The Handbook of Nanomedicine

Second Edition

The Handbook of Nanomedicine

Second Edition

Kewal K. Jain MD, FRACS, FFPM

Jain PharmaBiotech, Basel, Switzerland

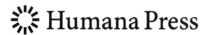

Humana Press

Kewal K. Jain
Jain PharmaBiotech
Blaesiring 7
Basel 4057
Switzerland

ISBN 978-1-61779-982-2 ISBN 978-1-61779-983-9 (eBook)
DOI 10.1007/978-1-61779-983-9
Springer New York Heidelberg Dordrecht London

Library of Congress Control Number: 2012940572

Printed on acid-free paper

Humana Press is a brand of Springer
Springer is part of Springer Science+Business Media (www.springer.com)

Foreword

The *Handbook of Nanomedicine* provides a thorough guide to this new and very important interdisciplinary area of science and technology. It provides both the basics and a classification system for nanomedicine. Important areas such as nanoarrays, nanofluidics, nanoparticles, nanogenomics, nanoproteomics, nanobiotechnology, nanomolecular diagnostics, and nanopharmaceuticals are evaluated. The role of biotechnology in biological therapies, and in particular oncology, is discussed. Nanodevices in surgery and medicine are also examined. Another important focus of the handbook is the role of nanomedicine in medical specialty areas—particularly in neurology, cardiology, dermatology, pulmonology, geriatrics, orthopedics, and ophthalmology. Nanomedicine in microbiology and in regenerative medicine and tissue engineering is also discussed. In addition, ethical, safety, regulatory, educational, and commercialization issues are discussed. Finally, the handbook concludes with an assessment of the future of nanomedicine, which is very bright.

Robert Langer Ph.D.
David H. Koch Institute Professor
Massachusetts Institute of Technology
Cambridge, MA, USA

Preface to the Second Edition

Considerable advances have taken place in nanomedicine since the first edition of the book in 2008. The basic plan of the book has been retained with some reorganization, but most of the material has been updated or replaced with new developments. Important classical references were left in while new ones have been added. Most of the advances have occurred in nanodiagnostics and nanopharmaceuticals, particularly drug delivery using nanobiotechnology. Nanooncology remains the major area of clinical application although considerable advances have been made in other therapeutic areas, particularly nanocardiology and nanoneurology. Several new products have been approved, and clinical applications of nanobiotechnology are progressing. This has required the discussion of some regulatory issues. Combination of diagnosis and therapy is facilitated by nanobiotechnology and fits in with concepts of personalized medicine, which is being increasingly accepted. As with the first edition, requirements of both physicians and scientists have been kept in mind. However, the description is kept simple enough to be understood by any educated layperson.

The author wishes to acknowledge the help and encouragement received from Patrick J. Marton, Senior Editor, Springer Protocols, Humana Press, in completion of the project. David Casey has done an excellent job of editing and organizing this book.

Basel, Switzerland Kewal K. Jain, MD, FRACS, FFPM.

Preface to the First Edition

Nanomedicine is application of nanobiotechnology to clinical medicine. However, new technologies do not always enter medical practice directly. Nanobiotechnologies are being used to research the pathomechanism of disease, refine molecular diagnostics, and help in the discovery, development, and delivery of drugs. In some cases, nanoparticles are the nanomedicines. The role is not confined to drugs before devices, and surgical procedures are refined by nanobiotechnology, referred to as nanosurgery.

This handbook covers the broad scope of nanomedicine. Starting with the basics, the subject is developed to potential clinical applications, many of which are still at an experimental stage. The prefix nano is used liberally and indicates the nanodimension of existing scientific disciplines and medical specialties. Two important components of nanomedicine are nanodiagnostics and nanopharmaceuticals and constitute the largest chapters.

Keeping in mind that the readers of the book will include nonmedical scientists, pharmaceutical personnel, as well as physicians, technology descriptions and medical terminology are kept as simple as possible. As a single author book, duplication is avoided. I hope that readers at all levels will find it a concise, comprehensive, and useful source of information.

There is voluminous literature relevant to nanomedicine. Selected references are quoted in the text.

Basel, Switzerland Kewal K. Jain, MD, FRACS, FFPM.

Contents

List of Figures

Fig. 1.1 Sizes of biologically entities relevant to the brain. (*Top row above scale bar*) From left to right: (**a**) X-ray crystal structure of Alzheimer's disease candidate drug, dehydroevodiamine HCl (DHED); (**b** and **c**) porous metal oxide microspheres being endocytosed by BV2 microglia cell (close-up and low magnification) SEM images; (**d** and **e**) SEM and fluorescence micrograph of DHED microcrystals (DHED is blue-green luminescent). (*Bottom row below the scale bar*) Left to right: small molecules, such as dopamine, minocycline, mefenamic acid, DHED, and heme, are ~1 nm or smaller. The lipid bilayer is a few nanometers thick. Biomolecules such as a microRNA and a protein are only a few nanometers in size. A single cell or neuron is tens or hundreds of microns in size. Size of human brain is tens of centimeters (Reproduced from: Suh WH, et al. Prog Neurobiol 2009;87:133–70 (by permission)) 2

Fig. 2.1 The core, branching, and surface molecules of dendrimers (Source: Starpharma Holding Ltd., by permission) 18

Fig. 2.2 Schematic representation of Dip Pen Nanolithography (*DPN*). A water meniscus forms between the atomic force microscope (*AFM*) tip coated with oligonucleotide (*ODN*) and the Au substrate. The size of the meniscus, which is controlled by relative humidity, affects the ODN transport rate, the effective tip-substrate contact area, and DPN resolution (Modified from Piner et al. 1999. © Jain PharmaBiotech) 35

Fig. 2.3 Surface plasmon resonance (SPR) technology (Reproduced by permission of Biacore) 57

Fig. 3.1 Concept of nanopore-based sequencing. When a current is applied across the membrane, charged biomolecules migrate through the pores. As each nucleotide passes through the

nanopore, an electric signature is produced that characterizes it because the size of the nanopore allows only a single nucleic acid strand to pass through it at one time 90

Fig. 4.1 Scheme of biobarcode assay. Schematic illustrating PSA (prostate-specific antigen) detection using the biobarcode assay. Antibody-coated magnetic beads capture and concentrate the protein targets. The captured protein targets are labeled with gold nanoparticle probes that are co-loaded with target-specific secondary antibodies and DNA barcodes. The resulting complexes are separated magnetically and washed to remove excess probe. The DNA barcodes are then released from the complex and detected via hybridization to a surface-immobilized DNA probe and an oligonucleotide-functionalized gold nanoparticle. The gold particles are enlarged through silver deposition, and the light scattered from the particles is detected using the Verigene Reader optical detection system. Increased detection sensitivity is derived from (1) capturing and concentrating protein targets with an antibody-coated magnetic bead, (2) releasing multiple DNA barcodes per captured protein target (hundreds of barcode are attached to a 30-nm-diameter gold particle), and (3) ultrasensitive DNA detection via silver-amplified gold nanoparticles (Courtesy of Nanosphere Inc.) 138

Fig. 4.2 Scheme of an optical mRNA biosensor. Sequence-specific molecular beacons are used as molecular switches. This biosensor detects single molecules in fluids and can be used to search for molecular biomarkers to predict the prognosis of disease 157

Fig. 5.1 Application of nanobiotechnology at various stages of drug discovery (© Jain PharmaBiotech) 172

Fig. 5.2 Bacteria plus nanoparticles for drug delivery into cells (Source: Akin et al. 2007) 191

Fig. 5.3 Schematic image of a lipid nanoparticle (© Springer Science+Business Media LLC) 212

Fig. 6.1 Nucleic acid delivery with lipid nanoparticle (LNP) technology (Courtesy of Tekmira Pharmaceutical Corporation) 243

Fig. 6.2 Nanocochleate-mediated drug delivery. Addition of calcium ions to small phosphatidylserine vesicles induces their collapse into discs, which fuse into large sheets of lipid. These lipid sheets rolled up into nano-crystalline structures called nanocochleates (Courtesy of BioDelivery Sciences International Inc.) 248

List of Tables

Abbreviations

AFM	Atomic force microscopy
BBB	Blood–brain barrier
BioMEMS	Biological Microelectromechanical Systems
CNS	Central nervous system
DNA	Deoxyribonucleic acid
DPN	Dip Pen Nanolithography
ELISA	Enzyme-linked immunosorbent assay
FDA	Food and Drug Administration (USA)
FRET	Fluorescence resonance energy transfer
LNS	Lipid nanosphere
MEMS	Microelectromechanical Systems
MNP	Magnetic nanoparticle
MRI	Magnetic resonance imaging
NCI	National Cancer Institute (USA)
NIH	National Institutes of Health (USA)
NIR	Near-infrared
NP	Nanoparticle
ODN	Oligodeoxynucleotide
PAMAM	Polyamidoamine (dendrimers)
PCR	Polymerase chain reaction
PEG	Polyethylene glycol

PEI	Polyethylenimine
PLA	Polylactides
PLGA	Poly(lactic-co-glycolic) acid
POC	Point-of-care
QD	Quantum dot
RLS	Resonance light scattering
RNA	Ribonucleic acid
SERS	Surface-enhanced Raman scattering
SNP	Single-nucleotide polymorphism
SPM	Scanning probe microscope
SPR	Surface plasmon resonance

Chapter 1
Introduction

Nanomedicine

Nanomedicine is defined as the application of nanobiotechnology to medicine. Its broad scope covers the use of nanoparticles and nanodevices in healthcare for diagnosis as well as therapeutics. Safety, ethical, and regulatory issues are also included.

Basics of Nanobiotechnology

Nanotechnology (Greek word nano means dwarf) is the creation and utilization of materials, devices, and systems through the control of matter on the nanometer-length scale, i.e., at the level of atoms, molecules, and supramolecular structures. Nanotechnology, as defined by the National Nanotechnology Initiative (http://www.nano.gov/), is the understanding and control of matter at dimensions of roughly 1–100 nm, where unique phenomena enable novel applications. Encompassing nanoscale science, engineering, and technology, nanotechnology involves imaging, measuring, modeling, and manipulating matter at this length scale. It is the popular term for the construction and utilization of functional structures with at least one characteristic dimension measured in nanometers – a nanometer is one billionth of a meter (10^{-9} m). This is roughly four times the diameter of an individual atom, and the bond between two individual atoms is 0.15 nm long. Proteins are 1–20 nm in size. The definition of "small," another term used in relation to nanotechnology, depends on the application but can range from 1 nm to 1 mm. Nano is not the smallest scale; further down the power of 10 are angstrom (=0.1 nm), pico, femto, atto, and zepto. By weight, the mass of a small virus is about 10 attograms. An attogram is one-thousandth of a femtogram, which is one-thousandth of a picogram, which is one-thousandth of a nanogram. Dimensions of various objects in nanoscale are shown in Table 1.1.

K.K. Jain, *The Handbook of Nanomedicine*, DOI 10.1007/978-1-61779-983-9_1,

Table 1.1 Dimensions of various objects in nanoscale

Object	Dimension (nm)
Width of a hair	50,000
Red blood cell	2,500
Vesicle in a cell	200
Bacterium	1,000
Virus	100
Exosomes (nanovesicles shed by dendritic cells)	65–100
Width of DNA	2.5
Ribosome	2–4
A base pair in human genome	0.4
Proteins	1–20
Amino acid (e.g., tryptophan, the largest)	1.2 (longest measurement)
Aspirin molecule	1
An individual atom	0.25

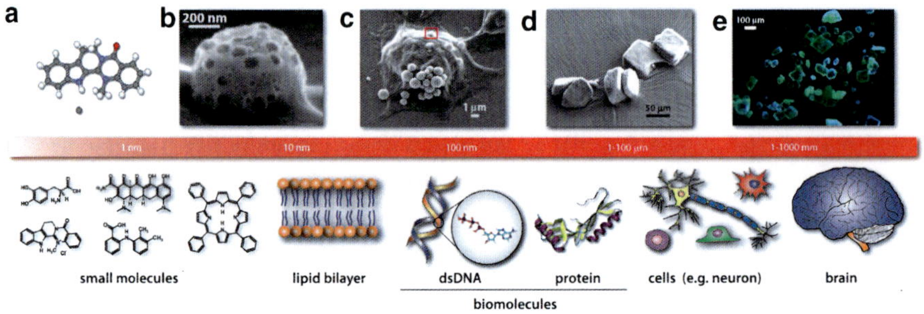

Fig. 1.1 Sizes of biologically entities relevant to the brain. (*Top row above scale bar*) From left to right: (**a**) X-ray crystal structure of Alzheimer's disease candidate drug, dehydroevodiamine HCl (DHED); (**b** and **c**) porous metal oxide microspheres being endocytosed by BV2 microglia cell (close-up and low magnification) SEM images; (**d** and **e**) SEM and fluorescence micrograph of DHED microcrystals (DHED is blue-green luminescent). (*Bottom row below the scale bar*) Left to right: small molecules, such as dopamine, minocycline, mefenamic acid, DHED, and heme, are ~1 nm or smaller. The lipid bilayer is a few nanometers thick. Biomolecules such as a microRNA and a protein are only a few nanometers in size. A single cell or neuron is tens or hundreds of microns in size. Size of human brain is tens of centimeters (Reproduced from: Suh WH, et al. Prog Neurobiol 2009;87:133–70 (by permission))

Given the inherent nanoscale functional components of living cells, it was inevitable that nanotechnology will be applied in biotechnology giving rise to the term nanobiotechnology. A brief introduction will be given to basic nanotechnologies from physics and chemistry, which are now being integrated into molecular biology to advance the field of nanobiotechnology. The aim is to understand the biological processes to improve diagnosis and treatment of diseases. Sizes of biologically entities relevant to the brain are shown in Fig. 1.1.

Relation of Nanobiotechnology to Nanomedicine

Technical achievements in nanotechnology are being applied to improve drug discovery, drug delivery, and pharmaceutical manufacturing. A vast range of applications have spawned many new terms, which are defined as they are described in various chapters. Numerous applications in the pharmaceutical industry can also be covered under the term "nanobiopharmaceuticals."

Landmarks in the Evolution of Nanomedicine

Historical landmarks in the evolution of nanomedicine are shown in Table 1.2.

Table 1.2 Historical landmarks in the evolution of nanomedicine

Year	Landmark
1905	Einstein published a paper that estimated the diameter of a sugar molecule at about 1 nm
1931	Max Knoll and Ernst Ruska discover the electron microscope – enables subnanomolar imaging
1959	Nobel Laureate Richard Feynman gave a lecture entitled "There's plenty of room at the bottom" at the annual meeting of the American Physical Society. He outlined the principle of manipulating individual atoms using larger machines to manufacture increasingly smaller machines (Feynman 1992)
1974	Start of development of molecular electronics by Aviram and Rattner (Hush 2003)
1974	Norio Taniguchi of Japan coined the word "nanotechnology"
1979	Colloidal gold nanoparticles used as electron-dense probes in electron microscopy and immunocytochemistry (Batten and Hopkins 1979)
1981	Conception of the idea of designing molecular machines analogous to enzymes and ribosomes (Drexler 1981)
1984	The first description the term dendrimer and the method of preparation of poly(amidoamine) dendrimers (Tomalia et al. 1985)
1985	Discovery of buckyballs (fullerenes) by Robert Curl, Richard Smalley, and Harold Kroto, which led to the award of Nobel Prize for chemistry in 1996 (Smalley 1985; Curl et al. 1997)
1987	Publication of the visionary book on nanotechnology potential "Engines of Creation" (Drexler 1987)
1988	Maturation of the field of supramolecular chemistry relevant to nanotechnology: construction of artificial molecules that interact with each other leading to award of the Nobel prize (Lehn 1988)
1990	Atoms visualized by the scanning tunneling microscope discovered in 1980s at the IBM Zürich Laboratory (Zürich, Switzerland), which led to award of a Nobel prize (Eigler and Schweizer 1990)
1991	Discovery of carbon nanotubes (Iijima et al. 1992)
1992	Principles of chemistry applied to the bottom-up synthesis of nanomaterials (Ozin 1992)
1994	Nanoparticle-based drug delivery (Kreuter 1994)
1995	FDA approved Doxil, a liposomal formulation of doxorubicin, as an intravenous chemotherapy agent for Kaposi's sarcoma. Drug carried by nanosize liposomes is less toxic with targeted delivery

(continued)

Table 1.2 (continued)

Year	Landmark
1997	Cancer targeting with nanoparticles coated with monoclonal antibodies (Douglas et al. 1997)
1997	Founding of the first molecular nanotechnology company – Zyvex Corporation
1998	First use of nanocrystals as biological labels, which were shown to be superior to existing fluorophores (Bruchez et al. 1998)
1998	Use of DNA-gelatin nanospheres for controlled gene delivery (Truong-Le et al. 1998)
1998	Use of the term "nanomedicine" in publications (Freitas 1998)
2000	Nanotechnology Initiative announced in the USA (Roco 2003)
2000	First FDA approval of a product incorporating the NanoCrystal® technology (Elan), solid-dose formulation of the immunosuppressant sirolimus – Rapamune® (Wyeth)
2003	Concept for nanolaser was developed at Georgia State University using nanospheres and nanolens system (Li et al. 2003)
2003	The US Senate passed the Nanotechnology Research and Development Act making the National Nanotechnology Initiative into law and authorized $3.7 billion over the next 4 years for the program
2005	FDA approved Abraxane™, a taxane based on nanotechnology, for the treatment of breast cancer. Nanoparticle form of the drug overcomes insolubility problems encountered with paclitaxel and avoids the use of toxic solvents

Nanomedicine as a Part of Evolution of Medicine

Medicine is constantly evolving, and new technologies are incorporated into the diagnosis and treatment of patients. This process is sometimes slow, and there can be a gap of years before new technologies are integrated in medical practice. The reasons for the delay are:

- Establishing the safety and efficacy of innovative treatments is a long process, particularly with clinical trials and regulatory reviews.
- The current generation of medical practitioners are still not well oriented toward biotechnology, and conservative elements of the profession may be slow in accepting and learning about nanobiotechnology, which is at the cutting edge of biotechnology.
- High cost of new technologies is a concern for the healthcare providers. Cost-benefit studies are needed to convince the skeptics that some of the new technologies may actually reduce the overall cost of healthcare.

Molecular medicine is already a recognized term. It should not be considered a subspecialty of medicine as molecular technologies would have an overall impact on the evolution of medicine. Recognition of the usefulness of biotechnology has enabled progress in the concept of personalized medicine, which again is not a branch of medicine but simply indicates a trend in healthcare and simply means the prescription of specific treatments and therapeutics best suited for an individual (Jain 2009). Various nanomachines and other nano-objects that are currently under investigation in medical research and diagnostics will soon find applications in the practice of medicine. Nanobiotechnologies are being used to create and study

Table 1.3 Nanomedicine in the twenty-first century

Nanodiagnostics
Extending limits of detection by refining currently available molecular diagnostic technologies
Development of new nanotechnology-based assays
Nanobiosensors
Nanoendoscopy
Nanoimaging
Nanopharmaceuticals
Nanoparticulate formulations of drugs
Nanotechnology-based drug discovery
Nanotechnology-based drug delivery
Regenerative medicine
Use of nanotechnology for tissue engineering
Transplantation medicine
Exosomes from donor dendritic cells for drug-free organ transplants
Nanomedicine specialties
Nanocardiology
Nanodermatology
Nanodentistry
Nanogerontology
Nanohematology
Nanoimmunology
Nanomicrobiology
Nanonephrology
Nanoneurology
Nanooncology
Nanoophthalmology
Nanoorthopedics
Implants
Bioimplantable sensors that bridge the gap between electronic and neurological circuitry
Durable rejection-resistant artificial tissues and organs
Implantations of nanocoated stents in coronary arteries to elute drugs and to prevent reocclusion
Implantation of nanoelectrodes in the brain for functional neurosurgery
Implantation of nanopumps for drug delivery
Nanosurgery
Minimally invasive surgery: miniaturized nanosensors implanted in catheters to provide real-time data
Nanosurgery by integration of nanoparticles and external energy
Nanorobotic treatments
Vascular surgery by nanorobots introduced into the vascular system
Nanorobots for detection and destruction of cancer

models of human disease, particularly immune disorders. Introduction of nanobiotechnologies in medicine will not create a separate branch of medicine but simply implies improvement of diagnosis as well as therapy.

Current research is exploring the fabrication of designed nanostructures, nanomotors, microscopic energy sources, and nanocomputers at the molecular scale, along with the means to assemble them into larger systems, economically and in great numbers. Some of the applications of nanobiotechnology in medicine are shown in Table 1.3.

Chapter 2
Nanotechnologies

Introduction

This chapter will focus on nanobiotechnologies that are relevant to applications in biomedical research, diagnostics, and medicine. Invention of the microscope revolutionized medicine by enabling the detection of microorganisms and study of histopathology of disease. Microsurgery was a considerable refinement over crude macrosurgery and opened the possibilities of procedure that were either not carried out previously or had high mortality and morbidity. Nanotechnologies, by opening up the world beyond microscale, will have a similar impact on medicine and surgery. Various nanobiotechnologies are described in detail in a special report on this topic (Jain 2012). Those relevant to understanding of diseases, diagnosis, and development of new drugs as well as management of diseases are described briefly in this chapter.

Classification of Nanobiotechnologies

It is not easy to classify the vast range of nanobiotechnologies. Some just represent motion on a nanoscale, but most of them are based on nanoscale structures, which come in a variety of shapes and sizes. A few occur in nature but most are engineered. The word nano is prefixed to just about anything that deals with nanoscale. It is not just biotechnology but many other disciplines such as nanophysics, nanobiology, etc. A simplified classification of basic nanobiotechnologies is shown in Table 2.1. Some technologies such as nanoarrays and nanochips are further developments.

K.K. Jain, *The Handbook of Nanomedicine*, DOI 10.1007/978-1-61779-983-9_2,

Table 2.1 Classification of basic nanomaterials and nanobiotechnologies

Nanoparticles
Fluorescent nanoparticles
Fullerenes
Gold nanoparticles
Lipoparticles
Magnetic nanoparticles
Nanocrystals
Nanoparticles assembly into micelles
Nanoshells
Paramagnetic and superparamagnetic nanoparticles
Polymer nanoparticles
Quantum dots
Silica nanoparticles
Nanofibers
Nanowires
Carbon nanofibers
Dendrimers
Polypropylenimine dendrimers
Composite nanostructures
Cochleates
DNA-nanoparticle conjugates
Nanoemulsions
Nanoliposomes
Nanocapsules enclosing other substances
Nanoshells
Nanovesicles
Nanoconduits
Nanotubes
Nanopipettes
Nanoneedles
Nanochannels
Nanopores
Nanofluidics
Nanostructured silicon
Nanoscale motion and manipulation at nanoscale
Cantilevers
Femtosecond laser systems
Nanomanipulation
Surface plasmon resonance
Visualization at nanoscale
Atomic force microscopy
Magnetic resonance force microscopy and nanoscale MRI
Multiple single-molecule fluorescence microscopy
Nanoparticle characterization by Halo™ LM10 technology
Nanoscale scanning electron microscopy
Near-field scanning optical microscopy
Optical imaging with a silver superlens
Partial wave spectroscopy
Photoactivated localization microscopy
Scanning probe microscopy
Super-resolution microscopy for in vivo cell imaging
Ultrananocrystalline diamond
Visualizing atoms with high-resolution transmission electron microscopy

Nanoparticles

Nanoparticles (NPs) form the bulk of nanomaterials. They can be made of different materials, e.g., gold. A NP contains tens to thousands of atoms and exists in a realm that straddles the quantum and the Newtonian. At those sizes, every particle has new properties that change depending on its size. As matter is shrunk to nanoscale, electronic and other properties change radically. NPs may contain unusual forms of structural disorder that can significantly modify materials properties and thus cannot solely be considered as small pieces of bulk material. Two NPs, both made of pure gold, can exhibit markedly different behavior – different melting temperature, different electrical conductivity, and different color – if one is larger than the other. That creates a new way to control the properties of materials. Instead of changing composition, one can change size. Some applications of nanoparticles take advantage of the fact that more surface area is exposed when material is broken down to smaller sizes. For magnetic NPs, the lack of blemishes produces magnetic fields remarkably strong considering the size of the particles. NPs are also so small that in most of them, the atoms line up in perfect crystals without a single blemish.

Zinc sulfide NPs a mere ten atoms across have a disordered crystal structure that puts them under constant strain, increasing the stiffness of the particles and probably affecting other properties, such as strength and elasticity. In similar semiconducting NPs, such as those made of cadmium selenide, slight differences in size lead to absorption and emission of different wavelengths of light, making them useful as fluorescent tracers. The dominant cause of such properties is quantum mechanical confinement of the electrons in a small package. But the disordered crystal structure now found in nanoparticles could affect light absorption and emission also. X-ray diffraction of single nanoparticles is not yet possible, and other methods are used to analyze X-ray diffraction images of nanoparticles so as to separate the effects of size from those of disordered structure.

It is beyond the scope of this handbook to describe all of the NPs. A few selected NPs relevant to nanomedicine are described briefly.

Gold Nanoparticles

Mass spectrometry analysis has determined the formula of gold nanocrystal molecules to be Au333(SR)79 (Qian et al. 2012). This metallic nanocrystal molecule exhibits fcc-crystallinity and surface plasmon resonance (SPR) at approximately 520 nm. Simulations have revealed that atomic shell closing largely contributes to the particular robustness of Au333(SR)79, albeit the number of free electrons is also consistent with electron shell closing based on calculations using a confined free electron model. This work clearly demonstrates that atomically precise nanocrystal molecules are achievable and that the factor of atomic shell closing contributes to their extraordinary stability compared to other sizes.

Ultrashort pulsed laser ablation in liquids represents a powerful tool for the generation of pure gold nanoparticles avoiding chemical precursors and thereby making them useful for biomedical applications. However, there is a concern that their biochemical properties may change because of their properties of accepting electrons, which often adsorb onto the nanoparticles. A study has shown that cotransfection of plasmid DNA and laser-generated gold nanoparticles does not disturb the bioactivity of GFP-HMGB1 fusion protein – either uptake of the vector through the plasma membrane or protein accumulation in the nucleus (Petersen et al. 2009). Thus, laser-generated gold nanoparticles provide a good alternative to chemically synthesized nanoparticles for use in biomedical applications.

DNA molecules are attached to gold nanoparticles, which tangle with other specially designed pieces of DNA into clumps that appear blue. The presence of lead causes the connecting DNA to fall apart. That cuts loose the individual gold nanoparticles and changes the color from blue to red. Gold nanoparticles are also used as a connecting point to build biosensors for detection of disease. A common technique for a diagnostic test consists of an antibody attached to a fluorescent molecule. When the antibody attaches to a protein associated with the disease, the fluorescent molecule lights up under ultraviolet light. Instead of a fluorescent molecule, a gold nanoparticle can be attached to the antibody, and other molecules such as DNA can be added to the nanoparticle to produce bar codes. Because many copies of the antibodies and DNA can be attached to a single nanoparticle, this approach is much more sensitive and accurate than the fluorescent molecule tests used currently.

Cubosomes

Methods and compositions for producing lipid-based cubic phase nanoparticles were first discovered in the 1990s. Since then, a number of studies have described properties such as particle size, morphology, and stability of cubic phase dispersions, which can be tuned by composition and processing conditions. Stable particle dispersions with consistent size and structure can be produced by a simple processing scheme comprising a homogenization and heat treatment step. Because of their unique microstructure, they are biologically compatible and capable of controlled release of solubilized active ingredients such as drugs and proteins (Garg et al. 2007). As a drug delivery vehicle, high drug payloads, stabilization of peptides or proteins, and simple preparation process are also advantages of a cubosome (Wu et al. 2008). The ability of cubic phase to incorporate and control release of drugs of varying size and polar characteristics, and biodegradability of lipids make it a versatile drug delivery system for various routes of administration, including oral, topical (or mucosal), and intravenous. Furthermore, proteins in cubic phase appear to retain their native conformation and bioactivity and are protected against chemical and physical inactivation.

Fluorescent Nanoparticles

Microwave plasma technique has been used to develop fluorescent nanoparticles. In a second reaction, a layer of organic dye is deposited, and the final step is an outer cover of polymer, which protects the nanoparticles from exposure to environments.

Each layer has characteristic properties. The size of the particles varies, and these are being investigated for applications in molecular diagnostics. Fluorescent nanoparticles can also be used as labels for immunometric assays.

Switchable fluorescent silica nanoparticles have been prepared by covalently incorporating a fluorophore and a photochromic compound inside the particle core (May et al. 2012). The fluorescence can be switched reversibly between an on- and off-state via energy transfer. The particles were synthesized using different amounts of the photoswitchable compound (spiropyran) and the fluorophore (rhodamine B) in a size distribution between 98 and 140 nm and were characterized in terms of size, switching properties, and fluorescence efficiency by TEM, and UV\Vis and fluorescence spectroscopy.

Fullerenes

Fullerene technology derives from the discovery in 1985 of carbon-60, a molecule of 60 carbon atoms that form a hollow sphere one nanometer in diameter. The molecule was named buckyball or fullerene or buckminsterfullerene, because of its similarity to the geodesic dome designed by Buckminster Fuller. Subsequent studies have shown that fullerenes actually represent a family of related structures containing 20, 40, 60, 70, or 84 carbons. C60, however, is the most abundant member of this family. Fullerenes are entirely insoluble in water, but suitable functionalization makes the molecules soluble. Initial studies on water-soluble fullerene derivatives led to the discovery of the interaction of organic fullerenes with DNA, proteins, and living cells. Subsequent studies have revealed interesting biological activity aspects of organic fullerenes owing to their photochemistry, radical quenching, and hydrophobicity to form one- to three-dimensional supramolecular complexes. In these areas of research, synthetic organic chemistry has played an important role in the creation of tailor-made molecules.

Upon contact with water, under a variety of conditions, C60 spontaneously forms a stable aggregate with nanoscale dimensions (25–500 nm), termed "nano-C60," that is both soluble and toxic to bacteria (Fortner et al. 2005). This finding challenges conventional wisdom because buckyballs are notoriously insoluble by themselves, and most scientists had assumed they would remain insoluble in nature. The findings also raise questions about how the aggregates will interact with other particles and living things in natural ecosystems.

A method of application of C60 to cultured cells has been described that does not require water-solubilization techniques (Levi et al. 2006). Normal and malignant cells take up C60 and the inherent photoluminescence of C60 is detected within multiple cell lines. Treatment of cells with up to 200 mg/ml (200 ppm) of C60 does not alter morphology, cytoskeletal organization, and cell cycle dynamics nor does it inhibit cell proliferation. This study shows that pristine C60 is nontoxic to the cells and suggests that fullerene-based nanocarriers may be used for biomedical applications. Fullerenes have important applications in treatment of various diseases such as cancer and as an antioxidant neuroprotective for neurodegenerative disorders in addition to use as contrast agent for brain imaging (Miller et al. 2007).

Lipoparticles

Lipoparticles are nanometer-sized spheres surrounded by a lipid bilayer and embedded with conformationally intact integral membrane proteins. Interactions with integral membrane proteins have been particularly difficult to study because the proteins cannot be removed from the lipid membrane of a cell without disrupting the structure and function of the protein. The ability to solubilize integral membrane proteins has applications for microfluidics, biosensors, high-throughput screening, antibody development, and structural studies of complex receptors.

Nanoparticles Assembly into Micelles

Assembly of gold and silver nanoparticle building blocks into larger structures is based on a novel method that goes back to one of nature's oldest known chemical innovations, i.e., the self-assembly of lipid membranes that surround every living cell (Zubarev et al. 2006). The method makes use of the hydrophobic effect, a biochemical phenomenon that all living creatures use to create membranes, ultrathin barriers of fatty acids that form a strong, yet dynamic, sack around the cell, sealing it from the outside world. Cell membranes are one example of a micelle, a strong bilayer covering that is made of two sheets of lipid-based amphiphiles, molecules that have a hydrophilic end and a hydrophobic end. Like two pieces of cellophane tape being brought together, the hydrophobic sides of the amphiphilic sheets stick to one another, forming the bilayered micelle. All micelles form in three shapes: spheres, cylinders, and sack-like vesicles. By varying the length of the polystyrene arm, the solvents used, and the size of the gold particles, it is possible to form spheres and vesicles and vary the diameter of their cylinders, some of which grew to well over 1,000 nm in length. This method may enable creation of a wide variety of useful materials, including potent cancer drugs and more efficient catalysts for the chemical industry.

Nanoshells

Nanoshells are ball-shaped structures measuring ~100 nm and consist of a core of nonconducting glass that is covered by a metallic shell, which is typically gold or silver. Nanoshells possess highly favorable optical and chemical properties for biomedical imaging and therapeutic applications. These particles are also effective substrates for surface-enhanced Raman scattering (SERS) and are easily conjugated to antibodies and other biomolecules. By varying the relative the dimensions of the core and the shell, the optical resonance of these nanoparticles can be precisely and systematically varied over a broad region ranging from the near-UV to the mid-infrared. This range includes the NIR wavelength region where tissue transmissibility peaks, which forms the basis of absorbing nanoshells in NIR thermal therapy of tumors. In addition to spectral tunability, nanoshells offer other advantages over conventional organic dyes including improved optical properties and reduced susceptibility to chemical/thermal denaturation. Furthermore, the same conjugation protocols used to bind biomolecules to gold colloid are easily modified for nanoshells. The core/shell ratio and overall size of a gold nanoshell influence its scattering and absorption properties.

Gold Nanoshells (Spectra Biosciences) possess physical properties similar to gold colloid, in particular a strong optical absorption due to the collective electronic response of the metal to light. The optical absorption of gold colloid yields a brilliant red color, which is very useful in consumer-related medical products such as home pregnancy tests. In contrast, the optical response of Gold Nanoshells depends dramatically on the relative sizes of the nanoparticle core and the thickness of the gold shell. Gold Nanoshells can be made either to absorb or scatter light preferentially by varying the size of the particle relative to the wavelength of the light at their optical resonance. Several potential biomedical applications of nanoshells are under development, including immunoassays, modulated drug delivery, photothermal cancer therapy, and imaging contrast agents.

Paramagnetic and Superparamagnetic Nanoparticles

Paramagnetic particles are important tools for cell sorting, protein separation, and single-molecule measurements. The particles used in these applications must meet the following requirements: uniform in size, highly paramagnetic, stable in physiological salt buffer, functionizable, and 100–1,000 nm in size. They have been used for the detection of model pathogens. Paramagnetic nanoparticles, which are linked to antibodies, enable highly specific biological cell separations.

Superparamagnetic iron oxide nanoparticles (SPION) with appropriate surface chemistry have been widely used experimentally for numerous in vivo applications such as magnetic resonance imaging (MRI) contrast enhancement, tissue repair, immunoassay, detoxification of biological fluids, hyperthermia, drug delivery, in cell separation, etc. These applications require that these nanoparticles have high magnetization values and size smaller than 100 nm with overall narrow particle size distribution, so that the particles have uniform physical and chemical properties. In addition, these applications need special surface coating of the magnetic particles, which has to not only be nontoxic and biocompatible but also allow a targetable delivery with particle localization in a specific area. Nature of surface coatings of the nanoparticles not only determines the overall size of the colloid but also plays a significant role in biokinetics and biodistribution of nanoparticles in the body. Magnetic nanoparticles can bind to drugs, proteins, enzymes, antibodies, or nucleotides and can be directed to an organ, tissue, or tumor using an external magnetic field or can be heated in alternating magnetic fields for use in hyperthermia. Magnetic labeling of cells provides the ability to monitor their temporal spatial migration in vivo by MRI. Various methods have been used to magnetically label cells using SPIONs. Magnetic tagging of stem cells and other mammalian cells has the potential for guiding future cell-based therapies in humans and for the evaluation of cell-based treatment effects in disease models.

Polymer Nanoparticles

Polymer nanoparticles, synthetic as well as biopolymers, are biocompatible, biodegradable, and nontoxic. They can be conjugated with other nanoparticles. Different types of polymer nanoparticles have been designed as drug delivery devices. Biodegradable polymeric nanoparticles are promising drug delivery devices because of their ability to deliver proteins, peptides, and genes as well as targeting therapeutics to specific organs/tissues. Although several synthetic polymers are available, natural polymers are still popular for drug delivery; these include acacia gum, chitosan, gelatin, and albumin. Examples of synthetic biodegradable polymers for controlled release drug delivery are polylactides (PLA), polyglycolides (PLG), and poly(lactide-co-glycolides) or PLGA.

Quantum Dots

Quantum dots (QDs) are nanoscale crystals of semiconductor material that glow or fluoresce when excited by a light source such as a laser. QD nanocrystals of cadmium selenide are 200–10,000 atoms wide and coated with zinc sulfide. The size of the QD determines the frequency of light emitted when irradiated with low energy light. The QDs were initially found to be unstable and difficult to use in solution. Multicolor optical coding for biological assays has been achieved by embedding different-sized QDs into polymeric microbeads at precisely controlled ratios. Their novel optical properties such as size-tunable emission and simultaneous excitation render these highly luminescent QDs ideal fluorophores for wavelength-and-intensity multiplexing. The use of ten intensity levels and six colors could theoretically code one million nucleic acid or protein sequences. Imaging and spectroscopic measurements indicate that the QD-tagged beads are highly uniform and reproducible, yielding bead identification accuracies as high as 99.99 % under favorable conditions. DNA hybridization studies demonstrate that the coding and target signals can be simultaneously read at the single-bead level. This spectral coding technology is expected to open new opportunities in gene expression studies, high-throughput screening, and medical diagnostics.

Latex beads filled with several colors of nanoscale semiconductor QDs can serve as unique labels for any number of different probes. When exposed to light, the beads identify themselves and their linked probes by emitting light in a distinct spectrum of colors – a sort of spectral bar code. The shape and size of QDs can be tailored to fluoresce with a specific color. Current dyes used for lighting up protein and DNA fade quickly, but QDs could allow tracking of biological reactions in living cells for days or longer.

QDs can also be placed in a strong magnetic field, which gives an electron on the dot two allowed energy states separated by an energy gap that depends on the strength of the field. The electron can jump the gap by absorbing a photon of precisely that energy, which can be tuned, by altering the field, to correspond with the energy of a far-infrared photon. Once it is excited by absorption of a photon, the electron can leap onto the terminal of a single-electron transistor, where it "throws the switch" and is detected.

Due to their sheer brightness and high photostability, QDs have the ability to act as molecular beacons. When attached to compounds or proteins of interest, QDs enable researchers to track movements within biological media or whole organisms, significantly impacting the way medical professionals study, diagnose, and treat diseases. Applications of QDs include the following:

- Life sciences research: tracking proteins in living cells
- Fluorescence detection: microscopy, biosensors, multicolor flow cytometry
- Molecular diagnostics
- Ex vivo live cell imaging
- In vivo targeting of cells, tissues, and tumors with monitoring by PET and MRI
- High-throughput screening
- Identification of lymph nodes in live animals by NIR emission during surgery

The new generations of QDs have far-reaching potential for the study of intracellular processes at the single-molecule level, high-resolution cellular imaging, long-term in vivo observation of cell trafficking, tumor targeting, and diagnostics. Best-known commercial preparation is Qdot™ (Life Technologies).

Silica Nanoparticles

In the case of silica, the formation of diatom shell or sponge spicule has attracted much attention in the last decade since it could provide key information to elaborate new hierarchically structured materials and nanodevices. The mineral phase is thought to be formed by the controlled assembly of nanoparticles generated in vivo from diluted precursor solutions, in the presence of biomolecular templates. Biomolecules present in silicifying organisms have been extracted and identified (Lopez et al. 2005). Silicon particles vary in size from 25 to 1,000 nm. Biomimetic approaches have led to the identification of several natural or synthetic molecules that are able to activate silica formation in conditions that closely resemble those found in the living organisms' intracellular compartments. Additionally, several of these systems are able to form silica nanoparticles whose size range and limited polydispersity reproduce colloidal biosilica. Extraction and characterization of biosilicifying molecules from living organisms, however, are still limited. Silicon nanoparticles have been used in drug delivery and gene therapy.

Bacterial Structures Relevant to Nanobiotechnology

Nanostructures Based on Bacterial Cell Surface Layers

Among the most commonly observed bacterial cell surface structures are monomolecular crystalline arrays of proteinaceous subunits termed S-layers, which are the simplest type of biological membrane developed during evolution. As an important component of the bacterial cell envelope, S-layers can fulfill various biological

Table 2.2 Applications of S-layers in nanobiotechnology

As a matrix for controlled immobilization of functional molecules
Binding of enzymes for bioanalytical biosensors
Immobilizing monoclonal antibodies for dipstick style immunoassays
Immobilizing antibodies for preparation of microparticles for ELISA
S-layers as carriers for conjugated vaccines
S-layer-coated liposomes
Immobilization of functional molecules on S-layer-coated liposomes
Entrapping of functional molecules for drug delivery
S-layer-coated liposomes with immobilized antigens and haptens for vaccines
Vehicles for producing fusion proteins
Vaccines
Biosensors
Diagnostics

functions and are usually the most abundantly expressed protein species in a cell (Pavkov-Keller et al. 2011). S-layer plays an important part in interactions of microbial cell with the environment. S-layers are generally 5–10 mm thick, and pores in the protein lattices are of identical size and morphology in the 2 to 8-nm range. S-layers have applications in nanobiotechnology as shown in Table 2.2.

Bacterial Magnetic Particles

Magnetic bacteria synthesize intracellular magnetosomes that impart a cellular swimming behavior referred to as magnetotaxis. The magnetic structures, magnetosomes, aligned in chains are postulated to function as biological compass needles allowing the bacterium to migrate along redox gradients through the Earth's geomagnetic field lines. Despite the discovery of this unique group of microorganisms several years ago, the mechanisms of magnetic crystal biomineralization have yet to be fully elucidated. A lipid bilayer membrane of approximately 2–4 nm in thickness encapsulates individual magnetosomes (50–100 nm in diameter). Magnetosomes are also referred to as bacterial magnetic particles (BMPs) to distinguish them from artificial magnetic particles (AMPs). The aggregation of BMPs can be easily dispersed in aqueous solutions compared with AMPs because of the enclosing organic membrane.

BMPs have potential applications in the interdisciplinary fields of nanobiotechnology, medicine, and environmental management. Through genetic engineering, functional proteins such as enzymes, antibodies, and receptors have been successfully displayed on BMPs, which have been utilized in various biosensors and bioseparation processes (Yoshino et al. 2010). The use of BMPs in immunoassays enables the separation of bound and free analytes by applying a magnetic field. Proteins can be attached covalently to solid supports such as BMPs that prevent desorption of antibodies during an assay. Large-scale production of functionally active antibodies or enzymes expressed on BMP membranes can be accomplished.

Carbon Nanotubes

Carbon nanotubes are rolled-up sheets of carbon atoms that appear naturally in soot, and are central to many nanotechnology projects. These nanotubes can go down in diameter to 1 nm, are stronger than any material in the universe, and can be of any length. These can be used as probes for AFMs that can image individual molecules in both wet and dry environments. This has enormous opportunities for application as conventional structure-based pharmaceutical design is hampered by the lack of high-resolution structural information for most protein-coupled receptors. It is possible to insert DNA into a carbon nanotube. Devices based on the DNA-nanotube combination could eventually be used to make electronics, molecular sensors, devices that sequence DNA electronically, and even gene delivery systems.

Medical Applications of Nanotubes

- Cyclic peptide nanotubes can act as a new type of antibiotic against bacterial pathogens.
- Cyclic peptide nanotubes can be used as artificial ion channels than open and close in response to electrical and chemical stimuli.
- It is easy to chemically functionalize the surfaces of template-synthesized nanotubes, and different functional groups can be attached to the inner versus outer surfaces of the tubes.
- Biomolecules, such as enzymes, antibodies, and DNA chains, can be attached to the nanotube surfaces to make biofunctionalized nanotubes.
- Template-synthesized nanotubes can be used as smart nanophase extraction agents, e.g., to remove drug molecules from solution.
- Template-synthesized nanotube membranes offer new approaches for doing bioseparations, e.g., of drug molecules.
- Nanoscale electromechanical systems (nanotweezers) based on carbon nanotubes have been developed for manipulation and interrogation of nanostructures within a cell.
- Carbon nanotubes can be used as tips for AFM.
- Lumen of a nanotube can carry payloads of drugs.
- Nanotubes can be used in biosensors.
- Blood-compatible carbon nanotubes, with heparin immobilized on the surface, are building blocks for in vivo nanodevices. Activated partial thromboplastin time and thromboelastography studies prove that heparinization can significantly enhance the blood compatibility of nanomaterials (Murugesan et al. 2006).

Studies of electrophoretic transport of ssRNA molecules through 1.5-nm-wide pores of carbon nanotube membranes reveal that RNA entry into the nanotube pores is controlled by conformational dynamics and exit by hydrophobic attachment of RNA bases to the pores. Differences in RNA conformational flexibility and hydrophobicity result in sequence-dependent rates of translocation, which is a prerequisite for nanoscale separation devices.

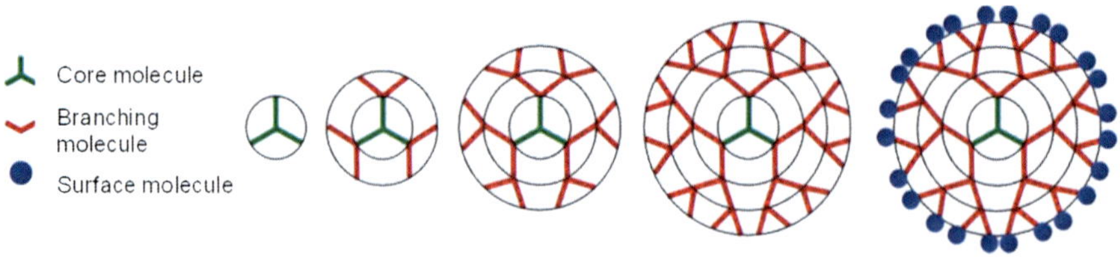

Fig. 2.1 The core, branching, and surface molecules of dendrimers (Source: Starpharma Holding Ltd., by permission)

The uptake of single-walled carbon nanotubes (SWCNTs) into cells appears to occur through phagocytosis. There are no adverse effects on the cells, and the nanotubes retained their unique optical properties suggesting that SWCNTs might be valuable biological imaging agents, in part because SWCNTs fluoresce in the NIR portion of the spectrum, at wavelengths not normally emitted by biological tissues. This may allow light from even a handful of nanotubes to be selectively detected in vivo. Although long-term studies on toxicity and biodistribution must be completed before nanotubes can be used in medical tests, but nanotubes are useful as imaging markers in laboratory in vitro studies, particularly in cases where the bleaching, toxicity, and degradation of more traditional markers are problematic.

Dendrimers

Dendrimers (dendri – tree, mer – branch) are a novel class of three-dimensional nanoscale, core-shell structures that can be precisely synthesized for a wide range of applications. Specialized chemistry techniques allow for precise control over the physical and chemical properties of the dendrimers. They are constructed generation by generation in a series of controlled steps that increase the number of small branching molecules around a central core molecule. Up to ten generations can be incorporated into a single dendrimer molecule. The final generation of molecules added to the growing structure makes up the polyvalent surface of the dendrimer (see Fig. 2.1). The core, branching, and surface molecules are chosen to give desired properties and functions.

As a result of their unique architecture and construction, dendrimers possess inherently valuable physical, chemical, and biological properties. These include:

- Precise architecture, size, and shape control. Dendrimers branch out in a highly predictable fashion to form amplified 3D structures with highly ordered architectures.
- High uniformity and purity. The proprietary stepwise synthetic process used produces dendrimers with highly uniform sizes (monodispersity) possessing precisely defined surface functionality and very low impurity levels.

- High loading capacity. Internal cavities intrinsic to dendrimer structures can be used to carry and store a wide range of metals, organic or inorganic molecules.
- High shear resistance. Through their 3D structure, dendrimers have a high resistance to shear forces and solution conditions.
- Low toxicity. Most dendrimer systems display very low cytotoxicity levels.
- Low immunogenicity when injected or used topically.

Properties

The surface properties of dendrimers may be manipulated by the use of appropriate "capping" reagents on the outermost generation. In this way, dendrimers can be readily decorated to yield a novel range of functional properties. These include:

- Polyvalency – The outer shell of each dendrimer can be manipulated to contain a large number of reactive groups. Each of these reactive sites has the potential to interact with a target entity, often resulting in polyvalent interactions.
- Flexible charge and solubility properties – Through use of appropriate capping groups on the dendrimer exterior, the charge and solubility of dendrimers can be readily manipulated.
- Flexible binding properties – By using appropriate capping groups on the dendrimer exterior, dendrimers can be designed to exhibit strong affinity for specific targets.
- Transfection – Dendrimers are able to move through cell boundaries and transport genetic materials into cell interiors.

Applications

Dendrimers, with their highly customizable properties, are basic building blocks with the promise of enabling specific nanostructures to be built to meet existing needs and solve evolving problems. Dendrimer research and development is currently making an impact on a broad range of fields as shown by exponential growth in the number of dendrimer-based publications. While the potential applications of dendrimers are unlimited, some of their current uses relevant to nanomedicine are shown in Table 2.3.

Advances in understanding of the role of molecular weight and architecture on the in vivo behavior of dendrimers, together with recent progress in the design of biodegradable chemistries, have enabled the application of dendrimers as antiviral drugs, tissue repair scaffolds, targeted carriers of chemotherapeutics, and optical oxygen sensors. Before such products can reach the market, however, the field must not only address the cost of manufacture and quality control of pharmaceutical-grade materials but also assess the long-term human and environmental health consequences of dendrimer exposure in vivo.

Table 2.3 Potential applications of dendrimers in nanomedicine

Diagnostics
Sensors
Imaging contrast agents
Drug delivery
Improved delivery of existing drugs
Improved solubility of existing drugs
Drug development
Polyvalent dendrimers interacting simultaneously with multiple drug targets
Development of new pharmaceuticals with novel activities
Improving pharmacological activity of existing drugs
Improving bioavailability of existing drugs
Medicine and surgery
Prevention of scar tissue formation after surgery

DNA Octahedron

DNA octahedron is a single strand of DNA that spontaneously folds into a highly rigid, nanoscale octahedron that is several million times smaller than the length of a standard ruler and about the size of several other common biological structures, such as a small virus or a cellular ribosome (Shih et al. 2004). The octahedron consists of 12 edges, 6 vertices, and 8 triangular faces. The structure is about 22 nm in diameter. Making the octahedron from a single strand was a breakthrough. Because of this, the structure can be amplified with the standard tools of molecular biology and can easily be cloned, replicated, amplified, evolved, and adapted for various applications. This process also has the potential to be scaled up so that large amounts of uniform DNA nanomaterials can be produced. These octahedra are potential building blocks for new tools for basic biomedical science. With these we have biological control, and not just synthetic chemical control, over the production of rigid, wire frame DNA objects.

Potential Applications

Because all 12 edges of the octahedral structures have unique sequences, they are versatile molecular building blocks that could potentially be used to self-assemble complex higher-order structures. Possible applications include using these octahedra as artificial compartments into which proteins or other molecules could be inserted, something like a virus in reverse – DNA is on the outside and proteins on the inside. In nature, viruses are self-assembling nanostructures that typically have proteins on the outside and DNA or RNA on the inside. The DNA octahedra could possibly form scaffolds that host proteins for the purposes of X-ray crystallography, which depends on growing well-ordered crystals, composed of arrays of molecules.

Nanowires

The manipulation of photons in structures smaller than the wavelength of light is central to the development of nanoscale integrated photonic systems for computing, communications, and sensing. Assembly of small groups of freestanding, chemically synthesized nanoribbons and nanowires into model structures illustrates how light is exchanged between subwavelength cavities made of three different semiconductors. With simple coupling schemes, lasing nanowires can launch coherent pulses of light through ribbon waveguides that are up to a millimeter in length. Also, interwire coupling losses are low enough to allow light to propagate across several right-angle bends in a grid of crossed ribbons. Nanoribbons function efficiently as waveguides in liquid media and provide a unique means for probing molecules in solution or in proximity to the waveguide surface. These results lay the groundwork for photonic devices based on assemblies of active and passive nanowire elements. There are potential applications of nanowire waveguides in microfluidics and biology. Some nanowire-based nanobiosensors are in development.

Polymer Nanofibers

Polymer nanofibers, with diameters in the nanometer range, possess larger surface areas per unit mass and permit easier addition of surface functionalities compared with polymer microfibers. Research on polymer nanofibers, nanofiber mats, and their applications has seen a remarkable growth over the last few years. Among various methods of manufacture, electrospinning has been used to convert a large variety of polymers into nanofibers and may be the only process that has the potential for mass production. Although measurement of mechanical properties such as tensile modulus, strength, and elongation is difficult because of the small diameters of the fibers, these properties are crucial for the proper use of nanofiber mats. Owing to their high surface area, functionalized polymer nanofibers will find broad applications as drug delivery carriers, biosensors, and molecular filtration membranes.

Nanopores

Nanopores are tiny structures that occur in the cell in nature for specific functions. At the molecular level, specific shapes are created that enable specific chemical tasks to be completed. For example, some toxic proteins such as alpha-hemolysin can embed themselves into cell membranes and induce lethal permeability changes there due to its central pore. The translocation of polymers across nanometer scale apertures in cell membranes is a common phenomenon in biological systems. The first proposed application was DNA sequencing by measuring the size of nanopore, application of an electric potential across the membrane, and waiting for DNA to migrate through the pore to enable one to measure the difference between bases in the sequence (see Chap. 3). Protein engineering has been applied to ion channels and pores, and protein as well as nonprotein can be constructed. Potential applications of engineered nanopores are:

- Tools in basic cell biology
- Molecular diagnostics: sequencing
- Drug delivery
- Cryoprotection and desiccation of cells
- Components of nanodevices and nanomachines
- Nanomedicine

Nanoporous Silica Aerogel

Nanoporous silica aerogels have been used in nanotechnology devices such as aerogel nanoporous insulation blankets. Silica aerogel substrate enables stable formation of lipid bilayers that are expected to mimic real cell membranes. Typical bilayers are 5 nm in thickness, and the silica beads in aerogel are approximately 10–25 nm in diameter. Silica aerogels have a unique structure and chemistry that allow for the transformation of nanosized liposomes into continuous, surface-spanning lipid bilayers. These lipid bilayers adsorb to the aerogel surface and exhibit the characteristic lateral mobility of real cell membranes. The high (98 %) porosity of aerogel substrates creates an underlying "water well" embedded in the aerogel pore structure that allows these membrane molecules to carry out normal biological activities including transport across the membrane. This porosity could potentially accommodate the movement of membrane proteins or other membrane-extruding molecules.

This aerogel is an improvement over conventional substrates for synthetic biomembranes as it is porous, thus minimizing nonphysiological interaction between membrane proteins and a hard substrate surface. This prevents the proteins from becoming immobilized, denatured, and eventually losing their biological functions. Applications of lipid bilayers are:

- Model biological membranes for research.
- Biosensors and lab-on-chip devices (microfluidic systems, analyte detector, etc.).
- Bio-actuating devices.
- Arrays for use in screening arrays of compounds for membrane-associated drug targets. Lipid bilayer system has been used in immunological screening for drug targets.
- Display libraries of compounds.
- Patterned lipid bilayers can be used for tissue culture and engineering (micropatterns of lipid membranes direct discriminative attachment or growth of living cells).

Advantages of aerogel biomembrane are:

- Best able to mimic the lateral mobility of molecules in real cell membranes.
- Enable membrane transport studies due to liquid permeability of aerogels.
- Both sides of supported membranes are accessible compared to only one side in conventional solid support.
- Can be used to design functional membranes for different applications by incorporating organic, inorganic, polymeric, and/or biologically active components into the aerogel structures.
- Nonphysiological interaction of the membrane-associated components with the underlying support (compared to glass).
- Membranes on the aerogel maintain stability for weeks.

Nanostructured Silicon

Silicon has been used for implants in the human body for several years. Following nanostructuring, silicon can be rendered biocompatible and biodegradable. BioSilicon™ (pSiMedica Ltd.) contains nanosized pores measuring 100 nm. The "silicon skeleton" between the pores comprises tens of silicon atoms in width. Initial applications are in drug delivery. The kinetics of drug release from BioSilicon™ can be controlled by adjusting the physical properties of the matrix, including modifying the pore size. Other potential applications include nanospheres for targeted systemic and pulmonary drug delivery. Nanostructured silicon, as multilayered mirrors, can be used for subcutaneous implants for diagnostics. Nanostructures can be used as prostheses to improve adhesion to bone tissue.

Polymer Nanofibers

Polymer nanofibers, with diameters in the nanometer range, possess larger surface areas per unit mass and permit easier addition of surface functionalities compared with polymer microfibers. Research on polymer nanofibers, nanofiber mats, and their applications has seen a remarkable growth over the last few years. Among various methods of manufacture, electrospinning has been used to convert a large variety of polymers into nanofibers and may be the only process that has the potential for mass production. Although measurement of mechanical properties such as tensile modulus, strength, and elongation is difficult because of the small diameters of the fibers, these properties are crucial for the proper use of nanofiber mats. Owing to their high surface area, functionalized polymer nanofibers will find broad applications as drug delivery carriers, biosensors, and molecular filtration membranes in future.

Nanoparticle Conjugates

DNA-Nanoparticle Conjugates

DNA-DNA hybridization has been exploited in the assembly of nanostructures including biosensors and DNA scaffolds. Many of these applications involve the use of DNA oligonucleotides tethered to gold nanoparticles or nanoparticles may be hybridized with one another. Two types of DNA-nanoparticle conjugates have been developed for these purposes. Both types entail the coupling of oligonucleotides through terminal thiol groups to colloidal gold particles. In one case, the oligonucleotides form the entire monolayer coating the particles, whereas in the other case, the oligonucleotides are incorporated in a phosphine monolayer, and particles containing discrete numbers of oligonucleotides are separated by gel electrophoresis. A minimal length of 50 residues is required, both for separation by electrophoresis and hybridization with complementary DNA sequences. These limitations of shorter oligonucleotides are attributed to interaction between the DNA and the gold. In a new technique, glutathione monolayer-protected gold clusters were reacted with 19- or 20-residue thiolated oligonucleotides, and the resulting DNA-nanoparticle conjugates could be separated on the basis of the number of bound oligonucleotides by gel electrophoresis and assembled with one another by DNA-DNA hybridization (Ackerson et al. 2005). This approach overcomes previous limitations of DNA-nanoparticle synthesis and yields conjugates that are precisely defined with respect to both gold and nucleic acid content.

Networks of Gold Nanoparticles and Bacteriophage

Biological molecular assemblies are excellent models for the development of nanoengineered systems with desirable biomedical properties. A spontaneous biologically active molecular network has been fabricated that consists of bacteriophage (phage) directly assembled with gold (Au) nanoparticles and termed Au-phage (Souza et al. 2006). When the phage is engineered so that each phage particle displays a peptide, such networks preserve the cell surface receptor binding and internalization attributes of the displayed peptide. The spontaneous organization of these targeted networks can be manipulated further by incorporation of imidazole (Au-phage-imid), which induces changes in fractal structure and near-infrared optical properties. The networks can be used as labels for enhanced fluorescence and dark-field microscopy, surface-enhanced Raman scattering detection, and near-infrared photon-to-heat conversion. Together, the physical and biological features within these targeted networks offer convenient multifunctional integration within a single entity with potential for nanotechnology-based biomedical applications such as biological sensors and cell-targeting agents. This genetically programmable nanoparticle with a biologically compatible metal acts as a nanoshuttle that can target specific locations in the body. For example, it could potentially locate damaged areas on arteries that have been caused by heart disease, and then deliver stem cells to the site that can build new blood vessels. It may be able to locate specific tumors, which could then be treated by either heating the gold particles with laser light and/or using the nanoparticles to selectively deliver a drug to destroy the cancer.

Protein-Nanoparticle Combination

Proteins come in many handy shapes and sizes, which make them major players in biological systems. Chaperonins are ring-shaped proteins found in all living organisms where they play an essential role in stabilizing proteins and facilitating protein folding. A chaperonin can be adapted for technological applications by coaxing it to combine with individual luminescent semiconductor nanoparticles. In bacteria, this chaperonin protein takes in and refolds denatured proteins in order to return them to their original useful shapes. This ability would make the proteins good candidates for drug carriers.

Cadmium sulfite nanoparticles emit light as long as they are isolated from each other; encasing the nanoparticles in the protein keeps the tiny particles apart. The biological fuel molecule ATP releases the nanoparticles from the protein tubes, freeing the particles to clump together, which quenches the light. The protein-nanoparticle combination could be used to detect ATP. This blend of nanotechnology and molecular biology could lead to new bioresponsive electronic nanodevices and biosensors very different from the artificial molecular systems currently available. By adding selective binding sites to the solvent-exposed regions of the chaperonin, the protein-nanoparticle bioconjugate becomes a sensor for specific targets (Xie et al. 2009).

Nanomaterials for Biolabeling

Nanomaterials are suitable for biolabeling. Nanoparticles usually form the core in nanobiomaterials. However, in order to interact with biological target, a biological or molecular coating or layer acting as an interface needs to be attached to the nanoparticle. Coatings that make the nanoparticles biocompatible include antibodies, biopolymers, or monolayers of small molecules. A nanobiomaterial may be in the form of nanovesicle surrounded by a membrane or a layer. The shape is more often spherical but cylindrical, platelike, and other shapes are possible. The size and size distribution might be important in some cases, e.g., if penetration through a pore structure of a cellular membrane is required. The size is critical when quantum-sized effects are used to control material properties. A tight control of the average particle size and a narrow distribution of sizes allow creating very efficient fluorescent probes that emit narrow light in a very wide range of wavelengths. This helps with creating biomarkers with many and well-distinguished colors. The core itself might have several layers and be multifunctional. For example, combining magnetic and luminescent layers one can both detect and manipulate the particles.

The core particle is often protected by several monolayers of inert material, e.g., silica. Organic molecules that are adsorbed or chemisorbed on the surface of the particle are also used for this purpose. The same layer might act as a biocompatible material. However, more often an additional layer of linker molecules is required that has reactive groups at both ends. One group is aimed at attaching the linker to the nanoparticle surface and the other is used to bind various biocompatible substances such as antibodies depending on the function required by the application.

Efforts are being made to improve the performance of immunoassays and immunosensors by incorporating different kinds of nanostructures. Most of the studies focus on artificial, particulate marker systems, both organic and inorganic. Inorganic nanoparticle labels based on noble metals, semiconductor QDs, and nanoshells appear to be the most versatile systems for these bioanalytical applications of nanophotonics. The underlying detection procedures are more commonly based on optical techniques. These include nanoparticle applications generating signals as diverse as static and time-resolved luminescence, one- and two-photon absorption, Raman and Rayleigh scattering as well as surface plasmon resonance, and others. All efforts are aimed at achieving one or more of the following goals:

- Lowering of detection limits (if possible, down to single-molecule level)
- Parallel integration of multiple signals (multiplexing)
- Signal amplification by several orders of magnitude

Potential benefits of using nanoparticles and nanodevices include an expanded range of label multiplexing. Different types of fluorescent nanoparticles and other nanostructures have been promoted as alternatives for the fluorescent organic dyes that are traditionally used in biotechnology. These include QDs, dye-doped polymer, and silica nanoparticles (Dosev et al. 2008). Various nanomaterials for biolabeling are shown in Table 2.4.

DNA Nanotags

Bright fluorescent dye molecules can be integrated with DNA nanostructure to make nanosized fluorescent labels – DNA nanotags, which improve the sensitivity for fluorescence-based imaging and medical diagnostics. DNA nanotags are useful for detecting rare cancer cells within tissue biopsies. In addition, they offer the opportunity to perform multicolor experiments. This feature is extremely useful for imaging applications because the multiple colors can be seen simultaneously, requiring only one experiment using one laser and one fluorescence-imaging machine. Fluorescent DNA nanotags have been used in a rolling circle amplification immunoassay based as a versatile fluorescence assay platform for highly sensitive protein detection (Xue et al. 2012).

Table 2.4 Nanomaterials for biolabeling

Label/reporter	Characteristics	Function/applications
Dendrimer/silver nanocomposites	Water-soluble, biocompatible, fluorescent, and 3–7-nm diameter stable nanoparticles	Biomarkers for in vitro cell labeling
Electrogenerated chemiluminescence	Tris(2,2′-bipyridyl)ruthenium(II) molecular labels	Nanoscale bioassay
Europium(III)-chelate-doped nanoparticles	Combined with selection of high-affinity monoclonal antibodies coated on label particles and microtitration wells	The sensitivity for virus particle detection is improved compared to immunofluorometric assays
Fluorescent color-changing dyes	3-hydroxychromone derivatives that exhibit two fluorescence bands as a result of excited-state intramolecular proton transfer reaction	Biosensors
Fluorescent lanthanide nanorods	Retain their fluorescent properties after internalization into cells	Multiplexed imaging of molecular targets in living cells
Luminescent core/shell nanohybrid	Luminescent rare earth ions in a nanosized Gd2O3 core (3.5 nm) and FITC molecules entrapped within in a polysiloxane shell (2.5–10 nm)	Two different luminescence emissions: (1) FITC under standard illumination and (2) Tb^{3+} under high-energy source giving highly photostable luminescence
Magnetic nanotags (MNTs)	Alternative to fluorescent labels in biomolecular detection assays	Multiplex protein detection of cancer biomarkers at low concentrations
Nanogold® labels (Nanoprobes Inc.)	Unlike nanogold particles, gold labels are uncharged molecules, which are cross-linked to specific sites on biomolecules	Nanogold® labels have a range and versatility, which is not available with colloidal nanogold particles
Nanophosphors	Nanophosphors contain embedded lanthanide ions, like europium or terbium	Nanophosphor signals hardly fade and can also be used for multiplex testing
Plasmon-resonant nanoparticles	Scatter light with tremendous efficiency	Ultrabright nanosized labels for biological applications, replacing other labeling methods such as fluorescence
QD end-labeling	Multicolor fluorescence microscopy using conjugated QDs	Detection of single DNA molecules
SERS (Surface-enhanced Raman Scattering)-based nanotags	A metal nanoparticle where each type of tag exploits the Raman spectrum of a different small molecule, and SERS bands are 1/50th the width of fluorescent bands	Enables greater multiplexed analyte quantification than other fluorescence-based quantitation tags

Fluorescent Lanthanide Nanorods

Inorganic fluorescent lanthanide (europium and terbium) orthophosphate nanorods can be used as a novel fluorescent label in cell biology. These nanorods, synthesized by the microwave technique, retain their fluorescent properties after internalization into human umbilical vein endothelial cells or renal carcinoma cells. The cellular internalization of these nanorods and their fluorescence properties have been characterized by fluorescence spectroscopy, differential interference contrast microscopy, confocal microscopy, and transmission electron microscopy. At concentrations up to 50 ug/ml, the use of 3H-thymidine incorporation assays, apoptosis assays, and trypan blue exclusion demonstrates the nontoxic nature of these nanorods, a major advantage over traditional organic dyes (Patra et al. 2006). These nanorods can be used for the detection of cancer at an early stage, and functionalized nanorods are potential vehicles for drug delivery.

Magnetic Nanotags

Magnetic nanotags (MNTs) are a promising alternative to fluorescent labels in biomolecular detection assays, because minute quantities of MNTs can be detected with inexpensive sensors. Probe sensors are functionalized with capture antibodies specific to the chosen analyte. During analyte incubation, the probe sensors capture a fraction of the analyte molecules. A biotinylated linker antibody is subsequently incubated and binds to the captured analyte, providing binding sites for the streptavidin-coated MNTs, which are incubated further. The nanotag-binding signal, which saturates at an analyte concentration-dependent level, is used to quantify the analyte concentration. However, translation of this technique into easy-to-use and multiplexed protein assays, which are highly sought after in molecular diagnostics such as cancer diagnosis and personalized medicine, has been challenging. Multiplex protein detection of potential cancer biomarkers has been demonstrated at subpicomolar concentration levels (Osterfeld et al. 2008). With the addition of nanotag amplification, the analytic sensitivity extends into the low femtomolar concentration range. The multianalyte ability, sensitivity, scalability, and ease of use of the MNT-based protein assay technology make it a strong contender for versatile and portable molecular diagnostics in both research and clinical settings.

Molecular Computational Identification

Molecular computational identification, based on molecular logic and computation, has been applied on nanoscale. Examples of populations that need encoding in the microscopic world are cells in diagnostics or beads in combinatorial chemistry. Taking advantage of the small size (about 1 nm) and large "on/off" output ratios of molecular logic gates and using the great variety of logic types, input chemical combinations, switching thresholds, and even gate arrays in addition to colors, unique identifiers have been produced for small polymer beads (about 100 μm) used for synthesis of combinatorial libraries. Many millions of distinguishable tags become available. This method should be extensible to far smaller objects, with the only requirement being a "wash and watch" protocol. The basis of this approach is converting molecular science into technology concerning analog sensors and digital logic devices. The integration of molecular logic gates into small arrays has been a growth area in recent years (de Silva 2011).

Nanophosphor Labels

Nanostructures based on inorganic phosphors (nanophosphors) are a new emerging class of materials with unique properties that make them very attractive for labeling. The molecular lattice of phosphors contains individual embedded lanthanide ions, like europium or terbium. The crystal lattice or sometimes "activator ions" – such as cerium ions used especially for this purpose – absorbs the stimulating light and transfers the energy to the lanthanide ions, which are the true source of fluorescence. The color emitted depends mainly on the lanthanide ions used. Terbium, e.g., gives off a yellowish green color, while europium produces a red fluorescence. As shown by the "microparticles" in fluorescent lights, the cycle of stimulation and emission can be endlessly repeated, which means that the dye never fades.

Bayer scientists are developing nanophosphors, which many of the advantages of QDs and fewer disadvantages such as high cost and heavy metals content that may not be environmentally friendly. Nanophosphor signals hardly fade and can also be used for multiplex testing. And the major advantage they have over QDs is that the wavelength of their emitted light does not depend on particle size but on the type of lanthanide ions used. For this reason, their particle size, which is also no more than 10 nm, does not need to be monitored so precisely. As a result, the manufacturing process is simpler and less expensive. Moreover, most ions of lanthanides, also called rare earths, are considered less harmful to the environment, and this facilitates their manufacture and disposal.

Background fluorescence from biological components of cells makes it difficult to interpret the signal, e.g., the positive result of a diagnostic test for cancer. Nanophosphors are able to get round this problem because for many types of nanophosphor, the life span of the fluorescence, i.e., the time between stimulation and emission, extends to several milliseconds. Accordingly, when the nanophosphor is exposed to a brief impulse of light, the background fluorescence disappears before the test result is displayed. This considerably enhances the sensitivity of the fluorescent marker in its various applications. Another important advantage of the nanophosphor system, particularly where medical diagnostics are concerned, is its ability to transfer fluorescent energy to a closely related dye. This allows biochemical reactions, like the coupling between antibodies, to be detected without the need for any additional procedures. So the relevant antibodies in the patient's sample can be detected immediately after the dye has been added to the test solution.

Before the nanophosphors can be used to track down certain segments of DNA, e.g., in cancer tests, they themselves need to be attached to suitable DNA segments. It is always a major challenge to achieve stable coupling of small organic molecules or larger biomolecules with unique, inorganic nanoparticles. The particles have to be painstakingly adapted to the properties of the organic molecules and prevented from lumping together themselves in the process. If this can be done successfully, it will meet the demanding challenges of medical diagnostics in the future.

Photoluminescence imaging in vitro and in vivo has been shown by use of near-infrared to near-infrared (NIR-to-NIR) upconversion in nanophosphors. This NIR-to-NIR upconversion process provides deeper light penetration into biological specimen and results in high-contrast optical imaging due to absence of an autofluorescence background and decreased light scattering. Fluoride nanocrystals (20–30 nm size) codoped with the rare earth ions, Tm^{3+} and Yb^{3+}, have been synthesized and characterized by TEM, XRD, and photoluminescence spectroscopy (Nyk et al. 2009). In vitro cellular uptake was demonstrated with no apparent cytotoxicity. Subsequent animal imaging studies were performed using Balb-c mice injected intravenously with upconverting nanophosphors, demonstrating the high-contrast PL imaging in vivo by photoluminescence spectroscopy. Lanthanide-doped nanocrystals have also been used for imaging of cells and some deep tissues in animals. Polyethylenimine (PEI)-coated $NaYF^4$:Yb,Er nanoparticles produce very strong upconversion fluorescence when excited at 980 nm by an NIR laser, which is resistant to photobleaching and nontoxic to bone marrow stem cells (Chatterjee et al. 2008). The nanoparticles delivered into some cell lines or injected intradermally and intramuscularly into some tissues either near the body surface or deep in the body of rats showed visible fluorescence, when exposed to a 980-nm NIR laser.

Organic Nanoparticles as Biolabels

The use of organic nonpolymeric nanoparticles as biolabels was not considered to be promising or have any advantage over established metallic or polymeric probes. Problems include quenching of fluorescence in organic dye crystals, colloidal stability, and solubility in aqueous environments, but some of these can be circumvented. Labels have been constructed by milling and suspending a fluorogenic hydrophobic precursor, fluorescein diacetate, in sodium dodecyl sulfate (SDS). Thus, a negative surface charge is introduced, rendering the particles (500 nm) colloidally stable and minimizing leakage of fluorescein diacetate molecules into surrounding water. Now it has been shown that the polyelectrolyte multilayer architecture is not vital for the operability of this assay format. Instead of SDS and multilayers, the adsorption of only one layer of an amphiphilic polymeric detergent, e.g., an alkylated poly(ethyleneimine), is sufficient to stabilize the system and to provide an interface for the antibody attachment. This is the basis of a technology

"ImmunoSuperNova®" (invented by 8sens.biognostic AG, Germany). In this the reaction of the analyte molecule with the capture antibody is followed by an incubation step with the antibody-nanoparticle conjugate, which serves as detector. After some washing steps, an organic release solvent is added, dissolving the particle and converting fluorescein diacetate into fluorescein.

Quantum Dots as Labels

The unique optical properties of QDs make them appealing as in vivo and in vitro fluorophores in a variety of biological investigations, in which traditional fluorescent labels based on organic molecules fall short of providing long-term stability and simultaneous detection of multiple signals. The ability to make QDs water soluble and target them to specific biomolecules has led to promising applications in cellular labeling, deep-tissue imaging, assay labeling, and as efficient fluorescence resonance energy transfer donors.

DNA molecules attached to QD surface can be detected by fluorescence microscopy. The position and orientation of individual DNA molecules can be inferred with good efficiency from the QD fluorescence signals alone. This is achieved by selecting QD pairs that have the distance and direction expected for the combed DNA molecules. Direct observation of single DNA molecules in the absence of DNA-staining agents opens new possibilities in the study of DNA-protein interactions. This approach can be applied for the use of QDs for nucleic acid detection and analysis. CdSe QDs can also be used as labels for sensitive immunoassay of cancer biomarker proteins by electrogenerated chemiluminescence. This strategy has been successfully used as a simple, cost-effective, specific, and potential method to detect α-fetoprotein in practical samples (Liu et al. 2011a). In contrast to a QD that is selectively introduced as a label, an integrated QD is one that is present in a system throughout a bioanalysis, and has a simultaneous role in transduction and as a scaffold for biorecognition. The modulation of QD luminescence provides the opportunity for the transduction of these events via fluorescence resonance energy transfer (FRET), bioluminescence resonance energy transfer (BRET), charge transfer quenching, and electrochemiluminescence (Algar et al. 2010).

SERS Nanotags

Surface-enhanced Raman scattering (SERS) nanotags (Oxonica Inc.) are silica-coated gold nanoparticles that are active at the glass-metal interface, and are optically detectable tags. Each type of tag exploits the Raman spectrum of a different small molecule, and SERS bands are 1/50th the width of fluorescent bands. These enable a greater degree of multiplexing than current fluorescence-based quantitation tags. The spectral intensity of SERS-based tags is linearly proportional to the number of particles allowing them to be used for multiplexed analyte quantification.

Because they are coated with glass, attachment to biomolecules is straightforward. They can be detected with low-cost instrumentation. The particles can be interrogated in the near-IR, enabling detection in blood and other tissues. Another advantage of these particles is that they are stable and are resistant to photodegradation. Nanoplex biotags are capable of measuring up to 20 biomarkers in a single test without interference from biological matrices such as whole blood. SERS nanotags are also useful for point-of-care diagnostics. There is a great potential for multiplexed imaging in living subjects in cases in which several targeted SERS probes could offer better detection of multiple biomarkers associated with a specific disease (Zavaleta et al. 2009). The primary limitations of Raman imaging in humans are those also faced by other optical techniques, in particular limited tissue penetration. Over the last several years, Raman spectroscopy imaging has advanced significantly, and many critical proof-of-principle experiments have been successfully carried out. It is expected that imaging with Raman spectroscopy will continue to be a dynamic research field over the next decade (Zhang et al. 2010).

Silica Nanoparticles for Labeling Antibodies

Luminescent silicon dioxide nanoparticles with size of 50 nm containing rhodamine (R-SiO2) have been synthesized by sol-gel method. These particles can emit intense and stable room temperature phosphorescence signals. An immune reaction between goat-antihuman IgG antibody labeled with R-SiO2 and human IgG has been demonstrated on polyamide membrane quantitatively, and the phosphorescence intensity was enhanced after the immunoreactions (Jia-Ming et al. 2005). This is the basis of a room temperature phosphorescence immunoassay for the determination of human IgG using an antibody labeled with the nanoparticles containing binary luminescent molecules. This method is sensitive, accurate, and precise. Lissamine rhodamine B sulfonyl chloride and other dyes can be covalently bound to and contained in spherical silica nanoparticles (30–80 nm). Compared to organic molecular markers, these fluorophore hybrid silica particles exhibit superior photostability and detection sensitivity, e.g., for detecting trace levels of hepatitis B surface. Dye-doped fluorescent silica nanoparticles are highly efficient labels for glycans and applied to detect bacteria by imaging as well as to study carbohydrate-lectin interactions on a lectin microarray (Wang et al. 2011a).

Silver Nanoparticle Labels

Silver (Ag) nanoparticles have unique plasmon-resonant optical scattering properties that are useful for nanomedical applications as signal enhancers, optical sensors, and biomarkers. Sensitive electrochemical DNA hybridization detection assay uses Ag nanoparticles as oligonucleotide labeling tags. The assay relies on the hybridization of the target DNA with the Ag nanoparticle-oligonucleotide DNA probe, followed by the release of the Ag metal atoms anchored on the hybrids by oxidative metal dissolution and the indirect determination of the solubilized Ag ions by anodic stripping voltammetry. Liquid electrode plasma-atomic emission spectrometry requires no plasma gas and no high-power source, which makes it suitable for onsite portable analysis, and can be used for protein-sensing studies employing Ag nanoparticle labeling. Human chorionic gonadotropin (hCG) was used as a model target protein, and the immunoreaction in which hCG is sandwiched between two antibodies, one of which is immobilized on the microwell and the second is labeled with Ag nanoparticles, was performed (Tung et al. 2012). hCG was analyzed in the range from 10 pg/mL to 1 ng/mL. This detection method has a wide variety of promising applications in metal-nanoparticle-labeled biomolecule detection.

Micro- and Nanoelectromechanical Systems

The rapid pace of miniaturization in the semiconductor industry has resulted in faster and more powerful computing and instrumentation, which have begun to revolutionize medical diagnosis and therapy. Some of the instrumentation used for nanotechnology is an extension of MEMS (microelectromechanical systems), which refers to a key enabling technology used to produce microfabricated sensors and systems. The potential mass application of MEMS lies in the fact that miniature components are batch fabricated using the manufacturing techniques developed in the semiconductor microelectronics industry enabling miniaturized, low-cost, high-performance, and integrated electromechanical systems. The "science of miniaturization" is a much more appropriate name than MEMS, and it involves a good understanding of scaling laws and different manufacturing methods and materials that are being applied in nanotechnology.

MEMS devices currently range in size from one to hundreds of micrometers and can be as simple as the singly supported cantilever beams used in AFM or as complicated as a video projector with thousands of electronically controllable microscopic mirrors. NEMS devices exist correspondingly in the nanometer realm – nanoelectromechanical systems (NEMS). The concept of using externally controllable MEMS devices to measure and manipulate biological matter (BioMEMS) on the cellular and subcellular levels has attracted much attention recently. This is because initial work has shown the ability to detect single base pair mismatches of DNA and to quantifiably detect antigens using cantilever systems. In addition is the ability to controllably grab and manipulate individual cells and subsequently release them unharmed.

Surface nanomachining combines the processing methods of MEMS with the tools of electron beam nanofabrication to create 3D nanostructures that move (and thus can do new types of things). Ultrashort pulsed laser radiation, e.g., using femtolasers, is an effective tool for controlled material processing and surface nano/micromodification because of minimal thermal and mechanical damage. Surface nanomachining has potential applications in nanobiotechnology.

BioMEMS

Because BioMEMS involves the interface of MEMS with biological environments, the biological components are crucially important. To date, they have mainly been nucleic acids, antibodies, and receptors that are involved in passive aspects of detection and measurement. These molecules retain their biological activity following chemical attachment to the surfaces of MEMS structures (most commonly, thiol groups to gold), and their interactions are monitored through mechanical (deflection of a cantilever), electrical (change in voltage or current in the sensor), or optical (surface plasmon resonance) measurements. The biological components are in the nanometer range or smaller; therefore, the size of these systems is limited by the minimum feature sizes achievable using the fabrication techniques of the inorganic structures, currently 100 nm–1 μm. Commercially available products resulting from further miniaturization could be problematic because of the expanding cost and complexity of optical lithography equipment and the inherent slowness of electron beam techniques. In addition to size limitations, the effects of friction have plagued multiple moving parts in inorganic MEMS, limiting device speeds and useful lifetimes.

Microarrays and Nanoarrays

Arrays consist of orderly arrangements of samples, which, in the case of biochips, may be cDNAs, oligonucleotides, or even proteins. Macroarraying (or gridding) is a macroscopic scheme of organizing colonies or DNA into arrays on large nylon filters ready for screening by hybridization. In microarrays, however, the sample spot sizes are usually less than 200 μm in diameter and require microscopic analysis. Microarrays have sample or ligand molecules (e.g., antibodies) at fixed locations on the chip, while microfluidics involves the transport of material, samples, and/or reagents on the chip.

Microarrays provide a powerful way to analyze the expression of thousands of genes simultaneously. Genomic arrays are an important tool in medical diagnostic and pharmaceutical research. They have an impact on all phases of the drug discovery process from target identification through differential gene expression, identification, and screening of small molecule candidates to toxicogenomic studies for drug safety. To meet the increasing needs, the density and information content of the microarrays is being improved. One approach is fabrication of chips with smaller, more closely packed features – ultrahigh density arrays, which will yield:

- High information content by reduction of feature size from 200 μm to 50 nm
- Reduction in sample size
- Improved assay sensitivity

Nanoarrays are the next stage in the evolution of miniaturization of microarrays. Whereas microarrays are prepared by robotic spotting or optical lithography, limiting

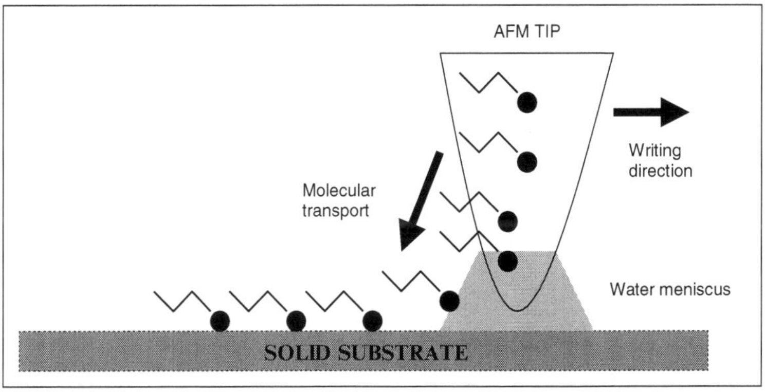

Fig. 2.2 Schematic representation of Dip Pen Nanolithography (*DPN*). A water meniscus forms between the atomic force microscope (*AFM*) tip coated with oligonucleotide (*ODN*) and the Au substrate. The size of the meniscus, which is controlled by relative humidity, affects the ODN transport rate, the effective tip-substrate contact area, and DPN resolution (Modified from Piner et al. 1999. © Jain PharmaBiotech)

the smallest size to several microns, nanoarrays require further developments in lithography strategies. Technologies available include the following:

- Electron beam lithography
- Dip Pen Nanolithography
- Scanning probe lithography
- Finely focused ion beam lithography
- Nanoimprint lithography

Dip Pen Nanolithography for Nanoarrays

Dip Pen Nanolithography™ (DPN™) commercialized by NanoInk Inc. employs the tip of an AFM to write molecular "inks" directly on a surface. Biomolecules such as proteins and viruses can be positioned on surfaces to form nanoarrays that retain their biological activity. DPN is schematically depicted in Fig. 2.2.

Advantages of DPN are as follows:

Ultrahigh resolution. DPN is capable of producing structures with line widths of less than 15 nm. This is compared to photolithography, which supports features of no less than 65-nm line width, and slower e-beam and laser lithography systems, which support features of 15-nm line width.

Flexibility. Direct fabrication is possible with many substances, from biomolecules to metals.

Accuracy. By leveraging existing highly accurate atomic force microscopy technology, DPN utilizes the best possible means for determining exactly where features are being placed on the substrate. This allows for the integration of multiple component nanostructures and for immediate inspection and characterization of fabricated structures.

Low capital cost. Techniques such as e-beam lithography that approach DPN-scale resolution are considerably more expensive to purchase, operate, and maintain.

Ease of use. DPN systems may be operated by nonspecialized personnel with minimal training. Further, DPN may be performed under normal ambient laboratory conditions with humidity control.

Speed. 100-nm spots can be deposited with a single DPN pen in less than a second. DPN can be used to fabricate arrays of a single molecule with more than 100,000 spots over 100 × 100 μm in less than an hour.

Applications of Dip Pen Nanolithography

Multiple allergen testing for high throughput and high sensitivity requires the development of miniaturized immunoassays that can be performed with minute amounts of test analyte that are usually available. Construction of such miniaturized biochips containing arrays of test allergens needs application of a technique able to deposit molecules at high resolution and speed while preserving its functionality. DPN is an ideal technique to create such biologically active surfaces, and it has already been successfully applied for the direct, nanoscale deposition of functional proteins, as well as for the fabrication of biochemical templates for selective adsorption. It has potential applications for detection of allergen-specific immunoglobulin E (IgE) antibodies and for mast cell activation profiling (Sekula-Neuner et al. 2012).

Protein Nanoarrays

High-throughput protein arrays allow the miniaturized and parallel analysis of large numbers of diagnostic biomarkers in complex samples. This capability can be enhanced by nanotechnology. DPN technique has been extended to protein arrays with features as small as 45 nm, and immunoproteins as well as enzymes can be deposited. Selective binding of antibodies to protein nanoarrays can be detected without the use of labels by monitoring small (5–15 nm) topographical height increases in AFM images.

Miniaturized microarrays, "mesoarrays," created by DPN with protein spots 400× smaller by area compared to conventional microarrays, were used to probe the ERK2-KSR binding event of the Ras/Raf/MEK/ERK signaling pathway at a physical scale below that previously reported (Thompson et al. 2011). This study serves as a first step toward an approach that can be used for analysis of proteins at a concentration level comparable to that found in the cellular environment.

Single-Molecule Protein Arrays

The ability to control the placement of individual protein molecules on surfaces could enable advances in many areas ranging from the development of nanoscale biomolecular devices to fundamental studies in cell biology. An approach that combines scanning probe block copolymer lithography with site-selective immobilization strategies has been used to create arrays of proteins down to the single-molecule level with arbitrary pattern control (Chai et al. 2011). Scanning probe block copolymer lithography was used to synthesize individual sub-10-nm single crystal gold nanoparticles to act as scaffolds for the adsorption of functionalized alkylthiol monolayers for facilitating the immobilization of specific proteins. The number of protein molecules that adsorb onto the nanoparticles depends on particle size; when the particle size approaches the dimensions of a protein molecule, each particle can support a single protein. This was demonstrated with both gold nanoparticle and QD labeling coupled with TEM imaging. The immobilized proteins remain bioactive, as demonstrated by enzymatic assays and antigen-antibody binding experiments.

Microfluidics and Nanofluidics

Microfluidics is the handling and dealing with small quantities (e.g., microliters, nanoliters, or even picoliters) of fluids flowing in channels the size of a human hair (~50 μm thick) even narrower. Fluids in this environment show very different properties than in the macroworld. This new field of technology was enabled by advances in microfabrication – the etching of silicon to create very small features. Microfluidics is one of the most important innovations of biochip technology. Typical dimensions of microfluidic chips are 1–50 cm^2 and have channels 5–100 μm. Usual volumes are 0.01–10 μl but can be less. Microfluidics is the link between microarrays and nanoarrays as we reduce the dimensions and volumes.

Microfluidics is the underlying principle of lab-on-a-chip devices, which carry out complex analyses, while reducing sample and chemical consumption, decreasing waste and improving precision and efficiency. The idea is to be able to squirt a very small sample into the chip, push a button and the chip will do all the work, delivering a report at the end. Microfluidics allows the reduction in size with a corresponding increase in the throughput of handling, processing, and analyzing the sample. Other advantages of microfluidics include increased reaction rates, enhanced detection sensitivity, and control of adverse events.

Drawbacks and limitations of microfluidics and designing of microfluidic chips include the following:

- Difficulties in microfluidic connections.
- Because of laminar flows, mixing can only be performed by diffusion.
- Large capillary forces.
- Clogging.
- Possible evaporation and drying up of the sample.

Applications of microfluidics include the following:

- DNA analysis
- Protein analysis
- Gene expression and differential display analysis
- Biochemical analysis

Nanotechnology on a Chip

Nanotechnology on a chip is a new paradigm for total chemical analysis systems. The ability to make chemical and biological information much cheaper and easier to obtain is expected to fundamentally change healthcare, food safety, and law enforcement. Lab-on-a-chip technology involves micro-total analysis systems that are distinguished from simple sensors because they conduct a complete analysis; a raw mixture of chemicals goes in and an answer comes out. Sandia National Laboratories is developing a handheld lab-on-a-chip that will analyze for airborne chemical warfare agents and liquid-based explosives agents. This development project brings together an interdisciplinary team with areas of expertise including microfabrication, chemical sensing, microfluidics, and bioinformatics. Although nanotechnology plays an important role in current efforts, miniaturized versions of conventional architecture and components such as valves, pipes, pumps, and separation columns are patterned after their macroscopic counterparts. Nanotechnology will provide the ability to build materials with switchable molecular functions that could provide completely new approaches to valves, pumps, chemical separations, and detection. For example, fluid streams could be directed by controlling surface energy without the need for a predetermined architecture of physical channels. Switchable molecular membranes and the like could replace mechanical valves. By eliminating the need for complex fluidic networks and microscale components used in current approaches, a fundamentally new approach will allow greater function in much smaller, lower-power total chemical analysis systems.

A new scheme for the detection of molecular interactions based on optical readout of nanoparticle labels has been developed. Capture DNA probes can be arrayed on a glass chip and incubated with nanoparticle-labeled target DNA probes, containing a complementary sequence. Binding are monitored by optical means, using reflected and transmitted light for the detection of surface-bound nanoparticles. Control experiments' significant influence of nonspecific binding on the observed contrast can be excluded. Distribution of nanoparticles on the chip surface can be demonstrated by scanning force microscopy.

BioForce Nanosciences has taken the technology of the microarray to the next level by creating the "nanoarray," an ultraminiaturized version of the traditional microarray that can actually measure interactions between individual molecules down to resolutions of as little as one nanometer. Here, 400 nanoarray spots can be placed in the same area as a traditional microarray spot. Nanoarrays are the next evolutionary step in the miniaturization of bioaffinity tests for proteins, nucleic acids, and receptor-ligand pairs. On a BioForce NanoArrayT, as many as 1,500 different samples can be queried in the same area now needed for just one domain on a traditional microarray.

Microfluidic Chips for Nanoliter Volumes

Nanoliter implies extreme reduction in quantity of fluid analyte in a microchip. The use of the word "nano" in nanoliter (nL) is in a different dimension than in nanoparticle, which is in nanometer (nm) scale. Chemical compounds within individual nanoliter droplets of fluid can be microarrayed on to glass slides at 400 spots/cm^2. Using aerosol deposition, subsequent reagents and water can be metered into each reaction center to rapidly assemble diverse multicomponent reactions without cross contamination or the need for surface linkage. Such techniques enable the kinetic profiling of protease mixtures, protease-substrate interactions, and high-throughput screening reactions. From one printing run that consumes <1 nanomole of each compound, large combinatorial libraries can be subjected to numerous separation-free homogeneous assays at volumes much smaller than current high-throughput methods. The rapid assembly of thousands of nanoliter reactions per slide using a small biological sample represents a new functional proteomics format implemented with standard microarraying and spot-analysis tools.

Use of Nanotechnology in Microfluidics

Construction of Nanofluidic Channels

Techniques such as nanoimprinting are used to construct large arrays of nanoscale grooves with efficiency and speed. Such grooves can be sealed with similar ease, to form nanofluidic channels. Laser-assisted direct imprint techniques enable the construction of millions of enclosed nanofluidic channels side by side on a single substrate, which is ideal for such parallel processing. By sputter-depositing silicon dioxide at an angle onto an array of prefabricated nanochannels imprinted into the surface of a biopolymer substrate, not only is an effective and uniform seal formed over the entire array, but the channels are further narrowed down to 10 nm, from an initial width of 55 nm. This process could be easily used for narrowing and sealing micro- and nanofluidic structures formed by other patterning techniques. By minimizing the hollow space in such structures, it could help surpass existing limitations in the spatial resolution of these techniques.

A chip-scale maze for combing out strands of DNA and inserting them into nanoscale channels was made using standard, inexpensive lithographic techniques. Their "gradient nanostructure" might be used to isolate and stretch DNA molecules for analysis, e.g., to look for bound proteins such as transcription factors along the strand. Such molecules would be obscured in normal solution because DNA, like any other linear polymer, collapses into a random coil as a featureless blob. The strands can, in principle, be straightened by feeding them into channels just a few tens of nanometers wide, using nanofluidic techniques.

Restriction mapping with endonucleases is a central method in molecular biology. Restriction mapping of DNA molecules can be performed using restriction endonucleases in nanochannels with diameters of 100–200 nm. It is based on the measurement of fragment lengths after digestion, while possibly maintaining the respective order. The location of the restriction reaction within the device is controlled by electrophoresis and diffusion of Mg^{2+} and EDTA. It is possible to measure the positions of restriction sites with precision using single DNA molecules with a resolution of 1.5 kbp within 1 min.

A review of nanofluidic systems reveals that these are divided into two large categories: top-down and bottom-up methods. The technology in the region of 1–10 nm is lacking and potentially can be covered by using the pulsed laser deposition method as a controlled way for thin film deposition (thickness of a few nanometers) and further structuring by the top-down method.

The benefits of operating in the nanoliter space include reduction of solvent, waste disposal costs, and human exposure by factors of 1,000×. New routine liquid handling capabilities include a purported 10× increase in MALDI sensitivity for analysis of proteins in proteomics work as demonstrated by various products such as nanoliter syringes based on induction-based fluidics technology, which uses electric fields to launch liquids to targets.

Nanoscale Flow Visualization

Most of the microscale flow visualization methods evolved from methods developed originally for macroscale flow. It is unlikely, however, that developed microscale flow visualization methods will be translated to nanoscale flows in a similar manner. Resolving nanoscale features with visible light presents a fundamental challenge. Although point-detection scanning methods have the potential to increase the flow measurement resolution on the microscale, spatial resolution is ultimately limited by the optical probe volume (length scale on the order of 100 nm), which, in turn, is limited by the wavelength of light employed. Optical spatially resolved flow measurements in nanochannels are difficult to visualize. There is a need for refinement of microscale flow visualization methods and the development of direct flow measurement methods for nanoflows.

Moving (Levitation) of Nanofluidic Drops with Physical Forces

The manipulation of droplets/particles that are isolated (levitated in gas/vacuum) from laboratory samples containing chemicals, cells, and bacteria or viruses is important both for basic research in physics, chemistry, biology, biochemistry, and colloidal science and for applications in nanotechnology and microfluidics. Various optical, electrostatic, electromagnetic, and acoustic methods are used for levitation.

Microfluidic drops can be moved with light – the lotus effect. On a super rough surface, when light shines on one side of a drop, the surface changes, the molecules switch, and the drop moves. This technology has potential applications in drug screening as it can be used for quickly analyzing and screening small amounts of biological materials. Called digital microfluidics, this approach enables one to quickly move small drops around by shining light on them. Hundreds of screens could be done on only one particular surface. The molecules, e.g., protein traces, do not interfere with movements of the drops because the surfaces are hydrophobic and the molecules have little contact with the surface.

The size of diamagnetic levitation devices can be reduced by using micron-scale permanent magnets to create a magnetic micromanipulation chip, which operates with femtodroplets levitated in air. The droplets used are one billion times smaller in volume than has been demonstrated by conventional methods. The levitated particles can be positioned with up to 300 nm accuracy and precisely rotated and assembled. Using this lab-on-a-chip, it might be possible to do the same thing with a large number of fluids, chemicals, and even red blood cells, bacteria, and viruses.

Electrochemical Nanofluid Injection

The ability to manipulate ultrasmall volumes of liquids is required in such diverse fields as cell biology, microfluidics, capillary chromatography, and nanolithography. In cell biology, it is often necessary to inject materials of high molecular weight such as DNA and proteins into living cells because their membranes are impermeable to such molecules. Currently used techniques for microinjection are limited by the relatively large injector size and poor control of the amount of injected material. An electrochemical attosyringe has been devised for electrochemical control of the fluid motion that enables one to sample and dispense attoliter-to-picoliter (10^{-18} to 10^{-12} l) volumes of either aqueous or nonaqueous solutions (Laforge et al. 2007). By changing the voltage applied across the liquid/liquid interface, one can produce a sufficient force to draw solution inside a nanopipette and then inject it into an immobilized biological cell. A high success rate was achieved in injections of fluorescent dyes into human cells in culture. The injection of femtoliter-range volumes can be monitored by video microscopy, and current/resistance-based approaches can be used to control injections from very small pipettes. Other potential applications of the electrochemical syringe include fluid dispensing in nanolithography and pumping in nanofluidic systems.

Nanofluidics on Nanopatterned Surfaces

A very thin layer of liquid behaves on a "nanopatterned" silicon surface, i.e., a surface etched with an ordered array of cavities, each only 20 nm deep. Watching how a liquid adsorbs on a nanopatterned surface is one way to study the basic properties of liquids that are confined in extremely tiny amounts within nanoscale structures. Understanding these properties will help in developing many useful fluid-based nanotechnologies. This work could help improve the "lab-on-a-chip."

Currently, the knowledge about the microscopic behavior of liquids on solid surfaces, known as "wetting" phenomena, is predominately based on measurements taken using structureless, flat surfaces. In those cases, the behavior of the liquid is based on the strength of attractive molecule-molecule forces known as "van der Waals interactions." But for a surface that contains a regular pattern of cavities, the shape of the surface influences how the liquid will fill those cavities. Analysis of the X-ray data reveals that a liquid layer builds up inside each nanocavity at a faster rate than on a flat surface of the same material. The wetting properties of the surface are considerably enhanced by the nanopatterning.

Nano-interface in a Microfluidic Chip

There are emerging experimental and conceptual platforms for probing living cells with nanotechnology-based tools in a microfluidic chip. Considerable advances have been made in measuring nanoscale mechanical, biochemical, and electrical interactions at the interface between biomaterials and living cells. By merging the fields of microfluidics, electrokinetics, and cell biology, microchips are capable of creating tiny, mobile laboratories. The challenge for the future of designing a nano-interface in a microfluidic chip to probe a living cell lies in seamlessly integrating techniques into a robust and versatile, yet reliable, platform. Potential benefits of nanosystems on a microchip result from real-time detection of numerous events in parallel. In addition to early detection of cell-level dysfunctions, these systems will enable broad screening that encompasses not just a large number of toxic stimuli and disease processes but also population subgroups. This will facilitate the development of personalized medicine. To reach this goal requires advancing the knowledge base of cellular and subcellular functions, perhaps by designing nanosystems that operate in the tissue milieu.

Nanofluidic Channels for Study of DNA

Nanofluidic channels enable molecular biologists to spot the association and dissociation of proteins on fluorescently labeled DNA. The simple system could even help researchers visualize induced tertiary structures such as loops, which push conventional optical or magnetic stretching methods to the limit. This silicon dioxide-glass nanochannel system, also referred to as nanoslit, requires no externally applied forces or fields (Krishnan et al. 2007). To unravel the molecules, scientists place a drop of solution containing DNA at one end of the nanochannel. Capillary action then draws the liquid into channels measuring 2–10 μm wide and 100 nm deep. After 1 min, a drop of buffer solution is added at the other end of the channel to equalize the pressure in the device and stop the flow. In channels of 100 nm depth or less, DNA molecules spontaneously adopt an extended state adjacent to the channel wall. The nanochannel geometry, however, physically confines polymer molecules to two spatial dimensions. Further reduction in configuration results in spontaneous axial stretching of molecules and appears to be electrostatically mediated. The physics for stretching a DNA molecule is built into the structure of the device. Fabrication of the channels and mass production of the unit are easy. Devices are made by first patterning a silicon substrate using laser lithography and then forming parallel channels 100 nm deep by either reactive ion etching in plasma or wet etching in HF. Cover glass is used to seal the channels from above.

Visualization and Manipulation on Nanoscale

4Pi Microscope

The most prominent restrictions of fluorescence microscopy are the limited resolution and the finite signal. Established conventional, confocal, and multiphoton microscopes resolve at best approximately 200 nm in the focal plane and only 500 nm in depth.

4Pi microscope (Leica Microsystems) uses a special phase- and wavefront-corrected double-objective imaging system linked to a confocal scanner to enable four- to sevenfold increased axial resolution over confocal and two-photon microscope. Even in living specimens, axial sections of ~100 nm are obtained. The system maintains all advantages of fast scanning, Acousto-Optical Beam Splitting (AOBS®), and spectral detection of the Leica TCS SP2 AOBS for routine operation. The first marked leap in resolution in commercial 3D fluorescence microscopy opens up new dimensions for research in cell and developmental biology. Colocalization studies of immunolabeled microtubules and mitochondria demonstrate the feasibility of 4Pi microscopy for routine biological measurements, in particular, to visualize the 3D entanglement of the two networks with unprecedented detail (Medda et al. 2006).

Atomic Force Microscopy

Basic AFM Operation

In its most basic form, atomic force microscopy (AFM) images topography by precisely scanning a probe across the sample to "feel" the contours of the surface. The interaction between the needle and the surface is measured, and an image is reconstructed from the data collected in this manner. With AFM, it is possible to reach an extremely high resolution. Because it can be applied under standard conditions in an aqueous environment, any significant perturbation of the sample can be avoided. In contrast to light microscopy and scanning electron microscopy, AFM provides the most optimal means to investigate the surface structures in three dimensions, with resolutions as high as 0.1–0.2 nm.

A key element of the AFM is its microscopic force sensor, or cantilever. The cantilever is usually formed by one or more beams of silicon or silicon nitride that is 100–500 μm long and about 0.5–5 μm thick. Mounted on the end of the cantilever is a sharp tip that is used to sense a force between the sample and tip. For normal topographic imaging, the probe tip is brought into continuous or intermittent contact with the sample and raster-scanned over the surface.

Advantages of AFM

In addition to its superior resolution and routine three-dimensional measurement capability, AFM offers several other clear advantages over traditional microscopy techniques. For example, scanning and transmission electron microscopy (SEM, TEM) image biologically inactive, dehydrated samples and generally require extensive sample preparation such as staining or metal coating. AFM eliminates these requirements and, in many cases, allows direct observation of native specimens and ongoing processes under native or near-native conditions.

Further adding to its uniqueness, the AFM can directly measure nanoscale interactive forces, e.g., ligand-receptor binding. Samples can be examined in ambient air or biological fluids without the cost and inconvenience of vacuum equipment. Sample preparation is minimal and allows the use of standard techniques for optical microscopy. The MultiMode AFM provides maximal resolution, while the BioScope AFM integrates the best of optical and atomic force microscopy to help life scientists explore new frontiers.

The ability of the AFM to create 3D micrographs with resolution down to the nanometer and Angstrom scales has made it an essential tool for imaging surfaces in applications ranging from semiconductor processing to cell biology. In addition to this topographical imaging, however, the AFM can also probe nanomechanical and other fundamental properties of sample surfaces, including their local adhesive or elastic (compliance) properties.

Microscopic adhesion affects a huge variety of events, from the behavior of paints and glues, ceramics, and composite materials to DNA replication and the action of drugs in the human body. Elastic properties are similarly important, often affecting the structural and dynamic behavior of systems from composite materials to blood cells. AFM offers a new tool to study these important parameters on the micron to nanometer scale using a technique that measures forces on the AFM probe tip as it approaches and retracts from a surface.

Force Sensing Integrated Readout and Active Tip

Force sensing integrated readout and active tip (FIRAT) is an extremely sensitive AFM technology that is capable of high-speed imaging 100 times faster than current AFM technology. Current AFM scans surfaces with a thin cantilever with a sharp tip at the end. An optical beam is bounced off the cantilever tip to measure the deflection of the cantilever as the sharp tip moves over the surface and interacts with the material being analyzed. FIRAT works a bit like a cross between a pogo stick and a microphone. In one version of the probe, the membrane with a sharp tip moves toward the sample, and just before it touches, it is pulled by attractive forces.

Much like a microphone diaphragm picks up sound vibrations, the FIRAT membrane starts taking sensory readings well before it touches the sample. And when the tip hits the surface, the elasticity and stiffness of the surface determine how hard the material pushes back against the tip. So rather than just capturing a topography scan of the sample, FIRAT can pick up a wide variety of other material properties.

FIRAT can capture additional measurements not possible before with AFM, including parallel molecular assays for drug screening and discovery, as well as material property imaging. This research breakthrough could prove invaluable for many types of nanoresearch, including translating into movies of molecular interactions in real time. FIRAT might eventually replace AFM.

AFM as Nanorobot

An AFM-based nanorobot has been introduced for biological studies (Xi et al. 2011). Using the AFM tip as an end effector, the AFM can be modified into a nanorobot that can manipulate biological objects at the single-molecule level. By functionalizing the AFM tip with specific antibodies, the nanorobot is able to identify specific types of receptors on the cell membrane. It is similar to the fluorescent optical microscopy but with higher resolution. By locally updating the AFM image based on interaction force information and objects' model during nanomanipulation, real-time visual feedback is obtained through the augmented reality interface. The development of the AFM-based nanorobotic system enables us to conduct in situ imaging, sensing, and manipulation simultaneously at the nanometer scale (e.g., protein and DNA levels). The AFM-based nanorobotic system offers several advantages and capabilities for studying structure-function relationships of biological specimens. As a result, many biomedical applications can be achieved by the AFM-based nanorobotic system.

Cantilever Technology

Cantilevers (Concentris) transform a chemical reaction into a mechanical motion on the nanometer scale. Measurements of a cantilever are length 500 μm, width 100 μm, thickness 25–500 μm, and deflection 10 nm. This motion can be measured directly by deflecting a light beam from the cantilever surface. Concentris uses an array of parallel VCSELs (vertical cavity surface emitting lasers) as stable, robust, and proven light source. A state-of-the-art position sensitive detector is employed as detection device.

The static mode is used to obtain information regarding the presence of certain target molecules in the sample substance. The surface stress caused by the adsorption of these molecules results in minute deflections of the cantilever. This deflection directly correlates with the concentration of the target substance. The dynamic mode allows quantitative analysis of mass loads in the subpicogram area.

Table 2.5 Applications of cantilever technology

Basic research
Study of chemical reactions or host-guest interactions on surfaces
Nanocalorimetry
Medical diagnostics
Parallel and label-free detection of disease markers, e.g., serum proteins or autoantibodies
Fast, label-free recognition of specific DNA sequences (SNPs, oncogenes, genotyping)
Detection of microorganisms and antimicrobial susceptibility
Drug discovery and life sciences research
Label-free biochemical assays and investigation of biomolecular interactions
Multiplexed assays
Process and quality control
Process monitoring
Purity analysis
Food analysis
Detection of trace contaminations, e.g., antibiotics, hormones, and pesticides
Detection of microorganisms
Identification and quality control
Environmental monitoring
Detection of heavy metal ions, pesticides, and air pollutants
Water analysis
Fragrance and flavor analysis
Using neural networks to analyze cantilever sensor array signals can identify and characterize complex chemical mixtures ("electronic nose" or "tongue")
Security devices
Detection of potentially hazardous chemicals and microorganisms
Workplace security

Source: Concentris GmbH

As molecules get adsorbed, minimal shifts in the resonance frequency of an oscillating cantilever can be measured and associated to reference data of the target substance. Both modes can also be operated simultaneously.

The controlled deposition of functional layers is the key to converting nanomechanical cantilevers into chemical or biochemical sensors. Inkjet printing is a rapid and general method to coat cantilever arrays efficiently with various sensor layers (Bietsch et al. 2004). Self-assembled monolayers of alkanethiols are deposited on selected Au-coated cantilevers and rendered them sensitive to ion concentrations or pH in liquids. The detection of gene fragments is achieved with cantilever sensors coated with thiol-linked single-stranded DNA oligomers on Au. A selective etch protocol proves the uniformity of the monolayer coatings at a microscopic level. A chemical gas sensor is fabricated by printing thin layers of different polymers from dilute solutions onto cantilevers. The inkjet method is easy to use, faster, and more versatile than coating via microcapillaries or the use of pipettes. In addition, it is scalable to large arrays and can coat arbitrary structures in noncontact.

The applications of cantilever technology, Cantosens (Concentris), are listed in Table 2.5 and discussed further in Chap. 3.

Further research continues at academic laboratories to develop nanoscale cantilevers, which would be smaller than the wavelength of light and make laser detection more difficult. When these devices are developed, they could be incorporated into AFMs, which are currently designed for standard size cantilevers.

AFM cantilevers have been actuated using a microheater at the bottom and integrated with deflection sensor as well as microactuator for imaging of soft biological samples in fluid (Fantner et al. 2009). Influence of the water was investigated on the cantilever dynamics, the actuation and the sensing mechanisms, as well as the crosstalk between sensing and actuation. Successful imaging of yeast cells in water using the integrated sensor and actuator shows the potential of the combination of this actuation and sensing method. This constitutes a major step toward the automation and miniaturization required to establish AFM in routine biomedical diagnostics and in vivo applications.

CytoViva® Microscope System

Specifically designed to support research in nanotechnology and infectious disease, the CytoViva Microscope System (CytoViva Inc.) employs a proprietary dark-field-based optical illumination technology, which dramatically improves contrast and signal-to-noise ratio. This transmitted-light illumination system enables scientists to observe a wide range of nanomaterials quickly and easily, without any special preparation. In addition, live cells and pathogens can also be viewed at a level of detail not possible with traditional optical imaging techniques such as phase contrast or differential interference contrast. When using the CytoViva Dual Mode Fluorescence system, researchers can also observe the interactions between fluorescently labeled nanoparticles or bacteria and live unlabeled cells. This unique capability can eliminate the need to create computer-enhanced overlay images, which require two different illumination methods and advanced software programs. Finally, when combined with the CytoViva Hyperspectral Imaging system, this high-contrast microscopy method enables researchers to secure spectral data from these images.

Fluorescence Resonance Energy Transfer

Fluorescence resonance energy transfer (FRET) is a process by which energy that would normally be emitted as a photon from an excited fluorophore can be directly transferred to a second fluorophore to excite one of its electrons. This, on decay, then generates an even longer wavelength photon. The extent of FRET is critically dependent on the distance between the two fluorophores as well as their spectral overlap. Thus, FRET is a powerful reporter of the separation of the two fluorophores. FRET is a simple but effective tool for measurements of protein-protein interactions.

It is one of the few techniques that are capable of giving dynamic information about the nanometer-range proximity between molecules, as opposed to simply the subcellular colocalization that is provided by fluorescence microscopy.

Magnetic Resonance Force Microscopy and Nanoscale MRI

IBM has been working over a decade to develop nanoscale magnetic resonance imaging technology called magnetic resonance force microscopy (MRFM). The central feature of MRFM is a silicon "microcantilever" that looks like a miniature diving board and is 1,000 times thinner than a human hair. It vibrates at a frequency of about 5,000 times a second, and a tiny but powerful magnetic particle attached to the tip attracts or repels individual electrons. The company claimed a breakthrough in nanoscale MRI by directly detecting for the first time a faint magnetic signal from single electrons buried inside solid samples. MRFM has been combined with 3D image reconstruction to achieve MRI with resolution <10 nm (Degen et al. 2009). The image reconstruction converts measured magnetic force data into a 3D map of nuclear spin density, taking advantage of the unique characteristics of the "resonant slice" that is projected outward from a nanoscale magnetic tip. The basic principles are demonstrated by imaging the 1H spin density within individual tobacco mosaic virus particles sitting on a nanometer-thick layer of adsorbed hydrocarbons. This result, which represents a 100-millionfold improvement in volume resolution over conventional MRI, demonstrates the potential of MRFM as a tool for 3D, elementally selective imaging on the nanometer scale.

With further progress in resolution and sample preparation, force-detected MRI techniques could have significant impact on the imaging of nanoscale biological structures, even down to the scale of individual molecules. Achieving resolution of 1 nm appears to be realistic because the current apparatus operates almost a factor of 10 away from the best demonstrated force sensitivities and field gradients. Even with a resolution >1 nm, MRFM may enable the basic structure of large molecular assemblies to be elucidated. MRFM image contrast can be enhanced beyond the basic spin-density information by using techniques similar to that developed for clinical MRI and NMR spectroscopy. Such contrast may include selective isotopic labeling, selective imaging of different chemical species, relaxation-weighted imaging, and spectroscopic imaging that reflects the local chemical environment.

The development represents a major milestone in the creation of a microscope that can make 3D images of molecules with atomic resolution. Such a device could have a major impact on the study of materials, ranging from proteins and pharmaceuticals to integrated circuits for which a detailed understanding of the atomic structure is essential. The ability to image the detailed atomic structure of proteins directly would also aid the development of new drugs. This new capability should ultimately lead to fundamental advances in nanotechnology and biology.

Multiple Single-Molecule Fluorescence Microscopy

Fitting the image of a single molecule to the point spread function of an optical system greatly improves the precision with which single molecules can be located. In nanometer-localized multiple single-molecule (NALMS) fluorescence microscopy, short duplex DNA strands are used as nanoscale "rulers" for validation. Nanometer accuracy of this microscope has been demonstrated for 2–5 single molecules within a diffraction-limited area. NALMS microscopy will greatly facilitate single-molecule study of biological systems because it covers the gap between fluorescence resonance energy transfer-based (<10 nm) and diffraction-limited microscopy (>100 nm) measurements of the distance between two fluorophores. NALMS microscopy has been applied to DNA mapping with <10-nm resolution.

Near-Field Scanning Optical Microscopy

Near-field scanning optical microscopy (NSOM) was the first technique that has overcome the limits of light microscopy by about one order of magnitude. Typically, the resolution range below 100 nm is accessed for biological applications. Using appropriately designed scanning probes allows for obtaining an extremely small near-field light excitation volume (some tens of nanometers in diameter). Because of the reduction of background illumination, high-contrast imaging becomes feasible for light transmission and fluorescence microscopy. The height of the scanning probe is controlled by atomic force interactions between the specimen surface and the probe tip. The control signal can be used for the production of a topographic (nonoptical) image that can be acquired simultaneously.

Scattering near-field microscopy (s-SNOM) can determine infrared "fingerprint" spectra of individual poly(methyl methacrylate) nanobeads and viruses as small as 18 nm. Amplitude and phase spectra are found surprisingly strong, even at a probed volume of only 10^{-20} l, and robust in regard to particle size and substrate. This makes infrared spectroscopic s-SNOM a versatile tool for chemical and protein-secondary-structure identification.

Nanosized Light Source for Single Cell Endoscopy

A nanosized light source is capable of emitting coherent light across the visible spectrum. Among the potential applications of this nanosized light source are single cell endoscopy and other forms of subwavelength bio-imaging, integrated circuitry for nanophotonic technology, and new advanced methods of cyber cryptography. Working with individual nanowires, scientists have developed the first electrode-free, continuously tunable coherent visible light source that is compatible with physiological environments. It was shown that nanowires with diameters as small as 20 nm and aspect ratios of more than 100 can be trapped and transported in 3D, enabling the construction of nanowire architectures that may function as active photonic devices (Pauzauskie et al. 2006). They have also demonstrated that it is possible to trap and manipulate single nanowires with optical tweezers, a critical capability not only for bio-imaging but also for wiring together nanophotonic circuitry. This nanowire light source is like a tiny flashlight that can scan across a living cell, enabling visualization of the cell and at the same time, mechanically interacting with it.

Nanoparticle Characterization by Halo™ LM10 Technology

Halo™ LM10 (NanoSight Ltd.) is based on the laser illumination of a specially designed optical element on to which sample is simply placed manually or allowed to flow across the surface. This is the first nanoparticle characterization tool, specifically designed for liquid phase sizing of individual nanoparticles, with the use of a conventional light microscope. Particles as small as 20 nm have been successfully visualized by this method, each particle being seen as an individual point of light moving under Brownian motion within the liquid. The intensity of light scattered by a particle varies as the sixth power of its radius. By doubling the diameter of the particle, 64-fold more light is scattered by the particle. This has significant implications for the early and simple detection of aggregation, flocculation, and dimerization of particulates at the nanometer scale.

Use of a shorter wavelength laser source capable of exciting fluorescent labels enables specific components within the sample to be distinguished from nonspecific background particles. The image can be analyzed by suitable software allowing changes in individual particle position to be followed furnishing real-time information about particle diffusion and particle-to-particle interactions. In the fluorescence mode, correlation techniques can be used to derive information by use of the technique known as fluorescence correlation spectroscopy. Halo™ LM10 is supported by Halo™ GS10 software.

The laser source need only be a few mW in power and can be delivered to the optical device via fiber optic connection, or the laser diode can be coupled directly to the edge of the optical element. The optical element can be manufactured in optical quality plastic or in glass or silica. The optical element need only be a few mm square and 2–5 mm in depth.

Larger volumes of sample containing dilute concentrations of particles of interest can be analyzed by being configured within a flow cell. Fabrication of the optical element is by industry standard metal coating techniques such as those found in the electronics and optical devices industries. Applications relevant to nanobiotechnology are shown in Table 2.6.

Table 2.6 Applications of optical nanoscopy

Molecular diagnostics
Detection of viral particles
DNA analysis
Mycoplasma detection in animal cell culture
Contaminant monitoring
Drug delivery
Drug carriers
Monitoring drug efficacy in body fluids
Biofilm production and implants
Nanoparticles
Environmental
Biodefense
Airborne contaminants such as asbestos particles
Medical
Clinical diagnosis of viral diseases, e.g., cerebrospinal fluid
Cancer cell detection, e.g., metastases

Source: Nanosight Ltd.

Nanoscale Scanning Electron Microscopy

Pharmaceutical enterprises require a range of imaging products that provide high-quality information, allowing them to reach their own targets on technology, productivity, and ultimately profitability. With the increasing expectations upon drug delivery systems for efficient and controlled delivery of the active material, there is a matching need for accurate information on these mechanisms. One of the most effective instruments in this area is a scanning electron microscope (SEM) from Carl Zeiss that has a unique ability to provide high-resolution images of a specimen under investigation. One example of its application in EVO® EP instrument (Carl Zeiss) is in manufacture of aspirin. The interaction of water with soluble aspirin demonstrates the mechanisms by which tablets lose mechanical strength and stability and hence release the active material. This process can be observed in real time in the SEM by introducing water vapor into the chamber at sufficiently high pressures that liquid water is condensed onto the specimen. During the wetting phase, the particle absorbs water and fragments. During the drying phase, the reverse processes can be followed in detail. The latest product from Carl Zeiss, ULTRA 55 field emission SEM, features a totally new "complete detection system," which enables simultaneous surface, compositional, and crystallographic imaging down to the nanometer level with high signal contrast and unsurpassed clarity.

Use of SEM to Reconstruct 3D Tissue Nanostructure

3D structural information is important in biological research. Excellent methods are available to obtain structures of molecules at atomic, organelles at electron microscopic, and of tissue at light-microscopic resolution. However, there is a need to reconstruct 3D tissue structure with a nanoscale resolution to identify small organelles such as synaptic vesicles. Such 3D data are essential to understand cellular networks that need to be completely reconstructed throughout a substantial spatial volume, particularly in the nervous system. Datasets meeting these requirements can be obtained by automated block-face imaging combined with serial sectioning inside the chamber of a SEM. Backscattering contrast is used to visualize the heavy-metal staining of tissue prepared using techniques that are routine for TEM. The resolution is sufficient to trace even the thinnest axons and to identify synapses. Stacks of several hundred sections, 50–70 nm thick, have been obtained at a lateral position jitter of typically under 10 nm. This opens up the possibility of automatically obtaining the electron-microscope-level 3D datasets needed to completely reconstruct the neuronal circuits.

Optical Imaging with a Silver Superlens

A superlens has been created using a thin film of silver as the lens and ultraviolet light that can overcome a limitation in physics that has historically constrained the resolution of optical images. The superlens has been used to record the images of an array of nanowires at a resolution of about 60 nm, whereas current optical microscopes can only make out details down to 400 nm. This work has a far-reaching impact on the development of detailed biomedical imaging. With current optical microscopes, scientists can only make out relatively large structures within a cell, such as its nucleus and mitochondria. With a superlens, optical microscopes could reveal the movements of individual proteins traveling along the microtubules that make up a cell's skeleton. SEM and AFM are now used to capture detail down to a few nanometers. However, such microscopes create images by scanning objects point by point, which means they are typically limited to nonliving samples, and image capture times can take up to several minutes. Optical microscopes can capture an entire frame with a single snapshot in a fraction of a second, opening up nanoscale imaging to living materials, which can help biologists better understand cell structure and function in real time and ultimately help in the development of new drugs to treat human diseases.

Photoactivated Localization Microscopy

Photoactivated localization microscopy (PALM) enables scientists peering inside cells to discern individual proteins at nanometer ~2 to 25 nm resolution (Betzig et al. 2006). The basic concepts behind this technology are simple: The researchers label the molecules they want to study with a photoactivatable probe, and then expose those molecules to a small amount of violet light. The light activates fluorescence in a small percentage of molecules, and the microscope captures an image of those that are turned on until they bleach. The process is repeated approximately 10,000 times, with each repetition capturing the position of a different subset of molecules. When a final image is created that includes the center of each individual molecule, it has a resolution previously only achievable with an electron microscope. Unlike electron microscopy, however, the new technique allows for more flexibility in labeling molecules of interest. The method is demonstrated in thin sections by imaging specific target proteins in lysosomes and mitochondria and in fixed, whole cells by imaging vinculin at focal adhesions, actin within a lamellipodium, and the distribution of the retroviral protein Gag at the plasma membrane. A great feature of PALM is that it can be readily used with electron microscopy, which produces a detailed image of very small structures, but not proteins, in cells. By correlating a PALM image showing protein distribution with an electron microscope image showing cell structure of the same sample, it becomes possible to understand how molecules are individually distributed in a cellular structure at the molecular scale. Correlative PALM unites the advantages of light and electron microscopy, producing a revolutionary new approach for looking at the cell in molecular detail. As the PALM technology advances, it may prove to be a key factor in unlocking at the molecular level secrets of intracellular dynamics that are unattainable by other methods. However, the time needed to collect the thousands of single-molecule images that go into each PALM picture is cumbersome. With the camera snapping one to two pictures each second, it can take 2–12 h to image a single sample. Activating more molecules per frame would reduce the number images that must be collected, and making the molecules brighter would reduce the time needed to take each image. Either would help to speed the PALM process. The technique is still undergoing refinements with an aim to developing a practical tool for use by biologists.

Scanning Probe Microscopy

The scanning probe microscope (SPM) system is an important tool for nonintrusive interrogation of biomolecular systems in vitro. Its particular merit is that it retains complete functionality in a biocompatible fluid environment and can track the dynamics of cellular and molecular processes in real time and real space at nanometer resolution, as an imaging tool, and with pN force sensing/imposing resolution, as an interaction tool. The capability may have relevance as a test bed for monitoring cellular response to environmental stimuli and pharmaceutical intervention. Best-known contributions of SPM are toward explanatory and predictive descriptions of biomolecular interactions at surfaces and interfaces, and there are some attempts to reconfigure the SPM platform for demonstration of novel biodevice applications. SPM enables high resolution without any of the drawbacks of electron microscopy, which can damage sensitive molecules by electrons. SPM enables investigation of biomolecules in fluid environments under physiological conditions and is useful for study of biology on nanoscale.

Scanning ion-conductance microscopy (SICM) is part of the larger family of SPM. It was specially designed for the submicrometer resolution scanning of soft nonconductive materials that are bathed in electrolyte solution. It consists of an electrically charged glass micro- or nanopipette probe filled with electrolyte lowered toward the surface of the sample (which is nonconducting for ions) in an oppositely charged bath of electrolyte. As the tip of the micropipette approaches the sample, the ion conductance and therefore current decreases since the gap through which ions can flow is reduced in size. SICM is a suitable tool for imaging surfaces of living cells in a contact-free manner and enables one to trace the outlines of entire cell soma and to detect changes in cellular shape and volume. SICM can also be employed to quantitatively observe cellular structures such as cell processes of living cells as well as cell soma of motile cells within hours (Happel and Dietzel 2009).

Partial Wave Spectroscopy

An optical microscopy technique, partial wave spectroscopy (PWS), is capable of quantifying statistical properties of cell structure at the nanoscale (Subramanian et al. 2009). PWS has been used to show for the first time the increase in the disorder strength of the nanoscale architecture not only in tumor cells but also in the microscopically normal-appearing cells outside of the tumor. Although genetic and epigenetic alterations have been previously observed in the field of carcinogenesis, these cells were considered morphologically normal. PWS can show organ-wide alteration in cell nanoarchitecture. This seems to be a general event in carcinogenesis, which is supported by data in three types of cancer: colon, pancreatic, and lung. These results have important implications in that PWS can be used as a new method to identify patients harboring malignant or premalignant tumors by interrogating easily accessible tissue sites distant from the location of the lesion. Once optimized, PWS could be used to detect cell abnormalities early and help physicians assess who might be at risk for developing cancer. Like a pap smear of the cervix, a simple brushing of cells is all that is needed to get the specimen required for testing. PWS can look inside the cell and see those critical building blocks, which include proteins, nucleosomes, and intracellular membranes, and detect changes to this nanoarchitecture. Conventional microscopy cannot do this, and other techniques that can (to some degree) are expensive and complex. PWS is simple, inexpensive, and minimally invasive.

Super-Resolution Microscopy for In Vivo Cell Imaging

Super-resolution microscopy comprises a variety of new approaches such as structured illumination (3D-SIM), localization microscopy (PALM, STORM), and stimulated emission depletion (STED) that have been developed to surpass the limits of conventional optical microscopes. These methods allow precise visualization and measurement of features that are below the diffraction limit. 3D-SIM (Applied Precision Inc./GE Healthcare) projects a structured light pattern onto the sample. The illumination pattern interacts with the fluorescent probes in the sample to generate interference patterns know as moiré fringes. By modulating the illumination pattern and collecting and reconstructing the subsequent images, super-resolution images with double the lateral and axial resolution are obtained.

Normal microscopes enable visualization of cells contents that are >200 nm in size but would be unable to detect a very small molecules such as insulin, which is about 10 nm in size. Super-resolution microscopy enables nanoscale observation of cells in vivo. In vivo study of cells provides invaluable information for study of pathomechanism of disease at cell level and for cell-based drug discovery. Electron microscopes have similar resolution to a super-resolution microscope, but they do not allow observation of cells in vivo. Exploration of cellular functions at nanoscale enables a better understanding of the processes that occur in a dysfunctional cell. Super-resolution microscopes have been used to study how the HIV virus penetrates cells and provide information for developing new drugs.

Ultrananocrystalline Diamond

A common problem in AFM is the deterioration of the tip apex as surfaces are scanned. To overcome this problem, an ultrananocrystalline diamond (UNCD) was used to fabricate a hard, low-wear probe for contact-mode writing techniques such as Dip Pen Nanolithography. Diamond, the hardest known material, is probably the optimal tip material for many applications. In addition to hardness, diamond is stiff, biocompatible, and wear resistant. Diamond tips with radii down to 30 nm were obtained through growth of UNCD films followed by selective etching of the silicon template substrate. The probes were monolithically integrated with diamond cantilevers and subsequently integrated into a chip body obtained by metal electroforming. The probes were characterized in terms of their mechanical properties, wear, and atomic force microscopy imaging capabilities. The developed probes performed exceptionally well in DPN molecular writing/imaging mode. Furthermore, the integration of UNCD films with appropriate substrates and the use of directed microfabrication techniques are particularly suitable for fabrication of one- and two-dimensional arrays of probes that can be used for massive parallel fabrication of nanostructures by the Dip Pen Nanolithography method. The technology can be employed for a variety of AFM scanning modes, from regular surface scanning in air or fluids to conductive AFM. It can also be employed as a nanofabrication tool. Examples include nanopatterning of biomolecules (for sequencing, synthesis, and drug discovery) and scanning probe electrochemistry (scanning electrode imaging, localized electrochemical etching or deposition of materials, and nanovoltametry). Potential markets include those industries where it is pivotal to preserve the performance of the tips or that require two-dimensional arrays for high throughput in which the cost of manufacturing is such that minimum possible tip wear is paramount. These include the chemical and biological sensor industry where high throughput and spatial resolution are important.

Visualizing Atoms with High-Resolution Transmission Electron Microscopy

The characterization of nanostructures down to the atomic scale is essential to understand some physical properties. Such a characterization is possible today using direct imaging methods such as aberration-corrected high-resolution transmission electron microscopy (HRTEM), when iteratively backed by advanced modeling produced by theoretical structure calculations and image calculations (Bar Sadan et al. 2008). Aberration-corrected HRTEM is therefore extremely useful for investigating low-dimensional structures, such as inorganic fullerene-like particles and inorganic nanotubes. The atomic arrangement in these nanostructures can lead to new insights into the growth mechanism or physical properties, where imminent commercial applications are unfolding. HRTEM study combined with modeling reveals new information regarding the chirality of the different shells and provides a better understanding of their growth mechanism. The next frontier will be seeing atoms in 3D.

Surface Plasmon Resonance

Surface plasmon resonance (SPR) is an optical-electrical phenomenon involving the interaction of light with the electrons of a metal. Light is coupled into the surface plasmon by means of either a prism or a grating on the metal surface. Depending on the thickness of a molecular layer at the metal surface, the SPR phenomenon results in a graded reduction in intensity of the reflected light. The optical-electronic basis of SPR is the transfer of the energy carried by photons of light to a group of electrons (a plasmon) at the surface of a metal.

In Biacore systems (Fig. 2.3), SPR arises when light is reflected under certain conditions from a conducting film at the interface between two media of different refractive index. The media are the sample and the glass of the sensor chip, and the conducting film is a thin layer of gold on the chip surface. SPR causes a reduction in the intensity of reflected light at a specific angle of reflection (the SPR angle). When molecules in the sample bind to the sensor surface, the concentration, and therefore the refractive index, at the surface changes, and an SPR response is detected. Plotting the response against time during the course of an interaction provides a quantitative measure of the progress of the interaction. This plot is called a sensorgram.

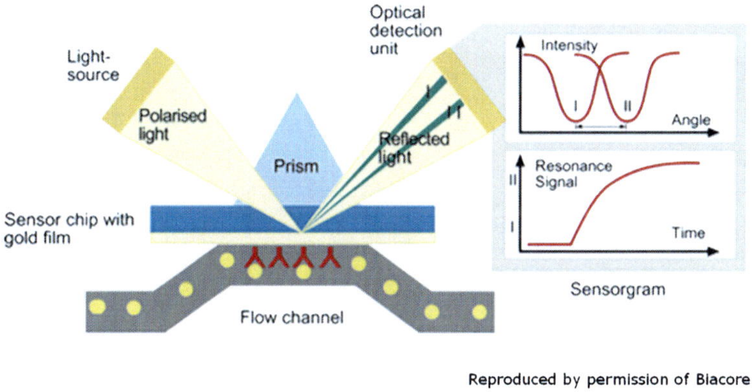

Fig. 2.3 Surface plasmon resonance (SPR) technology (Reproduced by permission of Biacore)

Biomedical applications take advantage of the exquisite sensitivity of SPR to the refractive index of the medium next to the metal surface, which makes it possible to measure accurately the adsorption of molecules on the metal surface and their eventual interactions with specific ligands. Applications of this technique include the following:

- Measurement in real time of the kinetics of ligand-receptor interactions
- Screening of lead compounds in the pharmaceutical industry
- Measurement of DNA hybridization
- Enzyme-substrate interactions
- Polyclonal antibody characterization
- Protein conformation studies
- Label-free immunoassays

Chapter 3
Nanotechnologies for Basic Research Relevant to Medicine

Introduction

Life sciences are the testing ground for many new biotechnologies for applications in medicine. Nanobiotechnology is a good example. Despite the remarkable speed of development of nanoscience, relatively little is known about the interaction of nanoscale objects with living systems. Much of the research in life sciences is directly relevant to applications described in the following chapters. Because of this overlap, some of the applications are indicated in this chapter and some of the research in life sciences is described along with applications. Important areas of research in life sciences where nanotechnologies are applied and that are relevant to applications in health sciences are:

- Role of nanotechnology in biological research
- Genomics and proteomics
- Gene sequencing
- Bioinformatics
- Assays

Nanotechnology and Biology

Investigative methods of nanotechnology have made inroads into uncovering fundamental biological processes, including self-assembly, cellular processes, and systems biology (such as neural systems). Key advances have been made in the ability to make measurements at the subcellular level and in understanding the cell as a highly organized, self-repairing, self-replicating, information-rich molecular machine. Single-molecule measurements are shedding light on the dynamics and mechanistic properties of molecular biomachines, both in vivo and in vitro, allowing the direct investigation of molecular motors, enzyme reactions, protein dynamics, DNA transcription, and cell signaling. It has also been possible to measure the chemical composition within a single cell in vivo. Micro total analysis systems, using molecular manipulation on nanoscale, offer the potential for highly efficient,

K.K. Jain, *The Handbook of Nanomedicine*, DOI 10.1007/978-1-61779-983-9_3,

simultaneous analysis of a large number of biologically important molecules in genomic, proteomic, and metabolic studies.

The physical sciences offer tools for synthesis and fabrication of devices for measuring the characteristics of cells and subcellular components and of materials useful in cell and molecular biology; biology offers a window into the most sophisticated collection of functional nanostructures that exists. Particles made of semiconductors at the nanoscale are already used in the electronic and information technology industries. For example, the active part of a single transistor on a Pentium silicon chip is a few tenths of a nanometer in size. The semiconductor laser used to read digital information on a CD or DVD has an active layer of similar dimensions. Creating the ability to import such electronic functions into the cell and meshing them with biological functions could open tremendous new possibilities, both for basic biological sciences and for medical and therapeutic applications. QDs are used by life science researchers as tiny beacons or biomarkers, allowing them to easily see individual genes, nucleic acids, proteins, or small molecules.

Nanosystems Biology

Systems biology is defined as the biology of dynamic interacting networks. It is also referred to as pathway, network, or integrative biology. An analysis of the structure and dynamics of network of interacting elements provides insights that are not obvious from analysis of the isolated components of the system. Proteomics plays an important role in systems biology because most biological systems involve proteins. Systems biology is providing new challenges for advancing science and technology. Analyses of pathways may provide new insight into the understanding of disease processes, developing more efficient biomarkers and understanding mechanisms of action of drugs.

Nanosystems biology is the application of nanobiotechnology, microfluidics, and molecular imaging methods, in the study of systems biology. It will play an important role in understanding biology of disease by:

- Providing refined tools for the study of proteomics.
- Nanotechnology provides real-time single-particle tracing in living cells.
- Nanotechnology will facilitate dissecting of signaling pathways.

The goal of nanosystems biology is to develop a suite of nanotechnology tools – ranging from integrated microfluidics to nanoelectronics to nanomechanical devices – that will enable a large-scale, systems-biology-driven, multiparameter analysis within a clinical setting, that is, every patient, every visit. Such an analysis constitutes an "informative diagnosis" of disease. Nanobiosensors are currently being developed for cancer biomarkers for early detection of the disease.

Nanobiology and the Cell

It is perhaps superfluous to use the term "nanobiology" because cell is the smallest unit alive. Molecules in the cell are organized in nanometer-scale dimensions. Visualizing the dynamic change in these molecules and studying the function of cells are among the challenges in nanobiology. A single molecule is the ultimate nanostructure. Single-molecule microscopy and spectroscopy are some of the techniques used to study single molecules.

The objective is to gain a detailed knowledge about biochemical processes occurring locally in the cell nucleus, which is a prerequisite for a comprehensive understanding of genome function. The combination of a resolution range of a few nanometers, high penetrating power, analytical sensitivity, and compatibility with wet specimens allows X-ray microscopy and scanning X-ray microscopy of whole cells. The spatial resolution of 20 nm reached up to now is much better than that of current conventional light microscopes. To further our understanding of chromosome biology and nuclear function, it is essential to develop techniques that enable the measurement of structures inside the living cell with a spatial resolution down to the scale of 10 nm.

The ability to work with an individual cell using nanotechnology is very promising, and there are already several research groups working on this. The single cell is an ideal sensor for detecting various chemical and biochemical processes, and the genetic manipulation of cells could be better done through mechanical rather than biochemical means.

Using a water droplet as a sort of nanoscale test tube, analysis and experimentation have been done at unprecedented tiny scales (Chiu 2010). The microfluidic device captures a single cell or even a small subcellular structure called an organelle within a droplet. The device has water in one channel and oil in an adjoining channel. The target is placed at the interface between the oil and water so the target is encapsulated as the water droplet is formed. It then employs a powerful laser microscope to study the contents and examine chemical processes, and a laser beam is used to manipulate the cell or even just a few molecules, combining them with other molecules to form new substances. This nanoscale laboratory enables one to do an experiment in the droplet on one cell or even a few molecules. The new approach makes it easier to get a wide range of information about a cell and to study structure and form simultaneously.

Biosensing of Cellular Responses

Cells represent the minimum functional and integrating communicable unit of living systems. Cultured cells both transduce and transmit a variety of chemical and physical signals, that is, production of specific substances and proteins, throughout their life cycle within specific tissues and organs. Such cellular responses might be usefully employed as parameters to obtain chemical information for both pharmaceutical and chemical safety, and drug efficacy profiles in vitro as a screening tool. However, such cellular signals are very weak and not easily detected with conventional analytical methods. By using micro- and nanobiotechnology methods integrated on-chip, a higher sensitivity and signal amplification has been developed for cellular biosensing. Nanotechnology is rapidly evolving to open new combinations of methods with improved technical performance, helping to resolve challenging bioanalytical problems including sensitivity, signal resolution, and specificity by interfacing these technologies in small volumes in order to confirm specific cellular signals. Integration of cell signals in both rapid time and small space and, importantly, between different cell populations (communication and systems modeling) will permit many more valuable measurements of the dynamic aspects of cell responses to various chosen stimuli and their feedback. This represents the future for cell-based biosensing.

Genetically encoded nanosensors have been developed to monitor glutamate levels inside and at the surface of living cells using the fluorescent indicator protein for glutamate (FLIPE), which consists of the glutamate/aspartate binding protein ybeJ from *Escherichia coli* fused to two variants of the green fluorescent protein. Such sensors respond to extracellular glutamate with a reversible concentration-dependent decrease in FRET efficiency. FLIPE nanosensors can be used for real-time monitoring of glutamate metabolism in living cells, in tissues, or in intact organisms, providing tools for studying metabolism or for drug discovery.

Single-cell imaging of sodium dynamics has been limited due to the narrow selection of fluorescent sodium indicators that are available currently. Fluorescent nanosensors that measure sodium in real time have been used for the detection of spatially defined sodium activity during action potentials in cells (Dubach et al. 2009). They are reversible and are completely selective over other cations such as potassium that were used to image sodium previously. The use of the nanosensors in vitro has been validated by determining drug-induced activation in heterologous cells transfected with the voltage-gated sodium channel NaV1.7. Spatial information of sodium concentrations during action potentials will provide insight at the cellular level on the role of sodium and how slight changes in sodium channel function can affect the entirety of an action potential. Regulation of sodium flux across the cell membrane plays a vital role in the generation of action potentials and regulation of membrane excitability in cells such as cardiomyocytes and neurons. Alteration of sodium channel function has been implicated in diseases such as epilepsy, long QT syndrome, and heart failure.

Control of T Cell Signaling Activity

The immunological synapse is a specialized cell-cell junction that is defined by large-scale spatial patterns of receptors and signaling molecules yet remains largely enigmatic in terms of formation and function. The marriage of inorganic nanotechnology with organic molecules and cells enables the scientists to go inside a living cell and physically move around its signaling molecules with molecular precision. Supported bilayer membranes and nanometer-scale structures fabricated onto the underlying substrate have been used to directly control T cell signaling activity.

Analysis of the resulting alternatively patterned synapses reveals a causal relation between the radial position of T cell receptors (TCRs) and signaling activity, with prolonged signaling from TCR microclusters that had been mechanically trapped in the peripheral regions of the synapse. These results are consistent with a model of the synapse in which spatial translocation of TCRs represents a direct mechanism of signal regulation.

Measuring Mass of Single Cells

A silicon cantilever mass sensor has been used to measure the mass of single cells with high accuracy (Park et al. 2008). HeLa cells injected into microfluidic channels were subsequently captured on the cantilevers and then cultured in a microfluidic environment. The resonant frequencies of the cantilevers were measured. The mass of a single HeLa cell was extracted from the resonance frequency shift of the cantilever and was found to be close to the mass value calculated from the cell density from the literature and the cell volume obtained from confocal microscopy. This technique could enable development of inexpensive, portable diagnostic devices and might also offer a unique glimpse into how cells change as they undergo cell division. Unlike conventional methods, this technique allows cells to remain in fluid while they are being measured, opening up a new realm of possible applications. In addition to weighing cells, the technology can be used to weigh nanoparticles or submonolayers of biomolecules with a resolution in solution that is six orders of magnitude more sensitive than commercial mass sensor methods. One application is mass-based flow cytometry, a way to weigh and count specific cells.

Nanostructures Involved in Endocytosis

Electron microscopy has been used to uncover new structures, 40 nm in size, which are involved in the very first step of particle and nutrient uptake into cells. Cells require a constant flux of nutrients and other chemicals for survival, and it is vitally important to understand how these materials reach the inside of the cell. Endocytosis, the process of regulated uptake by the cell, is vitally important as it occurs continuously, and a cell virtually consumes its entire covering every 30 min. Endocytosis can be hijacked by viruses to enter the cell, and so understanding this process can provide avenues to stop some viral infections. In addition, endocytosis can be used to deliver drugs into cells. The discovery of this pathway presents unexplored avenues for the development of new drugs to fight certain viral infections, as well as opening up new possibilities for gene therapy.

Clathrin-mediated endocytosis (CME) plays a fundamental role in many cellular activities including receptor downregulation, nutrient uptake, and maintenance of signal transmission across nerve cell junctions. Disturbances in CME are implicated in cancer and neurodegenerative diseases. Live-cell imaging and a novel fluorescence assay have been used to visualize the formation of clathrin-coated vesicles at single clathrin-coated pits (CCP) with a time resolution of seconds (Merrifield et al. 2005). This revealed how proteins linked to the actin, a part of the molecular motor, are transported to sites of coated pit. Disturbing actin polymerization with the toxin latrunculin B, a toxin found in Red Sea sponge, drastically reduces the efficiency of membrane scission and affects many aspects of CCP dynamics. The novel assay used in this study can be applied for drug screening. It has been shown that particles in the size range of tens to hundreds of nanometers can enter or exit cells via wrapping even in the absence of clathrin or caveolin coats, and an optimal particle size exists for the smallest wrapping time (Gao et al. 2005).

Nanotechnology-Based Live-Cell Single-Molecule Assays

Advances in nanotechnology are being harnessed to analyze individual cells. Many enzymatic reactions in the cell are coupled, that is, the product of one enzyme is passed on for use by the nearest enzyme to form a complex network, which functions somewhat like an assembly line. In these networks, proper functioning and control is provided by both space and time parameters. Many traditional techniques for analysis require "fixing" cell samples prior to analysis. This process often destroys the precise intracellular architecture governing the network. Thus, the rate of a biochemical reaction occurring in a test tube could be quite different from that observed for the same reaction inside a cell. DNA, mRNA, and some proteins exist in low numbers within cells, making them difficult to detect because they often remain hidden among all the other molecules. Measuring gene and protein expression levels may involve detecting a single molecule when an individual cell is used as the reaction vessel.

Quantum Dots for Stem Cell Labeling

A study has evaluated the in vivo multiplex and long-term imaging of mouse ESCs labeled with six different QDs and delivered with Qtracker (Lin et al. 2007). ESC viability, proliferation, and differentiation were not adversely affected by QDs compared with nonlabeled control cells. Labeled ESCs, injected subcutaneously in athymic nude mice, could be imaged with one single excitation wavelength and good contrast. QD 800 offered greater fluorescent intensity than the other QDs tested. With further improvements, QDs could gain greater utility for the long-term tracking of stem cells within deeper tissues. This could provide an innovative tool for noninvasive in vivo imaging of stem cell therapy.

Quantum Dots for Study of Apoptosis

Apoptosis (programmed cell death) is characterized by multiple biochemical and morphological changes in different organelles, including nuclei, mitochondria, and lysosomes. It is the dynamics of the spatiotemporal changes in the signaling and morphological adaptations which will ultimately determine the "shape" and fate of the cell. 3D reconstruction of nuclei and intracellular lipid peroxidation in cells exposed to oxidative stress induced by QDs enables a study of this process. This approach is also applicable more generally to investigating changes in organelle morphology in response to therapeutic interventions, stressful stimuli, and internalized nanoparticles. Moreover, the approach provides quantitative data for such changes, which will help us to better integrate compartmentalization of subcellular events and to link morphological and biochemical changes with physiological outcomes.

Single-Cell Injection by Nanolasers

The problem with previous methods of single-cell injection by nanolasers was low cell viability and low efficacy. To refine this procedure, transient membrane permeabilization of single living bovine aortic endothelial cells (BAEC) was studied as an effect of the incident laser intensity focused femtosecond (1 billionth of 1 millionth of a second) near-infrared laser pulses that created a pore or opening in the cellular wall of living cells and encouraged the cell to take in different molecules (Peng et al. 2007a). The rate of dye uptake by the cells was analyzed using time-lapse imaging. Membrane permeabilization occurs for laser intensities higher than $4.0 \times 1{,}012$ W/cm^2. For laser intensity above $3.3 \times 1{,}013$ W/cm^2, the cell disintegrates. Within these two limits, the rate of dye uptake increases logarithmically with increasing laser intensity. This functional dependence is explained by considering the Gaussian intensity distribution across the laser focal spot. Cell membrane permeabilization is explained by the creation of plasma within the laser focal spot. The physical understanding of the relationship between dye uptake, pore characteristics, and laser intensity allows control of the concentrations of molecules delivered into cells through the control of pore characteristics. The findings could serve as a set of guidelines for future research that requires precise microinjection of live single cells. The technique will enable researchers to use unprecedented precision to microinject cells or even perform nanosurgery on cells.

Study of Complex Biological Systems

Nanotechnology holds great promise for the analysis of complex processes inside living cells. It is anticipated to provide new tools to study the responses of different naturally occurring and genetically altered cell types and extend the approaches for monitoring cell behavior and activity in embryos, differentiated tissues, and organs as well as physiological systems. In addition to biological sensors that will be able to measure single-molecule behavior, nanodevices are presently being developed that can be used to relocate various components inside the cell nucleus. This will allow different regions in the nucleus to be probed and manipulated to study various processes, such as their permissiveness for transcription. This will likely open direct approaches for investigating structure-function relationships by perturbing the local organization of the genome and determining its effect on function.

A most promising method for nanomanipulation in living cells is the use of magnetic nanoparticles that are microinjected into the nucleus of living cells. Such particles can be functionalized by the covalent attachment of selected molecules, for example, specific proteins. Recently developed magnetic tweezers, in combination with high-resolution microscopy, would allow one to move such nanoparticles at will inside living cells, thereby changing local genome structure. Nanoprobes in the nucleus could be used to monitor changes in chromosome arrangement associated with changes in gene expression.

The study of complex biological systems requires methods to perturb the system in complex yet controlled ways to elucidate mechanisms and dynamic interactions and to recreate in vivo conditions in flexible in vitro setups. Nanotechnologies have been applied to the study of complex biological systems and provide advantages in these two areas. Particularly useful for controlling the chemical and mechanical microenvironments of cells is a set of techniques called soft lithography, whereby elastomeric materials are used to transfer and generate micro- and nanoscale patterns. Examples of some of the capabilities of soft lithography include the use of elastomeric stamps to generate micropatterns of protein and the use of elastomeric channels to localize chemicals with subcellular spatial resolutions. These nanotechnologies combined with mathematical modeling will considerably improve our understandings of cellular and subcellular physiology.

Molecular Motors

A molecular-level machine may be defined as an assembly of a number of molecules that are designed to perform movements. Molecular self-assembly is an important route toward the construction of artificial molecular-level devices. These devices are characterized by the energy source, the nature of the movement, the way it may be controlled, its repeatability, its purpose, and the time scale of the nanometer-scaled conformational changes. They play an important role in cell function.

A remarkable little rotary motor located at the base of every flagellum of *E. coli* bacteria, which spins to propel the organism, is being studied. This remarkable device is driven by proton gradients, has its own shaft seals through the cell membrane, can achieve ~50 % efficiency, and can spin bidirectionally up to a couple thousand rpm. A microrotary motor has been described, which is composed of a 20-μm-diameter silicon dioxide rotor driven on a silicon track by the gliding bacterium *Mycoplasma mobile* (Hiratsuka et al. 2006). This motor is fueled by glucose and inherits some of the properties normally attributed to living systems.

Every cell in the body has "a dynamic city plan" comprised of molecular highways, construction crews, street signs, cars, fuel, and exhaust. Maintenance of this highly organized structure is fundamental to the development and function of all cells, and much of it can be understood by figuring out how the functional units in the cell, molecular motors or biological nanomachines, do the work to keep cells orderly. Their function depends on catalytic activity of their constituent proteins. The miniature motor that drives our cells is the enzyme ATP synthase that converts food or light into ATP, the energy currency of the cell. Spinning at several thousand revolutions per minute, the detailed internal workings of the tiny motor are tough to decipher; high-speed imaging has been used to snap freeze-frames of the spinning shaft. Linear micromotors that move the cell, voltage-gated ion channels, DNA replication complexes, and countless other structures are quite complex, and their functions are still not well understood. Lasers, detectors, and optics have been used to study how two protein machines, myosin and kinesin, move about in vitro. The results show how multiple motors compete for activity on their protein scaffolds. Kinesin and another motor, dynein, operate in large groups to produce more than 10 times the speed that is expected from a single molecule. Use of a technique called fluorescence imaging with one-nanometer accuracy (FIONA), which has a spatial resolution of 1.5 nm, shows that myosin V, kinesin, and myosin VI all move in hand-over-hand fashion. Further studies with FIONA have shown that average step size is approximately 8 nm for both dynein and kinesin and that these work together rather than against each other in vivo producing up to 10 times the in vitro speed (Kural et al. 2005).

One of the unknowns about dynein was that the molecular site where chemical energy is initially released from ATP is very far away from where the mechanical force occurs and how the mechanical force was transmitted over a large distance. Using a variety of modeling techniques that allowed resolution at the level of atoms, scientists have identified a flexible, springlike "coiled coil" region within dynein, which couples the motor protein to the distant ATP site (Serohijos et al. 2006). It allows a very rapid transduction of chemical energy into mechanical energy. Conversion to mechanical energy allows dynein to transport cellular structures such as mitochondria that perform specific jobs such as energy generation, protein production, and cell maintenance. Dynein also helps force apart chromosomes during cell division. Although the research offers no immediate application to human disease, the authors noted that mutations of dynein have been implicated in some neurodegenerative and kidney disorders. Disruption of dynein's interaction with a particular regulator protein causes defects in nerve cell transmission and mimics the symptoms of patients with amyotrophic lateral sclerosis.

Proteins that function as molecular motors are surprisingly flexible and agile, able to navigate obstacles within the cell (Ross et al. 2006). These observations could lead to better ways to treat motor neuron diseases such as amyotrophic lateral sclerosis. Using a specially constructed microscope that allows researchers to observe the action of one macromolecule at a time, the team found that a protein motor is able to move back and forth along a microtubule – a molecular track – rather than in one direction, as previously thought. The proteins in this motor, dynein and dynactin, are the "long-distance truckers" of the cell: working together, they are responsible for transporting cellular cargo from the periphery of a cell toward its nucleus. Mutation in dynactin leads to degeneration of motor neurons, the hallmark of motor neuron disease. They found that a mutation decreases the efficiency of the dynein-dynactin motor in "taking out the trash" of the cell and thus leads to the accumulation of misfolded proteins in the cell, which may in turn lead to the degeneration of the neuron.

Another type of molecular motor provides the rigidity needed by the tiny sensors in the inner ear in order to respond to sound. This motor creates the proper amount of tension in the sensors and anchors itself to maintain that tension. A motor able to create structural changes by taking up slack in proteins and clamping down so that they remain in a rigid position may help explain many intricacies of cellular organization, such as how chromosomes line up and separate during cell division. Molecular motors called myosins are proteins that carry out cellular motion by attaching to and "walking" along fibers of actin. The interaction of actin and myosin is the mechanism behind cell actions such as muscle contractions, the pinching off of two daughter cells from a mother cell during division, and the hauling of cargo molecules around in a cell. Of the various types of myosin molecules, myosin VI is considered to be responsible for setting the tension for stereocilia, actin-filled rods on the sound-sensing hair cells of the inner ear. A defect in myosin VI results in deafness. Although it was known that myosin moves along actin fibers, it had never previously been demonstrated how myosin could function as an anchor or a clamp. It was enabled by optical tweezers, a focused laser that allows the manipulation of microscopic beads. This study involved looking at a single molecule and how it behaves, but there are very few proteins in biology that have been analyzed and understood down to this level. Such studies that provide an understanding of the biological nanostructures would be stimulating for nanobiotechnology. Designers of nanomachines can learn much from nanobiology as many nanostructures in biology can be used as tools in nanobiotechnology.

Biological molecular motors have a number of unique advantages over artificial motors, including efficient conversion of chemical energy into mechanical work and the potential for self-assembly into larger structures, as is seen in muscle sarcomeres and bacterial and eukaryotic flagella. The development of an appropriate interface between such biological materials and synthetic devices should enable us to realize useful hybrid micromachines. When the scientists discover how to design and mass-produce molecular motors artificially, it will be a major stride in the era of nanotechnology.

Nanomotor Made of Nucleic Acids

Although protein machines are abundant in biology, it has recently been proposed that nucleic acids could also act as nanomolecular machines in model systems. Several types of movements have been described with DNA machines: extension-contraction movement, rotation, and "scissors-like" opening and closing. The simple and robust device described is composed of a single 21-base oligonucleotide and relies on a duplex-quadruplex equilibrium that may be fueled by the sequential addition of DNA single strands, generating a DNA duplex as a by-product. The interconversion between two well-defined topological states induces a 5-nm two-stroke, linear motor-type movement, which is detected by fluorescence resonance energy transfer spectroscopy. This system could be used to obtain precise control of movements on the nanometer scale.

Nucleic acids are increasingly used to build nanometer-scale structures that may be used in future nanotechnology devices. There are two key goals in this area: to perform controlled mechanical movements and to produce complex structures from simple molecular building blocks. Several areas, such as error correction and scaled-up self-assembly, require particular attention if the potential of nucleic acid-based nanotechnology is to be fulfilled.

A DNA nanomechanical device enables the positional synthesis of products whose sequences are determined by the state of the device. Such a machine can emulate the translational capabilities of the ribosome. Ribosomes are miniature biological machines that weld together amino acids to form the enzymes that modulate body chemistry and the structural materials, like collagen. DNA was twisted and bent to build a structure that is approximately 110 nm long and 30 nm and 2 nm thick, roughly the same size as a ribosome, though not as complex. The DNA machine can swivel into four geometric positions and can be locked into any one of them by another fragment of DNA, which provides the instructions. Locking the machine into position dials in the sequence of very short DNA strands that it recognizes and positions for welding. The welding itself is performed by an enzyme that links DNA molecules. In the absence of the machine, this enzyme would create many different combinations of DNA strands; in its presence, only a single, preprogrammed combination results. The device has potential applications that include designer polymer synthesis, encryption of information, and use as a variable-input device for DNA-based computation.

phi29 DNA-Packaging Nanomotor

The condensation of bacteriophage (a virus that infects bacteria) phi29 genomic DNA into its preformed procapsid requires the DNA-packaging motor, which is the strongest known biological motor. The virus uses the motor to package DNA and move it into the capsid, a shell made of proteins, as part of the viral reproduction process. The packaging motor is an intricate ring-shaped protein/RNA complex, and its function requires an RNA component called packaging RNA (pRNA). Current structural information on pRNA is limited, which hinders studies of motor function. Site-directed spin labeling has been used to map the conformation of a pRNA 3-way junction that bridges binding sites for the motor ATPase and the procapsid (Zhang et al. 2012). The studies were carried out on a pRNA dimer, which is the simplest ring-shaped pRNA complex and serves as a functional intermediate during motor assembly. The studies establish a new method for mapping global structures of complex RNA molecules and provide information on pRNA conformation that aids investigations of phi29 packaging motor and developments of pRNA-based nanomedicine and nanomaterial.

The controllable DNA-packaging nanomotor has been constructed and is driven by synthetic ATP-binding pRNA monomers. When fed a supply of ATP fuel, the RNA strands kick against the axle in succession, much like pistons in a combustion engine. This phenomenon explains how RNA powers viral assembly. The nanomotor can stuff DNA into a protein shell and can be turned on and off. The motor components form regular arrays, which have potential applications in medicine and nanotechnology. Fusing pRNA with receptor-binding RNA aptamer, folate, short interfering RNA (siRNA), ribozyme, or another chemical group does not disturb dimer formation or interfere with the function of the inserted moieties. The motor pRNA can deliver siRNA, ribozyme, or other therapeutic molecules to specific cells and destroy them. This has been demonstrated in various cancer cells as well as in cells infected with hepatitis B virus. The use of such a nanomotor/pRNA/siRNA complex could extend the short half-life of therapeutic small molecules in vivo and overcome the delivery problems of molecules larger than 100 nm.

The nanoscale size range of phi29 motor is ideal for delivery inside the body. Anything smaller would be filtered out through the kidneys too quickly to be effective, and larger molecules would not be able to enter cells. The nanomotor can be attached to a lipid sheet, cell membrane, and liposome, which would take the place of the capsid into which the phi29 biological motor pumps DNA. The nanomotor, once embedded into the outer wall of the liposome or cell membrane, would pump DNA, drugs, or other therapeutic molecules into the liposome pocket's open space or directly into the cell through a controlled mechanism. The long-term goal is to place thousands of nanomotors in an array assembled on a porous surface, such as silicon, and to have them function for use in biosensing. Nanomotors are being studied further for potential use in the diagnosis and treatment of diseases such as cancer, AIDS, hepatitis B, and influenza. Nanomotors will be used to package and deliver therapeutic DNA or RNA to disease-causing cells. The aim is to create a medical tool using a device that mimics a natural biological structure. This biomimetic tool will be a hybrid of natural biological structures and synthetic structures that will operate on the nanoscale.

Light-Activated Ion Channel Molecular Machines

Light-regulated molecular machines, such as the SPARK (synthetic photoisomerizable azobenzene-regulated K^+) channels, may also have applications in nanobiotechnology. A molecular tether that attaches to the ion channel can change its shape when exposed to different wavelengths of light. The tether's long form will block the channel, but its short form will leave the channel open. When open, these channels allow positively charged ions to flow out of the neuron, which silences its activity. These engineered channels improve on previous attempts to selectively manipulate activity in a particular set of neurons and can be used for rapid, reversible, and precise silencing of neural firing. This new technique has potential applications in dissecting neural circuits and controlling activity downstream from sites of neural damage or degeneration

Application of AFM for Biomolecular Imaging

AFM has become a well-established technique for imaging single biomolecules under physiological conditions. The exceptionally high spatial resolution and signal-to-noise ratio of the AFM enables the substructure of individual molecules to be observed. In contrast to other methods, specimens prepared for AFM remain in a plastic state, which enables direct observation of the dynamic molecular response, creating unique opportunities for studying the structure-function relationships of proteins and their functionally relevant assemblies.

The combination of single-molecule imaging with other techniques to monitor topographical, biochemical, and physical parameters simultaneously is a powerful, interesting, and unique application of AFM. This correlation of biochemical and physical information can provide new insights into fundamental biological processes. Several different approaches for obtaining multiple parameters during AFM imaging have been developed, including AFM in combination with optical microscopy, patch-clamp electrophysiology, and ion-conductance pipettes. Used as a sensor, the AFM tip can also probe the charges of biological surfaces immersed in a buffer solution. So far, such approaches have successfully characterized protein interactions, but in the future they could be applied to imaging and detecting multiple parameters on a single molecule simultaneously.

AFM has been used to study and characterize gene complexes composed of plasmid DNA and cationic lipids. Surface morphology of spherical complexes with diameters of approximately 200–300 nm can be examined. However, AFM technique does not enable any conclusions to be drawn regarding the architecture of the inner core of the lipoplex as only the morphology of the surface structures can be obtained reliably.

One of the critical limitations of AFM is its inability to recognize the specific chemical composition of a molecule. AFM uses a tiny, highly sensitive probe that, when pulled across the surface of a sample, maps its topography down to the nanometer scale. Until recently, it could not identify exactly what the proteins on its map are as all proteins look the same in an AFM image. A solution to this limitation is use of an even tinier polymer thread that attached antibodies designed to bind to individual proteins to the tip of the AFM probe. When the antibody binds to the target protein, it creates a variance in the microscope's reading. This is a technique for identifying any antigen in a complex sample on the nanometer scale and has applicability that goes far beyond sorting out AFM images. It could be used to read arrays on the nanometer scale to enable mapping of the entire interaction potential landscape between a receptor and a ligand.

Future Insights into Biomolecular Processes by AFM

AFM has become a well-established technique for imaging individual macromolecules at a spatial resolution of <1 nm. The next step will be to establish new AFM methods to investigate structure-function relationships among the variety of molecular machines. Such results will provide insights into how such machines work at the molecular level and drive the understanding of common principles that govern them. The next challenge will be to study the behavior of individual molecular machines in heterogeneous assemblies and to understand how different machines form small functional entities. Here, again, AFM promises to be an important tool as it enables individual molecules to be imaged at sufficient resolution for their behavior within macromolecular complexes to be characterized.

4Pi Microscopy to Study DNA Double-Strand Breaks

DNA double-strand breaks caused by cellular exposure to genotoxic agents or produced by inherent metabolic processes initiate a rapid and highly coordinated series of molecular events resulting in DNA damage signaling and repair. Phosphorylation of histone (a spool around which DNA is wound) H2AX to form gamma-H2AX is one of the earliest of these events and is important for coordination of signaling and repair activities. An intriguing aspect of H2AX phosphorylation is that gamma-H2AX spreads a limited distance up to 1–2 Mbp from the site of a DNA break in mammalian cells. Generally light microscopy has limited resolution. By manipulating how light waves behave, however, biophysicists are expanding the limits of light microscopy, and one of the latest advances – the 4Pi microscope – provides never-before-seen views of cellular components, including structures within the nucleus. Identification of previously undescribed H2AX chromatin structures has been reported by successful application of 4Pi microscopy to visualize endogenous nuclear proteins (Bewersdorf et al. 2006). These observations suggest that H2AX is not distributed randomly throughout bulk chromatin, rather it exists in distinct clusters that themselves are uniformly distributed within the nuclear volume. These data support a model in which the size and distribution of H2AX clusters define the boundaries of gamma-H2AX spreading and also may provide a platform for the immediate and robust response observed after DNA damage. 4Pi microscopy allows researchers to actually see the response in three dimensions, at resolutions down to 100 nm. Therefore, the role of the physical structures in various processes within the nucleus can now be visualized.

Multi-isotope Imaging Mass Spectrometry

Secondary-ion mass spectrometry (SIMS) is an important tool for investigating isotopic composition in the chemical and materials sciences, but its use in biology has been limited by technical considerations. Multi-isotope imaging mass spectrometry (MIMS), which combines a new generation of SIMS instrument with sophisticated ion optics, labeling with stable isotopes, and quantitative image-analysis software, was developed to study biological materials. A beam of ions is used to bombard the surface atoms of the biological sample, and a fraction of the atoms are emitted and ionized. These "secondary ions" can then be manipulated with ion optics – in the way lenses and prisms manipulate visible light – to create an atomic mass image of the sample. The new instrument enables the production of mass images of high lateral resolution (down to 33 nm) as well as the counting or imaging of several isotopes simultaneously. As MIMS can distinguish between ions of very similar mass, it enables the precise and reproducible measurement of isotope ratios and thus of the levels of enrichment in specific isotopic labels, within volumes of less than a cubic micrometer. The sensitivity of MIMS is at least 1,000 times that of 14C autoradiography. The depth resolution can be smaller than 1 nm because only a few atomic layers are needed to create an atomic mass image (Lechene et al. 2006). Thus, the technology represents an imaging revolution.

MIMS can generate quantitative 3D images of proteins, DNA, RNA, sugar, and fatty acids at a subcellular level in tissue sections or cells and follow the fate of these molecules when they go into cells, where they go, and how quickly they are replaced. The method does not need staining or use of radioactive labeling. Instead, it is possible to use stable isotopes to track molecules. MIMS has been used to image unlabeled mammalian cultured cells and tissue sections; to analyze fatty-acid transport in adipocyte lipid droplets using 13C-oleic acid; to examine nitrogen fixation in bacteria using 15N gaseous nitrogen; to measure levels of protein renewal in the cochlea and in postischemic kidney cells using 15N-leucine; to study DNA and RNA codistribution and uridine incorporation in the nucleolus using 15N-uridine and 81Br of bromodeoxyuridine or 14C-thymidine; to reveal domains in cultured endothelial cells using the native isotopes 12C, 16O, 14N, and 31P; and to track a few 15N-labeled donor spleen cells in the lymph nodes of the host mouse, suggesting that MIMS may be highly useful in immunology and cancer research. MIMS makes it possible for the first time to both image and quantify molecules labeled with stable or radioactive isotopes within subcellular compartments, suggesting that MIMS may have applications in tracking stem cells and in understanding why some organ transplants are rejected.

Applications of Biomolecular Computing in Life Sciences

Early biomolecular computer research focused on laboratory-scale, human-operated computers for complex computational problems. Now, simple molecular-scale autonomous programmable computers have been shown to enable both input and output information in molecular form. Such computers, using biological molecules as input data and biologically active molecules as outputs, could produce a system for "logical" control of biological processes. An autonomous biomolecular computer can, at least in vitro, logically analyze the levels of mRNA species and in response produce a molecule capable of affecting levels of gene expression. This approach might be applied in vivo to biochemical sensing, genetic engineering, and even medical diagnosis and treatment. Such computers have been programmed to identify and analyze mRNA of disease-related genes associated with models of small-cell lung cancer and prostate cancer and to produce a single-stranded DNA molecule modeled after an anticancer drug. To have therapeutic value, such computers must be installed inside cells and protected from cellular defense mechanisms once they get there.

Using computer modeling, six phases of high-density nano-ice were predicted to form within CNTs at high pressure. High-density nano-ice self-assembled within smaller-diameter CNT exhibits a double-walled helical structure where the outer wall consists of four double-stranded helixes, which resemble a DNA double helix with the inner wall as a quadruple-stranded helix (Bai et al. 2006). This discovery could have major implications for scientists in other fields who study the protein structures that cause diseases such as Alzheimer's and bovine spongiform encephalopathy.

Molecular systems based on DNA computing and strand displacement circuitry can exhibit autonomous brain-like behaviors. Using a simple DNA gate architecture that allows experimental scale-up of multilayer digital circuits, arbitrary linear threshold circuits were transformed into DNA strand displacement cascades that function as small neural networks (Qian et al. 2011). Four fully connected artificial neurons, after training in silico, were demonstrated to remember four ssDNA patterns and recall the most similar one when presented with an incomplete pattern. These results suggest that DNA strand displacement cascades could be used to endow autonomous chemical systems with the capability of recognizing patterns of molecular events, making decisions, and responding to the environment.

Microbial Nanomaterials

Use of Bacteria to Construct Nanomachines

Single-molecule and super-resolution fluorescence imaging provide powerful tools for the biochemical, structural, and functional characterization of cells as biological nanomachines (Chiu and Leake 2011). Data obtained from single bacterial cells have been used successfully to make tiny bioelectronic circuits. Microbes can serve as templates for fabricating nanoscale structures and might obviate the need for the tedious and time-consuming construction of devices at the smallest scale. This may also form the basis for a new class of biosensors for real-time detection of dangerous biological agents, including anthrax and other microbial pathogens.

Exploiting the complex topography of the bacterial cell surface and microbial interactions with antibodies, more complex nanoscale structures can be constructed because of the natural ability of cells to dock with different kinds of molecules. This would be easier than the painstaking manipulation of individual nanosized components, such as the microscopic wires and tubes that comprise the raw materials of nanotechnology. Bacteria can be considered as nature's nanowires that can be easily grown and manipulated. One can attach microscopic gold particles to the shell of the bacterium, making it more like a nanoscale gold wire. The ability to capture and analyze individual microbes might make it easier for scientists to retrieve specified cells from a complex mixture and have applications in cell therapy.

Bacteriophage Nanoshells

The shell of bacteriophages protects the viral DNA during host-to-host transfer and serves as a high-pressure container storing energy for DNA injection into a host bacterium. A study of the mechanical properties of nanometer-sized bacteriophage shells (ϕ 29 nm) by applying point forces shows that empty shells withstand nanonewton forces while being indented up to 30% of their height (Ivanovska et al. 2004). The elastic response varied across the surface, reflecting the arrangement of shell proteins. The capsid of bacteriophage is remarkably dynamic yet resilient and tough enough to easily withstand the known packing pressure of DNA (≈60 atmospheres). These capsids, thus, provide not only a chemical shield but also significant mechanical protection for their genetic contents. Viral shells are a remarkable example of nature's solution to a challenging materials engineering problem: they self-assemble to form strong shells of precisely defined geometry by using minimum amount of different proteins. These results design principles for the construction of man-made nanoscale containers.

Natural Nanocomposites

Natural materials such as bone and tooth are nanocomposites of proteins and minerals with superior strength. Nanocomposites in nature exhibit a generic mechanical structure in which the nanometer size of mineral particles is selected to ensure optimum strength and maximum tolerance of flaws. The widely used engineering concept of stress concentration at flaws is no longer valid for nanomaterial design. Natural nanocomposites can be used as a guide to enhance the properties of artificial nanocomposite materials, which eventually could excel compared to their biological counterparts (Fleischli et al. 2008).

Nanotechnology in Biological Research

Invention of the scanning tunneling microscope opened up new realms in study of biology. A whole family of scanning probe instruments has been developed, extending biological to the scale of atoms and molecules. Such instruments are especially useful for imaging of biomolecular structures because they can produce topographic images with submolecular resolution in aqueous environments. Instruments with increased imaging rates, lower probe-specimen force interactions, and probe configurations not constrained to planar surfaces are being developed, with the goal of imaging processes at the single-molecule level, not only at surfaces but also within 3D volumes. New development in nanotechnology is facilitating the study of biological processes at nanoscale and providing an understanding of many life processes that are relevant to healthcare. Single-molecule nanobiology can be used as a tool for understanding the working principles of biological nanosystems in live cells.

AFM has been extensively used not only to image nanometer-sized biological samples but also to measure their mechanical properties by using the force curve mode of the instrument. The presence of specific receptors on the living cell surface has been mapped by this method. The force to break the co-operative 3D structure of globular proteins or to separate a double-stranded DNA into single strands has been measured. Extension of the method for harvesting functional molecules from the cytosol or the cell surface for biochemical analysis has been reported. There is a need for the development of biochemical nanoanalysis based on AFM technology.

Near-infrared (NIR) laser microscopy enables optical micromanipulation, piconewton force determination, and sensitive fluorescence studies by laser tweezers. Fluorescence images with high spatial and temporal resolution of living cells and tissues can be obtained via nonresonant fluorophore excitation with multiphoton NIR laser scanning microscopes. Furthermore, NIR femtosecond laser pulses at can be used to realize noninvasive contact-free surgery of nanometer-sized structures within living cells and tissues. These novel versatile NIR laser-based tools can be used for the determination of motility forces, coenzyme and chlorophyll imaging, 3D multigene detection, noninvasive optical sectioning of tissues (optical biopsy), functional protein imaging, and nanosurgery of chromosomes.

QDs for Biological Research

QDs are used in a variety of assays such as immunohistochemistry, flow cytometry, Western blotting, and plate-based assays. QDs have been used to study mechanism of protein trafficking. The early stages of receptor tyrosine kinase (RTK)-dependent signaling in living cells have been imaged using continuous confocal laser scanning microscopy and flow cytometry. Epidermal growth factor (EGF)-QDs are highly specific and potent in the binding and activation of the EGF receptor (erbB1), being rapidly internalized into endosomes that exhibit active trafficking and extensive fusion. It is expected that QD ligands will find widespread use in basic research and biotechnological developments.

A unique coating for inorganic particles at the nanoscale has been developed that may be able to disguise QDs as proteins. This process enables particles to function as probes that can penetrate the cell and light up individual proteins inside and create the potential for application in a wide range of drug development, diagnostic tools, and medications (Michalet et al. 2005).

The organic coatings – short chains of peptides – can be used to disguise QDs, quantum rods, and quantum wires so effectively that the cells mistake them for proteins, even when the coatings are used on particles that are inorganic and possibly even toxic. These peptide coatings trick the live cell into thinking that the nanoparticles are benign, protein-like entities. Therefore, one can use these coated particles to track the proteins in a live cell and conduct a range of studies at the molecular level. Using the new coatings, the UCLA team has been able to solubilize and introduce into the cell different-color quantum dots that can all be excited by a single blue light source. The color-encoding method is similar to the encoding of information that is sent down an optical fiber, called wavelength division multiplexing. The peptide-coating technology could, in principle, color encode biology itself, by painting different proteins in the cell with different-color quantum dots. The scientists are developing methods to attach QDs of specific colors to the different proteins on cells' surface and inside cells. By painting a subset of proteins in the cell with different-color QDs, one can follow the molecular circuitry, the dynamic rearrangement of circuit nodes, and the molecular interactions. In addition to the capacity to paint and observe many different proteins with separate colors, quantum dots can be used for the ultimate detection sensitivity: observing a single molecule. Until now, tracking and following a single protein in the cell has been extremely difficult.

By observing with a fluorescence microscope and high-sensitivity imaging cameras, researchers can track a single protein tagged with a fluorescent QD inside a living cell in 3D and within a few nanometers of accuracy. This process is, in some ways, the molecular equivalent of using the global positioning system (GPS) to track a single person anywhere on earth. Researchers can use optical methods to track several different proteins tagged with different-color QDs, measure the distances between them, and use those findings to better understand the molecular interactions inside the cell.

Particles disguised with the peptide coatings can enter a cell without affecting its basic functioning. Since the peptide-coated QDs are small, they have easy and rapid entry through the cell membrane. In addition, since multiple peptides of various lengths and functions could be deposited on the same single QD, it would be feasible to create smart probes with multiple functions.

This work on coatings was inspired by nature. Some plants and bacteria cells evolved unique capabilities to block toxic heavy-metal ions as a strategy to clean up the toxic environment in which they grow. These organisms synthesize peptides, called phytochelatins that reduce the amount of toxic-free ions by strongly binding to inorganic nanoparticles made of the sequestered toxic salts and other products.

The peptide coating bridges the inorganic chemistry world with the organic world on the nanometer scale. These coatings will be used to provide electrical contact between nanoscale inorganic electronic devices and functional proteins, which would lead to the evolution of novel and powerful "smart drugs," "smart enzymes," "smart catalysts," "protein switches," and many other functional hybrids of inorganic-organic substances. It might enable the creation of a hybrid nanoparticle that could be specifically targeted to identify and destroy cancer cells in the body.

Molecular Biology and Nanotechnology

Structural DNA Nanotechnology

DNA is an ideal molecule for building nanometer-scale structures. Strands of DNA interact in the most programmable way. Their enormous variability provides ample scope for designing molecules. DNA scaffolds could hold guest molecules in orderly arrays for crystallography. Nanometer-scale DNA machines can function by having part of their structure change from one DNA conformation to another. These movements can be controlled by chemical means or by the use of special DNA strands.

Structural DNA nanotechnology uses the basic chemical units of DNA – C, T, A, or G – to self-fold into a number of different building blocks that can further self-assemble into patterned structures. Unusual DNA motifs can be combined by specific structurally well-defined cohesive interactions (primarily sticky ends) to produce target materials with predictable 3D structures. This effort has generated DNA polyhedral catenanes, robust nanomechanical devices, and a variety of periodic arrays in 2D. This is a good example of artificial nanostructures that can be replicated using the machineries in live cells. The system has been used to produce specific patterns on the mesoscale through designing and combining specific DNA strands, which are then examined by AFM. The combination of these constructions with other chemical components is expected to contribute to the development of nanoelectronics, nanorobotics, and smart materials (Seeman 2007).

Three different structurally robust end states can be obtained in one system, all resulting from the addition of different set strands to a single floppy intermediate as an extension of the PX-JX2 DNA device (Chakraborty et al. 2008). The three states are related to each other by three different motions, a twofold rotation, a translation of ≈2.1–2.5 nm, and a twofold screw rotation, which combines these two motions. The transitions were demonstrated by gel electrophoresis, by FRET, and by AFM.

Many new tertiary interactions are being discovered, and some of these are being used for the purpose of generating new nucleic acid-based materials. These may ultimately lead to a new generation of capabilities for structural nucleic acid nanotechnology. As more knowledge is gained about the metabolism of DNA, new motifs may be discovered that are currently exploited by living systems and that can be used by the materials sciences to generate new materials. Structural nucleic acid nanotechnology is in its infancy, but it seems to be capable of remarkable versatility in the organization of matter on the nanoscale.

Despite the dramatic evolution of DNA nanotechnology, a versatile method that replicates artificial DNA nanostructures with complex secondary structures remains an appealing target. Previous success in replicating DNA nanostructures enzymatically in vitro suggests that a possible solution could be cloning these nanostructures by using viruses. A system has been reported where ssDNA nanostructure is inserted into a phagemid, a virus-like particle that infects a bacteria cell (Lin et al. 2008). Once inside the cell, the phagemid uses the cell just like a photocopier machine to reproduce millions of copies of the DNA. By theoretically starting with just a single phagemid infection and a single milliliter of cultured cells, the cells can churn out trillions of the DNA junction nanostructures. The DNA nanostructures produced in the cells were found to fold correctly, just like the previously built test tube structures. The simplicity, efficiency, and fidelity of nature are fully reflected in this system. UV-induced psoralen cross-linking is used to probe the secondary structure of the inserted junction in infected cells. These data suggest the possible formation of the immobile four-arm junction in vivo.

Another future goal is incorporation of DNA devices into nanorobotics. Nanoelectronic components, such as metallic nanoparticles or carbon nanotubes, will need to be combined with DNA molecules in compatible systems. It may be possible to have DNA-based replicating machines in a few decades.

RNA Nanotechnology

RNA has an important role in nanoscale fabrication due to its amazing diversity of function and structure. RNA molecules can be designed and manipulated with a level of simplicity characteristic of DNA while possessing versatility in structure and function similar to that of proteins. RNA helicases are a large family of molecular motors that utilize nucleoside triphosphates to unwind RNA duplexes and to remodel RNA protein complexes. These enzymes have a potential application in controlling conformational changes in nanoassemblies that contain RNA.

Assembly and folding principles of natural RNA can be used to build potentially useful artificial structures at the nanoscale. Reliable prediction and design of the 3D structure of artificial RNA building blocks have been achieved to generate molecular jigsaw puzzle units called tectosquares, which can be programmed with control over their geometry, topology, directionality, and addressability to algorithmically self-assemble into a variety of complex nanoscopic fabrics with predefined periodic and aperiodic patterns and finite dimensions. Such studies emphasize the modular and hierarchical characteristics of RNA by showing that small RNA structural motifs can code the precise topology of large molecular architectures. They demonstrate that fully addressable materials based on RNA can be synthesized and provide insights into self-assembly processes involving large populations of RNA molecules.

The ability of RNA to fold into a variety of rigid structural motifs can provide potential modules for supramolecular engineering. Nanogrids may eventually be used as a starting point to generate nanochips, nanocircuits, and nanocrystals with potential applications in nanotechnology and materials science. Healthcare applications include the development of medical implants, regeneration of organs, and nanodiagnostics. The most recent development in exploration of RNA nanoparticles is for pathogen detection, drug/gene delivery, and therapeutic application.

A conference on RNA nanotechnology concluded (Shukla et al. 2011):

- Applications and the impact of RNA nanotechnology in the assembly and delivery of siRNAs, pRNA, therapeutic ribozymes, RNA aptamers, and riboswitches are becoming a reality.
- RNA nanotechnology is already making a significant impact in drug delivery and therapeutics as many siRNA-based products are in the pipeline for potential treatment of viral infections, liver cancer, and Huntington's disease.
- Collaborative endeavors between academia, government, and industry are likely required to advance RNA nanotechnology into a practical tool for drug discovery and delivery in the near future.

Genetically Engineered Proteins for Nanobiotechnology

With the development of nanoscale engineering in physical sciences and the advances in molecular biology, it is possible to combine genetic tools with synthetic nanoscale constructs to create a novel methodology (Tamerler and Sarikaya 2009).

Peptides/proteins can now be genetically engineered to specifically bind to selected inorganic compounds for applications in nanobiotechnology. These genetically engineered proteins for inorganics (GEPIs) can be used in the assembly of functional nanostructures. Based on the three fundamental principles – molecular recognition, self-assembly, and DNA manipulation – GEPI has been used successfully in nanotechnology.

Organization and immobilization of inorganic nanoparticles in 2D or 3D is fundamental to the use of nanoscale effects. It would be desirable to use GEPIs that specifically recognize inorganics for nanoparticle assembly. An advantage of this approach is that GEPI can be genetically or synthetically fused to other functional biomolecular units or ligands to produce multifunctional molecular entities. Proteins may be useful in the production of tailored nanostructures, and the recognition activity of the protein could provide an ability to control the particle distribution, and particle preparation conditions could allow size control.

Self-assembled GEPI monolayers could open up new avenues for designing and engineering novel surfaces for a wide variety of nano- and biotechnology applications, for example, a GEPI recognizing and assembling on the surface of a therapeutic device could be fused to a human protein to enhance biocompatibility or used for drug delivery through colloidal inorganic particles. Coupled with a molecular motor, a GEPI may provide a critical step toward creating dynamic nanostructures. Ultimately, using nanopatterned multimetallic or multisemiconducting particles and localized surface plasmon effects, several different GEPI molecules could serve as specific linkers in creating nanoscale platforms for rapid development of nanoarrays for proteomics. Based on the insights achieved through these studies in the coming decade, and following the lead of molecular biology, a road map could be developed in which GEPI could be used as a versatile molecular linker and could open new avenues in the self-assembly of molecular systems in nanobiotechnology.

Single-Molecule Studies

Optical Trapping and Single-Molecule Fluorescence

Two of the mainstay techniques in single-molecule research are optical trapping and single-molecule fluorescence. Previous attempts to combine these techniques in a single experiment and on a single macromolecule of interest have met with little success, because the light intensity within an optical trap is more than 10 orders of magnitude greater than the light emitted by a single fluorophore. Instead, the two techniques have been employed sequentially, or spatially separated by distances of several micrometers within the sample, imposing experimental restrictions that limit the utility of the combined method. An instrument capable of true, simultaneous, spatially coincident optical trapping and single-molecule fluorescence has been developed (Lang et al. 2003). This opens the door to many types of experiment that employ optical traps to supply controlled external loads while fluorescent molecules report concurrent information about macromolecular structure. The combination of these two biophysical techniques in a single assay offers a powerful tool for studying molecular systems by allowing direct correlations to be made between nanoscale structural changes.

3D Single-Molecular Imaging by Coherent X-Ray Diffraction Imaging

Coherent X-ray diffraction imaging is a rapidly advancing form of microscopy: diffraction patterns, measured using the latest third-generation synchrotron radiation sources, can be inverted to obtain full 3D images of the interior density within nanocrystals. A 3D image of a nanocrystal, obtained by inversion of the coherent X-ray diffraction, shows the expected faceted morphology but in addition reveals a real-space phase that is consistent with the 3D evolution of a deformation field (Pfeifer et al. 2006). This method of measuring and inverting diffraction patterns from nanocrystals represents a vital step toward the ultimate goal of atomic resolution single-molecule imaging that is a major justification for development of X-ray free-electron lasers. It is hoped that one day this will be applied to determine the structure of single protein molecules placed in the femtosecond beam of a free-electron laser.

Nanochemistry

Nanochemistry involves the utilization of a chemical synthesis approach to make new materials with at least one physical dimension straddling the molecular (nanoscale) and macroscopic world (Ozin et al. 2009). In practice, nanochemistry deals with the production and the reactions of nanoparticles and their compounds.

A new generation of spectroscopic dyes is gradually becoming available to biological researchers, from an unexpected source: materials chemists who study the synthesis and properties of nanosized inorganic objects. Research into tailoring the optical properties, surface chemistry, and biocompatibility of metallic and semiconductor nanoparticles is fulfilling the promise of these nanostructures as customizable substitutes for organic molecular probes. Chemists have reported synthetic routes toward semiconductor "quantum dots" for fluorescent tagging, metal nanoparticles with extraordinarily high extinction coefficients for labeling in colorimetric and surface plasmon resonance assays, and elongated "nanorods" for measuring anisotropy. Nanoparticles are also providing alternatives to organic and organometallic probes for other (nonoptical) biological applications, such as paramagnetic particles for magnetic resonance contrast imaging and metal particles for thermal probing of specific biomolecular interactions. The chemical synthesis of most of these nanostructures is typically achieved with just one reaction, involving a chemical transformation of a precursor source of inorganic material followed by a nanocrystallization process in the same vessel. Fine control over synthetic conditions is responsible for the reproducible range of novel properties these materials exhibit.

As a result, nanostructured biological probes are easier to make and might eventually be less expensive to buy than organic dyes. In addition, the specific photochemical reactions that cause organic probes to photobleach or cross-link nonspecifically with biological samples are far less common for inorganic nanostructures. Admittedly, there are still limitations to using nanostructures as biological probes. Nanoparticles are significantly larger than molecular dyes, and a bound nanostructure might sterically block access to the active sites of a biomolecule or affect its diffusion. In addition, the bioconjugation chemistry of some nanomaterials is still not fully refined. Nevertheless, the control that synthetic chemists have demonstrated over the structure and properties of nanoparticle materials is impressive and naturally points toward their biological applications.

Nanoscale pH Meter

An all-optical nanoshell coated with pH-sensitive molecules that functions as a stand-alone nanoscale pH meter has been constructed, which monitors its local environment through the pH-dependent surface-enhanced Raman scattering (SERS) spectra of the adsorbate molecules (Bishnoi et al. 2006). The complex spectral output is reduced to a simple device characteristic by application of a locally linear manifold approximation algorithm. This study presents biologists with the first accurate method of measuring accurate pH changes inside living tissue and cells, including tumors, in real-time measurements.

Nanolaser Applications in Life Sciences

A nanolaser is a tiny laser which emits a coherent beam of light through the vibration of a single electron rather than the space-consuming optical pumping process of a traditional laser. The nanolasers were developed by growing semiconductor nanowires. The line widths, wavelengths, and power dependence of the nanowire emission characterize the nanowires as active optical cavities. Current leading solid-state lasers, often made of gallium arsenide or gallium nitride, are made of multilayer thin films and measure several micrometers in size. The nanowire laser is 1,000 times smaller, allowing localized optical illumination. It can be tuned to emit light of different wavelengths from the infrared to the deep ultraviolet by simply changing the diameter or composition of the nanowire.

One of the smallest lasers ever made, nanowire nanolaser, is too small to be seen even with the aid of the most powerful optical microscope. The nanowire nanolasers are pure crystals of zinc oxide that grow vertically in aligned arrays like the bristles on a brush. These crystal nanowires range from 2 to 10 μm in length, depending upon how long the growth process was allowed to proceed. The nanowire nanolaser, which emits flashes of ultraviolet light, measures just less than 100 nm. The individual ZnO nanowires have uniform diameters ranging from 10 to

300 nm. Under optical excitation, each individual ZnO nanowire serves as a Fabry-Perot optical cavity, and together they form a highly ordered nanowire ultraviolet laser array.

It is the small area of illumination that holds near-term potential for the nanolaser. Near-term products could include ultrahigh-resolution photolithography for next-generation microchips, as well as laser-powered biochips. Other potential applications for the nanolaser include high-density information storage, high-definition displays, photonics, optocommunications, and chemical analysis on microchips. Nanolaser spectroscopy has also been used to study very small biological structures. This technology has also been applied for biphotonic detection of cancer in single cells (see Chap. 4).

Nanoelectroporation

Whereas cells precisely control the passage of substances in and out of their membranes, efforts to transport materials into cells in human employ clumsier techniques, which often damage cells and provide little control, if any, over the material delivered. In nanoelectroporation, paired microchannels are connected by a nanochannel through which materials can move into cells and might help solve this problem (Boukany et al. 2011).

Nucleic acids are usually introduced into cells by bulk electroporation, in which suspended cells and reagents are placed together in a vessel, and an electric field is applied to increase the permeability of the cell membrane. The technique is simple, but it also destroys many cells and leaves many untransfected. Another technique, microfluidics-based electroporation, positions cells next to small openings that focus electric fields on only a small section of the cell membrane, which results in lower rates of cell death and higher rates of transfection, but it does not allow control over how much material is delivered. The new device enables electroporation on nanometer scale. Not only is the electric field applied to an area one-hundredth the size of those used in microfluidic-based methods, the volume of material delivered to cells can be precisely controlled through the duration and number of electric pulses. Bulk electroporation does not allow investigation of effects of dose levels. Microinjection can deliver precise dose levels, but this method works best with large cells, which are less easily damaged by the injection needle. A method enabling precisely controlled transfection for small cells would overcome this limitation.

This device described is made of a series of paired microchannels, each connected by an even tinier nanochannel with a diameter of about 90 nm. This nanochannel is made by laying gold-coated DNA strands into a low-viscosity resin into which microchannels have been stamped and then etching out the strands' impression. Cells are placed in one microchannel and the transfection material in the other. A voltage pulse creates a tiny pore in the cell membrane through which a precise amount of material can be driven.

The device was tested by transfecting cells with 18-mer oligonucleotides attached to a fluorescent marker as well as with an RNA-based molecular beacon (a probe designed to fluoresce upon hybridization to a target RNA or DNA molecule). Dose control was demonstrated by transfecting cancer cells with varying levels of a siRNA targeting a protein that inhibits apoptosis, demonstrating that varying dose levels affected cell viability. The technique also enabled controlled delivery of specified numbers of quantum dots and large DNA molecules into cells. With other forms of electroporation, nanoparticles tend to get stuck in the cell membrane, but nanochannel electroporation allows the particles to reach interior of cells. The device also works with large nucleic acids (larger than 4,000,000 Da, or about 6.6 kilobases), which are difficult to transfect using existing methods.

Currently, only a handful of cells can be transfected at a time with the nanochannel electroporation device because cells are loaded into the microchannel using optical tweezers. However, work is in progress on a second-generation device that would allow parallel transfection of 100,000 cells. The device can be used for studying fundamental biological problems, but the most important applications will be for modifying cells in gene therapy and reprogramming them. Current techniques can result in overdosing and other transfection-caused toxicity, but a precisely controlled high dose delivered to the precursor cells can be successful with low chance of forming cancerous cells.

Nanomanipulation

AFM enables the imaging and manipulation of biological systems at the nanometer scale. Examples include the following:

- Extraction of chromosomal DNA for genetic analysis
- Disruption of antibody-antigen bonds
- Dissection of biological membranes
- Nanodissection of protein complexes
- Controlled modulation of protein conformations

Nanomanipulation by Combination of AFM and Other Devices

An instrument constructed by combining an objective-type total internal reflection fluorescence microscope with an AFM can detect and confirm the result of cellular level manipulations made with the AFM, partly through the detection system of the highly sensitive fluorescence microscope. In this combination, manipulations are now possible from the nanometer to the micrometer scales, and the fluorescence detection system is sensitive enough even for localizing single molecules.

Nanomanipulation and nanoextraction on a scale close to and beyond the resolution limit of light microscopy are needed for many modern applications in biological research. For the manipulation of biological specimens, a combined microscope enabling ultraviolet (UV) microbeam laser manipulation together with manipulation by an AFM has been used. In a one-step procedure, human metaphase chromosomes can be dissected optically by the UV-laser ablation and mechanically by AFM manipulation. With both methods, sub-400-nm cuts are achieved routinely. Thus, the AFM is an indispensable tool for in situ quality control of nanomanipulation.

Design of a compact nanomanipulator that can be operated inside the sample chamber of a SEM for biological sample manipulation is based on that of AFM (Iwata et al. 2011). A self-sensitive cantilever is used to realize the compact body. Using this system, the scientists accomplished nanodissection of biological samples as well as AFM imaging under SEM observation. They then fabricated the surface of a rat renal glomerulus by scan scratching and succeeded in making a small hole on the wall of a blood capillary. As a result, it was possible to observe the internal structure of the capillary, which had been hidden beneath the surface wall. Furthermore, using two AFM units on the sample stage of the SEM, they successfully dissected the lens fiber cells taken from a rat eye in a multiprobe operation using the two cantilevers. This system is expected to become a very useful tool for micro- and nanometer-scale anatomical manipulations.

Surgery on Living Cells Using AFM with Nanoneedles

A tool has been developed for performing surgical operations on living cells at nanoscale resolution using AFM and a modified AFM tip (Obataya et al. 2005). The AFM tips are sharpened to ultrathin needles of 200–300 nm in diameter using focused-ion-beam etching. Force-distance curves obtained by AFM using the needles indicated that the needles penetrated the cell membrane following indentation to a depth of 1–2 μm. The force increase during the indentation process was found to be consistent with application of the Hertz model. A three-dimensional image generated by laser scanning confocal microscopy directly revealed that the needle penetrated both the cellular and nuclear membranes to reach the nucleus. This technique enables the extended application of AFM to analyses and surgery of living cells.

A nanoknife, fabricated from a commercial AFM cantilever by focused-ion-beam etching technique, has been proposed for single-cell cutting (Shen et al. 2011). The material identification of the nanoknife was determined using the energy dispersion spectrometry method. The buffering beam was used to measure the cutting force based on its deformation. The spring constant of the beam was calibrated based on a referenced cantilever by using a nanomanipulation approach. The cutting force and the sample slice angle for various nanoknives were evaluated. It was shown that the compression to the cell can be reduced when using the nanoknife with a small edge angle 5°. Consequently, the nanoknife was capable for in situ single-cell cutting tasks.

Optoelectronic Tweezers

The ability to easily manipulate cells has many applications, for example, isolation and study of circulating fetal cells in a mother's blood sample to sort out abnormally shaped organisms from healthy ones. This sorting process is usually painstakingly done by hand. A technician finds the cell of interest under a microscope and literally cuts out the piece of glass where the cell is located, taking care not to harm the sample. The conventional manipulation techniques – including optical tweezers, electrokinetic forces, magnetic tweezers, acoustic traps, and hydrodynamic flows – cannot achieve high resolution and high throughput at the same time. Optical tweezers offer high resolution for trapping single particles but have a limited manipulation area owing to tight-focusing requirements; on the other hand, electrokinetic forces and other mechanisms provide high throughput but lack the flexibility or the spatial resolution necessary for controlling individual cells.

An optical image-driven dielectrophoresis technique permits high-resolution patterning of electric fields on a photoconductive surface for manipulating single particles (Chiou et al. 2005). Such optoelectronic tweezers (OETs) can produce instant microfluidic circuits without the need for sophisticated microfabrication techniques. With direct optical imaging control, multiple manipulation functions are combined to achieve complex, multistep manipulation protocols. Microscopic polystyrene particles suspended in a liquid are sandwiched between a piece of glass and the photoconductive material. Wherever light hits the photosensitive material, it behaves like a conducting electrode, while areas not exposed to light behave like a nonconducting insulator. Once a light source is removed, the photosensitive material returns to normal. Depending upon the properties of the particles or cells being studied, they are either attracted to or repelled by the electric field generated by the OET. It requires 100,000 times less optical intensity than optical tweezers. Parallel manipulation of 15,000 particle traps can be done on a $1.3 \times 1.0\text{-mm}^2$ area. The researchers are now studying ways to combine this technology with computer pattern recognition so that the sorting process could be automated. A program could be designed to separate cells by size, luminescence, texture, fluorescent tags, and basically any characteristic that can be distinguished visually.

By combining the manipulation capabilities of OET with other relevant biological techniques (such as cell lysis and electroporation), it is possible to realize a true parallel, single-cell diagnostic and stimulation tool. The usefulness of the OET device has been demonstrated by integrating it onto single-chip systems capable of performing in situ, electrode-based electroporation/lysis; individual cell, light-induced lysis; and light-induced electroporation (Valley et al. 2009).

Optical Manipulation of Nanoparticles

Mechanical pumps do not scale down well for moving objects at the nanoscale. Optofluidics uses the pressure of light to move and manipulate biological molecules. A beam of light can trap and move particles as small as 75 nm in diameter, including DNA molecules. This is possible because of the paradoxical dual nature of light, which is considered to be a stream of particles called photons that can exert a force or as waves of expanding and contracting electric and magnetic fields. If light is confined to a waveguide narrower than its wavelength, the wave overflows and can exert a force beyond the guide. In a slot waveguide, two parallel silicon bars, 60 nm apart, serve as two parallel waveguides (Yang et al. 2009). Light waves traveling along each guide expand beyond its boundaries, but because the parallel guides are so close together, the waves overlap and most of the energy is concentrated in the slot. In addition to creating a more intense beam, this structure allows a beam of light to be channeled through air or water. This device will help to bridge the gap between optical manipulation and nanofluidics. A tiny biological sample could be carried through microscopic channels for processing. This would enable portable, fast-acting detectors for disease organisms or food-borne pathogens and other tests that now take hours or days. Further development could make it possible to separate DNA molecules by length for rapid DNA sequencing.

Manipulation of DNA Sequence by Use of Nanoparticles as Laser Light Antennas

An optical technique has been developed for the parallel manipulation of nanoscale structures with molecular resolution (Csaki et al. 2006). Bioconjugated metal nanoparticles are positioned at the location of interest, such as certain DNA sequences along metaphase chromosomes, prior to pulsed laser light irradiation of the whole sample. Nanoparticles are designed to absorb the introduced energy highly efficiently, thus acting as nanoantenna. As result of the interaction, structural changes of the sample with subwavelength dimensions and nanoscale precision are observed at the location of the particles. The process leading to the nanolocalized destruction is caused by particle ablation as well as thermal damage of the surrounding material. The procedure is highly parallel and can be potentially multiplexed by addressing several different sequences (such as genes). Potential applications are in DNA analysis such as DNA fingerprinting or mutation analysis as well as single-molecule manipulation.

Nanomanipulation of Single Molecule

The development of nanomanipulation techniques has given investigators the ability to manipulate single biomolecules in the order of nanometers and to record mechanical events of biomolecules at the single-molecule level. The techniques can elucidate the mechanism of molecular motors by directly monitoring the unitary process of the mechanical work and the energy conversion processes by combining these techniques with the single-molecule imaging techniques. The results strongly suggest that the sliding movement of the actomyosin motor is driven by Brownian movement. Other studies have reported data that are more consistent with the lever arm model. These methods and imaging techniques enable monitoring of the behavior of biomolecules at work and can be applied to other molecular machines.

By monitoring the end-to-end extension of a mechanically stretched, supercoiled, single DNA molecule, it is possible to directly observe the change in extension associated with unwinding of approximately one turn of promoter DNA by RNA polymerase. This approach has been used to quantify the extent of unwinding and compaction, the kinetics of unwinding and compaction, and effects of supercoiling, sequence, and nucleotides. The approach should permit analysis of other nucleic acid-processing factors that cause changes in DNA twist and/or DNA compaction.

Fluorescence-Force Spectroscopy

Despite the recent advances in single-molecule manipulation techniques, purely mechanical approaches cannot detect subtle conformational changes in the biologically important regime of weak forces. A hybrid scheme combining force and fluorescence has enabled the examination of the effect of subpiconewton forces on the nanometer-scale motion of the Holliday junction (HJ) at 100-hertz bandwidth (Hohng et al. 2007). The HJ is a four-stranded DNA structure that forms during homologous recombination, for example, when damaged DNA is repaired. To better understand the mechanisms and functions of proteins that interact with the HJ, researchers must first understand the structural and dynamic properties of the junction itself. But purely mechanical measurement techniques cannot detect the tiny changes that occur in biomolecules in the regime of weak forces. Mechanical interrogation of the HJ in three different directions helped elucidate the structures of the transient species populated during its conformational changes. This method combines the exquisite force control of an optical trap and the precise measurement capabilities of single-molecule fluorescence resonance energy transfer for mapping 2D reaction landscapes at low forces. It is readily applicable to other nucleic acid systems and their interactions with proteins and enzymes.

Nanomanipulation for Study of Mechanism of Anticancer Drugs

Using single-molecule nanomanipulation by magnetic tweezers, investigators have shown that anticancer drug topotecan, a camptothecin, kills cancer cells by preventing the enzyme DNA topoisomerase I from uncoiling double-stranded DNA in those cells (Koster et al. 2007). The DNA becomes locked in tight twists, called supercoils, which bulge out from the side of the overwound DNA molecule. If these supercoils accumulate and persist while the cell is trying to separate the two strands of DNA to make exact copies of the chromosomes during cell division, the cells will die. In vivo experiments in the budding yeast verified the resulting prediction that positive supercoils would accumulate during transcription and replication as a consequence of camptothecin poisoning of topoisomerase I. Based on the results of these studies, the supercoil theory was developed to explain camptothecins' cytotoxic effect, which could help in the clinical development of these agents.

Nanotechnology in Genomic Research

Studies in genetics, genomics, and cell biology have provided a foundation on which to build an understanding of genome biology. Extending this knowledge will require merging these approaches with additional disciplines and new technologies such as nanotechnology.

Nanotechnology for Separation of DNA Fragments

A technology using core-shell-type nanospheres and nanoparticle medium in conjunction with a pressurization technique was used to carry out separations of a wide range of DNA fragments with high speed and high resolution during microchip electrophoresis (Tabuchi et al. 2004). DNA fragments up to 15-kilobase pairs were successfully analyzed within 100 s without observing any saturation in migration rates. Optimal pressure conditions and concentrations of packed nanospheres are considered to be important for achieving improved DNA separations.

Electrophoretic transport of ions and macromolecules within long, thin nanochannels is being studied. Electrophoresis in nanochannels is a method to exploit coupling of disparate physical forces at the nanoscale in microfluidic devices for fast and accurate electrophoresis and chromatography. Potential application of nanochannel electrophoresis is improvements in DNA separation and sequencing. Conventional methods of DNA electrophoresis make use of either a gel or a concentrated solution of hydrophilic polymers as a separating medium in which DNA molecules migrate in the presence of an electric field. Gel-less separation of DNA via nanochannel electrophoresis, for example, might offer a significant reduction in both cost and time across a wide range of basic research, medical, and forensics applications (Baldessari and Santiago 2006). In order to achieve this goal, it is important at this stage to develop a fundamental understanding of how each phenomenon is regulated and to observe how the coupling of these affects separations. There is need to experimentally probe the dynamics of electrophoretic separations in nanochannels to expand the currently limited knowledge base.

Nanotechnology-Based DNA Sequencing

An efficient, nanoliter-scale microfabricated bioprocessor, integrating all three Sanger sequencing steps, thermal cycling, sample purification, and capillary electrophoresis, has been developed and evaluated (Blazej et al. 2006). Hybrid glass-polydimethylsiloxane wafer-scale construction is used to combine 250-nl

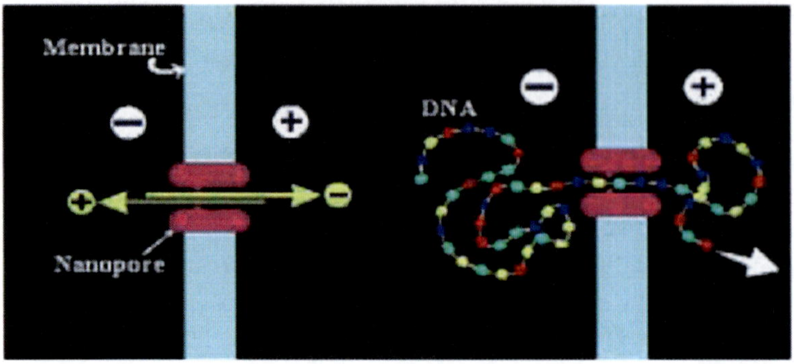

Fig. 3.1 Concept of nanopore-based sequencing. When a current is applied across the membrane, charged biomolecules migrate through the pores. As each nucleotide passes through the nanopore, an electric signature is produced that characterizes it because the size of the nanopore allows only a single nucleic acid strand to pass through it at one time

reactors, affinity-capture purification chambers, high-performance capillary electrophoresis channels, and pneumatic valves and pumps onto a single microfabricated device. Lab-on-a-chip-level integration enables complete Sanger sequencing from only 1 fmol of DNA template. Up to 556 continuous bases were sequenced with 99% accuracy, demonstrating read lengths required for de novo sequencing of human and other complex genomes. The performance of this miniaturized DNA sequencer provides a benchmark for predicting the ultimate cost and efficiency limits of Sanger sequencing.

A method for the single-molecule detection of specific DNA sequences involves hybridization of double-stranded DNA (dsDNA) with peptide nucleic acid (PNA) and threading the DNA through 4 to 5-nm pores in a silicon nitride membrane (Singer et al. 2010). The membrane is then placed between two small fluid chambers, and an electric current is applied across the membrane using a pair of silver electrodes. This current causes individual DNA molecules to move through the pore, unraveling as they enter the pore. When the matched DNA-PNA sequence passes through the pore, it produces a marked change in the electrical current passing between the two electrodes, which is easily distinguished from unaltered ssDNA, that is, DNA not duplexed with the PNA probe. The device is capable of analyzing one DNA molecule per second. This high-throughput, long-read length method can be used to identify key sequences embedded in individual DNA molecules, without the need for amplification or fluorescent/radio labeling. The concept of nanopore-based sequencing is shown in Fig. 3.1.

Role of Nanobiotechnology in Identifying Single Nucleotide Polymorphisms

Genetic analysis based on single nucleotide polymorphisms (SNPs) has the potential to enable identification of genes associated with disease susceptibility and to facilitate improved understanding and diagnosis of those diseases and should ultimately contribute to the provision of new therapies. To achieve this end, new technology platforms are required that can increase genotyping throughput while simultaneously reducing costs by as much as two orders of magnitude. Development of a variety of genotyping platforms with the potential to resolve this dilemma is already well advanced through research in the field of nanobiotechnology. Novel approaches to DNA extraction and amplification have reduced the times required for these processes to seconds. Microfluidic devices enable polymorphism detection through very rapid fragment separation using capillary electrophoresis and high-performance liquid chromatography, together with mixing and transport of reagents and biomolecules in integrated systems. The potential for application of established microelectronic fabrication processes to genetic analyses systems has been demonstrated (e.g., photolithography-based in situ synthesis of oligonucleotides on microarrays). Innovative application of state-of-the-art photonics and integrated circuitry is leading to improved detection capabilities. The diversity of genotyping applications envisaged in the future, ranging from the very-high-throughput requirements for drug discovery to rapid and cheap near-patient genotype analysis, suggests that several SNP genotyping platforms will be necessary to optimally address the different niches.

Nanobiotechnology for Study of Mitochondria

While autosomal nuclear DNA genes are confined to the nucleus, limited to two copies per cell, the mitochondrial DNA (mtDNA) genes are distributed throughout the cytoplasm and are present in numerous copies per cell. The mtDNA molecule is relatively small containing 16,569 nucleotide pairs. Mutations of mtDNA are responsible for several human diseases, for example, neurological diseases. Mitochondrial diseases are often underdiagnosed and therapies are inadequate. Nanobiotechnology has been applied to study of mitochondria (Weissig et al. 2007).

Nanomaterials for the Study of Mitochondria

Some of the nanomaterials for study of mitochondrial nanobiotechnology are shown in Table 3.1.

Study of Mitochondria with Nanolaser Spectroscopy

Because mitochondria are so small, averaging a few hundred nanometers, scientists have been unable to study them in vitro with the necessary precision. Nanolaser spectroscopy is useful for the study of mitochondria. Isolated mitochondria can be

Table 3.1 Nanomaterials for the study of mitochondria

Nanomaterial	Modification	Application
Biosomes	Self-assembly of mitochondriotropic bola-amphiphile (DQAsomes)	Selective delivery of DNA to mitochondria to enable mitochondrial gene therapy
Liposomes	Mitochondria-specific ligands with hydrophobic anchor to form mitochondria-targeted liposomes	Mitochondria-specific drug delivery
Nanoparticles	Mitochondria-specific ligands with linker to form mitochondria-targeted nanoparticles	Selective accumulation in mitochondria to probe or manipulate mitochondrial function
Quantum dots	Mitochondria-specific ligands with linker to form mitochondria-targeted quantum dots	To study the function and morphology of mitochondria

studied by nanolaser spectroscopy to assess the optical density of an individual mitochondrion. This measurement uniquely reflects the mitochondrion size and biomolecular composition. As such, the optical density is a powerful parameter that rapidly interrogates the biomolecular state of single cells and mitochondria. In normal cells, mitochondria are highly organized within the cytoplasm and highly scattering, yielding a highly correlated signal, whereas in cancer cells, mitochondria are more chaotically organized and poorly scattering (Gourley and McDonald 2007). These differences correlate with important bioenergetic disturbances that are hallmarks of many types of cancer. These optical methods may be useful for detecting cancer at an early stage.

Role of Nanotechnology in Proteomics Research

Study of Proteins by Atomic Force Microscopy

In contrast to other methods, specimens prepared for AFM remain in a plastic state, which enables direct observation of the dynamic molecular response, creating unique opportunities for studying the structure-function relationships of proteins and their functionally relevant assemblies. Electron crystallography and AFM enable the study of 2D membrane protein crystals. While electron crystallography provides atomic scale 3D density maps, AFM gives insight into the surface structure and dynamics at subnanometer resolution. Importantly, the membrane protein studied is in its native environment, and its function can be assessed directly. The approach allows both the atomic structure of the membrane protein and the dynamics of its surface to be analyzed.

Comparisons of AFM topographs with protein structures determined by electron microscopy and X-ray crystallography have shown excellent agreement within a lateral resolution of <1 nm and a vertical resolution of ~0.1 nm. AFM application in imaging and nanomanipulation include the extraction of chromosomal DNA for genetic analysis, the disruption of antibody-antigen bonds, the dissection of biological membranes, the nanodissection of protein complexes, and the controlled modulation of protein conformations.

Observing proteins in their native environment is a prerequisite for the proper assessment of function. Conformational changes of molecular assemblies can be observed by time-lapse AFM; however, such experiments lack the time resolution required to observe the turnover of most biological "machineries" owing to the relatively long time needed to record a topograph. One solution to this problem is to image the static conformations associated with different functional states of a biological macromolecule. Alternatively, the imaging speed of the AFM can be enhanced.

Single-Cell Nanoprobe for Studying Gene Expression of Individual Cells

The localization of specific mRNA generates cell polarity by controlling the translation sites of specific proteins. Although most of these events depend on differences in gene expression, no method is available to examine time-dependent gene expression of individual living cells. In situ hybridization (ISH) is a powerful and useful method for detecting the localization of mRNAs, but it does not allow a time-dependent analysis of mRNA expression in single living cells because the cells have to be fixed for mRNA detection. To overcome these issues, the extraction of biomolecules such as mRNAs, proteins, and lipids from living cells should be performed without severe damage to the cells. A single-cell nanoprobe (SCN) method to examine gene expression of individual living cells is available by using AFM without killing the cells. The SCN method has been compared with ISH (Uehara et al. 2007). Spatial beta-actin mRNA expression in single living cells was examined first, with the SCN method, and then the same cells were subjected to ISH for beta-actin mRNA. In the SCN method, quantity of beta-actin mRNA was analyzed by quantitative PCR, and in ISH, intensity of ISH was used as a parameter of concentration of beta-actin mRNA. The results showed that intensity of ISH as well as quantity of beta-actin mRNA detected by the SCN method is higher. Thus, SCN is a suitable and reliable method to examine mRNAs at medium or higher expression level.

Nanoproteomics

Nanoproteomics – application of nanobiotechnology to proteomics – improves on most current protocols including protein purification/display and automated identification schemes that yield unacceptably low recoveries with reduced sensitivity and speed while requiring more starting material. Low abundant proteins and proteins that can only be isolated from limited source material (e.g., biopsies) can be subjected to nanoscale protein analysis – nanocapture of specific proteins and complexes and optimization of all subsequent sample handling steps leading to mass analysis of peptide fragments. This is a focused approach, also termed "targeted proteomics," and involves examination of subsets of the proteome, for example, those proteins that are either specifically modified or bind to a particular DNA sequence or exist as members of higher-order complexes or any combination thereof. Some nanoproteomic technologies are described here briefly.

Dynamic Reassembly of Peptides

Nanofiber structures of some peptides and proteins as biological materials are being studied including their molecular mechanism of self-assembly and reassembly. An ionic self-complementary 16-residue peptide RADARADARADARADA (RADA16-I) has been reported to form a well-defined nanofiber scaffold that undergoes molecular self-assembly into nanofibers and eventually a scaffold hydrogel consisting of >99.5 % water (Yokoi et al. 2005). In this study, the nanofiber scaffold was sonicated into smaller fragments, and the mechanism of reassembly was studied by AFM. These sonicated fragments not only quickly reassemble into nanofibers that were indistinguishable from the original material but their reassembly also correlated with the rheological analyses showing an increase of scaffold rigidity as a function of nanofiber length. The disassembly and reassembly processes were repeated several times, and each time, the reassembly reached the original length. This reassembly process is important for the construction of new scaffolds for 3D cell culture, tissue repair, and regenerative medicine.

High-Field Asymmetric Waveform Ion Mobility Mass Spectrometry

An ion mobility technology – high-field asymmetric waveform ion mobility mass spectrometry (FAIMS) – has been introduced as online ion selection methods compatible with electrospray ionization (ESI). FAIMS uses ion separation to improve detection limits of peptide ions when used in conjunction with electrospray and nanoelectrospray MS. This facilitates the identification of low-abundance peptide ions often present at ppm levels in complex proteolytic digests and expands the sensitivity and selectivity of nanoLC-MS analyses in global and targeted proteomics approaches. This functionality likely will play an important role in drug discovery and biomarker programs for monitoring of disease progression and drug efficacy.

In nanoelectrospray ionization (nanoESI) techniques, the hydrophilic character of the emitters generally produces large bases for the Taylor cones, thereby generating relatively large droplet sizes and consequently reduced sensitivity. In order to minimize this wetting effect in nanoESI, a model hydrophobic polymer (an acrylic paint) was coated at the tip of commercial polyaniline (PANI)-coated emitters, and their performance was compared with that of unmodified PANI emitters using oxytocin and neuropeptide Y solutions. In experiments with oxytocin, the hydrophobic emitter produced higher signal intensities as well as higher signal-to-noise ratios than those from the unmodified PANI emitter. In addition, the hydrophobic emitter showed reusability and a slightly wider linear dynamic range than that from the unmodified PANI emitter (Choi and Wood 2007). In the case of neuropeptide Y, the hydrophobic emitter also enabled an approximately 350-fold overall increase in sensitivity than the unmodified PANI. The enhanced performance of the hydrophobic emitter clearly indicates potential for further increases in nanoESI sensitivity using emitters with tailored hydrophobic overcoatings.

Multiphoton Detection

A new detection technique called multiphoton detection (MPD) is in development at BioTrace Inc. and enables quantitation of subzeptomole amounts of proteins. It will be used for diagnostic proteomics, particularly for cytokines and other low-abundance proteins. BioTrace is developing supersensitive protein biochips to detect as low as 5 fg/ml (0.2 attomole/ml) concentration of proteins. Thus, this innovative type of the P-chips might permit about 1,000-fold better sensitivity than current protein biochips.

Nanoflow Liquid Chromatography

The use of liquid chromatography (LC) in analytical chemistry is well established, but relatively low sensitivity associated with conventional LC makes it unsuitable for the analysis of certain biological samples. Furthermore, the flow rates at which it is operated are not compatible with the use of specific detectors, such as electrospray ionization mass spectrometers. Therefore, due to the analytical demands of biological samples, miniaturized LC techniques were developed to allow for the analysis of samples with greater sensitivity than that afforded by conventional LC. In nanoflow LC (nanoLC), chromatographic separations are performed using flow rates in the range of low nanoliter per minute, which result in high analytical sensitivity due to the large concentration efficiency afforded by this type of chromatography. NanoLC, in combination to tandem mass spectrometry, was first used to analyze peptides and as an alternative to other mass spectrometric methods to identify gel-separated proteins. Gel-free analytical approaches based on LC and nanoLC separations have been developed, which are allowing proteomics to be performed in faster and more comprehensive manner than by using strategies based on the classical 2D gel electrophoresis approaches.

Protein identification using nanoflow liquid chromatography-mass spectrometry (MS)-MS (LC-MS-MS) provides reliable sequencing information for low femtomole level of protein digests. However, this task is more challenging for subfemtomole peptide levels.

Nanoproteomics for Study of Misfolded Proteins

Misfolding and self-assembly of proteins in nanoaggregates of different sizes and morphologies (nanoensembles, primarily nanofilaments and nanorings) are a complex phenomenon that can be facilitated, impeded, or prevented, by interactions with various intracellular metabolites, intracellular nanomachines controlling protein folding, and interactions with other proteins. A fundamental understanding of molecular processes leading to misfolding and self-aggregation of proteins involved in various neurodegenerative diseases will provide critical information to help identify appropriate therapeutic routes to control these processes. An elevated propensity of misfolded protein conformation in solution to aggregate with the formation of various morphologies impedes the use of traditional physical chemical approaches for studies of misfolded conformations of proteins. In an alternative approach, the protein molecules are tethered to surfaces to prevent aggregation and AFM force spectroscopy is used to probe the interaction between protein molecules depending on their conformations (Kransnoslobodtsev et al. 2005). It was shown that formation of filamentous aggregates is facilitated at pH values corresponding to the maximum of rupture forces. A novel surface chemistry was developed for anchoring of Aβ peptides at their N-terminal moieties. The use of the site-specific immobilization procedure allowed to measure the rupture of Aβ–Aβ contacts at single-molecule level. The rupture of these contacts is accompanied by the extension of the peptide chain detected by a characteristic elastomechanical component of the force-distance curves. Nanomechanical studies have potential applications for an understanding of the mechanisms of development of protein-misfolding diseases such as Alzheimer's disease.

Nanotube Electronic Biosensor for Proteomics

Single-walled carbon nanotubes can be used as a platform for investigating surface-protein and protein-protein binding and developing highly specific electronic biomolecule detectors. Nonspecific binding on nanotubes, a phenomenon found with a wide range of proteins, is overcome by immobilization of polyethylene oxide chains. A general approach is then advanced to enable the selective recognition and binding of target proteins by conjugation of their specific receptors to polyethylene oxide-functionalized nanotubes. These arrays are attractive because no labeling is required and all aspects of the assay can be carried out in solution phase. This scheme, combined with the sensitivity of nanotube electronic devices, enables highly specific electronic sensors for detecting clinically important biomolecules such as antibodies associated with human autoimmune diseases. These arrays are attractive because no labeling is required and all aspects of the assay can be carried out in solution phase. Interfacing novel nanomaterials with biological systems could therefore lead to important applications in disease diagnosis, proteomics, and nanobiotechnology in general.

Protein Nanocrystallography

Application of nanotechnology to structural proteomics can produce and characterize diffracting, stable, and radiation-resistant crystals of miniscule dimensions. Protein microcrystals obtained by nanotechnology-based protein thin-film template crystallization, as well as groundbreaking technology, such as AFM, nanogravimetry, and synchrotron microfocus, have enabled protein nanocrystallography to be defined as a unique technology capable of forming and characterizing stable protein microcrystals down to atomic resolution. A new route from art to science and technology has, therefore, been opened in protein crystallography, and it could be used to unravel the mysteries of many systems that remain unsolved.

Previous studies of symmetry preferences in protein crystals suggest that symmetric proteins, such as homodimers, might crystallize more readily on average than asymmetric, monomeric proteins. Proteins that are naturally monomeric can be made homodimeric artificially by forming disulfide bonds between individual cysteine residues introduced by mutagenesis. Furthermore, by creating a variety of single-cysteine mutants, a series of distinct synthetic dimers can be generated for a given protein of interest, with each expected to gain advantage from its added symmetry and to exhibit a crystallization behavior distinct from the other constructs. This strategy was tested on phage T4 lysozyme, a protein whose crystallization as a monomer has been studied exhaustively (Banatao et al. 2006). Experiments on three single-cysteine mutants, each prepared in dimeric form, yielded numerous novel crystal forms that cannot be realized by monomeric lysozyme. Six new crystal forms have been characterized. The results suggest that synthetic symmetrization may be a useful approach for enlarging the search space for crystallizing proteins.

QD-Protein Nanoassembly

An intein-based method for site-specific conjugation of QDs to target proteins in vivo has been described, which enables the covalent conjugation of any nanostructure and/or nanodevice to any protein and thus the targeting of such material to any intracellular compartment or signaling complex within the cells of the developing embryo (Charalambous et al. 2009). In vivo intein splicing produces fully functional conjugates that can be monitored in real time within live embryos. Use of near-infrared (NIR)-emitting QDs allows monitoring of QD conjugates within the embryo at depths where enhanced green fluorescent protein is undetectable demonstrating the advantages of QDs for this type of experiment.

Proteomics at Single-Molecule Level

Study of Protein Synthesis and Single-Molecule Processes

All life relies on the actions and reactions of single molecules within cells, but these molecules are so tiny that they have long eluded direct, real-time investigation using conventional light microscopes. Breakthrough technologies are enabling researchers to have an unprecedented view into the workings of individual molecules.

Single-molecule methods enable observation of the stepwise movement of aminoacyl-tRNA (aa-tRNA) into the ribosome during selection and kinetic proofreading using single-molecule fluorescence resonance energy transfer (smFRET).

Intermediate states in the pathway of tRNA delivery were observed using antibiotics and nonhydrolyzable GTP analogs. Three unambiguous FRET states have been identified corresponding to initial codon recognition, GTPase-activated and fully accommodated states (Blanchard et al. 2004a). The antibiotic tetracycline blocks progression of aa-tRNA from the initial codon recognition state, whereas cleavage of the sarcin-ricin loop impedes progression from the GTPase-activated state. These data support a model in which ribosomal recognition of correct codon-anticodon pairs drives rotational movement of the incoming complex of EF-Tu-GTP-aa-tRNA toward peptidyl-tRNA during selection on the ribosome. This is the basis of a mechanistic model of initial selection and proofreading.

Subsequently, tRNA molecules fluctuate between two configurations assigned as classical and hybrid states. The lifetime of classical and hybrid states, measured for complexes carrying aa-tRNA and peptidyl-tRNA at the A site, shows that peptide bond formation decreases the lifetime of the classical-state tRNA configuration by approximately sixfold. These data suggest that the growing peptide chain plays a role in modulating fluctuations between hybrid and classical states. smFRET was also used to observe aa-tRNA accommodation coupled with elongation factor G-mediated translocation (Blanchard et al. 2004b). Dynamic rearrangements in tRNA configuration are also observed subsequent to the translocation reaction. This work underscores the importance of dynamics in ribosome function and demonstrates single-particle enzymology in a system of more than two components.

Using these technologies, one can collect photons of light coming from a single molecule. This information reports on a biomolecule's location, its interaction with other molecules, and tiny motions within the molecule itself. These tools allow view of an enzymatic reaction from the very intuitive perspective of movements. Enzymes are molecular machines with moving parts, but these motions are on the order of a billionth of a centimeter. Nanoscale technologies will help to understand the mechanism of the ribosome, which is an assembly of about 60 different molecules working together to read the instructions for making new proteins coded in DNA. These instructions are presented to the ribosome in the form of messenger RNA (mRNA). The process of translating mRNA instructions into protein involves the selection by the ribosome of adaptor RNA molecules, called transfer RNA (tRNA). It is the selection of specific tRNA molecules that determines the relationship between the gene sequence and the sequence of the resulting protein. The reaction between tRNA and the ribosome is the basis of what is called the universal genetic code. Ribosome is important for cell function and human health. Because protein synthesis is crucial to the life cycle of all bacteria, roughly 50 % of antibiotics used today target ribosomal function. Ribosomal function is also a key to the success or failure of deadly viral infections such as HCV and HIV. Cancer cells, too, rely on protein synthesis to survive and multiply, so drugs that block ribosomal function in a cancer-specific manner might prove safe and effective chemotherapy. Genetic aberrations in the DNA-ribosome relationship can cause the enzyme to produce faulty proteins that trigger cystic fibrosis and other inherited illnesses. Understanding the mechanism of the ribosome may be a fundamental first step for developing antibiotics, cancer therapies, and antiviral drugs. The ribosome may one day even be a target for gene therapy.

Protein Expression in Individual Cells at the Single-Molecule Level

Measurements on single cells usually lack the sensitivity to resolve individual events of protein production. A microfluidic-based assay has been described that enables real-time observation of the expression of beta-galactosidase in living *E. coli* cells with single-molecule sensitivity (Cai et al. 2006). It revealed that protein production occurs in bursts, with the number of molecules per burst following an exponential distribution. The two key parameters of protein expression – the burst size and frequency – can be either determined directly from real-time monitoring of protein production or extracted from a measurement of the steady-state copy number distribution in a population of cells. Application of this assay to probe gene expression in individual budding yeast and mouse embryonic stem cells demonstrates its generality. Many important proteins are expressed at low levels and are thus inaccessible by current genomic and proteomic techniques. This microfluidic single-cell assay opens up possibilities for system-wide characterization of the expression of these low copy number proteins.

Single-Molecule Mass Spectrometry Using Nanotechnology

A 2D method for mass spectrometry (MS) in solution is based on the interaction between a nanometer-scale pore and analytes (Robertson et al. 2007). The technique involves creating a lipid bilayer membrane similar to those in living cells and "drilling" a pore in it with a protein (α-hemolysin) produced by the *Staphylococcus aureus* bacteria specifically to penetrate cell membranes. Charged molecules (such as single-stranded DNA) are forced one at a time into the nanopore, which is only 1.5 nm at its smallest point, by an applied electric current. As the molecules pass through the channel, the current flow is reduced in proportion to the size of each individual chain, allowing its mass to be easily derived. In this experiment, various-sized chains in solution of the uncharged polymer polyethylene glycol (PEG) were substituted for biomolecules. Each type of PEG molecule reduced the nanopore's electrical conductance differently as it moved through, allowing the researchers to distinguish one size of PEG chain from another. Because the dimensions of the lipid bilayer and the α-hemolysin pore, as well as the required amount of electrical current, are at the nanoscale level, the single-molecule mass spectrometry technology may one day be incorporated into "lab-on-a-chip" molecular analyzers and single-stranded DNA sequencers. This single-molecule analysis technique could prove useful for the real-time characterization of biomarkers (i.e., nucleic acids, proteins, or other biopolymers). With automated, unsupervised analytical and statistical methods, this technique should prove viable as a generalized analytical technique with nanopore arrays containing nanopores both with specific affinities for single biomarkers and with nonspecific transducers. In situ monitoring of cellular metabolism with such arrays should provide the sensitivity to monitor subtle changes observed through the release of biomarkers.

Nanoelectromechanical systems provide unparalleled mass sensitivity, which is now sufficient for the detection of individual molecular species in real time. In a nanoelectromechanical-mass spectrometry system, nanoparticles and protein species are introduced by electrospray injection from the fluid phase in ambient conditions into vacuum and are subsequently delivered to the system's detector by hexapole ion optics (Naik et al. 2009). Precipitous frequency shifts, proportional to the mass, are recorded in real time as analytes adsorb, one by one, onto a phase-locked, ultrahigh-frequency nanoelectromechanical resonator. These first nanoelectromechanical system-mass spectrometry spectra, obtained with modest mass sensitivity from only several hundred mass adsorption events, indicate the future capabilities of this approach. Substantial improvements are feasible in the near term, some of which are unique to nanoelectromechanical system-based mass spectrometry.

Biochips for Nanoscale Proteomics

Protein Biochips Based on Fluorescence Planar Waveguide Technology

The fluorescence planar waveguide (PWG) technology has demonstrated exceptional performance in terms of sensitivity, making it a viable method for detection in chip-based microarrays. Thin-film PWGs as used in ZeptoMARK™ protein microarrays (Zeptosens Bioanalytical Solutions) consist of a 150 to 300-nm thin film of a material with high refractive index, which is deposited on a transparent support with lower refractive index (e.g., glass or polymer). A parallel laser light beam is coupled into the waveguiding film by a diffractive grating, which is etched or embossed into the substrate. The light propagates within this film and creates a strong evanescent field perpendicular to the direction of propagation into the adjacent medium. It has been shown that the intensity of this evanescent field can be enhanced dramatically by increasing the refractive index of the waveguiding layer and equally by decreasing the layer thickness. Compared to confocal excitation, the field intensity close to the surface can be increased by a factor of up to 100. The field strength decays exponentially with the distance from the waveguide surface, and its penetration depth is limited to about 400 nm. This effect can be utilized to selectively excite only fluorophores located at or near the surface of the waveguide. By taking advantage of the high field intensity and the confinement of this field to the close proximity of the waveguide, PWG technology combines highly selective fluorescence detection with highest sensitivity.

For bioanalytical applications, specific capture probes or recognition elements for the analyte of interest are immobilized on the waveguide surface. The presence of the analyte in a sample applied to a PWG chip is detected using fluorescent reporter molecules attached to the analyte or one of its binding partners in the assay. Upon fluorescence excitation by the evanescent field, excitation and detection of fluorophores is restricted to the sensing surface, while signals from unbound molecules in the bulk solution are not detected. The result is a significant increase in the signal/noise ratio compared to conventional optical detection methods.

A variety of proteins can be immobilized on PWG microarrays as selective recognition elements for the investigation of specific ligand-protein interactions such as antigen-antibody, protein-protein, and protein-DNA interactions. Protein microarrays based on PWG allow the simultaneous, qualitative, and quantitative analysis of protein interactions with high sensitivity in a massively parallel manner. This method enables cost-effective determination of efficacy of drug candidates in a vast number of preclinical study samples.

Nanofilter Array Chip

Massachusetts Institute of Technology engineers have developed a microfabricated nanofilter array chip that can size-fractionate protein complexes and small DNA molecules based on the Ogston sieving mechanism (Fu et al. 2006). Nanofilter arrays with a gap size of 40–180 nm were fabricated and characterized. Millions of pores can be spread across a microchip the size of a thumbnail. In the model, proteins move through deep and shallow regions that act together to form energy barriers. These barriers separate proteins by size. The smaller proteins go through quickly, followed by increasingly larger proteins, with the largest passing through last. To date, the Ogston sieving model has been used to explain gel electrophoresis, even though no one has been able to unequivocally confirm this model in gel-based experiments. The performance of the current one-dimensional sieves matches the speed of one-dimensional gels, but it can be improved greatly. The sieves could potentially be used to replace 2D gels in the process of discovering disease biomarkers.

Role of Nanotechnology in Study of Membrane Proteins

Nanoparticles for Study of Membrane Proteins

As a large fraction of proteins are likely to be membrane-bound, technical improvements are needed in the analysis of membrane proteins. A method has been described for visualization of membrane proteins labeled with 10-nm gold nanoparticles in cells, using an all-optical method based on photothermal interference contrast (Cognet et al. 2003). The high sensitivity of the method and the stability of the signals allow 3D imaging of individual nanoparticles without the drawbacks of photobleaching and blinking inherent to fluorescent markers. The photothermal interference contrast method provides an efficient, reproducible, and promising way to visualize low amounts of proteins in cells by optical means.

Nanotube-vesicle networks with reconstituted membrane protein from cells and with interior activity have been defined by an injection of microparticles or molecular probes (Davidson et al. 2003). The functionality of a membrane protein after reconstitution was verified by single-channel ion-conductance measurements in excised inside-out patches from the vesicle membranes. The distribution of protein, determined by fluorescence detection, in the network membrane was homogeneous and could diffuse via a nanotube connecting two vesicles. The authors also show how injecting small unilamellar protein-containing vesicles can differentiate the contents of individual containers in a network. This system can model a variety of biological functions and complex biological multicompartment structures and might serve as a platform for constructing complex sensor and computational devices. This technology platform is now being developed by Nanoxis AB.

Study of Single Protein Interaction with Cell Membrane

The nanoscale research team at University of California (Davis, CA) is using aerogel to study single protein interaction with cell membrane. Aerogel, a semitransparent and lightweight silica substance, behaves like a living cell membrane. Aerogels, as model cell membranes, have potential applications in biosensors and lab-on-a-chip devices. Formation of fluid planar biomembranes has been reported on hydrophilic silica aerogels and xerogels. Scanning electron microscopy results showed the presence of interconnected silica beads of approximately 10–25 nm in diameter and nanoscale open pores of comparable size for the aerogel and grain size of approximately 36–104 nm with approximately 9–24-nm-diameter pores for the xerogel (Weng et al. 2004).

Aerogel has advantages over most artificial membranes that are used in current studies. Most artificial membranes lack functionality because they are one sided. Proteins can penetrate these membranes only on one side instead of both as in case of natural membranes in cells. The aerogel is unique in that it is accessible on both sides, just like the lipid bilayers that form actual membranes. This technique can be used for the detection of diseases in cases in which there is a need to study the specific protein involved. The researchers will not only test the traffic of molecules through their synthetic silica membranes but also develop aerogels out of new materials. Future aerogel materials will be nontoxic so that they may eventually be used in clinical trials. Possible materials used include aluminum oxide, carbon, and even sugar-based compounds.

Quantum Dots to Label Cell Surface Proteins

Antibody-based labeling has been used for targeting quantum dots (QDs) to cells, but there are two problems: the size of the QD conjugate after antibody attachment and the instability of many antibody-antigen interactions. One way to overcome these limitations is to tag the mammalian cell surface proteins with acceptor peptide (AP), which can be biotinylated by biotin ligase added to the medium, while endogenous proteins remain unmodified (Howarth et al. 2005). The biotin group then serves as a handle for targeting streptavidin-conjugated QDs. QDs have been targeted to cell surface cyan fluorescent protein and epidermal growth factor receptor in HeLa cells and to alpha-amino-3-hydroxy-5-methyl-4-isoxazolepropionate (AMPA) receptors in neurons. Labeling is specific for the AP-tagged protein and is highly sensitive. Thus, biotin ligase labeling provides a specific, rapid, and generally applicable method for detecting and tracking cell surface proteins.

Study of Single Membrane Proteins at Subnanometer Resolution

High-resolution atomic force microscopy (AFM) enables observation of substructures of single membrane proteins at subnanometer resolution as well as their conformational changes, oligomeric state, molecular dynamics, and assembly (Janovjak et al. 2006). Complementary to AFM imaging, single-molecule force spectroscopy experiments allow detection of molecular interactions established within and between membrane proteins. The sensitivity of this method makes it possible to detect the interactions that stabilize secondary structures such as transmembrane α-helices, polypeptide loops, and segments within. It has elucidated unfolding and refolding pathways of membrane proteins as well as their energy landscapes. These approaches will provide insights into membrane protein structure, function, and unfolding. They could help to answer key questions on the molecular basis of certain neuropathological disorders.

Nanoparticle-Protein Interactions

Due to their small size, nanoparticles have distinct properties compared with the bulk form of the same materials. In a biological fluid, proteins associate with nanoparticles, and the amount and presentation of the proteins on the surface of the particles lead to an in vivo response. Proteins compete for the nanoparticle "surface," leading to a protein "corona" that largely defines the biological identity of the particle. Thus, knowledge of rates, affinities, and stoichiometries of protein association with, and dissociation from, nanoparticles is important for understanding the nature of the particle surface seen by the functional machinery of cells. Approaches have been developed to study these parameters and apply them to plasma and simple model systems, albumin and fibrinogen (Cedervall et al. 2007). A series of copolymer nanoparticles are used with variation of size and composition (hydrophobicity). Isothermal titration calorimetry is suitable for studying the affinity and stoichiometry of protein binding to nanoparticles. The rates of protein association and dissociation are determined using surface plasmon resonance technology with nanoparticles that are thiol-linked to gold and through size exclusion chromatography of protein-nanoparticle mixtures. This method is less perturbing than centrifugation and is developed into a systematic methodology to isolate nanoparticle-associated proteins. The kinetic and equilibrium binding properties depend on protein identity as well as particle surface characteristics and size.

Protein Engineering on Nanoscale

Nanowires for Protein Engineering

When folded into their native structures, proteins in biological systems function as nanostructured machines. Although proteins provide many valuable properties, poor physical stability and poor electrical characteristics have prevented their direct use in electrical circuits. By contrast, some polypeptides tend to aggregate into other well-ordered structures, namely amyloid fibrils. Such well-ordered protein fibrils are attractive materials for nanobiotechnology because they self-associate through noncovalent bonds under controlled conditions. Self-assembling amyloid protein fibers have been used to construct nanowire elements. Such applications will potentially become one of the next trends in protein engineering and nanobiotechnology. Complex circuit schematics could be generated with these fibers, initiated by patterned surface modifications such as lithography, growth in flows or magnetic-field gradients, alignment by electrical fields, active patterning with optical tweezers, dielectrophoresis, and 3D patterning using hydrogels or microfluidic channels. These fibers were placed across gold electrodes, and additional metal was deposited by highly specific chemical enhancement of the colloidal gold by reductive deposition of metallic silver and gold from salts. The resulting silver and gold wires were approximately 100 nm wide. These biotemplated metal wires demonstrated the conductive properties of a solid metal wire, such as low resistance and ohmic behavior. With such materials, it should be possible to harness the extraordinary diversity and specificity of protein functions to nanoscale electrical circuitry. There is a great opportunity to expand further the potential interconnections in these circuits by exploiting the natural diversity and strength of protein-protein interactions.

A Nanoscale Mechanism for Protein Engineering

Proteins are switched on and off in living cells by a mechanism called allosteric control; proteins are regulated by other molecules that bind to their surface, inducing a change of conformation or distortion in the shape of the protein, making the protein either active or inactive. An artificial nanoscale mechanism of allosteric control is based on mechanical tension by chemically stringing a short piece of DNA around the protein and controlling the stiffness by inserting a molecular spring on the protein (Choi et al. 2005). The protein can be switched on and off by changing the stiffness of the DNA. By gluing together two disparate pieces of the cell's molecular machinery, a protein and a piece of DNA, this method creates a spring-loaded protein which can be turned on and off. This approach to protein engineering could lead to a new generation of targeted "smart" drugs that are active only in cells where a certain gene is expressed or a certain DNA sequence is present. Such drugs would have reduced side effects. Another application for the new molecules is as amplified molecular probes. Currently, it is difficult for scientists to study a single live cell and find what gene it is expressing, but with an amplified molecular probe, one could inject the probe into a single cell and detect that the cell is expressing a particular gene. An amplified molecular probe would make it possible to study the individuality of cells, with applications in stem cell research and early detection of disease.

Role of Nanoparticles in Self-Assembly of Proteins

In their physical dimensions, surface chemistry, and degree of anisotropic interactions in solution, CdTe nanoparticles are similar to proteins. How to direct and control the self-assembly of nanoparticles is a fundamental question in nanotechnology. A method has been discovered to induce nanocrystals in a fluid to assemble into free-floating sheets the same way some protein structures form in living organisms (Tang et al. 2006). The sheets, about 2 μm in width, which can appear colored under UV illumination from bright green to dark red depending on the nanoparticle size, are made from cadmium telluride crystals, a material used in solar cells. Computer simulation and concurrent experiments have demonstrated that the dipole moment, small positive charge, and directional hydrophobic attraction are the driving forces for the self-organization process. The data establish an important connection between two basic building blocks in biology and nanotechnology, that is, proteins and nanoparticles, and this is important for assembling materials from the bottom up for applications ranging from drug delivery to energy.

Role of Nanotechnology in Peptide Engineering

Several types of self-assembling peptide systems have been developed, for example, peptide nanotubes and nanovesicles. These self-assembling peptide systems are simple, versatile, and easy to produce and represent a significant advance in the molecular engineering. Protein design studies using coiled coils have illustrated the potential of engineering simple peptides to self-associate into polymers and networks. Although basic aspects of self-assembly in protein systems have been demonstrated, it remains a major challenge to create materials whose large-scale structures are well determined from design of local protein-protein interactions. A helical peptide has been designed, which uses phased hydrophobic interactions to drive assembly into nanofilaments and nanofibrils (Wagner et al. 2006). The nanostructures designed here are characterized by biophysical methods such as dynamic light scattering and atomic force microscopy to study their behavior on surfaces. The assembly of such structures can be predictably regulated by using various environmental factors such as pH and specifically designed "capping" peptides. This ability to regulate self-assembly is a critical feature in creating smart peptide biomaterials.

Manipulating Redox Systems for Nanotechnology

Redox proteins and enzymes carry out many key biological reactions, the underlying process of which is electron transfer. Protein-mediated electron transfer plays a key role in cellular processes such as respiration and photosynthesis. Redox proteins are attractive targets for nanobiotechnology because many can be detected both electrochemically and optically; they can perform specific reactions; they are capable of self-assembly; and their dimensions are in the nanoscale. There are several examples of novel approaches where redox proteins are "wired up" in efficient electron-transfer chains, are "assembled" in artificial multidomain structures, and are "linked" to surfaces in nanodevices for biosensing and nanobiotechnological applications. Redox proteins present advantages for applications in electronic devices and biological tools because their intrinsic dimensions are in nanoscale and they are capable of self-assembly.

Self-Assembling Peptide Scaffold Technology for 3D Cell Culture

Biomaterial scaffolds are components of cell-laden artificial tissues and transplantable biosensors. Some of the most promising new synthetic biomaterial scaffolds are composed of self-assembling peptides that can be modified to contain biologically active motifs. Peptide-based biomaterials can be fabricated to form 2D and 3D structures.

Until recently, 3D cell culture has required either animal-derived materials, with their inherent reproducibility and cell signaling issues, or much larger synthetic scaffolds, which fail to approximate the physical nanometer-scale and chemical attributes of native ECM. PuraMatrix™ (3DM, Inc.) nanometer-sized fibers provide a scaffold encapsulating cells in 3D and allow defined cell culture conditions, cell migration, nutrient diffusion, and cell harvesting. The hydrogels are composed of short strand of standard amino acids and 99.5 % water, which then self-assemble into very fine fibers resembling a bare extracellular matrix (ECM). Researchers in the fields of cancer biology, stem cell biology, and tissue engineering have been the first to realize that ECM and a tuned 3D microenvironment are critical for the proper understanding required for drug discovery, cell biology, and cell therapy development. Mesenchymal stem cells have been shown to differentiate into mature osteoblasts to form mineralized matrices within the synthetic self-assembling hydrogel scaffold (Hamada et al. 2008).

For the first time, the cell biology and drug discovery communities now have a biocompatible bare ECM which can be combined with relevant proteins and growth factors to more closely resemble in vivo milieu and which is compatible with cGMP requirements for cell therapy, medical device, and bioproduction applications. The PuraMatrix gels have undergone extensive in vivo toxicology safety testing.

There is a growing trend in 3D cell culture, moving cell biology research away from flat 2D cell cultures in traditional petri dishes. Some of the pioneers in 3D cell culture lend their observations that 3D microenvironments can radically alter cell behavior, enabling cells to mimic in vivo responses to drug targets and medical therapies much more accurately. Given the growing body of literature, drug discovery efforts at major pharmaceutical and biotechnology companies are beginning to adopt 3D culture techniques in their cell-based assays, especially in the context of high-content screening. Using products like PuraMatrix in order to create synthetic ECM scaffolds and tuned 3D microenvironments has proven to yield better data while also reducing the number of animals used for expensive in vivo testing.

Nanobiotechnology and Ion Channels

Ion channels provide the basis for the regulation of electrical excitability in the central and peripheral nervous systems. They are proteins that are equipped with a membrane-spanning ion-conducting pore. Disturbances in ion channels are the cause of many neurological disorders. Ion channels were traditionally studied by the patch-clamp technique, which was derived from the conventional voltage-clamp method. This method is now supplemented by nanotechniques, which are being used to study cell membranes and their proteins. Nanoscopy is the characterization of the membrane channels by techniques that resolve their morphological and physical properties and dynamics in space and time in the nanorange. These techniques make the study of structure and function of single-channel molecules in living cells possible and are currently being developed for automated and high-throughput measurements and fluorescence. Nanopores and their applications were described in Chap. 2 and earlier in this chapter.

AFM for Characterization of Ion Channels

AFM has been combined directly with the patch-clamp technique for the characterization of biological electromechanical transduction channels in living inner and outer hair cells of the cochlea. Using an AFM stylus with a tip diameter of only a few nanometers, it is possible to displace individual stereocilia of cochlear hair cells, resulting in opening of single transduction channels. In contrast to the outside-out and the inside-out patch-clamp configuration, this technique enables investigation of single mechanosensitive ion channels in entire cells.

Aquaporin Water Channels

Aquaporin (AQP) water channels are proteins that enable water to move rapidly into and out of cells. The atomic structure of mammalian AQPs illustrates how this family of proteins is freely permeated by water but not protons (hydronium ions, H_3O^+). AQP4 water channels can assemble in cell plasma membranes in orthogonal arrays of particles, which can be visualized by QD single-particle tracking (Crane et al. 2008).

Definition of the subcellular sites of expression predicted their physiological functions and potential clinical disorders. Analysis of several human disease states has confirmed that aquaporins are involved in these including abnormalities of kidney function, loss of vision, onset of brain edema, starvation, and arsenic toxicity. Research in this area requires nanoscale both in size and time scales. In molecular dynamics simulations of water in short (0.8 nm) hydrophobic pores, the water density in the pore fluctuates on a nanosecond time scale. In long simulations (460 ns in total) at pore radii ranging from 0.35 to 1.0 nm, researchers have quantified the kinetics of oscillations between a liquid-filled and a vapor-filled pore. One cannot assume that the behavior of water within complex biological pores may be determined by extrapolation from the knowledge of the bulk state or short simulations alone. Simulations aimed at collective phenomena such as hydrophobic effects may require simulation times >50 ns.

Remote Control of Ion Channels Through Magnetic-Field Heating of Nanoparticles

An approach to remotely activate temperature-sensitive cation channels in cells is based on radio-frequency magnetic-field heating of nanoparticles (Huang et al. 2010). Superparamagnetic ferrite nanoparticles were targeted to specific proteins on the plasma membrane of cells expressing TRPV1 and heated by a radio-frequency magnetic field. Using fluorophores as molecular thermometers, it was shown that the induced temperature increase is highly localized. Thermal activation of the channels triggers action potentials in cultured neurons without observable toxic effects. This approach can be adapted to stimulate other cell types and, moreover, may be used to remotely manipulate other cellular machinery for novel therapeutics.

Role of Nanobiotechnology in Engineering Ion Channels

Biological ionic channels contain precisely arranged arrays of amino acids that can efficiently recognize and guide the passage of K^+ or Na^+ across the cell membrane. However, the design of inorganic channels with novel recognition mechanisms that control the ionic selectivity remains a challenge. A design for a controllable ion-selective nanopore (molecular sieve) is based on a single-walled carbon nanotube with specially arranged carbonyl oxygen atoms modified inside the nanopore, which was inspired by the structure of potassium channels in membrane-spanning proteins (Gong et al. 2010). Molecular dynamics simulations show that the remarkable selectivity of this nanopore is attributed to the hydration structure of K^+ or Na^+ confined in the nanochannels, which can be precisely tuned by different patterns of the carbonyl oxygen atoms.

Supported membrane nanodevices can be based on natural or artificial ion channels embedded in a lipid membrane deposited on a chip wafer. Membrane conductance is modulated by biorecognitive events, with the use of intrinsic binding sites of the ion channel or via artificial sites fused to the channel protein. Artificial ion gates are constructed by coupling a specific ligand for the analyte near the channel entrance or a site important to triggering channel conformation. The binding event leads to the closure of the ion channel or induces a conformational change of the channel, reducing the ion flux. The signal transduced from the device is the decrease in the ion flux-induced electron current at a silver-silver chloride electrode at ultimate single-molecule sensitivity.

Among the natural ion channels, gramicidin A, a transport antibiotic, was found to be most suitable and thus was used to set up prototypes of membrane biochips, using self-association of the dimer. Covalent dimerization-based devices make use of the downregulation of the permanently open membrane-spanning bisgramicidin ion channel. The reactive group at the C-terminus, a hydroxy group, allows precise coupling of the analyte-binding moiety in gramicidin as well as bisgramicidin. The device is set up with bilayer membranes deposited on apertures of a hydrophobic frame structure produced via microlithography, facing an aqueous or hydrogel microenvironment on both sides, constructing black lipid membranes or patch-clamp devices "on chip." The setup of the device needs gel membrane supports that allow membrane formation and contribute to the stability of the bilayer by exposure of functional groups that promote electrostatic interaction and formation of hydrogen bridges and enable the introduction of covalent spacers and anchors. Photo-cross-linked polyvinylpyrrolidone and polyacrylamide, electropolymerized polydiaminobenzene and coated agarose, as well as various chemical modifications of these polymers were employed as membrane supports. With optimized assemblies, the membrane support enables the formation of stable bilayer membranes. Supports with and without hydrophilic and hydrophobic anchors were studied with reference to promoting the formation of a self-assembled membrane, to their electric resistance, and to the capability to insert functional ionophores. All components, including novel chemically engineered ion channels, novel amphiphilic lipids, a microlithographically designed chip, isolating polymer frames, and a hydrogel membrane support, are combined in the new bionanodevice. Sensitivity and specificity were proved, for example, with the use of an antibody-antigen couple downregulating the ion flux through the membrane channel. Single ion channels incorporated in the supported lipid bilayer gave stable signals at an operational stability of several hours, which is already sufficient to test and screen for membrane receptors but still insufficient to use this device as a sensor for off-site application. Further optimization to increase operational and storage stability is done by a number of groups to allow a broad application of these devices.

Synthetic nanopore membranes can mimic the function of ligand-gated ion channels. It has been shown that transmembrane ion current in a hydrophobic alumina nanopore membrane can be switched from an "off" state to an "on" state by exposure of the membrane to hydrophobic ionic surfactants. However, in biological channels there are no electrodes, and the ion current is driven by an electrochemical potential difference across the cell membrane. This function of the ligand-gated ion channel can be mimicked by applying a porous battery cathode film to one face of the hydrophobic alumina membrane and a porous battery anode film to the other face. Similar to the naturally occurring channel, such a membrane has a built-in electrochemical potential difference across the membrane. In the absence of the ligand, the membrane is in its "off" state, and the electrochemical potential difference cannot be utilized to drive a transmembrane ion current. In contrast, when the ligand is detected, the membrane switches to its "on" state and the transmembrane battery discharges, producing a corresponding transmembrane ion current.

Nanotechnology and Bioinformatics

Bioinformatics, also referred to as computational biology, is the use of highly sophisticated computer databases to store, analyze, and share biological information. This is a discipline at the interface of computer science, mathematics, and biology. The tremendous amount of data generated by new biotechnologies requires bioinformatic tools for analysis. Analyzing how multiple genes function together can produce terabytes of data. But as nanotech enables greater sensing and collecting of data, the information flow could become measured in petabytes, or quadrillion bytes, of information. Bioinformatics is essential for microarray data analysis and even more so for nanoarray data. In nanotechnology, the investigation of behavior and properties across a wide range of length scales is vital. Over the past two decades, computational techniques have evolved to the point that they now cover all length and time scales from the electronic to the macroscale, the realm of nanotechnology.

In the near future, it may be possible to fully model an individual cell's structure and function by computers connected to nanobiotechnology systems. Such a detailed virtual representation of how a cell functions might enable scientists to develop novel drugs with unprecedented speed and precision without any experiments in living animals. An example of application of this approach is construction of 3D nanomap of synapse.

SyNAPSE (Systems of Neuromorphic Adaptive Plastic Scalable Electronics) is a DARPA (Defense Advanced Research Projects Agency)-funded program to develop electronic neuromorphic machine technology that scales to biological levels (http://www.artificialbrains.com/darpa-synapse-program). It is an attempt to build a new kind of computer with similar form and function to the mammalian brain; such artificial brains would be used in robots. The program started in 2008 and is scheduled to run until 2016.

3D Nanomap of Synapse

Researchers have constructed a new detailed nanomap of the 3D terrain of a neuronal synapse, which shows the tiny spines and valleys resolved at nanometer scale (Coggan et al. 2005). It is already changing the conventional views of the synaptic landscape. A biologically accurate computer simulation of synaptic function combines 3D electron microscope maps with physiological measurements from real neurons. The textbook view of the synapse describes it as a place where rifle-like volleys of neurotransmitter are launched from one defined region of the sending neuron to another defined target on the receiving neuron. In contrast, the new data suggest that synapse can act like a shotgun, firing buckshot-like bursts of neurotransmitter to reach receptors arrayed beyond the known receiving sites. This method was applied to study the chick ciliary ganglion, which is a cluster of neurons that connect the brain to the iris of the eye. It launches the neurotransmitter acetylcholine from saclike vesicles across the synapse to two types of receptors, called $\alpha7$ and $\alpha3$ nicotinic receptors. The image of this ganglion is not one of a simple synapse with a single release site but multiple release sites. And it shows $\alpha3$ receptors within the postsynaptic region but $\alpha7$ receptors outside this region. This model showed that if the neurotransmitter were released only from vesicles in active zones, as previously believed, it would be a poor match to actual properties of the neuron. But according to the new model of neurotransmitter release, called ectopic release, the location of $\alpha 7$ receptors can match the actual properties of the synapse very accurately. The new 3D modeling technique could offer a powerful tool for understanding neurological disease, such as myasthenia gravis. The model can also be used as a tool for drug discovery. Drug discovery efforts can be focused on the site of the anomaly.

Chapter 4
Nanomolecular Diagnostics

Introduction

Clinical application of molecular technologies to elucidate, diagnose, and monitor human diseases is referred to as molecular diagnosis. It is a broader term than DNA (deoxyribonucleic acid) diagnostics and refers to the use of technologies that use DNA, RNA (ribonucleic acid), genes, or proteins as bases for diagnostic tests. The scope of the subject is much wider and includes in vivo imaging and diagnosis at the single-molecule level. A more detailed description of molecular diagnostics is presented elsewhere (Jain 2012a).

Because of the small dimension, most of the applications of nanobiotechnology in molecular diagnostics fall under the broad category of biochips/microarrays but are more correctly termed nanochips and nanoarrays. Nanotechnology-on-a-chip is a general description that can be applied to several methods. Some of these do not use nanotechnologies but merely have the capability to analyze nanoliter amounts of fluids.

Biochips, constructed with microelectromechanical systems on a micron scale, are related to micromanipulation, whereas nanotechnology-based chips on a nanoscale are related to nanomanipulation. Even though microarray/biochip methods employing the detection of specific biomolecular interactions are now an indispensable tool for molecular diagnostics, there are some limitations. DNA microarrays and ELISA rely on the labeling of samples with a fluorescent or radioactive tag – a highly sensitive procedure that is time-consuming and expensive.

The chemical modification and global amplification of the nucleic acid samples are achieved by polymerase chain reaction (PCR), which can introduce artifacts caused by the preferential amplification of certain sequences. Alternative label-free methods include surface plasmon resonance (SPR) and quartz crystal microbalance, which rely on mass detection. Nanotechnologies also provide label-free detection. Nanotechnology is thus being applied to overcome some of the limitations of biochip technology. This chapter focuses on the application of nanotechnologies to nucleic acid as well as protein diagnostics.

K.K. Jain, *The Handbook of Nanomedicine*, DOI 10.1007/978-1-61779-983-9_4,

Nanodiagnostics

Nanomolecular diagnostics is the use of nanobiotechnology in molecular diagnostics and can be termed "nanodiagnostics" (Jain 2003). Numerous nanodevices and nanosystems for sequencing single molecules of DNA are feasible. It seems quite likely that there will be numerous applications of inorganic nanostructures in biology and medicine as biomarkers. Given the inherent nanoscale of receptors, pores, and other functional components of living cells, the detailed monitoring and analysis of these components will be made possible by the development of a new class of nanoscale probes. Biological tests measuring the presence or activity of selected substances become quicker, more sensitive, and more flexible when certain nanoscale particles are put to work as tags or labels. Nanotechnology will improve the sensitivity and integration of analytical methods to yield a more coherent evaluation of life processes.

It is difficult to classify such a variety of technologies, but various nanotechnologies with potential applications in molecular diagnostics are listed in Table 4.1. Nanotechnology-on-a-chip was described in Chap. 2. Some of the other technologies will be described briefly in the following text using examples of commercial products. Applications in clinical laboratory have been reviewed elsewhere (Jain 2007).

Rationale of Nanotechnology for Molecular Diagnostics

Numerous nanodevices and nanosystems for sequencing single molecules of DNA are feasible. It is likely that there will be numerous applications of inorganic nanostructures in biology and medicine as markers:

- Nanoscale probes would be suitable for detailed analysis of receptors, pores, and other components of living cells that are on a nanoscale.
- Nanoscale particles, used as tags or labels, increase the sensitivity, speed, and flexibility of biological tests measuring the presence or activity of selected substances.
- Nanotechnology will improve the sensitivity and integration of analytical methods to yield a more coherent evaluation of life processes.

Nanoarrays for Molecular Diagnostics

Several nanoarrays and nanobiochips are in development (Jain 2012b). Some of these will be reviewed here.

Table 4.1 Nanotechnologies with potential applications in molecular diagnostics

Nanotechnology to improve polymerase chain reaction (PCR)
Nanotechnology-on-a-chip
Microfluidic chips for nanoliter volumes: NanoChip
Optical readout of nanoparticle labels
Nanoarrays
Protein nanoarrays
Nanotechnology-based cytogenetics
Study of chromosomes by atomic force microscopy (AFM)
Quantum dot fluorescent in situ hybridization (FISH)
Nanoscale single-molecule identification
Nanoparticle technologies
Gold particles
Nanobarcodes
Magnetic nanoparticles: ferrofluids, supramagnetic particles combined with MRI
Quantum dot technology
Nanoparticle probes
Nanowires
Nanopore technology
Measuring length of DNA fragments in a high-throughput manner
DNA fingerprinting
Haplotyping
DNA nanomachines for molecular diagnostics
Nanoparticle-based immunoassays
DNA-protein and nanoparticle conjugates
Nanochip-based single-molecular interaction force assays
Resonance light scattering technology
Nanosensors
Cantilever arrays
Living spores as nanodetectors
Nanopore nanosensors
Quartz nanobalance DNA sensor
PEBBLE (probes encapsulated by biologically localized embedding) nanosensors
Nanosensor glucose monitor
Photostimulated luminescence in nanoparticles
Optical biosensors: for example, surface plasmon resonance technology

Nanofluidic/Nanoarray Devices to Detect a Single Molecule of DNA

One of the more promising uses of nanofluidic devices is isolation and analysis of individual biomolecules, such as DNA, which could lead to new detection schemes for cancer. One of these devices entails first constructing silicon nanowires on a substrate, or chip, using standard photolithographic and etching techniques, followed by a chemical oxidation step that converts the nanowires into hollow nanotubes (Fan et al. 2005). Using this process, the investigators can reliably create nanotubes with diameters as small as 10 nm, though devices used for biomolecule isolation contain nanotubes with a diameter of 50 nm. To trap DNA molecules requires a device consisting of a silicon nanotube connecting two parallel microfluidic channels. Electrodes provide a source of current used to drive DNA into the nanotubes. Each time a single DNA molecule moves into the nanotube, the electrical current changes suddenly. The current returns to its baseline value when the DNA molecule exits the nanotube. On average, a DNA molecule remains within the nanotube for about 7.5 ms, which should be sufficient to make a variety of analytical measurements that could reveal cancer-associated mutations. The investigators are now adding optical and electrical circuitry to probe the trapped DNA molecules.

The nanoAnalyzer System (BioNano Genomics) is designed to enable direct visualization and linear analysis of multi-megabase genomic DNA at the single-molecule level with high feature resolution in massive parallel fashion. The platform is also anticipated to significantly reduce the cost and time needed for the extensive data and integrative analyses that have hindered widespread use of whole genome studies to date. It is expected to have broad application in systems biology, personalized medicine, pathogen detection, drug development, and clinical research.

Protein Nanoarrays

Protein microarrays provide a powerful tool for the study of protein function. However, they are not widely used, in part because of the challenges in producing proteins to spot on the arrays. Protein microarrays can be generated by printing complementary DNAs onto glass slides and then translating target proteins with mammalian reticulocyte lysate. Epitope tags fused to the proteins allowed them to be immobilized in situ. This obviates the need to purify proteins, avoided protein stability problems during storage, and captured sufficient protein for functional studies. This technology has been used to map pairwise interactions among 29 human DNA replication initiation proteins, recapitulate the regulation of Cdt1 binding to select replication proteins, and map its geminin-binding domain.

NanoArray Assay System™ (NanoInk) enables detection, identification, and quantitation of clinically relevant, low abundance proteins from a wide variety of sample types for applications such as biomarker analysis, translational medicine, and toxicology. NanoInk assays consume much smaller sample and reagent volumes than do traditional ELISA and bead-based assays, generating more proteomic data with less starting material and lowering assay costs for tests that are typically expensive to run.

Protein Nanobiochip

Nanotechnology Group of the NEC Corporation has developed a prototype protein analysis technology that can analyze samples about 20 times faster than conventional techniques. This technology can complete an analysis of a blood sample in about 60–70 min, compared to a day or so required for such analysis by conventional methods.

Biomarker proteins as early warning signs for diseases such as cancer can be identified for diagnostic purposes by finding their isoelectric points and their molecular weights. Isoelectric points are chemical features that refer to the electrical state of a molecule when it has no net charge. Conventional protein chips use a gel across which an electric current is applied to find the targeted protein's isoelectric points. In the new process, instead of being filtered through a block of gel, the protein molecules are separated by their isoelectric points by a capillary action as the proteins flow in a solution along channels in the chip. A test chip by NEC measures 21 mm^2 and contained four sets of tiny channels in which the capillary action takes place. The protein molecules are then dried and irradiated by a laser. Their molecular weights are then measured by a mass spectrometer. The laser helps the proteins leave the chip, and the mass spectrometer is used to judge the molecular weights of the protein molecules in the samples by measuring how early they reach a detector. In the mass spectrometer, light molecules fly faster than heavy ones in an electric field. The mass spectrometer judges the weight of the molecules by monitoring the timing of when each molecule reaches a detector. In addition to being faster than techniques that use gel blocks, the new method also needs blood samples of about 1 μL compared to about 20 μL or more that are needed using gel-based techniques. The company should commercialize the technology, and the technique could be used for health checks that might cost as little as $100. The current status of this product is not known.

Fullerene Photodetectors for Chemiluminescence Detection on Microfluidic Chip

Solution-processed thin-film organic photodiodes have been used for microscale chemiluminescence (Wang et al. 2007). The active layer of the photodiodes comprised a blend of the conjugated polymer poly(3-hexylthiophene) and a soluble derivative of fullerene C60. The devices had an active area of 1 × 1 mm and a broadband response from 350 to 700 nm, with an external quantum efficiency of more than 50 % between 450 and 550 nm. The photodiodes have a simple layered structure that allows integration with planar chip-based systems. To evaluate the suitability of the organic devices as integrated detectors for microscale chemiluminescence, a peroxyoxalate-based chemiluminescence reaction (PO-CL) was monitored within a poly(dimethyl-siloxane) (PDMS) microfluidic device. Quantitation of hydrogen peroxide indicated excellent linearity and yielded a detection limit of 10 microM, comparable with previously reported results using micromachined silicon microfluidic chips with integrated silicon photodiodes. The combination of organic photodiodes with PDMS microfluidic chips offers a means of creating compact, sensitive, and potentially low-cost microscale CL devices with wide-ranging applications in chemical and biological analysis and clinical diagnostics.

AFM for Molecular Diagnostics

Nanofountain AFM Probe

Nanofountain AFM probe (NFP) has been used for nanofabrication of protein dot and line patterns (Loh et al. 2008). Biomolecules are continuously fed in solution through an integrated microfluidic system and deposited directly onto a substrate. Deposition is controlled by application of an electric potential of appropriate sign and magnitude between the probe reservoir and substrate. Submicron dot and line molecular patterns were generated with resolution that depended on the magnitude of the applied voltage, dwell time, and writing speed. By using an energetic argument and a Kelvin condensation model, the quasi-equilibrium liquid-air interface at the probe tip was determined. The analysis revealed the origin of the need for electric fields in achieving protein transport to the substrate and confirmed experimental observations suggesting that pattern resolution is controlled by tip sharpness and not overall probe aperture. As such, the NFP combines the high-resolution of dip-pen nanolithography with the efficient continuous liquid feeding of micropipettes while enabling scalability to 1D and 2D probe arrays for high throughput.

AFM for Immobilization of Biomolecules in High-Density Microarrays

Nanoscale resolution is an important step in the preparation of nanoarrays and placement of probe biomolecules. A flexible procedure has been described for simultaneous spatially controlled immobilization of functional biomolecules with submicrometer resolution by molecular ink lithography using AFM (Breitenstein et al. 2010). Bottom-up fabrication of surface bound nanostructures enables the immobilization of different types of biomolecules. The method works on transparent as well as on opaque substrates. The spatial resolution is better than 400 nm and is limited only by the AFM's positional accuracy after repeated z-cycles since all steps are performed in situ without moving the supporting surface. The principle is demonstrated by hybridization to different immobilized DNA oligomers and was validated by fluorescence microscopy. This method not only enables deposition of DNA at submicrometer resolution but also proteins and other molecules of biological relevance that can be coupled to biotin.

AFM for Nanodissection of Chromosomes

Chromosomal dissection provides a direct advance for isolating DNA from cytogenetically recognizable region to generate genetic probes for fluorescence in situ hybridization (FISH), a technique that became very common in cytogenetics and molecular genetics research and diagnostics. Several methods for microdissection (glass needle or a laser beam) are available to obtain specific probes from metaphase chromosomes. There are limitations of the conventional methods of dissection because a large number of chromosomes are needed for the production of a probe. Moreover, these methods are not suitable for single chromosome analysis, because of the relatively large size of the microneedles. New dissection techniques are required for advanced research on chromosomes at the nanoscale level. Both AFM and scanning near-field optical microscopy (SNOM) have been used to obtain local information from G-bands and chromosomal probes.

AFM has been used as a tool for nanomanipulation of single chromosomes to generate individual cell-specific genetic probes. Molecular and nanomanipulation techniques have been combined to enable both nanodissection and amplification of chromosomal and chromatidial DNA (Di Bucchianico et al. 2011). Cross-sectional analysis of the dissected chromosomes reveals 20- and 40-nm-deep cuts. Isolated single chromosomal regions can be directly amplified and labeled by the degenerate oligonucleotide-primed polymerase chain reaction (DOP-PCR) and subsequently hybridized to chromosomes and interphasic nuclei. FISH, performed with the DOP-PCR products as test probes, has been tested successfully in avian microchromosomes and interphasic nuclei. Chromosome nanolithography, with a resolution beyond the limit of light microscopy, could be useful for the construction of chromosome band libraries and for molecular cytogenetic mapping in investigation of genetic diseases.

Nanoparticles for Molecular Diagnostics

Gold Nanoparticles

Bits of DNA and Raman-active dyes can be attached onto gold nanoparticles, which assemble onto a sensor surface only in the presence of a complementary target. If a patterned sensor surface of multiple DNA strands is used, the technique can detect millions of different DNA sequences simultaneously. The current non-optimized detection limit of this method is 20 femtomolars. Gold nanoparticles are particularly good labels for sensors because a variety of analytical techniques can be used to detect them, including optical absorption, fluorescence, Raman scattering, atomic or magnetic force, and electrical conductivity. Gold nanoparticles and Raman spectroscopy have been used to detect bacteria and viruses. This approach could replace PCR and fluorescent tags commonly used today. The detection system can also be used on biochips dotted with DNA. If the targeted disease exists in the sample, its DNA will bind onto the complementary strands of DNA on the chip and gold nanoparticle. The chip is treated with silver-based solution, which coats the nanoparticles. When exposed to a light scanner, the coating enhances the signal enough to detect minute amounts of DNA. Since the Raman band is narrower than the fluorescent band, it allows more dyes to detect more targets quickly. If the sequence of interest is present in the sample, it will bind to the DNA and cause the solution to change color. Labeling oligonucleotide targets with gold nanoparticle rather than fluorophore probes substantially alters the melting profiles of the targets from an array substrate. Nanoparticle-based DNA detection systems are more sensitive and specific than current genomic detection systems.

QDs for Molecular Diagnostics

There is considerable interest in the use of QDs as inorganic fluorophores, owing to the fact that they offer significant advantages over conventionally used fluorescent markers. For example, QDs have fairly broad excitation spectra – from ultraviolet to red – that can be tuned depending on their size and composition. At the same time, QDs have narrow emission spectra, making it possible to resolve the emissions of different nanoparticles simultaneously and with minimal overlap. Finally, QDs are highly resistant to degradation, and their fluorescence is remarkably stable. Advantages of QD technology are:

- Simple excitation – lasers are not required
- Simple instrumentation
- Availability of red/infrared colors enables whole-blood assays
- High sensitivity

QDs have been used as possible alternatives to the dyes for tagging viruses and cancer cells. A major challenge is that QDs have an oily surface, whereas the cellular environment is water-based. Attempts are being made to modify the surface chemistry of QDs so that they interact with water-friendly molecules like proteins and DNA. The current goal is to develop QDs that can target a disease site and light it up. This can someday lead to an integrated system that will also use the QDs to diagnose as well to deliver drug therapies to the disease site. QDs can be designed to emit light at any wavelength from the infrared to visible to ultraviolet. This enables the use of a large number of colors and thus multiplexed assays can be performed. Potential applications of QDs in molecular diagnostics can be summarized as follows:

- Cancer
- Genotyping
- Whole-blood assays
- Multiplexed diagnostics
- DNA mapping
- Immunoassays and antibody tagging
- Detection of pathogenic microorganisms

Quantum Dots for Detection of Pathogenic Microorganisms

QDs have been used as fluorescent labels in immunoassays for quantitative detection of foodborne pathogenic bacteria such as *Salmonella typhimurium.* QDs coated with streptavidin are added to react with biotin on the secondary antibody. Measurement of the intensity of fluorescence produced by QDs provides a quantitative method for microbial detection. QDs can be used for ultrasensitive viral detection of a small number of microorganisms.

Bioconjugated QDs for Multiplexed Profiling of Biomarkers

Bioconjugated QDs provide a new class of biological labels for evaluating biomarkers on intact cells and tissue specimens. In particular, the use of multicolor QD probes in immunohistochemistry is considered one of the most important and clinically relevant applications. At present, however, clinical applications of QD-based immunohistochemistry have achieved only limited success. A major bottleneck is the lack of robust protocols to define the key parameters and steps. Preliminary results and detailed protocols for QD-antibody conjugation, tissue specimen preparation, multicolor QD staining, image processing, and biomarker quantification have been published (Xing et al. 2007). The results demonstrate that bioconjugated QDs can be used for multiplexed profiling of biomarkers and ultimately for correlation with disease progression and response to therapy. In general, QD bioconjugation is completed within 1 day, and multiplexed molecular profiling takes 1–3 days depending on the number of biomarkers and QD probes used.

Imaging of Living Tissue with QDs

Tiny blood vessels, viewed beneath a mouse's skin with multiphoton microscopy appear so bright and vivid in high-resolution images that researchers can see the vessel walls ripple with each heartbeat. Capillaries, hundreds of microns below the skin of living mice, can be illuminated in an unprecedented detail using QDs circulating through the blood as fluorescent imaging labels. Although there are easier ways to take a mouse's pulse, this level of resolution with high signal-to-noise ratio illustrates how useful multiphoton microscopy with QDs can become in biological research for tracking cells and visualizing tissue structures deep inside living animals. Monitoring of vascular changes in malignant tumors is a potential application. This approach will pave the way for many new noninvasive in vivo imaging methods using QDs.

Carbohydrate-encapsulated QD can be used for medical imaging. Certain carbohydrates, especially those included on tumor glycoproteins, are known to have affinity for certain cell types, and this can be exploited for medical imaging. Conjugating luminescent QDs with target-specific glycans permits efficient imaging of the tissue to which the glycans bind with high affinity. Accurate imaging of primary and metastatic tumors is of primary importance in disease management. Second-generation QDs contain the glycan ligands and PEG of varying chain lengths. PEG modification produces QDs that maintain high luminescence while reducing nonspecific cell binding.

Procedures have been developed for using QDs to label live cells and to demonstrate their use for long-term multicolor imaging. Two approaches are endocytic uptake of QDs and selective labeling of cell-surface proteins with QDs conjugated to antibodies, which should permit the simultaneous study of multiple cells over long periods of time as they proceed through growth and development. Use of avidin permits stable conjugation of the QDs to ligands, antibodies, or other molecules that can be biotinylated, whereas the use of proteins fused to a positively charged peptide or oligohistidine peptide obviates the need for biotinylating the target molecule. Specific labeling of both intracellular and cell-surface proteins can be achieved by bioconjugation of QDs. For generalized cellular labeling, QDs not conjugated to a specific biomolecule may be used.

Fluorescent semiconductor QDs hold great potential for molecular imaging in vivo. However, the utility of existing QDs for in vivo imaging is limited because they require excitation from external illumination sources to fluoresce, which results in a strong autofluorescence background and a paucity of excitation light at nonsuperficial locations. QD conjugates that luminesce by bioluminescence resonance energy transfer in the absence of external excitation have been prepared by coupling carboxylate-presenting QDs to a mutant of the bioluminescent protein Renilla reniformis luciferase (So et al. 2006). The conjugates emit long-wavelength (from red to near infrared) bioluminescent light in cells and in animals, even in deep tissues, and are suitable for multiplexed in vivo imaging. Compared with existing QDs, self-illuminating QD conjugates have greatly enhanced sensitivity in small-animal imaging, with an in vivo signal-to-background ratio of $>10^3$ for 5 pmol of conjugate.

Several advances have recently been made using QDs for live cell and in vivo imaging, in which QD-labeled molecules can be tracked and visualized in 3D. QDs have been investigated for their use for multiplex immunohistochemistry and in situ hybridization which, when combined with multispectral imaging, has enabled quantitation and co-localization of gene expression in clinical tissue specimens (Byers and Hitchman 2011).

Use of Nanocrystals in Immunohistochemistry

A method has been described for simple convenient preparation of bright, negatively or positively charged, water-soluble CdSe/ZnS core/shell nanocrystals (NCs) and their stabilization in aqueous solution (Sukhanova et al. 2004). Single NCs can be detected using a standard epifluorescent microscope, ensuring a detection limit of one molecule coupled with an NC. NC-antibody (Ab) conjugates were tested in dot blots and exhibited retention of binding capacity within several nanogram of antigen detected. The authors further demonstrated the advantages of NC-Ab conjugates in the immunofluorescent detection and 3D confocal analysis of p-glycoprotein (p-gp), one of the main mediators of the multidrug resistance phenotype. The labeling of p-gp with NC-Ab conjugates was highly specific. Finally, the authors demonstrated the applicability of NC-Abs conjugates obtained by the method described to specific detection of antigens in paraffin-embedded formaldehyde-fixed cancer tissue specimens, using immunostaining of cytokeratin in skin basal carcinoma as an example. They concluded that the NC-Ab conjugates may serve as easy-to-do, highly sensitive, photostable labels for immunofluorescent analysis, immunohistochemical detection, and 3D confocal studies of membrane proteins and cells.

Magnetic Nanoparticles

Magnetic Nanoparticles for Bioscreening

Iron nanoparticles, 15–20 nm in size, having saturation magnetization, have been synthesized, embedded in copolymer beads of styrene and glycidyl methacrylate (GMA), which were coated with poly-GMA by seed polymerization (Maeda et al. 2006). The resultant Fe/St-GMA/GMA beads had diameters of 100–200 nm. By coating with poly-GMA, the zeta potential of the beads changed from −93.7 to −54.8 mV, as measured by an electrophoresis method. This facilitates nonspecific protein adsorption suppression, as revealed by gel electrophoresis method, which is a requisite for nanoparticles to be applied to carriers for bioscreening.

Magnetic nanoparticles can be used for detection of biomolecules and cells based on magnetic resonance effects using a general detection platform termed diagnostic magnetic resonance (DMR) technology, which covers numerous sensing principles, and magnetic nanoparticle biosensors have been designed to detect a wide range of targets including DNA/mRNA, proteins, enzymes, drugs, pathogens, and circulating tumor cells (Haun et al. 2010). The basic principle of DMR is the use of magnetic nanoparticles as proximity sensors that modulate the spin relaxation time of neighboring water molecules, which can be quantified using clinical MRI scanners or benchtop nuclear magnetic resonance (NMR) relaxometers. The capabilities of DMR technology have been advanced considerably with the development of miniaturized, chip-based NMR (μNMR) detector systems that are capable of performing highly sensitive measurements on microliter sample volumes and in multiplexed format. Thus, DMR biosensor technology holds considerable promise to provide a high-throughput, low-cost, and portable platform for large-scale molecular and cellular screening in clinical and point-of-care settings.

Monitoring of Implanted NSCs Labeled with Nanoparticles

Noninvasive monitoring of stem cells, using high-resolution molecular imaging, will be important for improving clinical neural transplantation strategies. Labeling of human neural stem cells (NSCs) grown as neurospheres with magnetic nanoparticles was shown to not adversely affect survival, migration, and differentiation or alter neuronal electrophysiological characteristics (Guzman et al. 2007). Using MRI, the authors demonstrated that human NSCs transplanted either to the neonatal, the adult, or the injured rodent brain respond to cues characteristic for the ambient microenvironment resulting in distinct migration patterns. Nanoparticle-labeled human NSCs survive long-term and differentiate in a site-specific manner identical to that seen for transplants of unlabeled cells. The impact of graft location on cell migration and MRI characteristics of graft cell death and subsequent clearance were also described. Knowledge of migration patterns and implementation of noninvasive stem cell tracking might help to improve the design of future clinical NSC transplantation.

Perfluorocarbon Nanoparticles to Track Therapeutic Cells In Vivo

Using perfluorocarbon nanoparticles 200 nm in size to label endothelial progenitor cells taken from human UCB enables their detection by MRI in vivo following administration (Partlow et al. 2007). The MRI scanner can be tuned to the specific frequency of the fluorine compound in the nanoparticles, and only the nanoparticle-containing cells are visible in the scan. This eliminates any background signal, which often interferes with medical imaging. Moreover, the lack of interference means one can measure very low amounts of the labeled cells and closely estimate their number by the brightness of the image. Since several perfluorocarbon compounds are available, different types of cells potentially could be labeled with different compounds, injected and then detected separately by tuning the MRI scanner to each one's individual frequency. This technology offers significant advantages over other cell-labeling technologies in development. Laboratory tests showed that the cells retained their usual surface markers and that they were still functional after the labeling process. The labeled cells were shown to migrate to and incorporate into blood vessels forming around tumors in mice. These could soon enable researchers and physicians to directly track cells used in medical treatments using unique signatures from the ingested nanoparticle beacons. They could prove useful for monitoring tumors and diagnosing as well as treating cardiovascular problems.

Superparamagnetic Nanoparticles for Cell Tracking

Magnetic nanoparticles are a powerful and versatile diagnostic tool in biology and medicine. It is possible to incorporate sufficient amounts of superparamagnetic iron oxide nanoparticles (SPIONs) into cells, enabling their detection in vivo using MRI. Because of their small size, they are easily incorporated into various cell types (stem cells, phagocytes, etc.) allowing the cells to be tracked in vivo, for example, to determine whether stem cells move to the correct target area of the body.

Superparamagnetic iron oxide nanoparticles (SPIONS), used clinically for specific magnetic sorting, can be used as a magnetic cell label for in vivo cell visualization. The fact that SPIONs coated with different commercially available antibodies can bind to specific cell types opens extensive possibilities for cell tracking in vivo. A study has investigated the biological properties, including proliferation, viability, and differentiation capacity of MSCs labeled with clinically approved SPIONs (Jasmin et al. 2011). Rat MSCs were isolated, cultured, and incubated with dextran-covered SPIONs (ferumoxide) alone or with poly-L-lysine (PLL) or protamine chlorhydrate. Whereas labeling of MSCs incubated with ferumoxide alone was poor, 95 % MSCs were labeled when incubated with ferumoxide in the presence of PLL or protamine. MSCs incubated with ferumoxide and protamine were efficiently visualized by MRI; they maintained proliferation and viability for up to 7 days and remained competent to differentiate. After 21 days, MSCs pre-treated with mitomycin C still showed a large number of ferumoxide-labeled cells. The efficient and long-lasting uptake and retention of SPIONs by MSCs using a protocol employing ferumoxide and protamine may be applicable to patients, since both ferumoxide and protamine are approved for human use.

Unfortunately, SPIONs are no longer being manufactured. Second-generation, ultrasmall SPIONs (USPIO), however, offer a viable alternative. Ferumoxytol (Feraheme™) is one USPIO composed of a nonstoichiometric magnetite core surrounded by a polyglucose sorbitol carboxymethylether coat. The colloidal, particle size of ferumoxytol is 17–30 nm. Ferumoxytol has been approved by the FDA as an iron supplement for treatment of iron deficiency in patients with renal failure. This agent has been used "off label" for stem cell labeling (Castaneda et al. 2011). This technique may be applied for noninvasive monitoring of stem cell therapies in preclinical and clinical settings.

SPIONs for Calcium Sensing

A family of calcium indicators for MRI is formed by combining a powerful SPION-based contrast mechanism with the versatile calcium-sensing protein calmodulin and its targets (Atanasijevic et al. 2006). Calcium-dependent protein-protein interactions drive particle clustering and produce up to fivefold changes in T2 relaxivity, an indication of the sensors' potency. Robust MRI signal changes are achieved even at nanomolar particle concentrations that are unlikely to buffer calcium levels. When combined with technologies for cellular delivery of nanoparticulate agents, these sensors and their derivatives may be useful for functional molecular imaging of biological signaling networks in live, opaque specimens.

Magnetic Nanoparticles for Labeling Molecules

Bound to a suitable antibody, magnetic nanoparticles are used to label specific molecules, structures, or microorganisms. Magnetic immunoassay techniques have been developed in which the magnetic field generated by the magnetically labeled targets is detected directly with a sensitive magnetometer. Binding of antibody to target molecules or disease-causing organism is the basis of several tests. Antibodies labeled with magnetic nanoparticles give magnetic signals on exposure to a magnetic field. Antibodies bound to targets can thus be identified as unbound antibodies disperse in all directions and produce no net magnetic signal.

SPIONs have been functionalized to identify *Mycobacterium avium spp.* paratuberculosis (MAP) through magnetic relaxation (Kaittanis et al. 2007). The results indicate that the MAP nanoprobes bind specifically to MAP and can quantify the bacterial target quickly in milk and blood with high sensitivity. The advantage of this approach is that detection is not susceptible to interferences caused by other bacteria. The use of these magnetic nanosensors is anticipated in the identification and quantification of bacteria in clinical and environmental samples.

Study of Living Cells by SPIONs

Technologies to assess the molecular targets of biomolecules in living cells are lacking. A technology called magnetism-based interaction capture (MAGIC) can identify molecular targets on the basis of induced movement of SPIONs inside living cells. Intracellular proteins can be painted with fluorescent materials and drugs embedded with SPIONs inserted into the cell. These nanoprobes captured the small molecule's labeled target protein and were translocated in a direction specified by the magnetic field. Use of MAGIC in genome-wide expression screening can identify multiple protein targets of a drug. MAGIC can also be used to monitor signal-dependent modification and multiple interactions of proteins. Internalized SPIONs can be moved inside cells by an external magnetic field. MAGIC can be useful in the development of diagnostics and biosensors. Its ultimate use would be for the analysis of interactions inside living cells of patients.

Imaging Applications of Nanoparticles

Molecular imaging now encompasses all imaging modalities including those used in clinical care: optical imaging, nuclear medical imaging, ultrasound imaging, MRI, and photoacoustic imaging. Molecular imaging always requires accumulation of contrast agent in the target site, often achieved most efficiently by steering nanoparticles containing contrast agent into the target. This entails accessing target molecules hidden behind tissue barriers, necessitating the use of targeting groups. For imaging modalities with low sensitivity, nanoparticles bearing multiple contrast groups provide signal amplification. The same nanoparticles can in principle deliver both contrast medium and drug, allowing monitoring of biodistribution and therapeutic activity simultaneously. Nanoparticles with multiple bioadhesive sites for target recognition and binding share functionalities with many subcellular organelles (ribosomes, proteasomes, ion channels, and transport vesicles), which are of similar sizes. The materials used to synthesize nanoparticles include natural proteins and polymers, artificial polymers, dendrimers, fullerenes and other carbon-based structures, lipid–water micelles, viral capsids, metals, metal oxides, and ceramics. Signal generators incorporated into nanoparticles include iron oxide, gadolinium, fluorine, iodine, bismuth, radionuclides, QDs, and metal nanoclusters (Debbage and Jaschke 2008). Diagnostic imaging applications, now appearing, include sentinel node localization and stem cell tracking.

There is rapid growth in the use of MRI for molecular and cellular imaging. Much of this work relies on the high relaxivity of nanometer-sized, ultrasmall dextran-coated iron oxide particles. Chemical modifications to nanosized virus particles may improve MRI. Attachment of a large number of gadolinium chelates, the chemical compound used in MRI contrast agents, onto the surface of the viral particles resulted in the generation of a very intense signal in a clinical MRI scanner (Anderson et al. 2006). Dendrimers, magnetic nanoparticles, QDs, and ferrofluids are examples of some of the nanoparticles that have been used along with imaging technologies for enhancement of contrast. Some of these are described in the preceding section.

Computer tomography (CT) is among the most convenient imaging/diagnostic tools used currently in terms of availability, efficiency, and cost. In contrast to other imaging modalities, such as PET, single-photon emission computed tomography (SPECT) and MRI, CT is not usually considered as a molecular imaging modality. However, development of nanoparticles as contrast agents is enabling specific applications of CT for molecular imaging. Current clinical CT contrast agents are predominantly based on the high atomic number iodine molecules ($Z=53$), which are effective in absorbing X-rays; but the small size of iodine molecules allows very short imaging times due to rapid clearance by the kidneys. Use of high-Z nanoparticles, for example, polymers, liposomes, and micelles as contrast agents may enable CT imaging at lower radiation doses and with improved sensitivity as well as specificity (Shilo et al. 2012). There is some concern about the toxicity of these nanoparticle contrast agents. Once the toxicity issues are resolved, clinical trials could be initiated in humans. Concomitant encapsulation of metal nanoparticles (diagnosis) and drug molecules (therapy) into carriers, such as liposomes, offers simultaneous in vivo imaging and tracking of drug efficacy, which will facilitate development of personalized medicine.

Dendritic Nanoprobes for Imaging of Angiogenesis

Angiogenesis is an important process in ischemia and cancer. A biodegradable positron-emitting dendritic nanoprobe targeted at αvβ3 integrin, a biomarker known to modulate angiogenesis, was developed for the noninvasive imaging of angiogenesis (Almutairi et al. 2009). The nanoprobe has a modular multivalent core-shell architecture consisting of a biodegradable heterobifunctional dendritic core chemoselectively functionalized with heterobifunctional polyethylene oxide (PEO) chains that form a protective shell, which imparts biological stealth and dictates the pharmacokinetics. Each of the eight branches of the dendritic core was functionalized for labeling with radiohalogens. Placement of radioactive moieties at the core was designed to prevent in vivo dehalogenation, a potential problem for radiohalogens in imaging and therapy. Targeting peptides of cyclic arginine-glycine-aspartic (RGD) acid motifs were installed at the terminal ends of the PEO chains to enhance their accessibility to αvβ3 integrin receptors. This nanoscale design enabled a 50-fold enhancement of the binding affinity to αvβ3 integrin receptors with respect to the monovalent RGD peptide alone Cell-based assays of the ^{125}I-labeled dendritic nanoprobes using αvβ3-positive cells showed a sixfold increase in αvβ3 receptor-mediated endocytosis of the targeted nanoprobe compared with the nontargeted nanoprobe, whereas αvβ3-negative cells showed no enhancement of cell uptake over time. In vivo biodistribution studies of ^{76}Br-labeled dendritic nanoprobes showed excellent bioavailability for the targeted and nontargeted nanoprobes. In vivo studies in a murine hind limb ischemia model for angiogenesis revealed high specific accumulation of ^{76}Br-labeled dendritic nanoprobes targeted at αvβ3 integrins in angiogenic muscles, allowing highly selective imaging of this critically important process.

Gadolinium-Loaded Dendrimer Nanoparticles for Tumor-Specific MRI

A target-specific MRI contrast agent for tumor cells expressing high affinity folate receptor was synthesized using a fifth-generation polyamidoamine dendrimer (Swanson et al. 2008). Surface-modified dendrimer was functionalized for targeting with folic acid, and the remaining terminal primary amines of the dendrimer were conjugated with the bifunctional NCS-DOTA (Dow Chemical) chelator that forms stable complexes with gadolinium. In xenograft tumors in immunodeficient mice induced with human epithelial cancer cells expressing folate receptor, 3D MRI results showed specific and statistically significant signal enhancement in tumors generated with targeted nanoparticle compared with signal generated by nontargeted contrast nanoparticle. The targeted dendrimer contrast nanoparticles infiltrated tumor and were retained in tumor cells up to 48 h following injection. The presence of folic acid on the dendrimer resulted in specific delivery of the nanoparticle to tissues and xenograft tumor cells expressing folate receptor in vivo. The specificity of the dendrimer nanoparticles for targeted cancer imaging with the prolonged clearance time compared favorably with the current clinically approved gadodiamide (Omniscan) contrast agent. Potential applications of this approach include determination of the folate receptor status of tumors and monitoring of drug therapy.

Gadonanotubes for MRI

More than 25 million patients in the USA undergo MRI each year, and contrast agents are used in about 30 % of these procedures. Gadolinium agents are the most effective and the most commonly used MRI contrast agents. Gadonanotubes are made of the same highly toxic metal gadolinium (Gd^{3+}) that is used in MRI currently, but the metal atoms are encased inside a carbon nanotube. The ultrashort nanotubes are only about 20–100 times longer than they are wide, and once inside the nanotubes, the gadolinium atoms naturally aggregate into tiny clusters of about 10 atoms each. Clustering causes the unexplained increases in magnetic and MRI effects. Gadonanotubes are at least 40–90 times more effective than Gd^{3+}-based MRI agents now in use. Shrouding the toxic metals inside the benign carbon is expected to significantly reduce or eliminate the metal's toxicity. Currently available methods of attaching disease-specific antibodies and peptides can be applied to gadonanotubes so they can be targeted to malignant and other diseased cells.

Gold Nanorods and Nanoparticles as Imaging Agents

Gold nanorods excited at 830 nm on a far-field laser-scanning microscope produced strong two-photon luminescence (TPL) intensities, and the TPL excitation spectrum can be superimposed on to the longitudinal plasmon band (Wang et al. 2005c). The TPL signal from a single nanorod is 58 times that of the two-photon fluorescence signal from a single rhodamine molecule. Gold nanorods can be used as imaging agents as demonstrated by in vivo imaging of single nanorods flowing in mouse ear blood vessels.

Nanoprobes Inc. reported that 1.9 nm gold nanoparticles may overcome many limitations to traditional X-ray contrast agents. Gold has higher X-ray absorption than iodine with less bone and tissue interference, thus achieving better contrast with lower X-ray dose. Because nanoparticles clear the blood more slowly than iodine agents, they permit longer imaging times. In studies in mice, a 5-mm tumor growing in one thigh was clearly evident from its increased vascularity and resultant higher gold content. The gold particles thus enable direct imaging, detection, and measurement of angiogenic and hypervascularized regions. The 1.9-nm gold nanoparticles were found to clear through the kidneys: a closer examination of the kidneys revealed a remarkably detailed anatomical and functional display, with blood vessels less than 100 μm in diameter delineated, thus enabling in vivo vascular casting. Toxicity was also low: mice intravenously injected with the gold nanoparticles survived over 1 year without signs of illness.

In Vivo Imaging Using Nanoparticles

Fluorescence provides remarkable results for in vivo imaging, but it has several limitations, particularly because of the need for tissue autofluorescence by external illumination and weak tissue penetration of low wavelength excitation light. An alternative optical imaging technique has been developed by using nanoparticles with persisting luminescence suitable for small-animal imaging (le Masne de Chermont et al. 2007). These nanoparticles can be excited before injection, and their in vivo distribution can be followed in real time for more than 1 h without the need for an external illumination source. Chemical modification of the nanoparticle surface can be done to target organs such as the lung or the liver or for prolonging luminescence during circulation of the nanoparticles in blood. Tumors have been identified by this technique.

A significant impediment to the widespread use of noninvasive in vivo vascular imaging techniques is the current lack of suitable intravital imaging probes. One strategy is the use of viral nanoparticles as a platform for the multivalent display of fluorescent dyes to image tissues deep inside living organisms. The bioavailable cowpea mosaic virus (CPMV) can be fluorescently labeled to high densities with no measurable quenching, resulting in exceptionally bright particles with in vivo dispersion properties that allow high-resolution intravital imaging of vascular endothelium for periods of at least 72 h. CPMV nanoparticles can be used to visualize the vasculature and blood flow in living mouse and chick embryos to a depth of up to 500 μm. Intravital visualization of human fibrosarcoma-mediated tumor angiogenesis using fluorescent CPMV provides a means to identify arterial and venous vessels for monitoring tumor neovascularization.

Manganese Oxide Nanoparticles as Contrast Agent for Brain MRI

A new MRI contrast agent using manganese oxide nanoparticles produces images of the anatomic structures of mouse brain which are as clear as those obtained by histological examination (Na et al. 2007). The new contrast agent will enable better research and diagnosis neurological disorders such as Alzheimer's disease, Parkinson's disease, and stroke. Furthermore, antibodies can be attached to the manganese oxide nanoparticles, which recognize and specifically bind to receptors on the surface of breast cancer cells in mouse brains with breast cancer metastases. The tumors were clearly highlighted by the antibody-coupled contrast agent. The same principle should allow other disease-related changes or physiological systems to be visualized by using the appropriate antibodies.

Nanoparticles Versus Microparticles for Cellular Imaging

Typically, millions of dextran-coated ultrasmall iron oxide particles (USIOPs) must be loaded into cells for efficient detection. Single, micrometer-sized iron oxide particles (MSIOPs) can be detected by MRI in vivo in living animals. Experiments studying effects of MRI resolution and particle size indicate that significant signal effects could be detected at resolutions as low as 200 μm. Cultured cells can be labeled with fluorescent MSIOPs in a way that single particles are present in individual cells and can be detected both by MRI and fluorescence microscopy. Single particles injected into single-cell-stage mouse embryos can be detected at later embryonic stages, demonstrating that even after many cell divisions, daughter cells still carry individual particles. These observations show that MRI can detect single particles and indicate that microparticle detection will be useful for cellular imaging for certain purposes and may be preferable to nanoparticles. MSIOPs will be useful in following the division of stem cells and in vivo labeling of cells.

Nanoparticles as Contrast Agent for MRI

The determination of brain tumor margins both during the presurgical planning phase and during surgical resection has long been a challenging task in the therapy of brain tumor patients. Multimodal (near-infrared fluorescent and magnetic) nanoparticles were used as a preoperative MRI contrast agent and intraoperative optical probes. Key features of nanoparticle metabolism, namely, intracellular sequestration by microglia and the combined optical and magnetic properties of the probe, allowed delineation of brain tumors both by preoperative MRI and by intraoperative optical imaging. This prototypical multimodal nanoparticle has unique properties that may allow radiologists and neurosurgeons to see the same probe in the same cells and may offer a new approach for obtaining tumor margins.

Alphanubeta3-targeted paramagnetic nanoparticles have been employed to noninvasively detect very small regions of angiogenesis associated with nascent melanoma tumors (Schmieder et al. 2005). Each particle was filled with thousands of molecules of the metal that is used to enhance contrast in conventional MRI scans. The surface of each particle was decorated with a substance that attaches to newly forming blood vessels, which are present at tumor sites. The goal was to create a high density of the glowing particles at the site of tumor growth so they are easily visible. Molecular MRI results were corroborated by histology. This study lowers the limit previously reported for detecting sparse biomarkers with molecular MRI in vivo when the growths are still invisible to conventional MRI. Earlier detection can potentially increase the effectiveness of treatment. This is especially true with melanoma, which begins as a highly curable disorder then progresses into an aggressive and deadly disease. A second benefit of the approach is that the same nanoparticles used to find the tumors could potentially deliver stronger doses of anticancer drugs directly to the tumor site with fewer side effects. Targeting the drugs to the tumor site in this way would also allow stronger doses without systemic toxicity than would be possible if the drug were injected or delivered in some other systemic way. The nanoparticles might also allow physicians to more readily assess the effectiveness of the treatment by comparing MRI scans before and after treatment. Other cancer types might be accessible to this approach as well, because all tumors recruit new blood vessels as they grow.

Optical Molecular Imaging Using Targeted Magnetic Nanoprobes

Dynamic magnetomotion of magnetic nanoparticles (MNPs) detected with magnetomotive optical coherence tomography (MM-OCT) represents a new method for contrast enhancement and therapeutic interventions in molecular imaging. In vivo imaging of dynamic functionalized iron oxide MNPs using MM-OCT was demonstrated in a preclinical mammary tumor model (John et al. 2010). Using targeted MNPs, in vivo MM-OCT images exhibit strong magnetomotive signals in mammary tumor, and no significant signals were measured from tumors of rats injected with nontargeted MNPs or saline. The results of in vivo MM-OCT are validated by MRI, ex vivo MM-OCT, Prussian blue staining of histological sections, and immunohistochemical analysis of excised tumors and internal organs. The MNPs are antibody functionalized to target the human epidermal growth factor receptor 2 (HER2 neu) protein. Fc-directed conjugation of the antibody to the MNPs aids in reducing uptake by macrophages in the reticuloendothelial system, thereby increasing the circulation time in the blood. These engineered magnetic nanoprobes have multifunctional capabilities enabling them to be used as dynamic contrast agents in MM-OCT and MRI.

QDs for Biological Imaging

Targeted QDs, coated with paramagnetic and pegylated lipids, have been developed for detection by MRI (Mulder et al. 2006). The QDs were functionalized by covalently linking v3-specific peptides, and the specificity was assessed and confirmed on cultured endothelial cells. The bimodal character, the high relaxivity, and the specificity of this nanoparticulate probe make it an excellent contrast agent for molecular imaging purposes. Among other applications, those in cancer are most important.

Accurate imaging of diseased cells (e.g., primary and metastatic tumors) is of primary importance in disease management. The NIH has developed carbohydrate-encapsulated QDs with detectable luminescent properties useful for imaging of cancer or other disease tissues. Certain carbohydrates, especially those included on tumor glycoproteins, are known to have affinity for certain cell types. One notable glycan used in this technology is the Thomsen-Friedenreich disaccharide (Galbeta1-3GalNAc) that is readily detectable in 90 % of all primary human carcinomas and their metastases. These glycans can be exploited for medical imaging. Encapsulating luminescent QDs with target- specific glycans permits efficient imaging of the tissue to which the glycans bind with high affinity.

Multifunctional nanoparticle probes based on semiconductor QDs have been used for cancer targeting and imaging in living animals. The structural design involves encapsulating luminescent QDs with an ABC triblock copolymer and linking this amphiphilic polymer to tumor-targeting ligands and drug-delivery functionalities. In vivo targeting studies of human prostate cancer growing in nude mice indicate that the QD probes accumulate at tumors both by the enhanced permeability and retention of tumor sites and by antibody binding to cancer-specific cell-surface biomarkers (Gao et al. 2004). Using both subcutaneous injection of QD-tagged cancer cells and systemic injection of multifunctional QD, sensitive and multicolor fluorescence imaging of cancer cells have been achieved under in vivo conditions. These results raise new possibilities for ultrasensitive and multiplexed imaging of molecular targets in vivo.

SPIONs Combined with MRI

Highly lymphotropic SPIONs measuring 2–3 nm on average, which gain access to lymph nodes by means of interstitial-lymphatic fluid transport, have been used in conjunction with high-resolution MRI to reveal small and otherwise undetectable lymph node metastases. In patients with prostate cancer who undergo surgical lymph node resection or biopsy, MRI with lymphotropic SPIONs can identify all patients with nodal metastases, which is not possible with conventional MRI alone, and has implications for the management. In men with metastatic prostate cancer, adjuvant androgen-deprivation therapy with radiation is the mainstay of management.

Sentinel lymph node (SLN) imaging and biopsy is an important part of the workup of some cancers in humans. The presence of lymph node metastases is an important factor in breast cancer patient prognosis. Therefore, the precise identification of SLNs in these patients is critical. Conventional methods have drawbacks including lack of depth, skin staining (blue dye), poor spatial resolution, and exposure to ionizing radiation. Among the newer methods, magnetic resonance lymphography, in which a gadolinium-labeled nanoparticle is injected and imaged, provides superior anatomic resolution and assessment of lymphatic dynamics, overcoming some of the drawbacks of other methods. Optical imaging employing various nanoparticles, including QDs, also provides the capability of mapping each lymphatic drainage in a different color. However, autofluorescence arising from normal tissues can compromise the sensitivity and specificity of in vivo fluorescence imaging using QDs by lowering the target-to-background signal ratio. Since bioluminescence resonance energy transfer QD (BRET-QD) nanoparticles can self-illuminate in NIR in the presence of the substrate, imaging using BRET-QDs does not produce any autofluorescence. These advantages of BRET-QD enable real-time, quantitative lymphatic imaging without image processing (Kosaka et al. 2011).

Use of lymphatic imaging agents will improve our understanding of the lymphatic system. It is conceivable that an anticancer drugs and a tumor vaccine can be incorporated into the imaging agent for the delivery of regional therapy (Ravizzini et al. 2009).

Concluding Remarks and Future Prospects of Nanoparticles for Imaging

Surface functionalization has expanded further the potential of nanoparticles as probes for molecular imaging. Ongoing research of nanoparticles for biomedical imaging focuses on increased selectivity and reduced nonspecific uptake with increased spatial resolution containing stabilizers conjugated with targeting ligands. Structural design of nanomaterials for biomedical imaging continues to expand and diversify. Synthetic methods aim to control the size and surface characteristics of nanoparticles to optimize distribution, half-life, and elimination. Although molecular imaging applications using nanoparticles are advancing into clinical applications, challenges such as storage stability and long-term toxicology should continue to be addressed (Nune et al. 2009).

Applications of Nanopore Technology for Molecular Diagnostics

Nanopore Technology for Detection of Single DNA Molecules

Nanopore sequencing was described in Chap. 3. Nanopores hold great promise as single-molecule analytical devices and biophysical model systems because the ionic current blockades they produce contain information about the identity, concentration, structure, and dynamics of target molecules. Nanopore technology can distinguish between and count a variety of different molecules in a complex mixture. For example, it can distinguish between hybridized or unhybridized unknown RNA and DNA molecules that differ only by a single nucleotide. Nanopore biosensors can enable direct, microsecond-scale nucleic acid characterization without the need for amplification, chemical modification, surface adsorption, or the binding of probes.

A mutant was constructed of porin MspA of *Mycobacterium smegmatis* that is capable of electronically detecting and characterizing single molecules of ssDNA as they are electrophoretically driven through the pore (Butler et al. 2008). A second mutant with additional exchanges of negatively charged residues for positively charged residues in the vestibule region exhibited a factor of ≈20 higher interaction rates, required only half as much voltage to observe interaction, and allowed ssDNA to reside in the vestibule ≈100 times longer than the first mutant. These results introduce MspA as a nanopore for nucleic acid analysis and highlight its potential as an engineerable platform for single-molecule detection and characterization applications.

Nanocytometry

Nanocytometry is a nanotechnology-based approach to flow cytometry. It incorporates previous work on a nanoelectronic technique for detecting the binding of unlabeled antibody-antigen pairs. Nanocytometry uses resistive-pulse sensing and artificial nanopores to detect and measure cell size, which is determined by the change in resistance when an individual cell passes through the pore (Carbonaro et al. 2008). As a proof of principle, it was shown that it was possible to measure the change in size when cells undergo apoptosis. The novel method has an integrated microfluidic chip, which can adapt to sort cancer and other types of cells based on their cell-surface protein expression.

A low-cost, flow-through nanocytometer has been presented that utilizes a colloidal suspension of nonfunctionalized magnetic nanoparticles for label-free manipulation and separation of microparticles (Kose and Koser 2012). The size-based separation is mediated by magnetically excited biocompatible ferrofluid particles with up to 99 % separation efficiency and a throughput of 3×10^4 particles/s per mm^2 of channel cross section. The device is readily scalable and applicable to live cell sorting offering competitive cytometer performance in a simple and inexpensive package.

Nanocytometry is a significant improvement over conventional flow cytometry, because the system permits label-free signal detection, extreme reproducibility and sensitivity, and cell separations using only a few cells. Conventional flow cytometry requires a large sample of cells and usually requires labeling. Nanocytometry could provide an important new technology applicable to cancer. For example, nanocytometry could be used to improve upon physicians' ability to detect minimal residual disease states as well as circulating tumor cells (CTCs) and upon a scientist's ability to study cell populations that occur in very small numbers such as stem cells.

DNA–Protein and Nanoparticle Conjugates

Semisynthetic conjugates composed of nucleic acids, proteins, and inorganic nanoparticles have been synthesized and characterized. For example, self-assembled oligomeric networks consisting of streptavidin and double-stranded DNA are applicable as reagents in immunoassays. Covalent conjugates of ssDNA and streptavidin are utilized as biomolecular adapters for the immobilization of biotinylated macromolecules at solid substrates via nucleic acid hybridization. This "DNA-directed immobilization" enables reversible and site-selective functionalization of solid substrates with metal and semiconductor nanoparticles or, vice versa, for the DNA-directed functionalization of gold nanoparticles with proteins, such as immunoglobulins and enzymes. This approach is applicable for the detection of chip-immobilized antigens. Moreover, covalent DNA-protein conjugates allow for their selective positioning along single-stranded nucleic acids and thus for the construction of nanometer-scale assemblies composed of proteins and/or nanoclusters. Examples include the fabrication of functional biometallic nanostructures from gold nanoparticles and antibodies, applicable as diagnostic tools in bioanalytics. Gold nanoparticles decorated with fluorescein-modified DNA enables improvement of the detection limit of ascorbic acid quantification by two orders of magnitude due to enhanced cleavage of DNA catalyzed by gold clusters (Malashikhina and Pavlov 2012).

Resonance Light Scattering Technology

Resonance light scattering (RLS) technology, developed at Genicon Sciences Corporation (now acquired by Life Technologies), offers uniquely powerful signal generation and detection capabilities applicable to a wide variety of analytical bioassay formats. RLS exploits submicroscopic metallic particles (e.g., gold and silver) of uniform diameter (in the nanometer range) which scatter incident white light to generate monochromatic colored light that appears as highly intense fluorescence Each RLS particle produces intense light scattering that can be viewed with the naked eye. Under low-power microscope magnification, individual 80-nm gold particles can be readily observed. The scattering produced by these particles creates a "halo" with an apparent 1-μm diameter. As a result, one can conduct ultrasensitive assays to define location and relative frequency of target molecules. RLS signal generation technology is up to 1,000,000 times more sensitive than current fluorescence signaling technology. Other advantages of RLS technology are that RLS signals do not require computer-enhanced imaging of data as they are so intense. Research applications of RLS technology are:

- Gene expression. Relative gene-expression studies on slide-based cDNA microarrays
- DNA sequencing. RLS-based DNA sequencing on sequence-by-hybridization biochips
- Microfluidics. RLS particles for solution-based assays in nanofluidic flow-through microarrays
- Immunohistology. Rapid in situ localization/quantitation of proteins in tissue sections using RLS-coupled antibodies
- Homogeneous. RLS particles for bimolecular, microvolume studies in solution

Clinical applications of RLS technology are:

- RLS technology is being used to score SNPs for discrimination of therapeutically relevant alleles.
- RLS technology provides ultrahigh-sensitivity probes for in situ hybridizations to quantitate therapeutically important DNA and RNA molecules.
- Antibody-coupled RLS particles can deliver increased sensitivity for detection of rare analytes in diagnostic assays.
- Nanoparticle-labeled bacterial RNA generates reproducible RLS signals that are at least 50 times more intense than state-of-the-art confocal-based fluorescence signals for detection of bacterial pathogens.

Nanobarcodes Technology

Metallic nanobarcodes have been produced with striping patterns prepared by sequential electrochemical deposition of metal ions. The differential reflectivity of adjacent stripes enables identification of the striping patterns by conventional light microscopy. This readout mechanism does not interfere with the use of fluorescence for detection of analytes bound to particles by affinity capture, as demonstrated by DNA and protein bioassays. Among other applications such as SNP mapping and multiplexed assays for proteomics, nanobarcodes can be used for population diagnostics and in point-of-care handheld devices. Multiplexed biodetection based on barcoded nanowires has been described with potential use in cancer detection (Brunker et al. 2007). Key performance advantages relative to existing encoded bead technologies include:

- The ability to use the widely installed base of optical microscopes for readout
- The ability to use multiple colors of fluorophores for quantitation
- The ability to generate hundreds to thousands of unique codes that can be distinguished at high speed

Nanobarcode Particle Technology for SNP Genotyping

Nanobarcode particle technology has been used in universal array for high-throughput SNP genotyping (Sha et al. 2006). The particles are encoded submicron metallic nanorods manufactured by electroplating inert metals such as gold and silver into templates and releasing the resulting striped nanoparticles. The power of this technology is that the particles are intrinsically encoded by virtue of the different reflectivity of adjacent metal stripes, enabling the generation of many thousands of unique encoded substrates. Using SNP found within the cytochrome P450 gene family, and a universal short oligonucleotide ligation strategy, simultaneous genotyping of

15 SNPs was demonstrated, a format requiring discrimination of 30 encoded nanowires (one per allele). To demonstrate applicability in practice, 160 genotypes were determined from multiplex PCR products from 20 genomic DNA samples.

QD Nanobarcode for Multiplexed Gene-Expression Profiling

QD nanobarcode-based microbead random array platform (Life Technologies) has been used for accurate and reproducible gene-expression profiling in a high-throughput and multiplexed format (Eastman et al. 2006). Four different sizes of QDs, with emissions at 525, 545, 565, and 585 nm, are mixed with a polymer and coated onto the magnetic microbeads (8-μm diameter) to generate a nanobarcoded QBeads. Twelve intensity levels for each of the four colors are used. Gene-specific oligonucleotide probes are conjugated to the surface of each spectrally nanobarcoded bead to create a multiplexed panel, and biotinylated cRNAs are generated from sample total RNA and hybridized to the gene probes on the microbeads. A fifth streptavidin QD (655 nm or infrared QD) binds to biotin on the cRNA, acting as a quantification reporter. The intensity of the 655-nm Qdot reflects the level of biotinylated cRNA captured on the beads and provides the quantification for the corresponding target gene. The system shows a level of sensitivity, which is better than that with a high-density microarray system, and approaches the level usually observed for quantitative PCR. The QBead nanobarcode system has a dynamic range of 3.5 logs, better than the 2–3 logs observed on various microarray platforms. The hybridization reaction is performed in liquid phase and completed in 1–2 h, at least 1 order of magnitude faster than microarray-based hybridizations. Detectable fold change is lower than 1.4-fold, showing high precision even at close to single copy per cell level. Reproducibility for this proof-of-concept study approaches that of Affymetrix GeneChip microarray. In addition, it provides increased flexibility, convenience, and cost-effectiveness in comparison to conventional gene-expression profiling methods.

Biobarcode Assay for Proteins

An ultrasensitive method for detecting protein analytes relies on nanoparticle probes that are encoded with DNA that is unique to the protein target of interest and antibodies. Magnetic separation of the complexed probes and target followed by dehybridization of the oligonucleotides on the nanoparticle probe surface allows the determination of the presence of the target protein by identifying the oligonucleotide sequence released from the nanoparticle probe. Because the nanoparticle probe carries with it a large number of oligonucleotides per protein-binding event, there is substantial amplification and PSA can be detected at 30 attomolar concentration. Alternatively, a PCR on the oligonucleotide barcodes can boost the sensitivity to 3 attomolar. Comparable clinically accepted conventional assays for detecting the

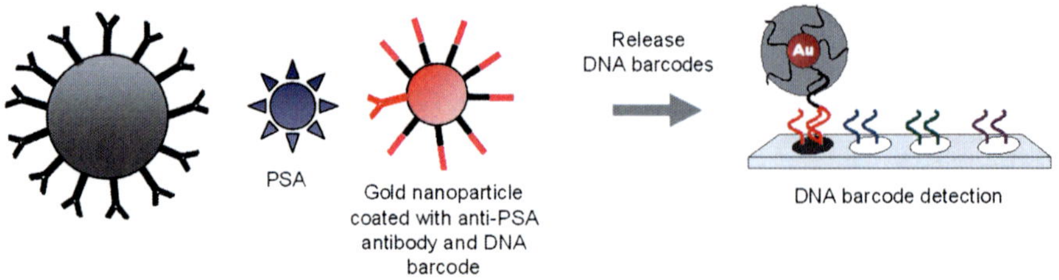

Fig. 4.1 Scheme of biobarcode assay. Schematic illustrating PSA (prostate-specific antigen) detection using the biobarcode assay. Antibody-coated magnetic beads capture and concentrate the protein targets. The captured protein targets are labeled with gold nanoparticle probes that are co-loaded with target-specific secondary antibodies and DNA barcodes. The resulting complexes are separated magnetically and washed to remove excess probe. The DNA barcodes are then released from the complex and detected via hybridization to a surface-immobilized DNA probe and an oligonucleotide-functionalized gold nanoparticle. The gold particles are enlarged through silver deposition, and the light scattered from the particles is detected using the Verigene Reader optical detection system. Increased detection sensitivity is derived from (1) capturing and concentrating protein targets with an antibody-coated magnetic bead, (2) releasing multiple DNA barcodes per captured protein target (hundreds of barcode are attached to a 30-nm-diameter gold particle), and (3) ultrasensitive DNA detection via silver-amplified gold nanoparticles (Courtesy of Nanosphere Inc.)

same target have sensitivity limits of 3 picomolars, six orders of magnitude less sensitive than what is observed with this method. Further development of this technology has resulted in a biobarcode assay with a 500 zeptomolar target DNA sensitivity limit (Nam et al. 2004). Magnetic separation and subsequent release of barcode DNA from the gold nanoparticles leads to a number of barcode DNA strands for every target DNA (see Fig. 4.1).

One reagent is a gold nanoparticle only 30 nm in diameter; the other is a 1-μm magnetic microparticle (MMP). During the assay, the two spheres capture and sandwich the analytes. The MMPs and whatever is bound to them are then captured using a magnet, and unreacted gold NPs are washed away. Thus, only those gold spheres that have captured the analyte remain. Each gold bead also bears an abundance of biobarcodes, custom oligonucleotides that uniquely identify the reaction. The system ultimately detects barcodes released from the beads by heating to 55 °C and not the analytes themselves. Chip-based barcode DNA detection can be done with PCR-like sensitivity but without the use of PCR.

A nanoparticle-based biobarcode assay (BCA) was used to measure the concentration of amyloid β-derived diffusible ligands (ADDLs) in the cerebrospinal fluid (CSF) as a biomarker for Alzheimer's disease (Georganopoulou et al. 2005). Commercial enzyme-linked immunoassays (ELISA) can only detect ADDLs in brain tissue where the biomarker is most highly concentrated. Studies of ADDLs in the CSF have not been possible because of their low concentration. The biobarcode amplification technology, which is a million times more sensitive than ELISA, can detect ADDLs in the CSF where the biomarker is present in very low concentrations. This study is a step toward a diagnostic tool, based on soluble pathogenic markers for Alzheimer's disease. The goal is to ultimately detect and validate the marker in blood.

Using the Verigene ID system (Nanosphere Inc.), one can quantify the barcodes using the kind of technology found in a flatbed scanner, providing results as clear as an at-home pregnancy strip test. Biobarcode system is extremely sensitive for protein detection. At 30 attomolar, it is five orders of magnitude more sensitive than is ELISA (peak sensitivity of around 3 picomolars). The system has enormous potential for multiplexing. It could hypothetically test for 415 different analytes simultaneously by tagging the different gold beads with different barcode sequences. The assay, however, and the fundamental issues with antibodies, such as cross-reactivity, nonspecific binding, and lot-to-lot variability, remain. Antibodies can distort, fall apart, or cling to the wrong analyte. These issues are being addressed. In 2007, the FDA cleared Verigene® Warfarin Metabolism nucleic acid test followed by clearance of Verigene® F5/F2/MTHFR nucleic acid test, which detects disease-associated gene mutations that can contribute to blood coagulation disorders and difficulties metabolizing folate (vitamin B12). Mutations in three specific genes can increase an individual's risk for dangerous blood clots and their leading complication, stroke. Patients that test positively for an increased risk of blood clots can be managed with anticoagulant therapy such as warfarin. Hypercoagulation tests for mutations associated with a predisposition to blood clots are currently among the most frequently conducted human genetic tests. The test is available in single and multitarget (multiplex) formats, allowing users to select the test cartridge that best fits the clinical indications for testing.

A modified form of the BCA called the surface-immobilized biobarcode assay (SI-BCA) is available in a microfluidic chip format (Goluch et al. 2009). The SI-BCA employs microchannel walls functionalized with antibodies that bind with the intended targets. Compared with the conventional BCA, it reduces the system complexity and results in shortened process time, which is attributed to significantly reduced diffusion times in the microscale channels. Raw serum samples, without any pretreatment, were evaluated with this technique. PSA in the samples was detected at concentrations ranging from 40 pM to 40 fM. The detection limit of the assay using buffer samples is 10 fM. The entire assay, from sample injection to final data analysis, was completed in 1 h 20 min. This ability to easily and quickly detect very low levels of PSA, not detectable by conventional assays, may enable diagnosis of prostate cancer recurrence years earlier than is currently possible. Furthermore, the effectiveness of postoperative treatment could be assessed by monitoring a patient's PSA levels. This level of sensitivity in detecting low concentrations of PSA will require revision of the normal laboratory values as currently written in reference manuals.

Single-Molecule Barcoding System for DNA Analysis

Molecular confinement offers new routes for arraying large DNA molecules, enabling single-molecule schemes aimed at the acquisition of sequence information. Such schemes can rapidly advance to become platforms capable of genome analysis if elements of a nascent system can be integrated at an early stage of development. Integrated strategies are needed for surmounting the stringent experimental requirements of nanoscale devices regarding fabrication, sample loading, biochemical labeling, and detection. Disposable devices featuring both micro- and nanoscale features have been shown to greatly elongate DNA molecules when buffer conditions are controlled for alteration of DNA stiffness (Jo et al. 2007). Analytical calculations that describe this elongation were presented. A complementary enzymatic labeling scheme was developed that tags specific sequences (barcodes) on elongated molecules within described nanoslit devices that are imaged via fluorescence resonance energy transfer. Collectively, these developments enable scalable molecular confinement approaches for genome analysis.

Nanoparticle-Based Colorimetric DNA Detection Method

Nucleic acid diagnostics is dominated by fluorescence-based assays that use complex and expensive enzyme-based target or signal-amplification procedures. Many clinical diagnostic applications will require simpler, inexpensive assays that can be done in a screening mode. Nanosphere Inc.'s Verigene™ platform uses a "spot-and-read" colorimetric detection method for identifying nucleic acid sequences based on optical properties of gold nanoparticles without the need for conventional signal or target amplification. Nucleic acid targets are recognized by DNA-modified gold probes, which undergo a color change that is visually detectable when the solutions are spotted onto an illuminated glass waveguide. Sensitivity of the spot test is improved by monitoring scattered light rather than reflected light from 40- to 50-nm-diameter gold particles. This scatter-based method enables detection of zeptomole quantities of nucleic acid targets without target or signal amplification when coupled to an improved hybridization method that facilitates probe-target binding in a homogeneous format. In comparison to a previously reported absorbance-based method, this method increases detection sensitivity by over four orders of magnitude and has been applied to the rapid detection of mecA in methicillin-resistant *Staphylococcus aureus* genomic DNA samples.

Nanoparticle assemblies interconnected with DNA triple helixes can be used to colorimetrically screen for triplex DNA binding molecules and simultaneously determine their relative binding affinities based on melting temperatures (Han et al. 2006). Nanoparticles assemble only when DNA triple helixes form between DNA from two different particles and a third strand of free DNA. In addition, the triple helix structure is unstable at room temperature and only forms in the presence of triplex DNA binding molecules which stabilize the triple helix. The resulting melting transition of the nanoparticle assembly is much sharper than the analogous triplex structure without nanoparticles. Upon nanoparticle assembly, a concomitant red-to-blue color change occurs. The assembly process and color change do not occur in the presence of duplex DNA binders and therefore provide a significantly better screening process for triplex DNA binding molecules compared to standard methods.

Rapid colorimetric analysis of a specific DNA sequence has been achieved by combining gold nanoparticles (AuNPs) with an asymmetric PCR (Deng et al. 2012). In the presence of the correct DNA template, the bound oligonucleotides on the surface of AuNPs selectively hybridize to form complementary sequences of ssDNA target generated from asymmetric PCR with a concomitant color change from ruby red to blue purple. It is a simple colorimetric method for specific nucleic acid sequence analysis with high specificity and sensitivity and has been used for the detection of *Bacillus anthracis* in clinical samples.

SNP Genotyping with Gold Nanoparticle

Conventional SNP detection techniques are mainly PCR-based. Nanosphere's Verigene technology enables multiplex SNP genotyping in total human genomic DNA without the need for target amplification by PCR. This direct SNP genotyping method requires no enzymes and relies on the high sensitivity of the gold nanoparticle probes.

A simple and rapid MS-based disulfide barcode method relies on magnifying the signal from a dual-modified gold nanoparticle and enables direct SNP genotyping of total human genomic DNA without the need for primer-mediated enzymatic amplification (Yang et al. 2010). Disulfides that are attached to the gold nanoparticle serve as a "barcode" that allows different sequences to be detected. Specificity is based on two sequential oligonucleotide hybridizations, which include two steps: the first is the capture of the target by gene-specific probes immobilized onto magnetic beads; the second is the recognition of gold nanoparticles functionalized with allele-specific oligonucleotides. The sensitivity of this method reaches down to the 0.1 fM range, thus approaching that of PCR. The feasibility of this method was demonstrated by applying it to genomic DNA samples representing all possible genotypes of the SNPs G2677T and C3435T in the human MDR1 gene.

Nanoparticle-Based Up-Converting Phosphor Technology

Up-converting phosphor technology (UPT) is a label detection technology that can be applied to the detection of minute quantities of various substances such as antigens, proteins, and DNA. UPT particles are small ceramic nanospheres composed of rare earth metals and have been shown to be 1,000 times more sensitive than current fluorescent technologies. This particle-based detection provides a stronger signal for each event detected and thereby enhances sensitivity in diagnostic assay systems. UPT has potential in a broad array of DNA testing applications including drug discovery, SNP analysis, and infectious disease testing. Employment of UPT, by bypassing target amplification, brings genetic-based testing a step closer to the point-of-care environment.

A rapid and quantitative UPT-based lateral-flow assay was developed for on-site quantitative detection of different Brucella species with high specificity, reproducibility, and stability (Qu et al. 2009). UPT-lateral flow IL-10 assay is a user-friendly, rapid alternative for IL-10 ELISAs, which is suitable for multiplex detection of different cytokines, and can be merged with antibody-detection assays for simultaneous detection of cellular and humoral immunity (Corstjens et al. 2011).

Surface-Enhanced Resonant Raman Spectroscopy

SERRS (surface-enhanced resonant Raman spectroscopy) beads bring various components of the technology into a single robust nanosized polymer-bead support with broad applications in molecular and immunodiagnostics. Focusing on organic fluorescent dyes, because of their strong excitation cross section, compounds are selected experimentally for strong affinity for the silver enhancing surfaces and good spectral resolution. Initially using four dyes, the possibilities for tens to hundreds of unique labels are currently under development. The chosen dyes also have excitation peaks that overlap with the metal plasmon frequency, thereby adding the all-important resonant amplification to the signal intensity.

At the core of the bead is the Raman-active substrate, where silver colloid, with defined physical characteristics, provides the surface-enhancement substrate and is combined with the dye or dyes for specific bead encoding. Control of the various parameters surrounding dye/colloid aggregate permits SERRS response to be modulated as desired.

To protect the SERRS-active complex from degradation, the aggregate is encapsulated in a polymer coating, a process that incorporates a multitude of dye/colloid particles into the same bead. This leads to highly sensitive beads with responses in excess of that achieved using the conformation of single dye molecules on an enhancing surface.

The polymer coating is treated further with a polymer shell to allow a variety of biologically relevant probe molecules (e.g., antibodies, antigens, nucleic acids) to be attached through standard bioconjugation techniques. While most of the development is focused on heterogeneous assays in a 96-well assay sample presentation, other designs include higher plate capacities (384 well) for higher throughput screening and microarray slide reading for DNA and proteomic analysis. Photonic crystal surfaces are used for enhancing the detection of SERR and the development of high-resolution photonic crystal-based laser biosensors, which can be used for gene-expression analysis, and protein biomarker detection (Cunningham 2010).

Near-Infrared (NIR)-Emissive Polymersomes

In vivo fluorescence imaging with near-infrared (NIR) light has enormous potential for a wide variety of molecular diagnostic applications. Because of its quantitative sensitivity, inherent biological safety, and relative ease of use, fluorescence-based imaging techniques are being increasingly used in small-animal research. Moreover, there is substantial interest in the translation of novel optical techniques into the clinic, where they will prospectively aid in noninvasive and quantitative screening, disease diagnosis, and posttreatment monitoring of patients. Effective deep-tissue fluorescence imaging requires the application of exogenous NIR-emissive contrast agents. Currently, available probes fall into two major categories: organic and inorganic NIR fluorophores (NIRFs). Various studies have used polymersomes (50-nm- to 50-μm-diameter polymer vesicles) for the incorporation and delivery of large numbers of highly emissive oligo(porphyrin)-based, organic NIRFs (Ghoroghchian et al. 2009). The total fluorescence emanating from the assemblies gives rise to a localized optical signal of sufficient intensity to penetrate through the dense tumor tissue of a live animal. Robust NIR-emissive polymersomes thus define a soft matter platform with exceptional potential to facilitate deep-tissue fluorescence-based imaging for in vivo diagnosis.

Nanobiotechnology for Detection of Proteins

Detection of proteins is an important part of molecular diagnostics. Uses of protein nanobiochips and nanobarcode technology for detection of proteins have been described in preceding sections. Other methods will be included in this section.

Captamers with Proximity Extension Assay for Proteins

Multivalent circular aptamers or "captamers" are formed through the merger of aptameric recognition functions with the DNA as a nanoscale scaffold. Whereas the sequence immobilized to the microtiter plate is termed captamer, the sequence used for detection is called detectamer. Aptamers are useful as protein-binding motifs for diagnostic applications, where their ease of discovery, thermal stability, and low cost make them ideal components for incorporation into targeted protein assays. Captamers are compatible with a highly sensitive protein detection method termed the "proximity extension" assay (Di Giusto et al. 2005). The circular DNA architecture facilitates the integration of multiple functional elements into a single molecule: aptameric target recognition, nucleic acid hybridization specificity, and rolling circle amplification. Successful exploitation of these properties is demonstrated for the molecular analysis of thrombin, with the assay delivering a detection limit nearly three orders of magnitude below the dissociation constants of the two contributing aptamer–thrombin interactions.

Use of liposomes as labels for aptamer-based assays and successful incorporation of cholesteryl–TEG DNA aptamers into liposomal lipid bilayers with subsequent successful function in target recognition further demonstrates the versatility of liposomes as signaling reagents and their potential as a standard platform technology for various analyses. Such an assay yields a limit of detection of 64 pM or 2.35 ng/mL, corresponding to 6.4 fmol or 235 pg, respectively, in a 100-μL volume (Edwards et al. 2010).

Real-time signal amplification, detection under isothermal conditions, specificity, and sensitivity would suggest potential application of captamer-based protein assay for further development of personalized medicine.

Nanobiosensors

Nanosensors are devices that employ nanomaterials, exploiting novel size-dependent properties, to detect gases, chemicals, biological agents, electric fields, light, heat, or other targets. The term "nanobiosensors" implies use of nanosensors for detection of chemical or biological materials. Nanomaterials are exquisitely sensitive chemical and biological sensors (Jain 2003a).

The sensors can be electronically gated to respond to the binding of a single molecule. Prototype sensors have demonstrated detection of nucleic acids, proteins, and ions. These sensors can operate in the liquid or gas phase, opening up an enormous variety of downstream applications. The detection schemes use inexpensive low-voltage measurement schemes and detect binding events directly, so there is no need for costly, complicated, and time-consuming labeling chemistries such as fluorescent dyes or the use of bulky and expensive optical detection systems. As a result, these sensors are inexpensive to manufacture and portable. It may even be possible to develop implantable detection and monitoring devices based on these detectors.

Some of the technologies that can be incorporated in biosensing are already covered in earlier sections. An example is nanopore technology, which can form the basis of nanosensors. Some of the biosensor devices are described in the following sections.

Cantilevers as Biosensors for Molecular Diagnostics

Cantilevers (Concentris) are small beams similar to those used in AFM to screen biological samples for the presence of particular genetic sequences. The surface of each cantilever is coated with DNA that can bind to one particular target sequence. On exposure of the sample to beams, the surface stress bends the beams by approximately 10 nm to indicate that the beams have found the target in the sample. This is considered biosensing.

Cantilever technology complements and extends current DNA and protein microarray methods because nanomechanical detection requires no labels, optical excitation, or external probes and is rapid, highly specific, sensitive, and portable. The nanomechanical response is sensitive to the concentration of oligonucleotides in solution, and thus one can determine how much of a given biomolecule is present and active. In principle, cantilever arrays also could quantify gene-expression levels of mRNA, protein–protein, drug-binding interactions, and other molecular recognition events in which physical steric factors are important. It can detect a single gene within a genome. Furthermore, fabricating thinner cantilevers will enhance the molecular sensitivity further, and integrating arrays into microfluidic channels will reduce the amount of sample required significantly. In contrast to SPR, cantilevers are not limited to metallic films, and other materials will be explored, for example, cantilevers made from polymers. In addition to surface-stress measurements, operating cantilevers in the dynamic mode will provide information on mass changes, and current investigations will determine the sensitivity of this approach. Currently, it is possible to monitor more than 1,000 cantilevers simultaneously with integrated piezoresistive readout, which in principle will allow high-throughput nanomechanical genomic analysis, proteomics, biodiagnostics, and combinatorial drug discovery.

Cantilevers in an array can be functionalized with a selection of biomolecules. Researchers at IBM, Zurich, Switzerland, reported the specific transduction, via surface-stress changes, of DNA hybridization and receptor-ligand binding into a direct nanomechanical response of microfabricated cantilevers. The differential deflection of the cantilevers was found to provide a true molecular recognition signal despite large nonspecific responses of individual cantilevers. Hybridization of complementary oligonucleotides shows that a single-base mismatch between two 12-mer oligonucleotides is clearly detectable. Similar experiments on protein A–immunoglobulin interactions demonstrate the wide-ranging applicability of nanomechanical transduction to detect biomolecular recognition. Microarray of cantilevers has been used to detect multiple unlabeled biomolecules simultaneously at nanomolar concentrations within minutes.

A specific test that uses micrometer-scale beams or "microcantilever" can detect prostate-specific antigen (PSA). PSA antibodies are attached to the surface of the microcantilever, which is applied to a sample containing PSA. When PSA binds to the antibodies, a change in the surface stress on the microcantilever makes it bend enough to be detected by a laser beam. This system is able to detect clinically relevant concentrations of PSA in a background of other proteins. The technique is simpler and potentially more cost-effective than other diagnostic tests because it does not require labeling and can be performed in a single reaction. It is less prone to false positives, which are commonly caused by the nonspecific binding of other proteins to the microcantilever.

Potential applications in proteomics include devices comprising many cantilevers, each coated with a different antibody, which might be used to test a sample rapidly and simultaneously for the presence of several disease-related proteins. One application is for detection of biomarkers of myocardial infarction such as creatine kinase at point-of-care. Other future applications include detection of disease by breath analysis, for example, presence of acetone and dimethylamine (uremia). Detection of a small number of Salmonella enterica bacteria is achieved due to a change in the surface stress on the silicon nitride cantilever surface in situ upon binding of bacteria. Scanning electron micrographs indicate that less than 25 adsorbed are required for detection.

Advantages of Cantilever Technology for Molecular Recognition

Cantilever technology has the following advantages over conventional molecular diagnostics:

- It circumvents the use of PCR.
- For DNA, it has physiological sensitivity and no labeling is required.
- In proteomics, it enables detection of multiple proteins and direct observation of proteins in diseases such as those involving the cardiovascular system.
- It enables the combination of genomics and proteomics assays.
- It is compatible with silicon technology.
- It can be integrated into microfluidic devices.

Antibody-Coated Nanocantilevers for Detection of Microorganisms

Nanocantilevers could be crucial in designing a new class of ultrasmall sensors for detecting viruses, bacteria, and other pathogens (Gupta et al. 2006). The cantilevers, coated with antibodies to detect certain viruses, attract different densities or quantity of antibodies per area depending on the size of the cantilever. The devices are immersed into a liquid containing the antibodies to allow the proteins to stick to the cantilever surface. Instead of simply attracting more antibodies, the longer cantilevers also contained a greater density of antibodies. The density is greater toward the free end of the cantilevers. The cantilevers vibrate faster after the antibody attachment if the devices have about the same nanometer-range thickness (~20 nm) as the protein layer. Moreover, the longer the protein-coated nanocantilever, the faster the vibration, which could only be explained if the density of antibodies increased with increasing lengths.

The cantilever's vibration frequency can be measured using an instrument called a laser Doppler vibrometer, which detects changes in the cantilever's velocity as it vibrates. This work may have broad impact on microscale and nanoscale biosensor design, especially when predicting the characteristics of nanobioelectromechanical sensors functionalized with biological capture molecules. The nanocantilevers could be used in future detectors because they vibrate at different frequencies when contaminants stick to them, revealing the presence of dangerous substances. Because of the nanocantilever's minute size, it is more sensitive than larger devices, promising the development of advanced sensors that detect minute quantities of a contaminant to provide an early warning that a dangerous pathogen is present. At the nanoscale, just adding the mass of one bacterium, virus or large molecule is enough to change the resonant frequency of vibration of the cantilever by a measurable amount, thereby signaling the presence of the pathogen. If one is trying to detect *E. coli*, other organisms in the fluid can weakly absorb on the detector by electrostatic forces. This is a problem in any biodetection and can be resolved by making the resonator vibrate from side to side. This will shake off loosely adhered materials, while whatever is tightly bound to an antibody will remain.

Cantilevers for Direct Detection of Active Genes

An innovative method for the rapid and sensitive detection of disease- and treatment-relevant genes is based on direct measurement of their transcripts (mRNA), which represent the intermediate step and link to protein synthesis (Zhang et al. 2006a). Short complementary nucleic acid segments (sensors) are attached to silicon cantilevers which are 450 nm thick and therefore react with extraordinary sensitivity. Binding of the targeted gene transcript to its matching counterpart on one of the cantilevers results in optically measurable mechanical bending.

Differential gene expression of the gene 1-8U, a potential marker for cancer progression or viral infections, could be observed in a complex background. The measurements provide results within minutes at the picomolar level without target amplification and are sensitive to base mismatches. An array of different gene transcripts can even be measured in parallel by aligning appropriately coated cantilevers alongside each other like the teeth of a comb. The new method complements current molecular diagnostic techniques such as the gene chip and real-time PCR. It could be used as a real-time sensor for continuously monitoring various clinical parameters or for detecting rapidly replicating pathogens that require prompt diagnosis. These findings qualify the technology as a rapid method to validate biomarkers that reveal disease risk, disease progression, or therapy response. Cantilever arrays have potential as a tool to evaluate treatment response efficacy for personalized medical diagnostics.

Carbon Nanotube Biosensors

Over the years, researchers have sought to tailor carbon nanotubes to detect chemicals ranging from small gas molecules to large biomolecules. The tubes' small size and unique electronic properties make them especially adept at detecting minute changes in the environment. Optical nanosensors can use single-walled carbon nanotubes that modulate their emission in response to the adsorption of specific biomolecules with two distinct mechanisms of signal transduction – fluorescence quenching and charge transfer. The nanotube-based chemical sensors developed so far generate an electric signal in the presence of a particular molecule. The basic design is widely applicable for such analytical tasks as detecting genes and proteins associated with diseases.

To test the feasibility of implanting the sensors in the body, oxidase- and ferricyanide-coated nanotubes were placed inside a sealed glass tube a centimeter long and 200 μm thick. The tube is riddled with pores large enough to let glucose enter but small enough to keep the nanotubes inside. The tube was then implanted in a sample of human skin, and the sensor could be excited with infrared light and detect its fluorescence.

Carbon Nanotube Sensors Coated with ssDNA and Electronic Readout

Nanoscale chemical sensors can be based on ssDNA as the chemical recognition site and single-walled carbon nanotube field-effect transistors (SWCN-FETs) as the electronic readout component (Staii et al. 2005). SWCN-FETs with a nanoscale coating of ssDNA respond to gas odors that do not cause a detectable conductivity change in bare devices. Responses of ssDNA/SWCN-FETs differ in sign and magnitude for different gases and can be tuned by choosing the base sequence of the ssDNA. ssDNA/SWCN-FET sensors detect a variety of odors, with rapid response and recovery times on the scale of seconds. The arrays of nanosensors could detect molecules on the order of one part per million. The sensor surface is self-regenerating: samples maintain a constant response with no need for sensor refreshing through at least 50 gas exposure cycles. The nanosensors could sniff molecules in the air or taste them in a liquid. This remarkable set of attributes makes sensors based on ssDNA decorated nanotubes promising for "electronic nose" and "electronic tongue" applications ranging from homeland security to disease diagnosis.

Carbon Nanotubes Sensors Wrapped with DNA and Optical Detection

SWCNs wrapped with DNA can be placed inside living cells and detect trace amounts of harmful contaminants using near-infrared light (Heller et al. 2006). The sensor is constructed by wrapping the double-stranded DNA around the surface of a single-walled carbon nanotube, in much the same fashion as a telephone cord wraps around a pencil. The DNA starts out wrapping around the nanotube with a certain shape that is defined by the negative charges along its backbone. Subtle rearrangement of an adsorbed biomolecule can be directly detected by such a carbon nanotube. At the heart of the new detection system is the transition of DNA secondary structure from the native, right-handed "B" form to the alternate, left-handed "Z" form. The thermodynamics that drive the switching back and forth between these two forms of DNA structure would modulate the electronic structure and optical emission of the carbon nanotube. When the DNA is exposed to ions of certain atoms such as calcium or mercury, the negative charges become neutralized and the DNA changes shape in a similar manner to its natural shape-shift from the B form to Z form. This reduces the surface area covered by the DNA, perturbing the electronic structure and shifting the nanotube's natural, near-infrared fluorescence to a lower energy. The change in emission energy indicates how many ions bind to the DNA. Removing the ions will return the emission energy to its initial value and flip the DNA back to the starting form, making the process reversible and reusable. The viability of this measurement technique was demonstrated by detecting low concentrations of mercury ions in whole blood, opaque solutions, and living mammalian cells and tissues where optical sensing is usually poor or ineffective. Because the signal is in the near infrared, a property unique to only a handful of materials, it is not obscured by the natural fluorescence of polymers and living tissues. The nanotube surface acts as the sensor by detecting the shape change of the DNA as it responds to the presence of target ions. This discovery opens the door to new types of optical sensors and biomarkers that exploit the unique properties of nanoparticles in living systems.

A pair of SWCNs provides at least four modes that can be modulated to uniquely fingerprint agents by the degree to which they alter either the emission band intensity or wavelength. This identification method was validated in vitro by demonstrating the detection of six genotoxic analytes, including chemotherapeutic drugs and reactive oxygen species, which are spectroscopically differentiated into four distinct classes, and also demonstrate single-molecule sensitivity in detecting hydrogen peroxide (Heller et al. 2009). SWCN sensor can be placed in living cells, healthy or malignant, and actually detect several different classes of molecules that damage DNA.

FRET-Based DNA Nanosensor

Rapid and highly sensitive detection of DNA is critical in diagnosing genetic diseases. Conventional approaches often rely on cumbersome, semiquantitative amplification of target DNA to improve detection sensitivity. In addition, most DNA detection systems (e.g., microarrays), regardless of their need for target amplification, require separation of unhybridized DNA strands from hybridized stands immobilized on a solid substrate and are thereby complicated by solution-surface binding kinetics. An ultrasensitive nanosensor is based on fluorescence resonance energy transfer (FRET) capable of detecting low concentrations of DNA in a separation-free format. This system uses quantum dots (QDs) linked to DNA probes to capture DNA targets (Zhang et al. 2005). The target strand binds to a dye-labeled reporter strand thus forming a FRET donor–acceptor ensemble. The QD also functions as a concentrator that amplifies the target signal by confining several targets in a nanoscale domain. Unbound nanosensors produce near-zero background fluorescence, but on binding to even a small amount of target DNA (~50 copies or less), they generate a very distinct FRET signal. A nanosensor-based oligonucleotide ligation assay has been demonstrated to successfully detect a point mutation typical of some ovarian tumors in clinical samples.

Ion-Channel Switch Biosensor Technology

The ion channel switch is a biosensor technology based upon a synthetic self-assembling membrane, which acts like a biological switch and is capable of detecting the presence of specific molecules and signaling their presence by triggering an electrical current. It has the ability to detect a change in ion flow upon binding with the target molecule resulting in a rapid result currently unachievable using existing technologies. An ion-channel biosensor comprised of gramicidin A channels embedded in a synthetic tethered lipid bilayer provides a highly sensitive and rapid detection method, for example, for influenza A in untreated clinical samples (Krishnamurthy et al. 2010).

Electronic Nanobiosensors

A signal-on, electronic DNA biosensor has been described that is label-free and achieves a subpicomolar detection limit (Xiao et al. 2006). The sensor, which is based on a target-induced strand displacement mechanism, is composed of a "capture probe" attached by its 5′ terminus to a gold electrode and a 5′ methylene blue-modified "signaling probe" that is complementary at both its 3′ and 5′ termini to the capture probe. In the absence of target, hybridization between the capture and signaling probes minimizes contact between the methylene blue and electrode surface, limiting the observed redox current. Target hybridization displaces the 5′ end of the signaling probe, generating a short, flexible single-stranded DNA element and producing up to a sevenfold increase in redox current. The observed signal gain is sufficient to achieve a demonstrated (not extrapolated) detection limit of 400 fM, which is among the best reported for single-step electronic DNA detection. Moreover, because sensor fabrication is straightforward, the approach appears to provide a ready alternative to the more cumbersome femtomolar electrochemical assays described to date.

Capacitors are critical elements in electrical circuits, and nanocapacitors are capacitors with electrodes spacing in the nano-order. When used with single-stranded DNA probes, target hybridization produces a measurable change in capacitance. When used in arrays, nanocapacitors can enable simultaneous detection of nucleic acids without labeling (Fortina et al. 2005).

Electrochemical Nanobiosensor

An electrochemical biosensor combining microfluidics and nanotechnology has been developed by GeneFluidics with 16 sensors in the array, each consisting of three single-layer gold electrodes – working, reference, and auxiliary. Each of the working electrodes contains one representative from a library of capture probes, which are specific for a clinically relevant bacterial urinary pathogen. The library included probes for *Escherichia coli, Proteus mirabilis, Pseudomonas aeruginosa, Enterococcus* spp., and the *Klebsiella-Enterobacter* group. A bacterial 16S rRNA target derived from single-step bacterial lysis was hybridized both to the biotin-modified capture probe on the sensor surface and to a second, fluorescein-modified detector probe. Detection of the target-probe hybrids is achieved through binding of a horseradish peroxidase (HRP)-conjugated anti-fluorescein antibody to the detector probe. Amperometric measurement of the catalyzed HRP reaction is obtained at a fixed potential of −200 mV between the working and reference electrodes. Species-specific detection of as few as 2,600 pathogenic bacteria in culture, inoculated urine, and clinical urine samples can be achieved within 45 min from the beginning of sample processing. In a feasibility study of this amperometric detection system using blinded clinical urine specimens, the sensor array had 100 % sensitivity for direct detection of gram-negative bacteria without nucleic acid purification or amplification (Liao et al. 2006). Identification was demonstrated for 98 % of gram-negative bacteria for which species-specific probes were available. When combined with a microfluidics-based sample preparation module, the integrated system could serve as a point-of-care device for rapid diagnosis of urinary tract infections.

Metallic Nanobiosensors

Fano resonances have been observed in the optical response of plasmonic nanocavities due to the coherent coupling between their superradiant and subradiant plasmon modes, and multiple Fano resonances occur as structure size is increased (Verellen et al. 2009). By putting together two specific nanostructures made of gold or silver, a prototype device can be constructed, which exhibits a highly sensitive ability to detect particular chemicals in the immediate surroundings once it is optimized. The nanostructures measure about 200 nm. One is shaped like a flat circular disk while the other looks like a doughnut with a hole in the middle. When brought together, they interact with light very differently to the way they behave on their own. When they are paired up, they scatter some specific colors within white light much less, leading to an increased amount of light passing through the structure undisturbed. This is distinctly different to how both structures scatter light separately. Metal nanostructures have been used as sensors but they interact very strongly with light due to so-called localized plasmon resonances. But this is the first time a pair with such a carefully tailored interaction with light has been created. This decrease in the interaction with light is in turn affected by the composition of molecules in close proximity to the structures. These nanosensors could be tailor-made to instantly detect the presence of particular molecules, for example, poisons or explosives in transport screening situations or proteins in patients' blood samples, with high sensitivity.

Quartz Nanobalance Biosensor

Single-strand DNA-containing thin films are deposited onto quartz oscillators to construct a device capable of sensing the presence of the complementary DNA sequences, which hybridize with the immobilized ones. DNA, once complexed with aliphatic amines, appears as a monolayer in a single-stranded form by X-ray small angle scattering. A quartz nanobalance is then utilized to monitor mass increment related to specific hybridization with a complementary DNA probe. The crystal quartz nanobalance, capable of high sensitivity, indeed appears capable of obtaining a prototype of a device capable of sensing the occurrence of particular genes or sequences in the sample under investigation.

Viral Nanosensor

Virus particles are essentially biological nanoparticles. Scientists at the Massachusetts General Hospital (Boston, MA) have used herpes-simplex virus (HSV) and adenovirus to trigger the assembly of magnetic nanobeads as a nanosensor for clinically relevant viruses. The nanobeads had a supramagnetic iron oxide core coated with dextran. Protein G was attached as a binding partner for antivirus antibodies. By conjugating anti-HSV antibodies directly to nanobeads using a bifunctional linker to avoid nonspecific interactions between medium components and protein G and using a magnetic field, the scientists could detect as few as five viral particles in a 10-mL serum sample. This system is more sensitive than ELISA-based methods and is an improvement over PCR-based detection because it is cheaper and faster and has fewer artifacts. Upon target binding, these nanosensors cause changes in the spin–spin relaxation times of neighboring water molecules, which can detect specific mRNA, proteins, and enzymatic activity by (NMR/MRI) techniques.

A QD-DNA nanosensor, based on fluorescence resonance energy transfer (FRET), has been used for the detection of the target DNA and single mismatch in hepatitis B virus (HBV) gene (Wang et al. 2010a). This DNA detection method is simple, rapid, and efficient due to the elimination of the washing and separation steps. In this study, oligonucleotides were attached to the QD surface to form functional QD-DNA conjugates. With the addition of DNA targets and Cy5-modified signal DNAs into the QD-DNA conjugates, sandwiched hybrids were formed leading to fluorescence from the acceptor by means of FRET on illumination of the donor. Oligonucleotide ligation assay was employed to efficiently detect single-base mutants in HBV gene. This simple method enables efficient detection that could be used for high throughput and multiplex detections of viral gene mutations.

PEBBLE Nanosensors

Scientists at the University of Michigan (Ann Arbor, Michigan) have developed PEBBLE (probes encapsulated by biologically localized embedding) nanosensors, which consist of sensor molecules entrapped in a chemically inert matrix by a microemulsion polymerization process that produces spherical sensors in the size range of 20–200 nm. These sensors are capable of real-time inter- and intracellular imaging of ions and molecules and are insensitive to interference from proteins. PEBBLE can also be used for early detection of cancer. PEBBLE nanosensors also show very good reversibility and stability to leaching and photobleaching, as well as very short response times and no perturbation by proteins. In human plasma, they demonstrate a robust oxygen sensing capability, little affected by light scattering and autofluorescence. PEBBLE has been developed further as a tool for diagnosis as well as treatment of cancer.

Detection of Cocaine Molecules by Nanoparticle-Labeled Aptasensors

Metallic or semiconductor nanoparticles (NPs) are used as labels for the electrochemical, photoelectrochemical, or surface plasmon resonance (SPR) detection of cocaine using a common aptasensor configuration (Golub et al. 2009). The aptasensors are based on the use of two anticocaine aptamer subunits, where one subunit is assembled on an Au support, acting as an electrode or a SPR-active surface, and the second aptamer subunit is labeled with Pt-NPs, CdS-NPs, or Au-NPs. In the different aptasensor configurations, the addition of cocaine results in the formation of supramolecular complexes between the NPs-labeled aptamer subunits and cocaine on the metallic surface, enabling quantitative analysis of cocaine. The supramolecular Au-NPs-aptamer-subunits–cocaine complex generated on the Au support allows the SPR detection of cocaine through the reflectance changes stimulated by the electronic coupling between the localized plasmon of the Au-NPs and the surface plasmon wave. All aptasensor configurations enable the analysis of cocaine with a detection limit in the range of 10^{-6} to 10^{-5} M. The major advantage of the sensing platform is the lack of background interfering signals.

Nanosensors for Glucose Monitoring

One of the main reasons for developing in vivo glucose sensors is the detection of hypoglycemia in people with insulin-dependent (type 1) diabetes. It is possible to engineer fluorescent micro/nanoscale devices for glucose sensing. Deployment of nanoparticles in the dermis may allow transdermal monitoring of glucose changes in interstitial fluid. Using electrostatic self-assembly, an example of nanotechnology for fabrication, two types of sensors are being studied: (1) solid nanoparticles coated with fluorescent enzyme-containing thin films and (2) hollow nanocapsules containing fluorescent indicators and enzymes or glucose-binding proteins. Nanoengineering of the coated colloids and nanocapsules allows precision control over optical, mechanical, and catalytic properties to achieve sensitive response using a combination of polymers, fluorescent indicators, and glucose-specific proteins. Challenges to in vivo use include understanding of material toxicity and failure modes and determining methods to overcome fouling, protein inactivation, and material degradation. Noninvasive glucose sensing will maximize acceptance by patients and overcome biocompatibility problems of implants. Near-infrared spectroscopy has been most investigated, but the precision needs to be improved for eventual clinical application.

The nanotube-based optical biosensor could free people with diabetes from the daily pinprick tests now required for monitoring blood sugar concentrations. Carbon nanotubes are coated with glucose oxidase, an enzyme that breaks down glucose molecules. Then ferricyanide, an electron-hungry molecule, is sprinkled onto the nanotubes' surfaces. Ferricyanide draws electrons from the nanotubes, quenching their capacity to glow when excited by infrared light. When glucose is present, it reacts with the oxidase, producing hydrogen peroxide. In turn, the hydrogen peroxide reacts with ferricyanide in a way that reduces that molecule's hunger for electrons. The higher the glucose level, the greater is the nanotube's infrared fluorescence.

Micromechanical detection of biologically relevant glucose concentrations can be achieved by immobilization of glucose oxidase (GOx) onto a microcantilever surface. The enzyme-functionalized microcantilever undergoes bending due to a change in surface stress induced by the reaction between glucose in solution and the GOx immobilized on the cantilever surface.

Nanobiosensors for Protein Detection

High-sensitivity biosensors for the detection of proteins have been developed using several kinds of nanomaterials. The performance of the sensors depends on the type of nanostructures with which the biomaterials interact. 1D structures such as nanowires, nanotubes, and nanorods are proven to have high potential for bio-applications. Different types of nanostructures that have attracted much attention by their performance as biosensors utilize materials such as polymers, carbon, and zinc oxide because of their sensitivity, biocompatibility, and ease of preparation (M et al. 2011). This publication describes the three stages in the development of biosensors: (1) fabrication of biomaterials into nanostructures, (2) alignment of the nanostructures, and (3) immobilization of proteins.

Optical Biosensors

Many biosensors that are currently marketed rely on the optical properties of lasers to monitor and quantify interactions of biomolecules that occur on specially derived surfaces or biochips. An integrated biosensor, based on phototransistor integrated circuits, has been developed for use in medical detection, DNA diagnostics, and gene mapping. The biochip device has sensors, amplifiers, discriminators, and logic circuitry on board. Integration of light-emitting diodes into the device is also possible. Measurements of fluorescent-labeled DNA probe microarrays and hybridization experiments with a sequence-specific DNA probe for HIV-1 on nitrocellulose substrates illustrate the usefulness and potential of this DNA biochip. A number of variations of optical biosensors offer distinct methods of sample application and detection in addition to different types of sensor surface. Surface plasmon resonance technology is the best-known example of this technology.

Laser Nanosensors

In a laser nanosensor, laser light is launched into the fiber, and the resulting evanescent field at the tip of the fiber is used to excite target molecules bound to the antibody molecules. A photometric detection system is used to detect the optical signal (e.g., fluorescence) originating from the analyte molecules or from the analyte-bioreceptor reaction. Laser nanosensors can be used for in vivo analysis of proteins and biomarkers in individual living cells (Vo-Dinh and Zhang 2011). The nanosensors are made of tapered optical fibers with distal ends having nanometer-sized diameters. Bioreceptors, such as antibody, peptides, and nucleic acids, are immobilized on the fiber tips and designed to be selective to target analyte molecules of interest. A laser beam is transmitted through the fiber and excites target molecules bound to the bioreceptor molecules. The resulting fluorescence from the analyte molecules is detected by a photodetection system. Nanosensors can provide minimally invasive tools to probe subcellular compartments inside individual living cells.

Physicists at University of Rochester have assembled a simple laser system to detect nanoparticles. They split a laser beam in two, sending one half to a sample. When the light hits a small particle, it is scattered back and recombines with the reserve half of the laser beam, producing a detectable interference pattern detectable only when a moving particle is present. This laser method works where others do not because it relies on the amplitude rather than intensity of light. The amplitude is the square root of intensity, so it decays much less than intensity as the particles get smaller. Single particles, as small as 7 nm in diameter, have been detected.

Researchers at the University of Twente (The Netherlands) have developed an ultrasensitive sensor that could potentially be used in a handheld device to detect various viruses and measure their concentration within minutes. It requires only a tiny sample of saliva, blood, or other body fluid. The device uses a silicon substrate containing channels that guide laser light. Light enters into the substrate at one end and is split into four parallel beams. When these beams emerge at the other end, they spread out and overlap with one another, creating a pattern of bright and dark bands, known as an interference pattern, which are recorded. A commercial version of the biosensor is being developed in collaboration with Paradocs Group BV (The Netherlands). Although the sensor has been shown to detect only the herpes-simplex virus, it could be used to quickly screen people at hospitals and emergency clinics for control of outbreaks of diseases such as SARS and avian flu.

Nanoshell Biosensors

Nanoshells can enhance chemical sensing by as much as 10 billion times. That makes them about 10,000 times more effective at Raman scattering than traditional methods. When molecules and materials scatter light, a small fraction of the light interacts in such a way that it allows scientists to determine their detailed chemical makeup. This property, known as Raman scattering, is used by medical researchers, drug designers, chemists, and other scientists to determine what materials are made of. An enormous limitation in the use of Raman scattering has been its extremely weak sensitivity. Nanoshells can provide large, clean, reproducible enhancements of this effect, opening the door for new, all-optical sensing applications. Each individual nanoshell can act as an independent Raman enhancer. That creates an opportunity to design all-optical nanoscale sensors – essentially molecular diagnostic instruments – that could detect as little as a few molecules of a target substance, which could be anything from a drug molecule or a key disease protein to a deadly chemical agent.

The metal cover of the nanoshell captures passing light and focuses it, a property that directly leads to the enormous Raman enhancements observed. Furthermore, nanoshells can be tuned to interact with specific wavelengths of light by varying the thickness of their shells. This tunability allows for the Raman enhancements to be optimized for specific wavelengths of light. The finding that individual nanoshells can vastly enhance the Raman effect opens the door for biosensor designs that use a single nanoshell, something that could prove useful for engineers who are trying to probe the chemical processes within small structures such as individual cells, or for the detection of very small amounts of a material, like a few molecules of a deadly biological or chemical agent. Nanoshells are already being developed for applications including cancer diagnosis, cancer therapy, testing for proteins associated with Alzheimer's disease, drug delivery, and rapid whole-blood immunoassays.

Plasmonics and SERS Nanoprobes

Surface plasmons are collective oscillations of free electrons at metallic surfaces. These oscillations can give rise to the intense colors of solutions of plasmon resonance nanoparticles and very intense scattering. While the use of plasmonic particle absorption-based bioaffinity sensing is now widespread throughout biological research, the use of their scattering properties is relatively less studied. Plasmon scatter can be used for long-range immunosensing and macromolecular conformation studies.

A variety of sensors, metallic nanostructured probes, metallic nanoshells and half-shells, and nanoarrays for SERS sensing have been developed at the Oak Ridge National Laboratory. The SERS technology can detect the chemical agents and biological species (e.g., spores, biomarkers of pathogenic agents) directly. A DNA-based technique based on surface-enhanced Raman gene (SERGen) probes can be also used to detect gene targets via hybridization to DNA sequences complementary to these probes. Advanced instrumental systems designed for spectral measurements and for multi-array imaging as well as for field monitoring (RAMiTS technology) have been constructed. Plasmonics and SERS nanoprobes are useful for biological sensing.

Optical mRNA Biosensors

mRNA quantification is important in molecular diagnostics. Traditional spectrophotometric method cannot distinguish DNA, rRNA, and tRNA species from mRNA. Scheme of an optical mRNA biosensor for examination of pathological samples is shown in Fig. 4.2.

Surface-Enhanced Micro-optical Fluidic Systems

The aim of the surface-enhanced micro-optical fluidic systems (SEMOFS) European project is to develop a new concept for biosensors: a polymer-based card

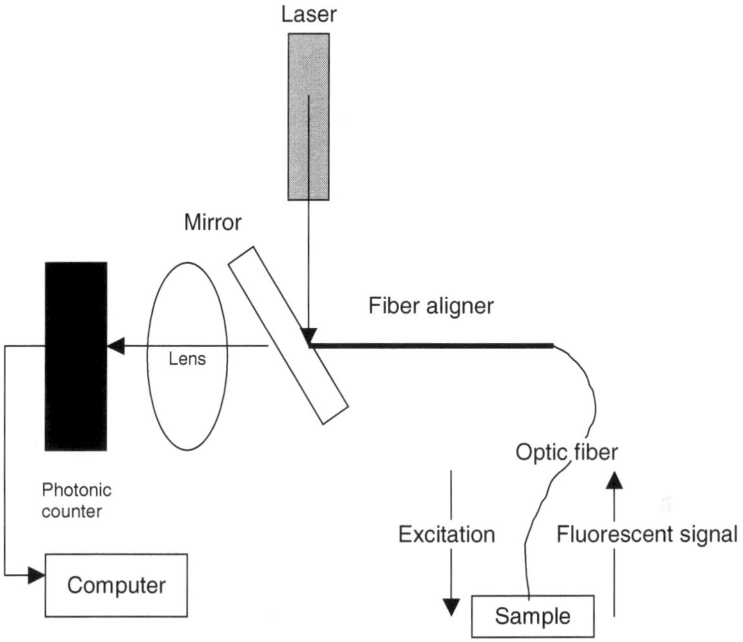

Fig. 4.2 Scheme of an optical mRNA biosensor. Sequence-specific molecular beacons are used as molecular switches. This biosensor detects single molecules in fluids and can be used to search for molecular biomarkers to predict the prognosis of disease

type integrated "plasmon-enhanced SPR" sensor. The card will combine biologically active surfaces with integrated optics (light source, detection) and biocompatible multichannel microfluidics. The project aims to achieve a significant breakthrough, since all functions will be totally integrated on a single polymer-based chip. The final product shall be manufactured with large-scale, mass production techniques. The card will therefore be extremely low cost and disposable while providing increased sensitivity and diagnosis possibilities. The project will focus on:

- Increasing detection sensitivity and access to new information of the biological sample
- Microfluidics on polymer substrate enabling multichanneling (further enhancing sensitivity by parallel analysis) and integrated fluid actuators
- Integrated optical detection concept based on organic light-emitting display (OLED)/waveguide/miniaturized spectrometer enabling card type integrated solution and multichanneling
- Hybrid micromachining to ensure compatibility of the mastering and replication protocols with constraints of industrial scale manufacturing
- Validation of expected applications and evaluation in clinical cancer diagnosis

Nanoparticle-Enhanced Sensitivity of Fluorescence-Based Biosensors

Sensitivity required for high-performance bioassays can be achieved using fluorescence-based techniques for biosensors. There is still a need for enhancement strategies, which can reduce limit of detection and increase sensitivity for the detection of low analyte concentrations in small sample volumes. Possible solutions include the use of SPR effect associated with metal nanostructures, each of which contains a high concentration of dye molecules (McDonagh et al. 2009). The degree of enhancement achieved is dependent on the nanoparticle, dye label, and nanoparticle deposition technique. For optimum assay enhancement, the antibody label must be located outside the quenching range and within the optimum distance from the metal nanoparticle. Nanoparticles with high brightness, low toxicity, biocompatibility, and ease of biomolecule conjugation are selected. For enhancement of bioassays, the nanoparticle is conjugated to the antibody, replacing the single dye label.

Nanowire Biosensors

Since their surface properties are easily modified, nanowires can be decorated with virtually any potential chemical or biological molecular recognition unit, making the wires themselves independent of the analyte. The nanomaterials transduce the chemical binding event on their surface into a change in conductance of the nanowire in an extremely sensitive, real-time, and quantitative fashion. Boron-doped silicon nanowires (SiNWs) have been used to create highly sensitive, real-time electrically based sensors for biological and chemical species. Biotin-modified SiNWs were used to detect streptavidin down to at least a picomolar concentration range. The small size and capability of these semiconductor nanowires for sensitive, label-free, real-time detection of a wide range of chemical and biological species could be exploited in array-based screening and in vivo diagnostics.

A novel approach to synthesizing nanowires (NWs) allows their direct integration with microelectronic systems for the first time, as well as their ability to act as highly sensitive biomolecule detectors that could revolutionize biological diagnostic applications. An interdisciplinary team of engineers in Yale University's Institute for Nanoscience and Quantum Engineering has overcome hurdles in NW synthesis by using a tried-and-true process of wet-etch lithography on commercially available silicon-on-insulator wafers. These NWs are structurally stable and demonstrate an unprecedented sensitivity as sensors for detection of antibodies and other biologically important molecules. According to researchers, not only can the NWs detect extremely minute concentrations (as few as 1,000 individual molecules in a cubic millimeter), they can do it without the hazard or inconvenience of any added fluorescent or radioactive detection probes. The study demonstrated ability of the NWs to monitor antibody binding and to sense real-time live cellular immune response using T-lymphocyte activation as a model. Within approximately 10 s, the NW could register T-cell activation as the release acid to the device. The basis for the sensors is the detection of hydrogen ions or acidity, within the physiological range of reactions in the body. Traditional assays for detection of immune system cells such as T cells or for antibodies usually take hours to complete.

When biological molecules bind to their receptors on the nanowire, they usually alter the current moving through the sensor and signal the presence of substance of interest. This direct detection dispenses with the time-consuming labeling chemistry and speeds up the detection process considerably. Nanowire biosensors are used for the detection of proteins, viruses, or DNA in a highly sensitive manner. They can be devised to test for a complex of proteins associated with a particular cancer and used for diagnosis as well as monitoring the progress of treatment.

Nanowire Biosensors for Detection of Single Viruses

Rapid, selective, and sensitive detection of viruses is crucial for implementing an effective response to viral infection, such as through medication or quarantine. Established methods for viral analysis include plaque assays, immunological assays, transmission electron microscopy, and PCR-based testing of viral nucleic acids. These methods have not achieved rapid detection at a single virus level and often require a relatively high level of sample manipulation that is inconvenient for infectious materials.

Direct, real-time electrical detection of single virus particles with high selectivity has been reported by using nanowire field-effect transistors (Patolsky et al. 2004). Measurements made with nanowire arrays modified with antibodies for influenza A showed discrete conductance changes characteristic of binding and unbinding in the presence of influenza A but not paramyxovirus or adenovirus. Simultaneous electrical and optical measurements using fluorescently labeled influenza A were used to demonstrate conclusively that the conductance changes correspond to binding/unbinding of single viruses at the surface of nanowire devices. pH-dependent studies further show that the detection mechanism is caused by a field effect and that the nanowire devices can be used to determine rapidly isoelectric points and variations in receptor-virus binding kinetics for different conditions. Larger arrays of reproducible nanowire devices might simultaneously screen for the presence of 100 or more different viruses. Finally, studies of nanowire devices modified with antibodies specific for either influenza or adenovirus show that multiple viruses can be selectively detected in parallel. The possibility of large-scale integration of these nanowire devices suggests potential for simultaneous detection of a large number of distinct viral threats at the single virus level.

Nanowires for Detection of Genetic Disorders

The surfaces of the silicon nanowire devices have been modified with peptide nucleic acid (PNA) receptors designed to recognize wild type versus the F508 mutation site in the cystic fibrosis transmembrane receptor gene (Hahm and Lieber 2004). Conductance measurements made while sequentially introducing wild type or mutant DNA samples exhibit a time-dependent conductance increase consistent with the PNA-DNA hybridization and enabled identification of fully complementary versus mismatched DNA samples. Concentration-dependent measurements show that detection can be carried out to at least the tens of femtomolar range. It provides more rapid results than current methods of DNA detection. This nanowire-based approach represents a step forward for direct, label-free DNA detection with extreme sensitivity and good selectivity and could provide a pathway to integrated, high-throughput, multiplexed DNA detection for genetic screening.

Nanowires Biosensor for Detecting Biowarfare\ Agents

A multi-striped biosensing nanowires system can be used for detecting biowarfare agents in the field (Tok et al. 2006). It is constructed from submicrometer layers of different metals including gold, silver, and nickel that act as "barcodes" for detecting a variety of pathogens ranging from anthrax, smallpox, and ricin to botulinum. Antibodies of specific pathogens are attached to the nanowires producing a small, reliable, sensitive detection system. The system could also be used during an outbreak of an infectious disease.

Concluding Remarks and Future Prospects of Nanowire Biosensors

A review has shown that nanowire biosensors modified with specific surface receptors represent a powerful nanotechnology-enabled diagnostic/detection platform for medicine and the life sciences (Patolsky et al. 2006). Key features of these devices include direct, label-free, and real-time electrical signal transduction, ultrahigh sensitivity, exquisite selectivity, and potential for integration of addressable arrays on a massive scale, which sets them apart from other sensor technologies that are currently available. Nanowire biosensors have unique capabilities for multiplexed real-time detection of proteins, single viruses, DNA, enzymatic processes, and small organic molecule-binding to proteins. Apart from their value as research tools, they have a significant impact on disease diagnosis, genetic screening, and drug discovery. They will facilitate the development of personalized medicine. Because these nanowire sensors transduce chemical/biological-binding events into electronic/digital signals, they have the potential for a highly sophisticated interface between nanoelectronic and biological information processing systems in the future.

Future Issues in the Development of Nanobiosensors

New biosensors and biosensor arrays are being developed using new materials, nanomaterials, and microfabricated materials including new methods of patterning. Biosensor components will use nanofabrication technologies. Use of nanotubes, buckminsterfullerenes (buckyballs), and silica and its derivatives can produce nanosized devices. Some of the challenges will be:

- Development of real-time noninvasive technologies that can be applied to detection and quantitation of biological fluids without the need for multiple calibrations using clinical samples.

- Development of biosensors utilizing new technologies that offer improved sensitivity for detection with high specificity at single-molecular level.
- Development of biosensor arrays that can successfully detect, quantify, and quickly identify individual components of mixed gases and liquid in an industrial environment.

It would be desirable to develop multiple integrated biosensor systems that utilize doped oxides, polymers, enzymes, or other components to give the system the required specificity. A system with all the biosensor components, software, plumbing, reagents, and sample processing is an example of an integrated biosensor. There is also a need for reliable fluid handling systems for "dirty" fluids and for relatively small quantities of fluids (nanoliter to attoliter quantities). These should be low cost, disposable, reliable, and easy to use as part of an integrated sensor system. Sensing in picoliter to attoliter volumes might create new problems in development of micro-reactors for sensing and novel phenomenon in very small channels.

Applications of Nanodiagnostics

Applications of nanotechnologies in clinical diagnostics have been expanding. Although some of these were mentioned along with individual technologies in the preceding section, other important applications will be identified here. Applications for diagnosis in special areas such as cancer are described in chapters dealing with these therapeutic areas.

Nanotechnology for Detection of Biomarkers

A biomarker is a characteristic that can be objectively measured and evaluated as an indicator of a physiological as well as a pathological process or pharmacological response to a therapeutic intervention. Classical biomarkers are measurable alterations in blood pressure, blood lactate levels following exercise, and blood glucose in diabetes mellitus. Any specific molecular alteration of a cell on DNA, RNA, metabolite, or protein level can be referred to as a molecular biomarker. From a practical point of view, the biomarker would specifically and sensitively reflect a disease state and could be used for diagnosis as well as for disease monitoring during and following therapy (Jain 2010). Currently available molecular diagnostic technologies have been used to detect biomarkers of various diseases such as cancer, metabolic disorders, infections, and diseases of the central nervous system. Nanotechnology has further refined the detection of biomarkers. Some biomarkers also form the basis of innovative molecular diagnostic tests.

The physicochemical characteristics and high surface areas of nanoparticles make them ideal candidates for developing biomarker harvesting platforms. Given the variety of nanoparticle technologies that are available, it is feasible to tailor nanoparticle surfaces to selectively bind a subset of biomarkers and sequestering them for later study using high-sensitivity proteomic tests (Geho et al. 2006). Biomarker harvesting is an underutilized application of nanoparticle technology and is likely undergo substantial growth. Functional polymer-coated nanoparticles can be used for quick detection of biomarkers and DNA separation.

DNA Y-junctions have been used as fluorescent scaffolds for EcoRII methyltransferase-thioredoxin fusion proteins and covalent links were formed between the DNA scaffold and the methyltransferase at preselected sites on the scaffold containing 5FdC (Singer and Smith 2006). The resulting thioredoxin-targeted nanodevice was found to bind selectively to certain cell lines but not to others. The fusion protein was constructed so as to permit proteolytic cleavage of the thioredoxin peptide from the nanodevice. Proteolysis with thrombin or enterokinase effectively removed the thioredoxin peptide from the nanodevice and extinguished cell line-specific binding measured by fluorescence. Potential applications for devices of this type include the ability of the fused protein to selectively target the nanodevice to certain tumor cell lines suggesting that this approach can be used to probe cell-surface receptors as biomarkers of cancer and may serve as an adjunct to immunohistochemical methods in tumor classification.

A magnetic nanosensor technology that is up to 1,000 times more sensitive than any technology now in clinical use is accurate regardless of which bodily fluid is being analyzed and can detect biomarker proteins over a range of concentrations three times broader than any existing method (Gaster et al. 2009). The nanosensor chip also can search for up to 64 different proteins simultaneously and has been shown to be effective in early detection of tumors in mice, suggesting that it may open the door to significantly earlier detection of even the most elusive cancers in humans. The magnetic nanosensor can successfully detect cancerous tumors in mice when levels of cancer-associated proteins are still well below concentrations detectable using the current standard method, ELISA. The sensor also can be used to detect biomarkers of diseases other than cancer.

Nanotechnology for Genotyping of Single-Nucleotide Polymorphisms

Nanoparticles for Detecting SNPs

There are two types of DNA-nanoparticles aggregation assays: one of the methods relies on cross-linking of the gold nanoparticle by hybridization, and the other is a non-cross-linking system (Sato et al. 2007). The cross-linking system has been used not only to detect target DNA sequences but also to detect metal ions or small molecules, which are recognized by DNAzymes. The non-cross-linking approach shows high performance in the detection of SNPs. These methods do not need special equipment and open up a new possibility of POC diagnoses.

The primer extension (PEXT) reaction is the most widely used approach to genotyping of SNPs. Current methods for analysis of PEXT reaction products are based on electrophoresis, fluorescence resonance energy transfer, fluorescence polarization, pyrosequencing, mass spectrometry, microarrays, and spectrally encoded microspheres. A dry-reagent dipstick method has been devised that enables rapid visual detection of PEXT products without instrumentation (Litos et al. 2007). The genomic region that spans each SNP of interest is amplified by PCR. Two primer extension reactions are performed with allele-specific primers (for one or the other variant nucleotide), which contain an oligo(dA) segment at the 5′-end. Biotin-dUTP is incorporated in the extended strand. The product is applied to the strip followed by immersion in the appropriate buffer. As the DNA moves along the strip by capillary action, it hybridizes with oligo(dT)-functionalized gold nanoparticles, such that only extended products are captured by immobilized streptavidin at the test zone, generating a red line. A second red line is formed at the control zone of the strip by hybridization of the nanoparticles with immobilized oligo(dA). The dipstick test is complete within 10 min. The described PEXT-dipstick assay is rapid and highly accurate; it shows 100 % concordance with direct DNA sequencing data. It does not require specialized instrumentation or highly trained technical personnel. It is appropriate for a diagnostic laboratory where a few selected SNP markers are examined per patient with a low cost per assay.

Nanopores for Detecting SNPs

Use of nanopore technology for sequencing was described earlier in this chapter. The focus in this section is the application for detection of SNPs. A voltage threshold has been discovered for permeation through a synthetic nanopore of dsDNA bound to a restriction enzyme that depends on the sequence (Zhao et al. 2007). Molecular dynamic simulations reveal that the threshold is associated with a nanonewton force required to rupture the DNA-protein complex. A single mutation in the recognition site for the restriction enzyme, that is, an SNP, can easily be detected as a change in the threshold voltage. Consequently, by measuring the threshold voltage in a synthetic nanopore, it may be possible to discriminate between two variants of the same gene (alleles) that differ in one base.

Nanobiotechnologies for Single-Molecule Detection

Various nanobiotechnologies for single-molecule detection are shown in Table 4.2. These have been described in preceding sections.

Protease-Activated QD Probes

QDs have been programmed to glow in presence of enzyme activity and give off NIR light only when activated by specific proteases. Altered expression of particular

Table 4.2 Nanobiotechnologies for single-molecule detection

Visualization of biomolecules by near nanoscale microscopy
Atomic force microscope
Scanning probe microscope
3D single-molecular imaging by nanotechnology
Near-field scanning optical microscope
Spectrally resolved fluorescence lifetime imaging microscopy
Nanolaser spectroscopy for detection of cancer in single cells
Nanoproteomics
Study of protein expression at single-molecule level
Detection of a single molecule of protein
Erenna™ Bioassay System: digital single-molecule detection platform
Nanofluidic/nanoarray devices: detection of a single molecule of DNA
Carbon nanotube transistors for genetic screening
Nanopore technology
Portable nanocantilever system for diagnosis
Nanobiosensors
QD-FRET nanosensors for single-molecule detection

proteases is a common hallmark of cancer, atherosclerosis, and many other diseases. NIR light also passes harmlessly through skin, muscle, and cartilage, so the new probes could detect tumors and other diseases at sites deep in the body without the need for a biopsy or invasive surgery. The probe's design makes use of a technique called "quenching" that involves tethering a gold nanoparticle to the QD to inhibit luminescence. The tether, a peptide sequence measuring only a few nanometers, holds the gold close enough to prevent the QD from giving off its light. The peptide tether used is one that is cleaved by the enzyme collagenase. The luminescence of the QDs is cut by more than 70 % when they are attached to the gold particles. They remain dark until the nanostructures were exposed to collagenase after which the luminescence steadily returns. The ultimate aim of the research is to pair a series of QDs, each with a unique NIR optical signature, to an index of linker proteases. This probe would be important for understanding and monitoring the efficacy of therapeutic interventions, including the growing class of drugs that act as protease inhibitors. An important feature of the protease imaging probes described in this study is the combination of the contrast enhancement achievable through a probe that can be activated and is combined with the brightness, photostability, and tunability of QDs.

Labeling of MSCs with QDs

QDs are useful for concurrently monitoring several intercellular and intracellular interactions in live normal cells and cancer cells over periods ranging from less than a second to over several days (several divisions of cells). QDs offer an alternative to organic dyes and fluorescent proteins to label and track cells in vitro and in vivo.

These nanoparticles are resistant to chemical and metabolic degradation, demonstrating long-term photostability. The cytotoxic effects of in vitro QD labeling on MSC proliferation and differentiation and use as a cell label in an in vitro cardiomyocyte coculture model have been investigated (Muller-Borer et al. 2007). Fluorescent QDs were shown to label MSC effectively, were easy to use, and showed a high yield as well as survival rate with minimal cytotoxic effects. Dose-dependent effects, however, suggest limiting MSC QD exposure.

The peptide CGGGRGD has been immobilized on CdSe–ZnS QDs coated with carboxyl groups by cross-linking with amine groups. These conjugates are directed by the peptide to bind with selected integrins on the membrane of hMSCs. Upon overnight incubation with optimal concentration, QDs effectively labeled all the cells. Long-term labeling of bone marrow-derived hMSCs with RGD-conjugated QDs was demonstrated during self-replication and differentiation into osteogenic cell lineages. Labeling of hMSCs with QDs has been carried out during self-replication and multilineage differentiations into osteogenic, chondrogenic, and adipogenic cells (Shah et al. 2007). QD-labeled hMSCs remained viable as unlabeled hMSCs from the same subpopulation suggesting the use of bioconjugated QDs as an effective probe for long-term labeling of stem cells.

Nanotechnology for Point-of-Care Diagnostics

Point-of-care (POC) or near patient testing means that diagnosis is performed in the doctor's office or at the bedside in case of hospitalized patients or in the field for several other indications including screening of populations for genetic disorders and cancer. POC involves analytical patient testing activities provided within the healthcare system but performed outside the physical facilities of the clinical laboratories. POC does not require permanent dedicated space but includes kits and instruments, which are either hand carried or transported to the vicinity of the patient for immediate testing at that site. The patients may even conduct the tests.

An example of POC test is CD4 T-cell count as guide to treatment of HIV/AIDS. The number of circulating CD4 T cells drops significantly when patients are infected with HIV/AIDS. CD4 counts assist in the decisions on when to initiate and when to stop the treatment, which makes this test so important at POC. While such testing is routine in Western countries and used repeatedly over the course of treatment to see if interventions are effective, it is unavailable to many people in the developing world, especially in rural areas. A cheap test for CD4+ T lymphocytes in the blood is in development using biosensor nanovesicles to enhance the signal.

After the laboratory and the emergency room, the most important application of molecular diagnostics is estimated to be at the POC. Nanotechnology would be another means of integrating diagnostics with therapeutics. Nanotechnology-based diagnostics provides the means to monitor drugs administered by nanoparticle carriers. A number of devices based on nanotechnology are among those with potential applications in POC testing.

Nanotechnology-Based Biochips for POC Diagnosis

The use of metal nanoparticles as labels represents a promising approach. They exhibit a high stability in signal and new detection schemes that would allow for robustness and low-cost readout in biochips. First examples of this kind have been established and are in the market, and more are in the development pipeline (Festag et al. 2008). Nanosphere Inc.'s Verigene™ platform will be suitable for development of POC testing.

Carbon Nanotube Transistors for Genetic Screening

Carbon nanotube network field-effect transistors (NTNFETs) have been reported that function as selective detectors of DNA immobilization and hybridization (Star et al. 2006). NTNFETs with immobilized synthetic oligonucleotides have been shown to specifically recognize target DNA sequences, including H63D SNP discrimination in the HFE gene, responsible for hereditary hemochromatosis, a disease in which too much iron accumulates in body tissues. The electronic responses of NTNFETs upon single-stranded DNA immobilization and subsequent DNA hybridization events were confirmed by using fluorescence-labeled oligonucleotides and then were further explored for label-free DNA detection at picomolar to micromolar concentrations. A strong effect of DNA counterions on the electronic response was observed, suggesting a charge-based mechanism of DNA detection using NTNFET devices. Implementation of label-free electronic detection assays using NTNFETs constitutes an important step toward low-cost, low-complexity, highly sensitive, and accurate molecular diagnostics. Label-free electronic detection of DNA has several advantages over state-of-the-art optical techniques, including cost, time, and simplicity. The sensitivity of the new device is good enough to detect a single-base mutation in an amount of DNA present in 1 mL of blood. This technology can bring to market handheld, POC devices for genetic screening, as opposed to laboratory methods using labor-intense labeling and sophisticated optical equipment. This device will be commercially developed by Nanomix Inc. as Sensation™ technology.

POC Monitoring of Vital Signs with Nanobiosensors

Researchers at the University of Arkansas (Fayetteville, AR) have worked with pentacene, a hydrocarbon molecule, and carbon nanotubes (CNTs) to develop the two types of nanobiosensors for vital signs: a temperature sensor and a strain sensor for respiration. The two similar but slightly different biosensors are integrated into "smart" fabrics – garments with wireless technology and will be able to monitor a patient's respiration rate and body temperature in real time. The addition of CNTs with pentacene increases biosensor sensitivity. As an organic semiconductor, pentacene is efficient and easy to control. Both biosensors were fabricated directly on flexible polymeric substrates. The strain sensor, which would monitor respiration rate, consisted of a Wheatstone bridge, an instrument that measures unknown electrical resistance, and a thin pentacene film that acted as a sensing layer. The system would work when a physiological strain, such as breathing, creates a mechanical deformation of the sensor, which then affects the electrical current's resistance. For the temperature sensor, the researchers used a thin-film transistor that helped them to observe electrical current in linear response to temperature change. Most importantly, in low-voltage areas, the current displayed the highest sensitivity to temperature changes. This device is useful for patients whose vital signs must be continuously monitored on bedside either at home or in hospital. The sensors and wireless networks can fit on garments such as undershirts. With this technology, the smart fabric can monitor vital signs and collect and send data to an information center in real time. The information can enable immediate detection of physiological abnormalities, which will allow physicians to begin treatment or prevent illness before full-blown disease manifestation.

Shri Lakshmi Nano Technologies Ltd. is collaborating with the University of Arkansas to optimize utilization of upcoming nanotechnologies to invent, design, and manufacture advanced conductive fabric incorporating a biosensor that will allow the monitoring of body temperature, blood pressure, ECG, heart rate, and other vital health signs.

Nanodiagnostics for the Battle Field and Biodefense

One of the areas of interest at the MIT's Institute for Soldier Nanotechnologies (Cambridge, MA) concerns ultrasensitive nanoengineered chemical detectors. Researchers have taken a major step toward making an existing miniature lab-on-a-chip fully portable, so the tiny device can perform hundreds of chemical experiments in any setting including the battlefield. This will make testing soldiers to see if they have been exposed to biological or chemical weapons much faster and easier. Neither of the previous approaches, such as mechanically forcing fluid through microchannels or capillary electroosmosis, offers portability. Within the lab-on-a-chip, biological fluids such as blood are pumped through channels about 10 μm wide. Each channel has its own pumps, which direct the fluids to certain areas of the chip so they can be tested for the presence of specific molecules. In the new system, known as a 3D AC electroosmotic pump, tiny electrodes with raised steps generate opposing slip velocities at different heights, which combine to push the fluid in one direction, like a conveyor belt. Simulations predict a dramatic improvement in flow rate, by almost a factor of 20, so that fast (mm/s) flows, comparable to pressure-driven systems, can be attained with battery voltages. If exposure to chemical or biological weapons is suspected, the device can automatically and rapidly test a miniscule blood sample, rather than sending a large sample to a laboratory and waiting for the results. The chips are so small and cheap to make that they could be designed to be disposable or they could be made implantable. Another project focuses on research to develop different approaches to sensing and characterization of materials, including toxins, with identifiable chemical signatures. Each project exploits manipulation of nanoscale features of materials to achieve one or more of specificity, spatial resolution, convenience of use, reduced power demand, or multifunctionality.

An Integrated Nanobiosensor

DINAMICS (DIagnostic NAnotech and MICrotech Sensors) is a Sixth Framework European project for develop an integrated cost-effective nanobiosensor assay for detection of bioterrorism and harmful environmental agents. The project (http://www.dinamics-project.eu/) started in 2007 and concluded at the end of 2011. The prime deliverable is an exploitable lab-on-a-chip device for detection of pathogens in water using on-the-spot recognition and detection based on the nanotechnological assembly of unlabeled DNA. DINAMICS has integrated DNA hybridization sensors with microfluidics and signal conditioning/processing both on silicon and polymer substrates avoiding the use of external apparatus for fluid handling, electrical signal generation, and processing, based on DNA hybridization. Measurements based on electrical signals and detection is through UV light absorption. The aim is development of a system where each sensing site in the microarray contains a UV microfabricated sensor. After DNA hybridization, the whole array is illuminated with UV light, and the absorption of each site is measured by the sensor. The project has culminated in an integrated multi-technology product that will be high tech, low-cost, and time-efficient sensing device applicable for use in the water industry. This will ensure a reliable source of cost reduction through a drastic shortening of the sensing pipeline and without the need of transferring the samples to an analytical laboratory.

Nanodiagnostics for Integrating Diagnostics with Therapeutics

Molecular diagnostics is an important component of personalized medicine. Improvement of diagnostics by nanotechnology has a positive impact on personalized medicine. Nanotechnology has potential advantages in applications in point-of-care (POC) diagnosis: on patient's bedside, self-diagnostics for use in the home, and integration of diagnostics with therapeutics. All of these will facilitate the development of personalized medicines.

Concluding Remarks About Nanodiagnostics

It is now obvious that direct analysis of DNA and protein could dramatically improve speed, accuracy, and sensitivity over conventional molecular diagnostic methods. Since DNA, RNA, protein, and their functional subcellular scaffolds and compartments are in the nanometer scale, the potential of single-molecule analysis approach would not be fully realized without the help of nanobiotechnology. Advances in nanotechnology are providing nanofabricated devices that are small, sensitive and inexpensive enough to facilitate direct observation, manipulation and analysis of single biological molecule from single cell. This opens new opportunities and provides powerful tools in the fields such as genomics, proteomics, molecular diagnostics, and high-throughput screening. A review of articles published over the past 10 years investigating the use of QDs, gold nanoparticles, cantilevers, and other nanotechnologies concluded that nanodiagnostics promise increased sensitivity, multiplexing capabilities, and reduced cost for many diagnostic applications as well as intracellular imaging. Further work is needed to fully optimize these diagnostic nanotechnologies for clinical laboratory setting and to address the issues of potential health and environmental risks related to QDs.

Various nanodiagnostics that have been reviewed will improve the sensitivity and extend the present limits of molecular diagnostics. Numerous nanodevices and nanosystems for sequencing single molecules of DNA are feasible. It seems quite likely that there will be numerous applications of inorganic nanostructures in biology and medicine as biomarkers. Given the inherent nanoscale of receptors, pores, and other functional components of living cells, the detailed monitoring and analysis of these components will be made possible by the development of a new class of nanoscale probes. Biological tests measuring the presence or activity of selected substances become quicker, more sensitive, and more flexible when certain nanoscale particles are put to work as tags or labels. Nanoparticles are the most versatile material for developing diagnostics.

Nanomaterials can be assembled into massively parallel arrays at much higher densities than is achievable with current sensor array platforms and in a format compatible with current microfluidic systems. Currently, quantum dot technology is the most widely employed nanotechnology for diagnostic developments. Among the recently emerging technologies, cantilevers are the most promising. This technology complements and extends current DNA and protein microarray methods, because nanomechanical detection requires no labels, optical excitation, or external probes and is rapid, highly specific, sensitive, and portable. This will have applications in genomic analysis, proteomics and molecular diagnostics. Nanosensors are promising for detection of bioterrorism agents that are not detectable with current molecular diagnostic technologies, and some have already been developed.

Future Prospects of Nanodiagnostics

Within the next decade, measurement devices based on nanotechnology, which can make thousands of measurements very rapidly and very inexpensively, will become available. The most common clinical diagnostic application will be blood protein analysis. Blood in systemic circulation reflects the state of health or disease of most organs. Therefore, detection of blood molecular fingerprints will provide a sensitive assessment of health and disease. Another important area of application will be cancer diagnostics. Molecular diagnosis of cancer including genetic profiling would be widely used by the year 2017. Nanobiotechnology would play an important part, not only in cancer diagnosis but also in linking diagnosis with treatment.

In the near future, nanodiagnostics would reduce the waiting time for the test results. For example, the patients with sexually transmitted diseases could give the urine sample when they first arrive at the outpatient clinic or physician's practice; the results could then be ready by the time they go in to see the doctor. They could then be given the prescription immediately, reducing the length of time worrying for the patient and making the whole process cheaper.

Future trends in diagnostics will continue in miniaturization of biochip technology to nanorange. The trend will be to build the diagnostic devices from bottom up starting with the smallest building blocks. Whether interest and application of nanomechanical detection will hold in the long range remains to be seen. Another trend is to move away from fluorescent labeling as miniaturization reduces the signal intensity, but there have been some improvements making fluorescent viable with nanoparticles.

Molecular electronics and nanoscale chemical sensors will enable the construction microscopic sensors capable of detecting patterns of chemicals in a fluid. Information from a large number of such devices flowing passively in the bloodstream allows estimates of the properties of tiny chemical sources in a macroscopic tissue volume. Such devices should be cabled to discriminate a single-cell-sized chemical source from the background chemical concentration in vivo, providing high-resolution sensing in both time and space. With currently used methods for blood analysis, such a chemical source would be difficult to distinguish from background when diluted throughout the blood volume and withdrawn as a blood sample.

Chapter 5
Nanopharmaceuticals

Introduction

The term "nanopharmaceuticals" covers discovery, development, and delivery of drug. The post-genomic era is revolutionizing the drug discovery process. The new challenges in the identification of therapeutic targets require efficient and cost-effective tools. Label-free detection systems use proteins or ligands coupled to materials, the physical properties of which are measurably modified following specific interactions. Among the label-free systems currently available, the use of metal nanoparticles offers enhanced throughput and flexibility for real-time monitoring of biomolecular recognition at a reasonable cost. This chapter will deal with the use of nanobiotechnologies for drug discovery and development, an important part of nanobiopharmaceuticals. Some technologies will accelerate target identification, whereas others will evolve into therapeutics. Use of nanobiotechnologies for drug delivery is an important part of nanomedicine.

Nanobiotechnology for Drug Discovery

Current drug discovery process needs improvement in several areas. Although many targets are being discovered through genomics and proteomics, the efficiency of screening and validation processes needs to be increased. Through further miniaturization, nanotechnology will improve the ability to fabricate massive arrays in small spaces using microfluidics and the time efficiency. This would enable direct reading of the signals from microfluidic circuits in a manner similar to a microelectronics circuit where one does not require massive instrumentation. This would increase the ability to do high-throughput drug screening. QDs and other nanoparticles (gold colloids, magnetic nanoparticles, nanobarcodes, nanobodies, dendrimers, fullerenes, and nanoshells) have received a considerable attention because of their unique properties that are useful for drug discovery (Jain 2005). Application of nanobiotechnologies to various stages of drug discovery is shown schematically in Fig. 5.1, and basic nanotechnologies applicable to drug discovery are listed in Table 5.1.

K.K. Jain, *The Handbook of Nanomedicine*, DOI 10.1007/978-1-61779-983-9_5,

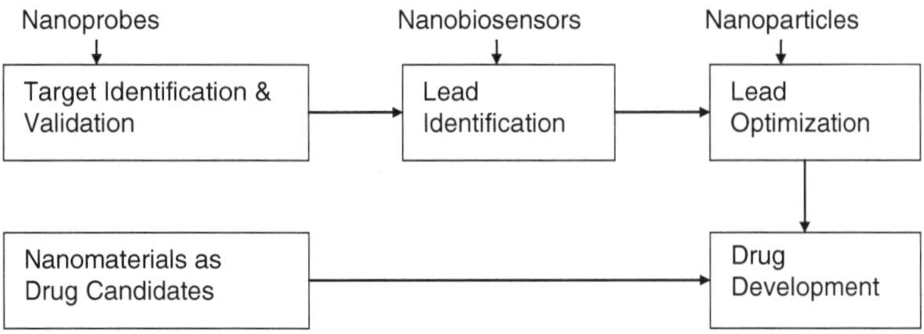

Fig. 5.1 Application of nanobiotechnology at various stages of drug discovery (© Jain PharmaBiotech)

Table 5.1 Basic nanobiotechnologies relevant to drug discovery

Nanoparticles
Gold nanoparticles
Lipoparticles
Magnetic nanoparticles
Micelles
Polymer nanoparticles
Quantum dots
Nanofibers
Nanowires
Carbon nanofibers
Nanoconduits
Nanotubes
Nanopipettes
Nanoneedles
Nanochannels
Nanopores
Nanofluidics
Nanobiotechnology applications in proteomics relevant to drug discovery
Nanoflow liquid chromatography
High-field asymmetric waveform ion mobility mass spectrometry
Use of nanotube electronic biosensor in proteomics
Fluorescence planar waveguide technology
Miscellaneous nanobiotechnologies
Visualization and manipulation at biological structures at nanoscale
Surface plasmon resonance (SPR)
Drug discovery through study of endocytosis on nanoscale
Nanosubstances as drug candidates
Dendrimers
Fullerenes
Nanobodies
Nanodevices
Nanobiosensors
Nanowire devices
Nanoarrays and nanobiochips
Cantilevers
Atomic force microscopy

Nanofluidic Devices for Drug Discovery

Development of nanofluidic devices with dimensions in the range of 1–100 nm provides opportunities for probing single molecules as fluids can no longer be considered as continua but rather as ensembles of individual molecules. Diffusion becomes an efficient mass transport mechanism on nanoscale, and these characteristics can be exploited to develop new analytical platforms for drug discovery and development.

As described in the preceding chapters, a nanopore can act as a single-molecule sensor to explore discrete molecular phenomena while operating at extremely high analytical throughput. The majority of nanopore-based studies involve the use of a protein channel that spontaneously inserts itself into a lipid membrane. A limitation of it is that it is not possible to control the pore diameter or to use it over a wide range of pH, salt concentration, temperature, and mechanical stress. An alternative to protein nanopores is the use of solid-state nanopores, which can be tuned in size with nanometer precision and display improved mechanical, chemical, and electrical stability. A novel approach for the optical detection of DNA translocation events through solid-state nanopores shows the potential for ultrahigh-throughput and parallel analysis at the single-molecule level (Chansin et al. 2007). In essence, each individual subwavelength pore acts as a waveguide for fluorescence excitation with a metallic layer on the freestanding membrane acting as an optical barrier between the illumination region and the analyte reservoir. This configuration allows for high-contrast imaging of single-molecule translocation events through multiple pores and with minimal background or noise (Hong et al. 2009).

Gold Nanoparticles for Drug Discovery

Tracking Drug Molecules in Cells

Gold nanoparticles have been used to demonstrate multiphoton-absorption-induced luminescence (MAIL), in which specific tissues or cells are fluorescently labeled using special stains that enable them to be studied. Gold nanoparticles can emit light so strongly that it is readily possible to observe a single nanoparticle at laser intensities lower than those commonly used for MAIL – sub-100-fs pulses of 790-nm light (Farrer et al. 2005). Moreover, gold nanoparticles do not blink or burn out, even after hours of observation. These findings suggest that metal nanoparticles are a viable alternative to fluorophores or semiconductor nanoparticles for biological labeling and imaging. Other advantages of the technique are that the gold nanoparticles can be prepared easily, have very low toxicity, and can readily be attached to molecules of biological interest. In addition, the laser light used to visualize the particles is a wavelength that causes only minimal damage to most biological tissues. This technology could enable tracking of a single molecule of a drug in a cell or other biological samples.

SPR with Colloidal Gold Particles

Conventional SPR is applied in specialized biosensing instruments. These instruments use expensive sensor chips of limited reuse capacity and require complex chemistry for ligand or protein immobilization. SPR has also been successfully applied with colloidal gold particles in buffered solution, which offers many advantages over conventional SPR. The support is cheap, easily synthesized, and can be coated with various proteins or protein–ligand complexes by charge adsorption. With colloidal gold, the SPR phenomenon can be monitored in any UV–vis spectrophotometer. For high-throughput applications, the technology has been adapted in an automated clinical chemistry analyzer. Among the label-free systems currently available, the use of metal nanocolloids offers enhanced throughput and flexibility for real-time biomolecular recognition monitoring at a reasonable cost.

Use of QDs for Drug Discovery

The use of QDs for drug discovery has been explored extensively. Both advantages and drawbacks have been investigated.

Advantages of the Use of QDs for Drug Discovery

- Enhanced optical properties as compared with organic dyes. QDs offer great imaging results that could not be achieved by organic dyes as they have narrowband emission together with large UV absorption spectra, which enables multiplexed imaging under a single light source.
- Multiple leads can be tested on cell culture simultaneously. Similarly, the absorption of several drug molecules can be studied simultaneously for a longer period of time.
- Using the surface functionalization properties of QDs, targeting capabilities can be added as well.
- Due to the inorganic nature of QDs, their interaction with their immediate environment at in vivo states can be minimal compared with their organic counterparts.

QDs carrying a surface-immobilized antagonist remain with nanomolar affinity on the cell surface, and particles carrying an agonist are internalized upon receptor binding. The receptor functions like a logic "and-gate" that grants cell access only to those particles that carry a receptor ligand "and" where the ligand is an agonist (Hild et al. 2010). Agonist- and antagonist-modified nanoparticles bind to several receptor molecules at a time. This multiligand binding leads to five orders of magnitude increased-receptor affinities, compared with free ligand, in displacement studies. More than 800 G protein-coupled receptors in humans provide an opportunity that targeting of a plethora of cells is possible for drug discovery and that switching from cell recognition to cell uptake is simply a matter of nanoparticle surface modification with the appropriate choice of ligand type.

Drawbacks of the Use of QDs for Drug Discovery

QDs have not been totally perfected, and some of the drawbacks are:

- Size variation during the synthesis of single color QDs is 2–4 %, which could create false results for applications such as capillary electrophoresis or gel electrophoresis. Therefore, QD synthesis techniques need to have improved quality control with respect to size distribution before they can be seriously utilized in drug discovery research.
- For ADME purposes, blue QDs (diameter of 3.7 nm) are the smallest class of the QD family, but they are considerably larger than organic dyes. Hence, the use of QDs for this purpose might not be desirable in special cases.
- Similarly, the number of functional groups attached to an organic dye is usually one, or it can be controlled very precisely. However, in the case of QDs, the functional groups usually decorate the entire surface and thus cause multiple attachments of target molecules.
- The transport of a large volume (due to multiple attachments of drug molecules to a single QD) across the membrane will be more difficult than a single molecule itself.
- To satisfy all the available surface groups, larger numbers of target molecules are needed; this could affect the cost of the experiment. Although several methods have been reported to reduce the number of surface groups around a single dot, each of these methods adds to the final size of the QDs, which might not be desired in many cases, especially in studies related to kinetics and transport of drug molecules.
- The "blinking" characteristics of QDs when they are excited with high-intensity light could be a limiting factor for fast scan systems such as flow cytometry.
- Under combined aqueous-UV excitation conditions, QDs demonstrate oxidation and release of Cd ions into the environment. This is a definite concern for in vivo applications. As an alternative, capping the surface of a core dot with a large bandgap semiconductor or proteins can eliminate or reduce the toxicity. But each additional step on the QDs will add to their final size and could even affect their final size distribution during these additional process steps.

QDs for Imaging Drug Receptors in the Brain

Cellular receptors are a critical target studied by scientists who develop new drug candidates for diseases including neurological disorders such as epilepsy and depression. More detailed understanding of the behavior of these receptors can open up new treatment options. Older imaging tools such as fluorescent dyes or polymer spheres are either too unstable or too big to effectively perform single-molecule tracking. Single-molecule properties in living cells can be tracked by using QD conjugates and produce photo resolutions up to eight times more detailed than the older imaging tools. QD conjugates are also an order of magnitude brighter than fluorescent dyes and can be observed for as long as 40 min compared to about 5 s for the dyes. Individual receptors of glycine (GlyRs), the main inhibitory neurotransmitter in the human CNS, and their dynamics in the neuronal membrane of living cells can be studied for periods ranging from milliseconds to minutes using QDs. Entry of GlyRs into the synapse by diffusion has been observed and confirmed by electron microscopy imaging of QD-tagged receptors. Length of observation time is critical for studying cellular processes, which change rapidly over a span of several minutes.

G-protein-coupled receptors (GPCRs) are the largest protein superfamily in the human genome; they comprise 30 % of current drug targets and regulate diverse cellular signaling responses. Role of endosomal trafficking in GPCR signaling regulation is significant, but this process remains difficult to study due to the inability to distinguish among many individual receptors because of simultaneously trafficking within multiple endosomal pathways. Accurate measurement of the internalization and endosomal trafficking of single groups of serotonin (5-hydroxytryptamine, 5-HT) receptors was shown by using single QD probes and quantitative colocalization (Fichter et al. 2010). Presence of a QD tag does not interfere with 5-HT receptor internalization or endosomal recycling. Direct measurements show simultaneous trafficking of the 5-HT1A receptor in two distinct endosomal recycling pathways. Single-molecule imaging of endosomal trafficking will significantly impact the understanding of cellular signaling and provide powerful tools to elucidate the actions of GPCR-targeted therapeutics.

Lipoparticles for Drug Discovery

Lipoparticle technology (Integral Molecular Inc) enables integral membrane proteins to be solubilized while retaining their intact structural conformation. Retaining the native structural conformation of membrane-bound receptors is essential during assay development for optimal lead selection and optimization. Lipoparticles can be paired with a multitude of detection systems, permitting the optimal detection system to be used depending on the target protein, the goal of the assay, and the preference of customers. Biosensors are one class of detection system currently being used with lipoparticles.

Biosensor for Drug Discovery with Lipoparticles

Interactions with integral membrane proteins have been particularly difficult to study because the receptors cannot be removed from the lipid membrane of a cell without disrupting the structure and function of the protein. Cell-based assays are the current standard for drug discovery against integral membrane proteins but are limited in important ways. Biosensors are capable of addressing many of these limitations. Biosensors are currently being used in target identification, validation, assay development, lead optimization, and ADMET (Absorption, Distribution, Metabolism, Excretion and Toxicity) studies but are best suited for soluble molecules. Integral is using lipoparticles to effectively solubilize integral membrane proteins for use in biosensors and other microfluidic devices.

A primary application of current biosensor technologies is the optimization of limited-scope drug libraries against specific targets. Paired with lipoparticle technology, biosensors can be used to address some of the most complex biological problems facing the drug discovery industry, including cell–cell recognition, cell adhesion, cell signaling, and lipid and protein–protein interactions:

- Where high-throughput screening of random libraries does not work
- Only weak ligands known, ultrasensitivity required
- When high-content information is needed (affinity, kinetics)
- Structure-based rational drug design
- ADMET: drug binding to cytochromes, serum proteins, lipid solubility
- Peptide-based ligand design where no ligand available

Lipoparticles provide a means for solubilizing integral membrane proteins that would lose their structure if extracted away from the lipid membrane. The methods for using lipoparticles range from traditional fluorescent detection technologies to emerging biosensor technologies. The optimal detection system can be used depending on the target protein, the goal of the assay, and the preference of customers.

Lipoparticles will be used for identification and optimization of chemical compounds and for antibody development. Lipoparticles will also be used to purify and concentrate structurally intact receptors from naturally occurring cell lines. This technology offers a "better-and-different" discovery platform for complex and difficult targets but can also be adapted to "faster-and-cheaper" detection systems.

Magnetic Nanoparticles Assays

Several assays are used for screening drug targets. Magnetic nanoparticles are used in many biochemical assays as labels for concentration, manipulation and, more recently, detection. Typically, one attaches the magnetic particles to the biochemical species of interest (target) using a chemically specific binding interaction. Once bound, the labels enable the manipulation of the target species through the application of magnetic forces. Spintronic sensors, specifically giant magnetoresistive and spin-dependent tunneling sensors, have been developed to detect and quantify labels in two main formats: flowing in a microfluidic channel and immobilized labels on a chip surface.

Analysis of Small Molecule–Protein Interactions by Nanowire Biosensors

Development of miniaturized devices that enable rapid and direct analysis of the specific binding of small molecules to proteins could be of substantial importance to the discovery of and screening for new drug molecules. A highly sensitive and label-free direct electrical detection of small-molecule inhibitors of ATP binding to

Abl has been reported by using silicon nanowire field-effect transistor devices (Wang et al. 2005a). Abl, which is a protein tyrosine kinase whose constitutive activity is responsible for chronic myelogenous leukemia, was covalently linked to the surfaces of silicon nanowires within microfluidic channels to create active electrical devices. Concentration-dependent binding of ATP and concentration-dependent inhibition of ATP binding by the competitive small-molecule antagonist Gleevec were assessed by monitoring the nanowire conductance. In addition, concentration-dependent inhibition of ATP binding was examined for four additional small molecules, including reported and previously unreported inhibitors. These studies demonstrate that the silicon nanowire devices can readily and rapidly distinguish the affinities of distinct small-molecule inhibitors and, thus, could serve as a technology platform for drug discovery.

Cells Targeting by Nanoparticles with Attached Small Molecules

Multivalent attachment of small molecules to nanoparticles can increase specific binding affinity and reveal new biological properties of such nanomaterial. Multivalent drug design has yielded antiviral and anti-inflammatory agents several orders of magnitude more potent than monovalent agents. Parallel synthesis of a library has been described, which is comprised of nanoparticles decorated with different synthetic small molecules (Weissleder et al. 2005). Screening of this library against different cell lines led to discovery of a series of nanoparticles with high specificity for endothelial cells, activated human macrophages, and pancreatic cancer cells. This multivalent approach could facilitate development of functional nanomaterials for applications such as differentiating cell lines, detecting distinct cellular states, and targeting specific cell types. It has potential applications in high-throughput drug discovery, diagnostics, and human therapeutics.

Role of AFM for Study of Biomolecular Interactions for Drug Discovery

An approach called TREC (topography and recognition imaging) uses any of a number of different ligands such as antibodies, small organic molecules, and nucleotides bound to a carefully designed AFM tip sensor which can, in a series of unbinding experiments, estimate affinity and structural data (Ebner et al. 2005). If a ligand is attached to the end of an AFM probe, one can simulate various physiological conditions and look at the strength of the interaction between the ligand and receptor under a wide range of circumstances. By functionalizing the tip, one can use it to probe biological systems and identify particular chemical entities on the surface of a biological sample. This opens the door to more effective use of AFM in drug discovery.

AFM has been used to study the molecular-scale processes underlying the formation of the insoluble plaques associated with Alzheimer's disease (AD). As one of a class of neurological diseases caused by changes in a protein's physical state, called "conformational" diseases, it is particularly well suited for study with AFM. Extensive data suggest that the conversion of the Aβ peptide from soluble to insoluble forms is a key factor in the pathogenesis of Alzheimer's disease (AD). In recent years, AFM has provided useful insights into the physicochemical processes involving Aβ morphology. AFM was the key in identifying the nanostructures which are now recognized as different stages of Aβ aggregation in AD and has revealed other forms of aggregation, which are observable at earlier stages and evolve to associate into mature fibrils. AFM can now be used to explore factors that either inhibit or promote fibrillogenesis, e.g., AFM can be used to compare monoclonal antibodies being studied as potential treatments for AD to select the one that does a better job of inhibiting the formation of these protofibrils. AFM not only can be reliably used to study the effect of different molecules on Aβ aggregation but it can also provide additional information such as the role of epitope specificity of antibodies as potential inhibitors of fibril formation.

Nanoscale Devices for Drug Discovery

Miniature devices are being used to study synthetic cell membranes in an effort to speed the discovery of new drugs for a variety of diseases, including cancer. Examples of these are "laboratories-on-a-chip" and Lab-on-Bead.

Laboratories-on-a-Chip

Microfluidic systems and nanoporous materials enable construction of miniature "laboratories-on-a-chip" that might contain up to a million test chambers, each capable of screening an individual drug. Such chips can be used to screen candidate compounds to find drugs to overcome anticancer drug resistance by deactivating the pumps in cell membranes that remove chemotherapy drugs from tumor cells, making the treatment less effective. The chips could dramatically increase the number of experiments that are possible with a small amount of protein.

Lab-on-Bead

A nanotechnology-based method for selecting peptide nucleic acid (PNA)-encoded molecules with specific functional properties from combinatorially generated libraries consists of three essential stages: (1) creation of a Lab-on-Bead library, a one-bead, one-sequence library that, in turn, displays a library of candidate molecules; (2) fluorescence microscopy-aided identification of single target-bound beads and the extraction – wet or dry – of these beads and their attached candidate molecules by a micropipette manipulator; and (3) identification of the target-binding candidate molecules via amplification and sequencing (Gassman et al. 2010). This novel integration of techniques harnesses the sensitivity of DNA detection methods and the multiplexed and miniaturized nature of molecule screening to efficiently select and identify target-binding molecules from large nucleic acid-encoded chemical libraries and has the potential to accelerate assays currently used for the discovery of new drug candidates by screening millions of chemicals simultaneously using nanosized plastic beads. One batch of nanoscopic beads can replace the work of thousands of conventional, repetitive laboratory tests. This process could be up to 10,000 times faster than current methods. By working at nanoscale, it will be possible to screen more than a billion possible drug candidates per day as compared to the current limit of hundreds of thousands per day.

Nanotechnology for Drug Design at Cellular Level

To create drugs capable of targeting some of the most devastating human diseases, one must first decode exactly how a cell or a group of cells communicates with other cells and reacts to a broad spectrum of complex biomolecules surrounding it. But even the most sophisticated tools currently used for studying cell communications suffer from significant deficiencies and typically can only detect a narrowly selected group of small molecules or, for a more sophisticated analysis, the cells must be destroyed for sample preparation. A nanoscale probe, the scanning mass spectrometry (SMS) probe, can capture both the biochemical makeup and topography of complex biological objects. SMS exploits an approach to electrospray ionization that enables continuous sampling from a highly localized picoliter volume in a liquid environment, softly ionizes molecules in the sample to render them amenable for mass spectrometric analysis, and sends the ions to the mass spectrometer (Kottke et al. 2010). The SMS probe can help map all those complex and intricate cellular communication pathways by probing cell activities in the natural cellular environment, which might lead to better disease diagnosis and drug design on the cellular level.

Nanobiotechnology-Based Drug Development

Dendrimers as Drugs

Dendrimers are a novel class of three-dimensional nanoscale, core–shell structures that can be precisely synthesized for a wide range of applications. Specialized chemistry techniques allow for precise control over the physical and chemical properties of the dendrimers. They are most useful in drug delivery but can also be used for the development of new pharmaceuticals with novel activities. Polyvalent dendrimers interact simultaneously with multiple drug targets. They can be developed into novel targeted cancer therapeutics. Polymer–protein and polymer–drug conjugates can be developed as anticancer drugs. These have the following advantages:

- Tailor-made surface chemistry
- Nonimmunogenic
- Inherent body distribution enabling appropriate tissue targeting
- Possibly biodegradable

Dendrimer conjugation with low molecular weight drugs has been of increasing interest recently for improving pharmacokinetics, targeting drugs to specific sites, and facilitating cellular uptake. Opportunities for increasing the performance of relatively large therapeutic proteins such as streptokinase (SK) using dendrimers have been explored in one study (Wang et al. 2007a). Using the active ester method, a series of streptokinase-poly(amido amine) (PAMAM) G3.5 conjugates were synthesized with varying amounts of dendrimer-to-protein molar ratios. All of the SK conjugates displayed significantly improved stability in phosphate buffer solution, compared to free SK. The high coupling reaction efficiencies and the resulting high enzymatic activity retention achieved in this study could enable a desirable way for modifying many bioactive macromolecules with dendrimers.

Glycodendrimers are carbohydrate-functionalized dendrimers for use in therapeutics, antigen presentation, and as biologically active compounds. GlycoSyn, a joint venture between Starpharma Holdings and Industrial Research Ltd, will provide manufacturing and specialized expertise in carbohydrate design, synthesis, and analysis. One of the first projects in the pipeline involves research undertaking cGMP manufacture of intermediates used in the production of Starpharma's vaginal microbicide – VivaGel, a polyvalent dendrimer-based pharmaceutical being developed to prevent the spread of HIV/AIDS and potentially other sexually transmitted infections including genital herpes.

Fullerenes as Drug Candidates

A key attribute of the fullerene molecules is their numerous points of attachment, allowing for precise grafting of active chemical groups in three-dimensional orientations. This attribute, the hallmark of rational drug design, allows for positional control in matching fullerene compounds to biological targets. In concert with other attributes, namely, the size of the fullerene molecules, their redox potential, and its relative inertness in biological systems, it is possible to tailor requisite pharmacokinetic characteristics to fullerene-based compounds and optimize their therapeutic effect.

Fullerene antioxidants bind and inactivate multiple circulating intracellular free radicals, giving them unusual power to stop free radical injury and to halt the progression of diseases caused by excess free radical production. Fullerenes provide effective defense against all of the principal damaging forms of reactive oxygen species. C-60 fullerene has 30 conjugated carbon–carbon double bonds, all of which can react with a radical species. In addition, the capture of radicals by fullerenes is too fast to measure and is referred to as "diffusion controlled," meaning the fullerene forms a bond with a radical every time it encounters one. Numerous studies demonstrate that fullerene antioxidants work significantly better as therapeutic antioxidants than other natural and synthetic antioxidants, at least for CNS degenerative diseases. In oxidative injury or disease, fullerene antioxidants can enter cells and modulate free radical levels, thereby substantially reducing or preventing permanent cell injury and cell death. Mechanisms of action of fullerene are as follows:

- Fullerenes can capture multiple electrons derived from oxygen free radicals in unoccupied orbitals.
- When an attacking radical forms a bond with fullerene creating a stable and relatively nonreactive fullerene radical.
- A tris-malonic acid derivative of the fullerene C60 molecule (C3) is capable of removing the biologically important superoxide radical.
- C3 localizes to mitochondria, suggesting that C3 functionally replaces manganese superoxide dismutase (SOD), acting as a biologically effective SOD mimetic.

Fullerenes have potential applications in the treatment of diseases where oxidative stress plays a role in the pathogenesis. These include the following:

- Degenerative diseases of the central nervous system including Parkinson's disease, Alzheimer's disease, and amyotrophic lateral sclerosis
- Multiple sclerosis
- Ischemic cardiovascular diseases
- Atherosclerosis
- Major long-term complications of diabetes
- Sun-induced skin damage and physical manifestations of aging

The first-generation antioxidant fullerenes are based on the C3 compound, produced by the precise grafting of three malonic acid groups to the C-60 fullerene surface. C3 has shown significant activity against a spectrum of neurodegenerative disorders in animal models. These animal models replicate many of the features of important human neurodegenerative diseases, including amyotrophic lateral sclerosis and Parkinson's disease.

The second-generation antioxidant fullerenes are based on DF-1, the dendrofullerene, produced by attaching a highly water-soluble conjugate to the C-60 fullerene core. In preclinical testing, C-60 has shown DF-1 to be highly soluble, nontoxic, and able to retain a high level of antioxidant activity in both cultured cells and animals.

A number of water-soluble C60 derivatives have been suggested for various medical applications. These applications include neuroprotective agents, HIV-1 protease inhibitors, bone disorder drugs, transfection vectors, X-ray contrast agents, photodynamic therapy agents, and a C60–paclitaxel chemotherapeutic.

Another possible application of fullerenes is to be found in nuclear medicine, in which they could be used as an alternative to chelating compounds that prevent the direct binding of toxic metal ions to serum components. This could increase the therapeutic potency of radiation treatments and decrease their adverse effect profile because fullerenes are resistant to biochemical degradation within the body.

Nanobodies

Nanobodies, derived from naturally occurring single-chain antibodies, are the smallest fragments of naturally occurring heavy-chain antibodies that have evolved to be fully functional in the absence of a light chain. The Nanobody technology (Ablynx) was originally developed following the discovery that camelidae (camels and llamas) possess a unique repertoire of fully functional antibodies that lack light chains. Like conventional antibodies, nanobodies show high target specificity and low inherent toxicity; however, like small-molecule drugs, they can inhibit enzymes and can access receptor clefts. Their unique structure consists of a single variable domain (VHH), a hinge region, and two constant domains (CH2 and CH3). The cloned and isolated VHH domain is a perfectly stable polypeptide harboring the full antigen-binding capacity of the original heavy chain. This newly discovered VHH domain is the basic component of Ablynx's Nanobodies. Ablynx's Nanobodies are naturally highly homologous to human antibodies. They can also be humanized to within 99 % sequence homology of human VH domains. Ablynx's Nanobody platform can quickly deliver therapeutic leads for a wide range of targets. Advantages of nanobodies are:

- They combine the advantages of conventional antibodies with important features of small-molecule drugs.
- Nanobodies can address therapeutic targets not easily recognized by conventional antibodies such as active sites of enzymes.
- Nanobodies are very stable.
- They can be administered by means other than injection.
- They can be produced cost-effectively on a large scale.
- Nanobodies have an extremely low immunogenic potential. In animal studies, the administration of nanobodies does not yield any detectable humoral or cellular immune response.

The cloning and selection of antigen-specific nanobodies obviate the need for construction and screening of large libraries and for lengthy and unpredictable in vitro affinity maturation steps. The unique and well-characterized properties enable nanobodies to excel conventional therapeutic antibodies in terms of recognizing uncommon or hidden epitopes, binding into cavities or active sites of protein targets, tailoring of half-life, drug format flexibility, low immunogenic potential, and ease of manufacture. Moreover, the favorable biophysical and pharmacological properties of nanobodies, together with the ease of formatting them into multifunctional protein therapeutics, leave them ideally placed as a new generation of antibody-based therapeutics. They have a potential as cancer therapeutic agents.

Another example of use of nanobodies as novel drugs is nanobody-conjugated human trypanolytic factor for treatment of human African trypanosomiasis (HAT). Normal human serum (NHS) contains apolipoprotein L-I (apoL-I), which lyses African trypanosomes except resistant forms such as *Trypanosoma brucei rhodesiense*, which expresses the apoL-I-neutralizing serum resistance-associated (SRA) protein, endowing this parasite with the ability to infect humans and cause HAT. A truncated apoL-I (Tr-apoL-I) has been engineered by deleting its SRA-interacting domain, which makes it lytic for *T. b. rhodesiense*. Tr-apoL-I has been conjugated with a nanobody that efficiently targets conserved cryptic epitopes of the variant surface glycoprotein of trypanosomes to generate a new type of immunotoxin with potential for trypanosomiasis therapy (Baral et al. 2006). Treatment with this engineered conjugate resulted in clear curative and alleviating effects on acute and chronic infections of mice with both NHS-resistant and NHS-sensitive trypanosomes. First-in-class' antithrombotic agent ALX-0081 (Ablynx NV) is in phase I trials (Van Bockstaele et al. 2009).

Role of Nanobiotechnology in the Future of Drug Discovery

None of the nanoparticles available is ideal for all requirements of drug discovery. The choice may depend on the needs. QDs can be used for high-throughput cell-based studies with the advantage of multiplexing (i.e., multiple leads can be tested at the same time). However, as discussed earlier, there are some limitations yet to be resolved for their use in the drug discovery studies, namely, toxicity, size variation, agglomeration, potential multiple drug attachment to a single QD, and blinking.

An increasing use of nanobiotechnology by the pharmaceutical and biotechnology industries is anticipated. Nanotechnology will be applied at all stages of drug development – from formulations for optimal delivery to diagnostic applications in clinical trials. In the near future, it may be possible to fully model an individual cell's structure and function by computers connected to nanobiotechnology systems. Such a detailed virtual representation of how a cell functions might enable scientists to develop novel drugs with unprecedented speed and precision without any experiments in living animals. In another promising area of application, non-biodegradable 3D scaffolds are being developed to hold stem cells for pharmaceutical and biological research. These tissue constructs can be used to test new drugs. Since tissues grow in three dimensions and not two, 3D growth would be more suitable for early drug screening.

Nanobiotechnology in Drug Delivery

Drug delivery is one of the important considerations in drug development and therapeutics. New technologies are applied for constructing innovative formulations and delivering them. The focus is on targeted drug delivery. This is important for delivery of biopharmaceuticals and treatment of diseases such as cancer and neurological disorders. In the pharmaceutical industry, there is potential to provide new formulations and routes of drug delivery. Among new technologies, nanobiotechnology has evoked considerable interest for application in the pharmaceutical industry. Applications of nanotechnology for drug delivery will be considered in this chapter.

Ideal Properties of Material for Drug Delivery

Properties of an ideal macromolecular drug delivery or biomedical vector are:

- Structural control over size and shape of drug or imaging-agent cargo-space
- Biocompatible, nontoxic polymer/pendant functionality
- Well-defined scaffolding and/or surface modifiable functionality for cell-specific targeting moieties
- Lack of immunogenicity or ability to evade the immune system
- Appropriate cellular adhesion, endocytosis, and intracellular trafficking to allow therapeutic delivery or imaging in the cytoplasm or nucleus
- Acceptable bioelimination or biodegradation
- Targeted delivery with binding to the target sites and accumulation in the target tissue with sparing of normal or nontarget tissues
- Controlled or triggerable drug release
- Molecular level isolation and protection of the drug against inactivation during transit to target cells
- Minimal nonspecific cellular and blood–protein binding properties
- Ease of consistent, reproducible, clinical grade synthesis

Nanobiotechnology fulfills many of these requirements for improved drug delivery. Nanoparticles as well as nanodevices are used for this purpose.

Improved Absorption of Drugs in Nanoparticulate Form

Micronization was in use prior to introduction of techniques for producing nanoparticles. Although several claims were made for increased absorption, no significant improvement was documented because the microparticle size was still above 3 μm (3,000 nm) and nanoparticle size could be as much as 30 times less.

Reduction of particle size from 5 μm to 200 nm increases the surface area of the particle by a factor of 25 with increase in solubility. As an example, reduction of iron phosphate to the nanoscale increases its absorption in the body.

Interaction of Nanoparticles with Human Blood

Nanoparticle size and plasma binding profile contribute to a particle's longevity in the bloodstream, which can have important consequences for therapeutic efficacy. Approximate doubling in nanoparticle hydrodynamic size was observed upon in vitro incubation of 30–50-nm colloidal gold nanoparticles in human plasma due to binding of plasma proteins to their surface (Dobrovolskaia et al. 2009).

Nanoscale Devices Delivery of Therapeutics

There are several requirements for developing a device small enough to efficiently leave the vasculature and enter cells to perform multiple, smart tasks. However, the major requirement involves size. Vascular pores limit egress of therapeutics to materials less than approximately 50 nm in diameter, and cells will not internalize materials much greater than 100 nm. As a result, the only currently available technology that fulfills these criteria consists of synthetic nanodevices. These are designed synthetic materials with structures less than 100 nm in size. Unlike fictional mechanical nanomachines, based on machines that have been "shrunken" to nanometer dimensions, several true nanomolecular structures have now been synthesized and applied to drug delivery, gene transfer, antimicrobial therapeutics, and immunodiagnostics.

Nanobiotechnology Solutions to the Problems of Drug Delivery

One of the major problems with drugs is solubility, which is an essential factor for drug effectiveness, independent of administration route. It is also a major challenge for pharmaceutical companies developing new pharmaceutical products since nearly half of new chemically based drugs are insoluble, or poorly soluble, in water. Many, otherwise promising, compounds never reach the market. Others reach the market but in a suboptimal formulation, possibly providing low or unpredictable bioavailability or posing an increased side effect risk. Enhanced solubility technology can be used to reformulate such drugs and increase their commercial potential. Nanobiotechnology provides the following solutions to the problems of drug delivery:

- The particle size is reduced to nanometer size range to increase the surface area, thereby increasing the rate of dissolution, e.g., Nanoedge technology (Baxter).
- Improving solubilization of the drug.

- Using noninvasive routes of administration eliminates the need for administration of drugs by injection.
- Development of novel nanoparticle formulations with improved stabilities and shelf lives.
- Development of nanoparticle formulations for improved absorption of insoluble compounds and macromolecules enables improved bioavailability and release rates, potentially reducing the amount of dose required and increasing safety through reduced side effects.
- Manufacture of nanoparticle formulations with controlled particle sizes, morphology, and surface properties would be more effective and less expensive than other technologies.
- Nanoparticle formulations that can provide sustained-release profiles up to 24 h can improve patient compliance with drug regimens.
- Direct coupling of drugs to targeting ligand restricts the coupling capacity to a few drug molecules, but coupling of drug carrier nanosystems to ligands allows import of thousands of drug molecules by means of one receptor-targeted ligand. Nanosystems offer opportunities to couple drugs with newly discovered disease-specific targets.

Nanosuspension Formulations

Nanosuspension formulations can be used to improve solubility of poorly soluble drugs. A large number of new drug candidates emerging from drug discovery programs are water insoluble, and therefore poorly bioavailable, leading to abandoned development efforts. These can now be rescued by formulating them into crystalline nanosuspensions. Techniques such as media milling and high-pressure homogenization have been used commercially for producing nanosuspensions. The unique features of nanosuspensions have enabled their use in various dosage forms, including specialized delivery systems such as mucoadhesive hydrogels. Nanosuspensions can be delivered by parenteral, peroral, ocular, and pulmonary routes. Currently, efforts are being directed to extending their applications in site-specific drug delivery. A large number of drugs are available as nanosuspensions. Advantages of nanosuspension are (Patel and Agrawal 2011):

- Higher drug loading can be achieved.
- Dose reduction is possible.
- Enhancement of physical and chemical stability of drugs.
- Suitable for hydrophilic drugs.

Baxter scientists have used Nanoedge technology to formulate the antifungal agent itraconazole as an intravenous nanosuspension. In studies on rats, formulation as a nanosuspension was shown to enhance efficacy of itraconazole relative to a solution formulation because of altered pharmacokinetics, leading to increased tolerability, permitting higher dosing and resultant tissue drug levels (Rabinow et al. 2007).

A study has compared the in vitro and in vivo antitumor efficacy as well as dose-dependent toxicity of camptothecin nanosuspension (Nano-CPT) with that of topotecan (TPT). Nano-CPT showed approximately six times in vitro cytotoxicity than TPT against cell lines MCF-7, and the same in vivo antitumor activity as TPT but with lower toxicity (Yao et al. 2012). The results indicate that Nano-CPT formulation has higher antitumor efficacy and lower toxicity than the conventional formulation of the drug.

Nanotechnology for Solubilization of Water-Insoluble Drugs

The Ubisol-Aqua™ (Zymes LLC) delivery system uses nanotechnology to enable the solubilization and reformulation of water-insoluble drugs, nutrients, and cosmetic ingredients. The capacity of Ubisol-Aqua™ to expand the usefulness of such compounds has been scientifically demonstrated with water-soluble formulations of coenzyme Q10 (HQO™) and antifungal antibiotics. Zymes LLC has successfully solubilized fish oil and omega-3 fatty acids (DHA/EPA/ALA) with an average particle size of 34 nm. Zymes offers its delivery system technology to industry partners in need of more effective ways of making their ingredients water soluble and thus more bioavailable.

Self-Assembled Nanostructures with Hydrogels for Drug Delivery

Drug delivery systems based on physical hydrogels with self-assembled nanostructures are attracting increasing attention as complements to chemically cross-linked hydrogels because of advantages of reduced toxicity, convenience of in situ gel formation, stimuli responsiveness, reversible sol-gel transition, and improved drug loading and delivery profiles. The driving forces of the self-assembly include hydrophobic interaction, hydrogen bonding, electrostatic interaction, and weak van der Waals forces. Stimuli-responsive properties of physical hydrogels include thermosensitivity and pH sensitivity. Fabrication of self-assembled nanostructures in drug delivery hydrogels, via physical interactions between polymer–polymer and polymer–drug, requires accurately controlled macro- or small molecular architecture and a comprehensive knowledge of the physicochemical properties of the therapeutics (Tang et al. 2011). Nanostructures within hydrogels, which interact with payloads, provide useful means to stabilize the drug form and control its release kinetics. Biocompatibles UK Ltd, a subsidiary of BTG plc, is developing these systems.

Nanomaterials and Nanobiotechnologies Used for Drug Delivery

Table 5.2 shows various nanomaterials and nanobiotechnologies used for drug delivery.

Table 5.2 Nanomaterials used for drug delivery

Structure	Size	Role in drug delivery
Bacteriophage NK97 (a virus that attacks bacteria)	~28 nm	Emptied of its own genetic material, NK97, which is covered by 72 interlocking protein rings, can act as a nanocontainer to carry drugs and chemicals to targeted locations
Canine parvovirus (CPV)-like particles	~26 nm	Targeted drug delivery: CPV binds to transferrin receptors, which are overexpressed in some tumors
Carbon magnetic nanoparticles	40–50 nm	For drug delivery and targeted cell destruction
Ceramics nanoparticles	~35 nm	Accumulate exclusively in the tumor tissue and allow the drug to act as sensitizer for PDT without being released
Cerasomes	60–200 nm	Cerasome is filled with C6 ceramide for use as an anticancer agent
Dendrimers	1–20 nm	Holding therapeutic substances such as DNA in their cavities
Gold nanoparticles	2–4 nm	Enable externally controlled drug release
HTCC nanoparticles	110–180 nm	Encapsulation efficiency is up to 90 %. In vitro release studies show a burst effect followed by a slow and continuous release
Micelle/nanopill	25–200 nm	Made from two polymer molecules – one water repellant and the other hydrophobic – that self-assemble into a sphere called a micelle that can deliver drugs to specific structures within the cell
Low-density lipoproteins	20–25 nm	Drugs solubilized in the lipid core or attached to the surface
Nanocochleates		Nanocochleates facilitate delivery of biologicals such as DNA and genes
Nanocrystals	<1,000 nm	NanoCrystal technology (Elan) can rescue a significant number of poorly soluble chemical compounds by increasing solubility
Nanodiamonds	550–800 nm	Biocompatibility and unique surface properties for drug delivery
Nanoemulsions	20–25 nm	Drugs in oil and/or liquid phases to improve absorption
Nanoliposomes	25–50 nm	Incorporate fullerenes to deliver drugs that are not water soluble and tend to have large molecules
Nanoparticle composites	~40 nm	Attached to guiding molecules such as MAbs for targeted drug delivery
Nanopore membrane		An implanted titanium device using silicone nanopore membrane can release encapsulated protein and peptide drugs
Nanospheres	50–500 nm	Hollow ceramic nanospheres created by ultrasound
Nanostructured organogels	~50 nm	Mixture of olive oil, liquid solvents, and adding a simple enzyme to chemically activate a sugar. Used to encapsulate drugs
Nanotubes	Single wall 1–2 nm Multiwall 20–60 nm	Resemble tiny drinking straws and that might offer advantages over spherical nanoparticles for some applications
Nanovalve	~500 nm	Externally controlled release of drug into a cell
Nanovesicles	25–100 nm	Bilayer spheres containing the drugs in lipids
Polymer nanocapsules	50–200 nm	Enclosing drugs
PEG-coated PLA nanoparticles	Variable size	PEG coating improves the stability of PLA nanoparticles in the gastrointestinal fluids and helps the transport of encapsulated protein across the intestinal and nasal mucus membranes
Quantum dots	2–10 nm	Combine imaging with therapeutics
Superparamagnetic iron oxide nanoparticles	10–100 nm	As drug carriers for intravenous injection to evade RES as well as to penetrate the very small capillaries within the body tissues and thus offer the most effective distribution in certain tissues

PEG poly(ethylene glycol), *PLA* poly(lactic acid), *HTCC* N-(2-hydroxyl) propyl-3-trimethyl ammonium chitosan chloride, *RES* reticuloendothelial system

Viruses as Nanomaterials for Drug Delivery

Specific targeting of tumor cells is an important goal for the design of nanotherapeutics for the treatment of cancer. Recently, viruses have been explored as nanocontainers for specific targeting applications, but these systems typically require modification of the virus surface using chemical or genetic means to achieve tumor-specific delivery. However, there is a subset of viruses with natural affinity for receptors on tumor cells that could be exploited for nanotechnology applications, e.g., the canine parvovirus for targeted drug delivery in cancer.

Bacteria-Mediated Delivery of Nanoparticles and Drugs into Cells

Nanoparticles and bacteria have been independently used to deliver genes and proteins into mammalian cells for monitoring or altering gene expression and protein production. Harmless strains of bacteria could be used as vehicles, harnessing bacteria's natural ability to penetrate cells and their nuclei. Researchers at Purdue University's Birck Nanotechnology Center have demonstrated the simultaneous use of nanoparticles and bacteria to deliver nucleic acid-based model drug molecules into cells in mice (Akin et al. 2007). In this approach, the gene or cargo is loaded onto the nanoparticles, ranging in size from 40 to 200 nm, which are attached to the bacteria with linker molecules. The bacteria successfully deliver the molecules, and the genes are released from the nanoparticles and expressed in cells. When the cargo-carrying bacteria attach to the recipient cell, they are engulfed by its outer membrane, forming "vesicles" or tiny spheres that are drawn into the cell's interior. Once inside the cell, the bacteria dissolve the vesicle membrane and release the cargo as shown in Fig. 5.2.

This technique may be used to deliver different types of cargo into a variety of cells and live animals for gene therapy without the need for complicated genetic manipulations. This delivery system also is more efficient than techniques using viruses as they usually incorporate only one copy of a gene cargo to virus particle. In this approach, bacteria can carry hundreds of nanoparticles, each of which can in turn carry hundreds of drug molecules, depending on the size of the nanoparticles. Released cargo can be designed to be transported to different locations in the cells to carry out disease detection and treatment simultaneously. The method might be used to take images of diseased tissues by inserting a cargo of fluorescent molecules into tumors that are ordinarily too small to be detected. It could enable insertion of relatively large structures, such as biosensors into the interiors of cells for the early detection of cancer and other diseases and to monitor the progress of disease as well as response to drug therapy. The carbon nanotubes could be delivered into diseased cells and then exposed to light, causing them to heat up and selectively kill only the diseased cells.

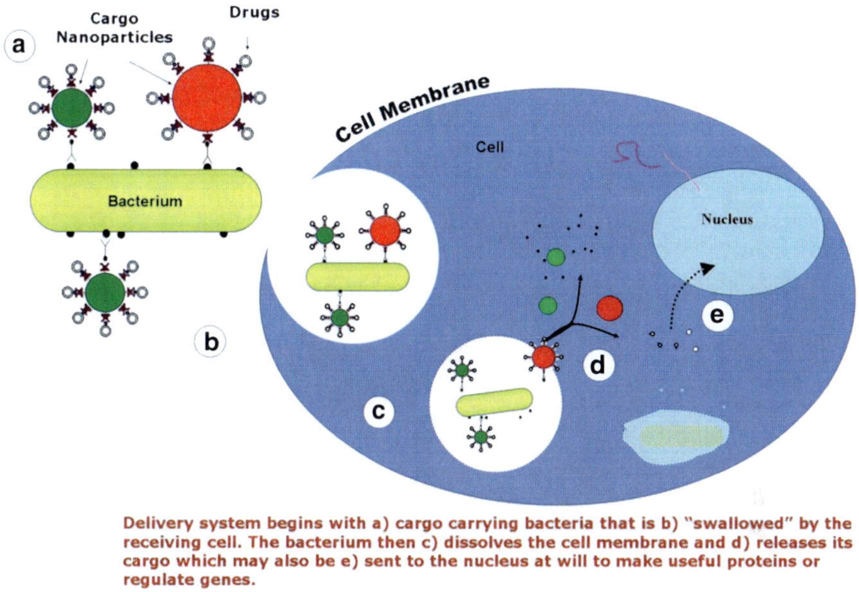

Fig. 5.2 Bacteria plus nanoparticles for drug delivery into cells (Source: Akin et al. 2007)

Cell-Penetrating Peptides

Cell-penetrating peptides (CPPs) are short basic peptide sequences that might display amphipathic properties. CPPs generally contain a small number (<20–30) of amino acids, among which are a great number of positively charged amino acids that confer cell internalization properties on those peptides. Originally derived from natural proteins, the number of designed CPPs with similar cell penetration properties has now expanded widely. These positively charged peptides internalize into all cell types, but with different efficiency. CPPs use all routes of pinocytosis to internalize, in addition to direct membrane translocation that requires interaction with lipid membrane domains. These differences in internalization efficiency according to the peptide sequence and cell type suggest that the CPPs interact with different molecular partners at the cell surface. The most popular CPPs are penetratin, Tat, and oligoarginine, which interact with carbohydrates and lipids. Cell surface composition influences cell internalization, and the interaction with molecules found in membranes reflects the internalization efficiency of the peptides (Walrant et al. 2012). For specific drug delivery, the exact molecular and chemical nature of membranes and their interactions with CPPs need to be identified.

Nanoparticle-Based Drug Delivery

Trend toward miniaturization of carrier particles had already started prior to the introduction of nanotechnology in drug delivery. As a part of introduction, microparticles and nanoparticles will be compared for their role as carriers of therapeutic substances

The suitability of nanoparticles for use in drug delivery depends on a variety of characteristics, including size and porosity. Acusphere Inc is creating porous particles that are smaller than red blood cells. Nanoparticles can be used to deliver drugs to patients through various routes of delivery. Nanoparticles are important for delivering drugs intravenously so that they can pass safely through the body's smallest blood vessels, for increasing the surface area of a drug so that it will dissolve more rapidly, and for delivering drugs via inhalation. Porosity is important for entrapping gases in nanoparticles, for controlling the release rate of the drug and for targeting drugs to specific regions.

It is difficult to create sustained-release formulations for many hydrophobic drugs because they release too slowly from the nanoparticles used to deliver the drug, diminishing the efficacy of the delivery system. Modifying water uptake into the nanoparticles can speed the release while retaining the desired sustained-release profile of these drugs. Water uptake into nanoparticles can be modified by adjusting the porosity of the nanoparticles during manufacturing and by choosing from a wide variety of materials to include in the shell. Different types of nanoparticles and nanotechnologies used for drug delivery will be mentioned briefly here, and specialized drug delivery for various disorders will be described in chapters dealing with those disorders.

Cationic Nanoparticles

Cationic nanoparticles built from drug, cationic lipid, and polyelectrolytes are excellent and active carriers of amphotericin B against *Candida albicans* (Vieira and Carmona-Ribeiro 2008). Assemblies of amphotericin B and cationic lipid, at extreme drug to lipid molar ratios, were wrapped by polyelectrolytes forming cationic nanoparticles of high colloid stability and fungicidal activity against *C. albicans*. Experimental strategy involved dynamic light scattering for particle sizing, zeta-potential analysis, determination of colloid stability, determination of AmB aggregation state by optical spectra, and determination of activity against *C. albicans* in vitro from cfu countings. The multiple assembly of antibiotic, cationic lipid and cationic polyelectrolyte consecutively nanostructured in each particle produced a strategical and effective attack against fungal infections.

Ceramic Nanoparticles

Ceramic (inorganic) particles with entrapped biomolecules have potential pharmaceutical applications in including drug delivery. Ceramic nanoparticles have several advantages such as:

- Manufacture processes are relatively similar to the well-known sol-gel process, require ambient temperature condition, and can be easily prepared with the desired size, shape, and porosity.
- Their small size (less than 50 nm) can help them to evade being trapped by the reticuloendothelial system of the body.
- There is no swelling or change in porosity with change in pH.
- These particles effectively protect doped molecules (enzymes, drugs, etc.) against denaturation induced by external pH and temperature.
- Such particles, including silica, aluminum, and titanium, are known for their compatibility with biological systems.
- Their surfaces can be easily modified for conjugation to monoclonal antibodies or ligands to target them to desired sites in vivo.

Cyclodextrin Nanoparticles for Drug Delivery

Cyclodextrins are a family of cyclic oligosaccharides with a hydrophilic outer surface and a lipophilic center. Cyclodextrin molecules are relatively large with a number of hydrogen donors and acceptors and, thus, in general they do not permeate lipophilic membranes. Cyclodextrins have mainly been used as complexing agents to increase aqueous solubility of poorly soluble drugs and to increase their bioavailability and stability.

Amphiphilic cyclodextrin nanoparticles, resulting from the esterification of primary hydroxyl groups by hydrocarbon chains varying from C6 to C14, are capable of forming spontaneously nanoparticles, which can be loaded with drugs. The drug can be released in a controlled manner following targeting delivery by the oral or the parenteral route. For injectable preparations, sterile filtration is not feasible since nanoparticle sizes are larger than the filter pore size and the yield after sterilization is very low. However, blank as well as drug-loaded cyclodextrin nanospheres and nanocapsules are capable of being sterilized by gamma irradiation with no effect on particle size, drug loading, and drug release properties.

Incorporation of cyclodextrins in poly(anhydride) nanoparticles improves their bioadhesive capability, the loading of lipophilic drugs, and has an effect on efflux membrane proteins as well as cytochrome P450. The combination between bioadhesive nanoparticles and P-gp inhibitors without pharmacological activity may be useful for promoting the oral bioavailability of drugs (Agüeros et al. 2011). CALAA01 (Calando Pharmaceuticals), a targeted, self-assembling nanoparticle system based on cyclodextrin complexed with siRNA, overcomes delivery problems of siRNAs and has been shown to be effective in phase I clinical trials.

Dendrimers for Drug Delivery

Well-characterized, commercially available dendritic polymers have been subjected to functionalization for preparing drug delivery systems of low toxicity, high loading capacity, ability to target specific cells, and transport through their membranes. This has been achieved by surface targeting ligands, which render the carriers specific to certain cells and PEG, securing water solubility, stability, and prolonged circulation. Moreover, transport agents facilitate transport through cell membranes, while fluorescent probes detect their intracellular localization. A common feature of surface groups is multivalency, which considerably enhances their binding strength with complementary cell receptors. To these properties, one should also add the property of attaining high loading of active ingredients coupled with controlled and/or triggered release (Paleos et al. 2010).

The unique properties of dendrimers such as high degree of branching and well-defined molecular weight make them ideal scaffolds for drug delivery. Advantages of dendrimers over linear polymers are:

- The large number of active functional groups on the surface of dendrimers allows them to be meticulously tailored and to act as nanoscaffolds or nanocontainers for various categories of drugs (Jain and Asthana 2007).
- Because of their well-defined molecular weight, they provide reproducible pharmacokinetic behavior compared to linear polymers containing fractions within a sample that vary greatly in molecular weight.
- The globular structure of dendrimers, as contrasted with the coil structure of most linear polymers, can modify their biological properties, enabling discovery of new effects related to macromolecular architecture.

Dendrimers are particularly useful for the delivery of anticancer drugs such as cisplatin and doxorubicin. Dendrimers are also agents for boron neutron capture therapy and photodynamic therapy for cancer. By adding stimuli-responsive properties to the dendrimers, dendritic polymers capable of controlled release can be produced (Kojima 2010). These stimuli-responsive dendrimers are potential next-generation drug carriers.

DNA-Assembled Dendrimers for Drug Delivery

A wide variety of nanoparticle drug delivery systems have been developed using DNA molecules to bind the dendrimers together. Nanometer-scaled dendrimers can be assembled in many configurations by using attached lengths of ssDNA molecules, which naturally bind to other DNA strands in a highly specific fashion. This approach enables targeting of a wide variety of molecules – drugs, contrast agents – to almost any cell. Nanoparticle complexes can be specifically targeted to cancer cells and are small enough to enter a diseased cell, either killing it from within or sending out a signal to identify it. However, construction of the particles is difficult and time-consuming.

Fullerenes for Drug Delivery

Amphiphilic Fullerene Derivatives

The amphiphilic fullerene monomer (AF-1) consists of a buckyball cage to which a Newkome-like dendrimer unit and five lipophilic C12 chains positioned octahedrally to the dendrimer unit are attached. An AF-1-based liposome termed "buckysome" has been described that is water soluble and forms stable spherical nanometer sized vesicles (Partha et al. 2007). Cryogenic electron microscopy (cryo-EM) indicates the formation of large (400 nm diameter) multilamellar, liposome-like vesicles and unilamellar vesicles in the size range of 50–150-nm diameter. In addition, complex networks of cylindrical, tube-like aggregates with varying lengths and packing densities were observed. Under controlled experimental conditions, high concentrations of spherical vesicles could be formed. In vitro results suggest that these supramolecular structures impose little to no toxicity. Ongoing studies are aimed at understanding cellular internalization of these nanoparticle aggregates. This delivery vector might provide promising features such as ease of preparation, long-term stability, and controlled release.

Fullerene Conjugates for Intracellular Delivery of Peptides

Cell walls, or membranes, form a protective covering around the cell's inner machinery and its DNA blueprints. Drugs are far more effective if they are delivered through the membrane directly into the cell, but this is difficult. A fullerene–peptide conjugate formed via the incorporation of a fullerene-substituted phenylalanine derivative, "bucky amino acid" (Baa), to a cationic peptide, acts as a passport for intracellular delivery, enabling transport of peptides that, in the absence of the fullerene amino acid, cannot enter the cell (Yang et al. 2007). Delivery of the fullerene species to either the cytoplasm or nucleus of the cell has been demonstrated. The hydrophobic nature of the fullerene assisting peptide transport is suggested by the effect of gamma-cyclodextrin in lowering the efficacy of transport. These data suggest that the incorporation of a fullerene-based amino acid provides a route for the intracellular delivery of peptides and as a consequence the creation of a new class of cell-penetrating peptides. The peptides were found effective at penetrating the defenses of both liver cancer cells and neuroblastoma cells.

Gold Nanoparticles as Drug Carriers

Gold nanoparticles (AuNPs), in addition to their applications in molecular diagnostics, can be conjugated with peptides, drugs, and other molecules for drug delivery as well as for thermal treatment of cancer. AuNPs have received considerable attention as model drug delivery platforms because of their surface characteristics that enable easy functionalization with chemicals and biological molecules and also due to their apparently low toxicity (Papasani et al. 2012). Efforts are being made to develop intelligent delivery systems by lining the walls of polymer "delivery-vehicle" particles with gold nanoparticles. By simply shining a laser on loaded delivery vehicles (i.e., particles filled with various contents, such as an enzyme or drug), the walls could be opened and the contents released. This technique has been used successfully for the release of an encapsulated enzyme on demand with a single nanosecond laser pulse. In contrast to the common approach for drug release by changes in the local environment at the site where drug delivery is needed, gold nanoparticle technology enables externally controlled drug release. In addition to drugs, these gold-coated vehicles could be used for the controlled delivery of a wide range of other substances including genes. There is no risk that the laser energy will be significantly absorbed by biological structures such as bodily organs because the absorption of the gold-coated delivery vehicles in the NIR region is intentionally engineered in the wavelength regime for which light has a maximum penetration depth in tissue.

Layered Double Hydroxide Nanoparticles

Layered double hydroxides (LDHs) were well known as catalyst and ceramic precursors, traps for anionic pollutants, catalysts, and additives for polymers. Their successful synthesis on the nanometer scale opened up a whole new application for delivery of drugs and other therapeutic/bioactive molecules (e.g., peptides, proteins, nucleic acids) to mammalian cells. LDH nanoparticles have advantages as well as disadvantages as carriers for nucleic acids and drugs, and some challenges need to be overcome before LDH nanoparticles can be used in a clinical setting (Ladewig et al. 2009). Size-dependent toxicity of LDH was examined in cultured human lung cells; 50-nm particles were determined to be more toxic than larger particles, while LDHs within the size range of 100–200 nm exhibited very low cytotoxicity in terms of cell proliferation, membrane damage, and inflammation response (Choi et al. 2008).

Nanocomposite Membranes for Magnetically Triggered Drug Delivery

Nanocomposite membranes based on thermosensitive, poly(N-isopropylacrylamide)-based nanogels and magnetite nanoparticles have been designed to achieve "on-demand" drug delivery upon the application of an oscillating magnetic field (Hoare et al. 2009). On–off release of sodium fluorescein over multiple magnetic cycles has been successfully demonstrated using prototype membrane-based devices. The total drug dose delivered was directly proportional to the duration of the "on" pulse. The membranes were noncytotoxic, were biocompatible, and retained their switchable flux properties after 45 days of subcutaneous implantation.

Nanocrystals

Nanocrystalline Silver

Silver has been valued for centuries for its medicinal properties. From ancient Greece to the American settlers, silver was used as a preservative for drinking water and other liquid storage. Decades ago, doctors would apply a thin layer of silver to large wounds to prevent infection and promote healing. Nucryst's silver nanocrystalline technology decreases the particle size, thus changing the physical and chemical properties. As the proportion of atoms on the surface increases, the result is a more powerful compound than conventional silver treatments. In vitro tests have demonstrated that active silver clusters of ions begin providing antimicrobial activity immediately and kill many organisms in 30 min, faster than other forms of silver.

Silcryst™ nanocrystals release sustained, uniform doses of silver. Silver nanocrystalline technology is capable of delivering a sustained release of active silver to the dressings over a longer period of time than any other silver treatment. Other treatments, such as silver sulfadiazine and silver nitrate, are characterized by the rapid depletion of active silver, forcing the regular scraping of creams from or applications of solutions to open wounds multiple times per day. This process is labor intensive and extremely traumatic for patients. Silver nanocrystalline technology dressings cover the wound providing sustained release of silver to the dressing, acting as a barrier to infection for up to 7 days. Acticoat™ (Smith & Nephew) dressings for burns and chronic wounds use Nucryst's proprietary Silcryst™ silver nanocrystalline technology. In vitro studies of Acticoat have demonstrated:

- Extensive antimicrobial spectrum of 150 different pathogens
- Rapid kill rates
- Effective against drug-resistant forms of bacteria, such as MRSA (methicillin-resistant *Staphylococcus aureus*) and VRE (vancomycin-resistant Enterococci), sometimes referred to as "superbugs"
- Fast-acting release of ionic silver to the dressing over a sustained period of time (effective for up to 7 days)

The company also is conducting preclinical studies on the use of nanocrystalline silver inhaled into the lungs for the treatment of serious lung infection or lung inflammation. In the future, the company plans to conduct research on the nanocrystalline structures of other metals, including gold, which is well known as a treatment for arthritis, and platinum, which is a well-known treatment for cancer, to determine if the behavior and performance of these metals also can be enhanced.

Elan's NanoCrystal Technology

NanoCrystal® (Elan) particles are small particles of drug substance, typically less than 1,000 nm in diameter, which are produced by milling the drug substance using a proprietary milling technique. The NanoCrystal® particles of the drug are stabilized against agglomeration by surface adsorption of selected GRAS (generally regarded as safe) stabilizers. The end result is a suspension of the drug substance that behaves like a solution – a NanoCrystal® colloidal dispersion, which can be processed into dosage forms for all routes of administration. NanoCrystal® technology is being used by Johnson & Johnson Pharmaceutical Research & Development in a phase III clinical trial of a long-acting injectable formulation of its paliperidone palmitate in patients with schizophrenia.

NanoCrystal® technology represents a valuable, enabling technology to evaluate new chemical entities which exhibit poor water solubility and is also a valuable tool for optimizing the performance of established drugs. NanoCrystal technology has the potential to rescue a significant number of poorly soluble chemical compounds. The drug in nanoform can be incorporated into common dosage forms, including tablets, capsules, inhalation devices, and sterile forms for injection, with the potential for substantial improvements to clinical performance. There are currently two pharmaceutical products that have been commercialized incorporating NanoCrystal technology, with several additional product launches anticipated over the near future. Advantages of this technology are:

- More rapid absorption of active drug substance
- Higher dose loading with smaller dose volume
- Aqueous based with no organic solvents needed
- Capability for sterile filtering
- Longer dose retention in blood and tumors for some compounds

Biorise System

Biorise system (Aptalis Pharmaceutical Technologies) creates new physical entities by physically breaking down a drug's crystal lattice. This results in drug nanocrystals and/or amorphous drug, which are then stabilized with biologically inert carriers. The carriers used in the Biorise system are biocompatible and readily disperse in the body's GI fluids. The final product is a free-flowing powder that can be incorporated into a variety of dosage forms to achieve the most effective delivery.

Aptalis uses three types of carriers: swellable microparticles, composite swellable microparticles, or cyclodextrins. When used in the Biorise system, all three carrier types improve both solubility and dissolution rates as well as the rate and overall percentage of drug absorption. The selection of the appropriate type of Biorise carrier is a critical step in the process and is dependent upon the drug delivery objective, drug carrier compatibility, and its drug loading capacity.

Aptalis has developed a number of activation systems that can convert a drug into its thermodynamically activated state. These systems provide flexibility and allow the technology to be applied to a range of compounds with differing characteristics. These systems include:

High Energy Mechanochemical Activation (HEMA). This system involves the application of friction and impact energy to the drug, thereby increasing its entropy and transforming the drug into its activated state. This system is a dry system and maintains the drug/carrier matrix in a powder form at all times.

Solvent-Induced Activation (SIA). This system is particularly suitable for thermolabile compounds and compounds with a low melting point. With this system, a drug can be solubilized in an appropriate solvent and layered onto swellable, cross-linked carriers. Controlled evaporation of the solvent and drying the material create nanoparticles and/or amorphous drug that is stabilized in a carrier.

Super Critical Fluid Activation (SCFA). A drug and carrier are placed in a solvent system within a soluble environment. The solvent is removed by controlled displacement using super critical fluids resulting in the precipitation of nanocrystalline and/or amorphous drug that is stabilized in a carrier.

Before Aptalis begins working on a compound, its experienced teams of scientists evaluate the compound and apply a mathematical model to predict the impact that Biorise will have on a drug. This model simulates an in vitro release profile and also determines the most appropriate carrier system as well as drug to carrier ratios. Modeling is a key component in the Biorise process as it helps to:

- Expedite development programs and accelerate the time to market
- Reduce the need for experimentation
- Speed up the rational screening process
- Rapidly predict the outcome of the project

Aptalis' Biorise system can be used to improve a product already on the market, a drug currently in development, as well as to rescue a drug that has been shelved due to solubility difficulties. The system also offers faster and more efficient processing times compared to other marketed technologies and is currently one of the few bioavailability enhancement technologies that is commercialized and being used in a marketed product. The Biorise system offers additional advantages including:

- No use of surfactants
- Produces a drug powder which can be incorporated into a variety of dosage forms including tablets and capsules
- Stable
- Cost-effective process
- Scaled-up, validated, approved by a regulatory agency and commercialized
- Ability to control and vary the ratio of nanocrystal and amorphous drug
- Uses GRAS materials

Nanodiamonds

Nanodiamonds are nanoparticles varying in size from 2 to 8 nm and can be used for diagnostic as well as therapeutic purposes. Water dispersion of previously insoluble drugs when complexed with nanodiamonds demonstrates great promise in expanding current drug delivery options (Lam and Ho 2009). Bovine insulin was noncovalently bound to detonated nanodiamonds via physical adsorption in an aqueous solution and demonstrated pH-dependent desorption in alkaline environments of sodium hydroxide (Shimkunas et al. 2009).

Carbon nanodiamonds are much more biocompatible than most other carbon nanomaterials, including carbon blacks, fullerenes, and carbon nanotubes. The noncytotoxic nature of nanodiamonds, together with their unique strong and stable photoluminescence, tiny size, large specific surface area and ease with which they can be functionalized with biomolecules, makes nanodiamonds attractive for various biomedical applications both in vitro and in vivo (Xing and Dai 2009).

Polymer Nanoparticles

Biodegradable polymer nanoparticles include chitosan (CS) nanoparticles, poly(ethylene glycol) (PEG) nanoparticles, and polylactide-co-glycolic acid (PLGA) nanoparticles.

Biodegradable PEG Nanoparticles for Penetrating the Mucus Barrier

Protective mucus coatings typically trap and rapidly remove foreign particles from the eyes, gastrointestinal tract, airways, nasopharynx, and female reproductive tract, thereby strongly limiting opportunities for controlled drug delivery at mucosal surfaces. A preparation of nanoparticles composed of a biodegradable diblock copolymer of polysebacic acid and polyethylene glycol (PSA-PEG), both of which are routinely used in humans, can diffuse through mucous membranes (Tang et al. 2010). In fresh undiluted human cervicovaginal mucus (CVM), which has a bulk viscosity approximately 1,800-fold higher than water at low shear, PSA-PEG nanoparticles diffused at an average speed only 12-fold lower than the same particles in pure water. In contrast, similarly sized biodegradable nanoparticles composed of PSA or PLGA diffused at least 3,300-fold slower in CVM than in water. PSA-PEG particles also rapidly penetrated sputum expectorated from the lungs of patients with cystic fibrosis, a disease characterized by hyperviscoelastic mucus secretions. Rapid nanoparticle transport in mucus is made possible by the efficient partitioning of PEG to the particle surface during formulation. Biodegradable polymeric nanoparticles capable of overcoming human mucus barriers and providing sustained drug release open significant opportunities for improved drug and gene delivery at mucosal surfaces. Beyond their potential applications for cystic fibrosis patients, the nanoparticles also could be used to help treat disorders such as lung and cervical cancer and inflammation of the sinuses, eyes, lungs, and gastrointestinal tract. Chemotherapy is typically given to the whole body and has many undesired side effects. If drugs are encapsulated in these nanoparticles and inhaled directly into the lungs of lung cancer patients, drugs may reach lung tumors more effectively, and improved outcomes may be achieved, especially for patients diagnosed with early stage NSCLC. PEG acts as a shield to protect the particles from interacting with proteins in mucus that would cause them to be cleared before releasing their contents. Nanoparticles can efficiently encapsulate several chemotherapeutics, and a single dose of drug-loaded particles is able to limit tumor growth in a mouse model of lung cancer for up to 20 days.

Additionally, PEG coating improves the stability of PLGA nanoparticles in the gastrointestinal fluids and helps the transport of the encapsulated protein, tetanus toxoid, across the intestinal and nasal mucous membranes. Furthermore, intranasal administration of these nanoparticles provided high and long-lasting immune responses.

PLGA-Based Nanodelivery Technologies

Polylactide-co-glycolic acid (PLGA) is a FDA-approved copolymer which is used in a host of therapeutic devices, owing to its biodegradability and biocompatibility. PLGA is synthesized by means of random ring-opening copolymerization of two different monomers. PLGA nanoparticles deliver molecules considered too large and complex to transport with known vectors. PLGA is nontoxic, does not illicit an immune response, causes comprehensive transfection, crosses the blood–brain barrier, and supports sustained drug release. PLGA has been successful as a biodegradable polymer because it undergoes hydrolysis in the body to produce the original monomers, lactic acid and glycolic acid. These two monomers under normal physiological conditions are by-products of various metabolic pathways in the body. Since the body effectively deals with the two monomers, there is very minimal systemic toxicity associated with using PLGA for drug delivery or biomaterial applications. Also, the possibility to tailor the polymer degradation time by altering the ratio of the monomers used during synthesis has made PLGA a common choice in the production of a variety of biomedical devices such as grafts, sutures, implants, and prosthetic devices. As an example, a commercially available drug delivery device using PLGA is Abbott's Lupron Depot® (leuprolide acetate) for the treatment of advanced prostate cancer.

Polymeric Micelles

Micelles are biocompatible nanoparticles varying in size from 50 to 200 nm in which poorly soluble drugs can be encapsulated, which represents a possible solution to the delivery problems associated with such compounds, and could be exploited to target the drugs to particular sites in the body, potentially alleviating toxicity problems. pH-sensitive drug delivery systems can be engineered to release their contents or change their physicochemical properties in response to variations in the acidity of the surroundings. One example of this is the preparation and characterization of novel polymeric micelles (PM) composed of amphiphilic pH-responsive poly(N-isopropylacrylamide) (PNIPAM) or poly(alkyl(meth)acrylate) derivatives. On one hand, acidification of the PNIPAM copolymers induces a coil-to-globule transition that can be exploited to destabilize the intracellular vesicle membranes. PNIPAM-based PMs, loaded with either doxorubicin or aluminum chloride phthalocyanine, are cytotoxic in murine tumor models. On the other hand, poly(alkyl(meth) acrylate) copolymers can be designed to interact with either hydrophobic drugs or polyions and release their cargo upon an increase in pH. The self-assembly of well-defined polypeptide-based diblock copolymers into micelles and stimuli-responsive behavior of polypeptides to pH and ionic strength is used to produce nanoparticles with controlled size and shape, which are particularly useful for encapsulation and delivery purpose at a controlled pH (Checot et al. 2007).

Chitosan Nanoparticles

Chitin, a polymer, is commercially extracted from shrimp shells and has several medical applications. Chitin is a very large sugar molecule with a large number of acetic acid molecules attached to it. Treatments with soda remove some of this acetic acid from the sugar backbone, converting chitin into biopolymer chitosan. Chitosan is prone to chemical and physical modifications and is very responsive to environmental stimuli such as temperature and pH. These features make chitosan a smart material with great potential for developing multifunctional nanocarrier systems to deliver large varieties of therapeutic agents administrated in multiple ways with reduced side effects. Chitosan modification with a variety of ligands specific for cell surface receptors can increase recognition and uptake of nanocarriers into cells through receptor-mediated endocytosis (Duceppe and Tabrizian 2010).

Chitosan nanoparticles are known for their ability to overcome biological barriers and facilitate the delivery of complex drugs such as insulin, vaccines, plasmid DNA, and genes. In the NanoBioSaccharides project which is financed by the European Union, scientists from universities and commercial companies in Germany, France, Spain, Denmark, and Italy will collaborate to optimize these technologies to further improve the delivery of macromolecules, e.g., insulin, via the nasal, pulmonary, and oral routes instead of via an injection into the blood vessels.

N-(2-hydroxyl) propyl-3-trimethyl ammonium chitosan chloride (HTCC) is water-soluble derivative of chitosan (CS), synthesized by the reaction between glycidyl-trimethyl-ammonium chloride and CS. HTCC nanoparticles have been formed based on ionic gelation process of HTCC and sodium tripolyphosphate (TPP). Bovine serum albumin (BSA), as a model protein drug, is incorporated into the HTCC nanoparticles measuring 110–180 nm in size with encapsulation efficiency up to 90 %. In vitro release studies showed a burst effect, and a slow and continuous release followed. Encapsulation efficiency was obviously increased with increase of initial BSA concentration.

Coating of PLGA nanoparticles with the mucoadhesive CS improves the stability of the particles in the presence of lysozyme and enhanced the nasal transport of the encapsulated tetanus toxoid. Nanoparticles made solely of CS are stable upon incubation with lysozyme. Moreover, these particles are very efficient in improving the nasal absorption of insulin as well as the local and systemic immune responses to tetanus toxoid, following intranasal administration.

By encapsulating drugs in CS cubes, the cells of the body are tricked into absorbing drugs that could not normally be transported across the cell membrane. After delivering its load directly into the cells affected, CS is broken down in the body and disappears without a trace. CS capsules can transport siRNA into the cell to switch off faulty genes selectively. A siRNA against the respiratory syncytial virus (RSV), RSV-NS1gene (siNS1), has been tested for its potential in decreasing RSV infection and infection-associated inflammation in rats (Kong et al. 2007). Plasmids encoding siNS1 were complexed with a chitosan nanoparticle delivery agent and administered intranasally. They were shown to be effective in reducing virus titers in the lung and in preventing the inflammation and airway hyperresponsiveness associated with the infection.

QDs for Drug Delivery

Engineered QDs with integrated targeting, imaging, and therapeutic functions are excellent materials for study of drug delivery in cells and experimental animals. They can provide important information for the rational design of biocompatible drug carriers and can serve as a superior alternative to magnetic and radioactive imaging contrast agents in preclinical drug screening, validation, and delivery research (Qi and Gao 2008). In a multifunctional QD coated with amphiphilic polymer, hydrophobic drugs can be trapped between the nanoparticle core and amphiphilic polymer coating layer, and hydrophilic targeting molecules can be immobilized on the outer surface.

To meet some challenges of drug delivery, the engineered QDs would be able to stabilize therapeutic compounds, increase their plasma circulation time while reducing the concentration of free drug to minimize unwanted side effects, and to release the drug with a well-controlled profile. In addition, it should be possible to covalently link the therapeutic compounds to the target on QD surface via cleavable chemical bonds, so that the bioconjugates are initially large enough to avoid renal filtration but, following cleavage of the ligands, are small enough to be cleared out of the body. With advances being made in the identification of new targeting ligands, the development of specialized nanoparticles and the discovery of better conjugation techniques, the QD-based drug delivery has excellent prospects. Role of QDs in drug delivery for cancer is described later in this chapter.

Special Procedures in Nanoparticle-Based Drug Delivery

Coated Nanoparticles for Penetrating Cell Membranes Without Damage

Most nanoscale objects are typically internalized by cells into membrane-bounded endosomes and fail to access the cytosolic cell machinery. A study has shown that gold nanoparticles coated with alternating bands of two different types of molecules can penetrate a cell quickly, passing into cells without damaging them, while those randomly coated with the same materials cannot (Verma et al. 2008). This is the first fully synthetic material that can pass through a cell membrane without rupturing it and is significant for drug delivery across biological membranes. In addition to the practical applications of such nanoparticles for drug delivery, they have been used to deliver fluorescent imaging agents to cells that could help explain how some biological materials such as peptides are able to enter cells.

Combinatorial Synthesis of Nanoparticles for Intracellular Delivery

Evaluation of a large library of structurally distinct nanoparticles with cationic cores and variable shells was carried out by using robotic automation. Nanoparticles were combinatorially cross-linked with a diverse library of amines, followed by measurement of molecular weight, diameter, RNA complexation, cellular internalization, and in vitro siRNA and pDNA delivery (Siegwart et al. 2011). Analysis revealed structure–function relationships and beneficial design guidelines. Cross-linkers optimally possessed tertiary dimethylamine or piperazine groups and potential buffering capacity. Covalent cholesterol attachment enabled intracellular delivery in vivo to liver hepatocytes in mice.

Drug Delivery Using "Particle Replication in Nonwetting Templates"

Most current techniques for particle formation are incompatible with organic materials because they involved baking, etching, or processing robust metals using with solvents that destroy fragile organic matter such as genes or drugs. The relatively simple process, called Particle Replication in Nonwetting Templates (PRINT), avoids creating films or "scum layers" that would clump particles together rather than allowing them to be harvested independent of one another. PRINT affords the simple, straightforward encapsulation of a variety of important bioactive agents, including proteins, DNA, and small-molecule therapeutics, which indicates that PRINT can be used to fabricate next-generation particulate drug delivery agents. The particles are so small they can be designed and constructed to measure <200 nm in diameter. The method avoids harsh treatment but also allows formation of uniform particles in any shape that designers choose: spheres, rods, cones, trapezoidal solids, etc. Besides drug delivery, this technology will have a profound impact on human health care in areas such as chemotherapy, gene therapy, and disease detection. Particles injected into the body can be designed to be biodegradable and incorporate as "cargo" any biological material that designers want to introduce into patients' bloodstreams for more efficient uptake by cells for diagnostic testing or therapy. Preliminary in vitro and in vivo studies have demonstrated future utility of PRINT particles as delivery vectors in nanomedicine (Gratton et al. 2007).

Encapsulating Water-Insoluble Drugs in Nanoparticles

Many of the most potent anticancer agents are poorly soluble in water, presenting a challenge for medicinal chemists who must develop methods of delivering these drugs in the watery environment of the human body. Nanoparticles appear to be perfectly suited to this task, and indeed, numerous research groups are developing nanoparticles specifically for delivering water-insoluble drugs to tumors. A fundamental understanding of particle size control in antisolvent precipitation is beneficial for designing mixing systems and surfactant stabilizers for forming nanoparticles of poorly water-soluble drugs with the potential for high dissolution rates.

Inverse emulsion photopolymerization is a method that uses light to create a well-defined polymeric nanoparticle with internal spaces that can provide a friendly environment to water-insoluble drugs and channels through which the entrapped drugs can escape into malignant cells (Missirlis et al. 2006). The investigators created these nanoparticles from two different polymers that cross-link to each other when exposed to light from an argon laser for 1 h. They then added the nanoparticles to a solution of doxorubicin and evaporated the solvent used to dissolve the anticancer drug. Nearly half of the drug in solution became encapsulated within the nanoparticles. The researchers note that the resulting nanoparticles contain a protein-repelling surface coating that should result in favorable pharmacokinetic behavior. Experiments to test the drug release characteristics of these nanoparticles showed that maximum release occurred at approximately 8 h and then remained close to that level for a week. The data imply that release occurs through a diffusion mechanism, i.e., drug travels through channels in the nanoparticle to the nanoparticle surface, as opposed to a disintegration mechanism in which the nanoparticle falls apart and releases drug. This novel colloidal system can be as a controlled delivery system for small hydrophobic drugs for cancer.

Filomicelles Versus Spherical Nanoparticles for Drug Delivery

Shape may be important in designing better nanotechnology-based drug delivery vehicles. A study in rodents has compared highly stable, polymer micelle assemblies known as filomicelles with spheres of similar chemistry and shown that filomicelles persisted in the circulation up to 1 week after intravenous injection (Geng et al. 2007). This is about ten times longer than their spherical counterparts and is more persistent than any known synthetic nanoparticle. Under fluid flow conditions, spheres and short filomicelles are taken up by cells more readily than longer filaments because the latter are extended by the flow. Preliminary results further demonstrate that filomicelles can effectively deliver the anticancer drug paclitaxel and shrink human-derived tumors in mice. Although these findings show that long-circulating vehicles need not be nanospheres, they also lend insight into possible shape effects of natural filamentous viruses.

Flash NanoPrecipitation

Flash NanoPrecipitation produces stable nanoparticles at high concentrations using amphiphilic diblock copolymers to direct self-assembly (Prudhomme et al. 2006). In NanoPrecipitation, two streams of liquid are directed toward one another in a confined area. The first stream consists of an organic solvent that contains the medicines and imaging agents, as well as long-chain molecules called polymers.

The second stream of liquid contains pure water. When the streams collide, the hydrophobic medicines, metal imaging agents, and polymers precipitate out of solution in an attempt to avoid the water molecules. The technique has been applied to the anticancer agent paclitaxel. The polymers immediately self-assemble onto the drug and imaging agent cluster to form a coating with the hydrophobic portion attached to the nanoparticle core and the hydrophilic portion stretching out into the water. By carefully adjusting the concentrations of the substances, as well as the mixing speed, the researchers are able to control the sizes of the nanoparticles. Uniform particles with tunable sizes from 50 to 500 nm can be prepared. The key to the process is the control of time scales for micromixing, polymer self-assembly, and particle nucleation and growth. The diffusion-limited assembly enables particles of complex composition to be formed. The stretched hydrophilic polymer layer keeps the particles from clumping together and prevents recognition by the immune system so that the particles can circulate through the bloodstream. The hydrophobic interior of the particles ensures that they are not immediately degraded by watery environments, though water molecules will, over time, break the particles apart, dispersing the medicine. Ideally, the particles would persist for 6–16 h after they are administered intravenously, which would allow enough time for the potent packages to slip into the solid tumor cells whenever they encounter them throughout the body.

Applications include controlled delivery of multiple drugs from nanoparticles as well as aerosol drug delivery. It enables the simultaneous encapsulation and controlled release of both hydrophobic and hydrophilic actives. The incorporation of gold nanoparticles and organic compounds into single nanoparticles enables simultaneous delivery and medical imaging. Finally, the ability to dry the nanoparticles by lyophilization or spray drying and to reconstitute them without aggregation greatly enhances the applicability of the technology.

These nanoparticles can deliver medicine deep into the lungs or infiltrate cancer cells while leaving normal ones alone. Only 100–300 nm wide, the particles can be loaded with medicines or imaging agents, like gold and magnetite, that will enhance the detection capabilities of CT scans and MRIs. The nanoparticles are too large to pass through the membrane of normal cells but will pass through larger defects in the capillaries in rapidly growing solid tumors.

Particles in this size range also could improve the delivery of inhaled drugs because they are large enough to remain in the lungs, but too small to trigger the body's lung-clearing defense systems. This trait could maximize the effectiveness of inhaled, needle-free vaccination systems. It has potential applications in the development of nanoparticle-based aerosol vaccines for tuberculosis and diphtheria. Because of their potential for use on a large scale at relatively low cost, these systems are particularly attractive for use in the developing world.

Magnetic Nanoparticles for Drug Delivery

These particles have a magnetic core with a polymer or metal coating, which can be functionalized. A therapeutic agent such as a cytotoxic drug or a therapeutic DNA can be attached to the particle. The particles are delivered by a catheter close to the site of action and further guided by external magnetics. Investigations of magnetic micro- and nanoparticles for targeted drug delivery began over 30 years ago, and major advances have been made in particle design and synthesis techniques, which have been reviewed (McBain et al. 2008). Although very few clinical trials have taken place, with this technique, it appears to be promising. nanoTherics is developing magnefect-nano™, which has the following advantages:

- Up to 1,000-fold higher transfection efficiencies at short transfection times when compared to cationic lipid reagents
- No adverse effects on cell viability
- Potential to target/penetrate physical barriers in vivo (e.g., mucous layers for cystic fibrosis gene transfection)
- Successful with "hard to transfect" cells/cell lines
- Cost-effective, saving time and materials
- Can be used with adherent and suspension cells
- Scalable for high-throughput screening

Nanoparticles Bound Together in Spherical Shapes

Altair Nanotechnologies Inc has developed unique micron-size structures (TiNano Spheres™) made by its patented "growth-in-film" nanotechnology. They consist of hundreds of nanoparticles bound together in spherical and near spherical shapes and are capable of carrying active pharmaceutical ingredients (API), biocides, or fungicides on either the interior or exterior surfaces. The nanoparticles have a very high surface area and when coated with an API delivers a very large amount of drug to biosystem interface. This larger interface could improve solubility and/or reaction rates.

Altair's nanotechnology is used to create competent porous microstructures consisting of high surface area nano-primary particles to enable new applications for hard to dissolve drugs. A sustained release of drugs is possible by applying the drug to the inside of the TiNano Spheres™. Dual action properties are possible by applying one drug to the inside and another to the outside of the TiNano Sphere. Altair has successfully deposited at least one of these drugs on the surface of TiNano Spheres™. Some of the many possible applications of TiNano Spheres™ are:

- Drug delivery by topical applications
- Sustained release of antibiotic and fungicides
- Sustained release of drugs for cholesterol lowering
- Pain and itch preparations with sustained-release action
- Sunscreen and after-sun care

Perfluorocarbon Nanoparticles for Imaging and Targeted Drug Delivery

Perfluorocarbon (PFC) nanoparticles are approximately 200 nm in diameter and are encapsulated in a phospholipid shell, which provides an ideal surface for the incorporation of targeting ligands, imaging agents, and drugs. PFC nanoparticles can serve as a platform technology for molecular imaging and targeted drug delivery applications. For molecular imaging, PFC nanoparticles can carry very large payloads of gadolinium to detect pathological biomarkers with MRI. A variety of different epitopes, including $\alpha v\beta 3$, tissue factor, and fibrin, have been imaged using nanoparticles formulated with appropriate antibodies or peptidomimetics as targeting ligands. Lipophilic drugs can also be incorporated into the outer lipid shell of nanoparticles for targeted delivery. Upon binding to the target cell, the drug is exchanged from the particle surfactant monolayer to the cell membrane through a novel process called "contact-facilitated drug delivery." By combining targeted molecular imaging and localized drug delivery, PFC nanoparticles provide diagnosis and therapy with a single agent and would facilitate the development of personalized medicine (Winter et al. 2007).

The contrast agents in development by Kereos Inc comprise tiny perfluorocarbon nanoparticles suspended in an emulsion. Agents such as technetium-99 m may be attached to the nanoparticles to provide the contrast that allows for imaging. In addition, nanoparticles are labeled with a specific ligand that causes the agent to target newly developing blood vessels. When injected into the body, the resulting agent will find and illuminate these vessels. Anticancer drugs and therapeutic radionuclides may also be incorporated into the nanoparticles to deliver therapy directly and selectively.

Prolonging Circulation of Nanoparticles by Attachment to RBCs

Polymeric nanoparticles are used as carriers for systemic and targeted drug delivery. They protect drugs from degradation until they reach their target and provide sustained release of drugs. However, applications of nanoparticles are limited by their short in vivo circulation lifetimes. They are quickly removed from the blood, sometimes in minutes, rendering them ineffective in delivering drugs. It is now possible to dramatically improve the in vivo circulation lifetime of polymeric nanoparticles by attaching them to the surface of red blood cells (RBCs) without affected their circulation (Chambers and Mitragotri 2007). The particles remain in circulation as long as they remain attached to RBCs, theoretically up to the circulation lifetime of a RBC, which is 120 days. Particles eventually detach from RBCs due to shear forces and cell–cell interactions and are subsequently cleared in the liver and spleen.

The researchers have learned that particles adhered to RBCs can escape phagocytosis because red blood cells have a knack for evading macrophages. Nanoparticles are not the first to be piggybacking on red blood cells; the strategy has already been adopted by certain bacteria, such as *Hemobartonella*, that adhere to RBCs and can remain in circulation for several weeks. Using RBCs to extend the circulation time of the particles avoids the need to modify the surface chemistry of the entire particle, which offers the potential to attach chemicals to the exposed surface for targeting applications. The exposed surface of the particles could be used to immobilize enzymes and improve their in vivo circulation lifetime. The enzyme would have direct access to plasma in the systemic circulation. RBC-mediated prolonged circulation may also be applied to gene delivery applications in which extended circulation times are difficult to achieve. Synthetic gene delivery vectors suffer from rapid clearance by the reticuloendothelial system, restricting transfection to the liver and lung. RBC attachment of gene vectors may provide a long-circulating depot, thereby increasing their residence time in blood.

RBC membrane-coated nanoparticles present a major breakthrough in drug delivery technology and show great promise for clinical applications (Fang et al. 2012). This technique could be applied for the delivery of drugs and circulating bioreactors in a wide variety of conditions such as cancer and heart disease.

Self-Assembling Nanoparticles for Intracellular Drug Delivery

EAK16-II, a self-assembling peptide, has been found to stabilize the hydrophobic anticancer agent ellipticine (EPT) in aqueous solution and form nanoparticles with an average size of ~100 nm (Bawa et al. 2011). This nanoformulation is cytotoxic to human lung carcinoma A549 cells that is comparable to EPT dissolved in dimethyl sulfoxide. It enhances EPT uptake significantly as compared to the microformulation. Promising therapeutic efficacy, specific delivery pathway, and intracellular distribution pattern discovered in this study may help further develop EPT as a nanoformulation for clinical applications.

Researchers at the University of Ulsan College of Medicine (Seoul, Korea) have developed self-assembling nanoparticles that can sense the low pH of endosomes and disintegrate, which not only releases their drug payload but enables it to exit the endosomes. Chitosan serves as the starting material for these self-assembling nanoparticles. The investigators modify the polymer by attaching a chemical derivative of the amino acid histidine to each of the sugar units in the chitosan backbone. At neutral pH, histidine is hydrophobic, or poorly soluble in water. The presence of multiple histidines on the water-soluble, or hydrophilic, chitosan backbone creates a molecule that naturally self-assembles into a structure that surrounds the hydrophobic histidines with a protective shell of hydrophilic chitosan. When added to cells grown in culture, the nanoparticles fuse with the cell membrane, forming endosomes inside the cell. At the low pH found inside an endosome, histidine takes on a positive charge and also becomes hydrophilic. As a result, the physical forces that held together the self-assembling nanoparticle no longer exist, and the nanoparticle falls apart. Any drug molecules entrapped within the nanoparticle is then released into the endosomes.

Trojan Nanoparticles

Trojan particles combine the drug release and delivery potential of nanoparticle systems with the ease of flow, processing, and aerosolization potential of large porous particle systems by spray drying solutions of polymeric and nonpolymeric nanoparticles into extremely thin-walled macroscale structures. These hybrid particles exhibit much better flow and aerosolization properties than the nanoparticles; yet, unlike the large porous particles, which dissolve in physiological conditions to produce molecular constituents, hybrid particles dissolve to produce nanoparticles with drug release and delivery. Formation of the large porous nanoparticle aggregates occurs via a spray-drying process that ensures the drying time of the sprayed droplet is sufficiently shorter than the characteristic time for redistribution of nanoparticles by diffusion within the drying droplet. Additional control over the physical characteristics is achieved by adding other components to the spray-dried solutions, including sugars, lipids, polymers, and proteins. The ability to produce large porous nanoparticles appears to be largely independent of molecular component type as well as the size or chemical nature of the nanoparticles.

These particles range in size from 25 nm to several hundred nms and can be used to deliver drugs to specific sites within the body. They are robust drug delivery systems that can be used to encapsulate drugs of varying chemistry and molecular weights. Trojan nanoparticles will be discussed further in connection with drug delivery across the BBB.

Therapeutic Protein Delivery from Nanoparticle–Protein Complexes

Scientists at the US Department of Energy's Brookhaven National Laboratory have attached gold nanoparticles to proteins to form sheets of protein–gold arrays (Hu et al. 2007). The nanoparticle–protein complexes can be used to identify functional parts of proteins and to construct new protein complexes, which can be used as precision vehicles for targeted drug delivery.

Therapeutic protein delivery from nanoparticle aggregates is being developed and is based on several independent variables including nanoparticle size and chemical composition of the particle. The nanoparticle aggregate technology allows remarkable versatility in protein loading and subsequent release. Optimization of a formulation can be achieved in a relatively short time for a given protein drug. Importantly, this technology can substantially reduce the "burst release" of the protein, which occurs with other delivery systems. In preclinical animal studies, the ability to control the release of the protein for periods of 3 months and greater has been established. Additional preclinical animal studies have shown that the materials used to produce the aggregates, which are included in several FDA-approved products, are biocompatible and therefore suitable for use in a drug delivery system. It represents one of the simplest delivery systems with great versatility for incorporating and delivering proteins. The abilities to load the protein in a cost-effective manufacturing process without using solvents or polymerization and to tightly control the drug release profile are potentially very significant advantages over other protein delivery technologies. Protein delivery is important as therapeutic proteins are being increasingly used to treat a wide variety of disease including cancer, infections, rheumatoid arthritis, and autoimmune diseases.

Triggered Release of Drugs from Nanoparticles

Controlling location and timing of the release enables use of potent drugs in a personalized manner so that the interaction with the right target is ensured. One method for this is externally triggered release of encapsulated compounds, which can be accomplished if drug delivery vehicles, such as liposomes or polyelectrolyte multilayer capsules, incorporate NP actuators. NPs can efficiently harvest energy from tissue penetrating, external stimuli, such as near infrared (NIR) light and alternating magnetic fields (AMFs), localize it by local field enhancement or convert it into heat, and thus trigger release of cargo from thermoresponsive vehicles. Cargo release can be externally triggered by magnetic or electromagnetic fields from vesicles loaded with superparamagnetic or metallic NPs. Control over the assembly of NPs into a responsive vesicle determines the vesicle stability, and the ability to control timing and dose of the cargo released from the vesicle. Thus, nondestructive, reversible changes in permeability of delivery vehicles enable pulsed cargo release, and therefore a close control over timing and dose of released cargo.

Liposomes

Basics of Liposomes

Liposome properties vary substantially with lipid composition, size, surface charge, and the method of preparation. They are therefore divided into three classes based on their size and number of bilayers:

1. Small unilamellar vesicles are surrounded by a single lipid layer and measure 25–50 nm in diameter.
2. Large unilamellar vesicles are a heterogeneous group of vesicles similar to and are surrounded by a single lipid layer.
3. Multilamellar vesicles consist of several lipid layers separated from each other by a layer of aqueous solution.

Lipid bilayers of liposomes are similar in structure to those found in living cell membranes and can carry lipophilic substances such as drugs within these layers in the same way as cell membranes. AFM is useful in evaluating the physical characteristics and stability of liposomes as drug delivery systems (Spyratou et al. 2009). The pharmaceutical properties of the liposomes depend on the composition of the lipid bilayer and its permeability and fluidity. Cholesterol, an important constituent of many cell membranes, is frequently included in liposome formulations because it reduces the permeability and increases the stability of the phospholipid bilayers. Until recently, the use of liposomes as therapeutic vectors was hampered by their toxicity and lack of knowledge about their biochemical behavior. The simplest use of liposomes is as vehicles for drugs and antibodies for the targeted delivery of anticancer agents. Furthermore, liposomes can be conjugated to antibodies or ligands to enhance target-specific drug therapy.

Stabilization of Phospholipid Liposomes Using Nanoparticles

A simple strategy of mixing phospholipid liposomes with charged nanoparticles and using sonication to mix them at low volume fraction produces particle-stabilized

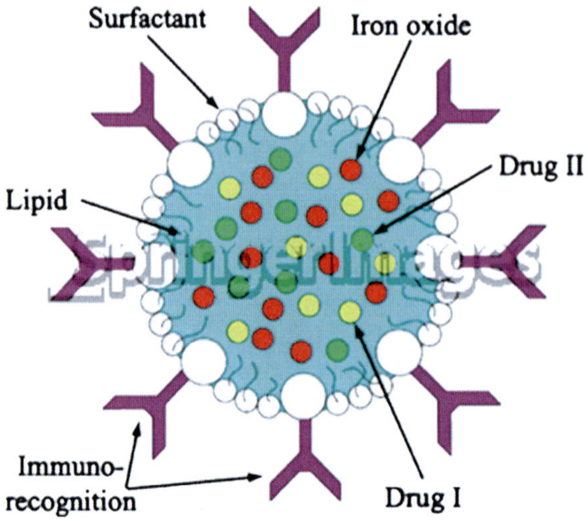

Fig. 5.3 Schematic image of a lipid nanoparticle (© Springer Science+Business Media LLC)

liposomes that repel one another and do not fuse (Zhang and Granick 2006). Subsequently, the volume fraction can be raised as high as 50 %, reversibly, still without fusion. The nanoparticles adhere to the capsules and prevent further growth, freezing them at the desired size. The lipid concentration can then be increased without limits. As proof of concept, fluorescent dyes were encapsulated within lipid capsules. No leakage occurred, and the lipids proved stable against further fusion. Although these particle-stabilized liposomes were stable against fusion, 75 % of the outer liposome surface remained unoccupied.

This opens the door to using particle-stabilized liposomes in various applications. The biocompatible containers could carry cargo such as enzymes, DNA, proteins, and drug molecules throughout living organisms. They could also serve as surrogate factories where enzyme-catalyzed reactions are performed. By attaching biomolecules to the capsule's surface, novel colloidal-size sensors could be produced. An additional use for stabilized lipid capsules is the study of behavior of a drug contained in this nanoenvironment.

Lipid Nanoparticles

Lipid nanoparticles are made up of lipids that assemble on their own into spherical particles or liposomes. This technology differs from conventional liposome technology in that the lipids used contain polymerizable functional groups that are cross-linked when exposed to UV light. As a result, the surface of the polymerized nanoparticle more closely resembles a nanosphere or bead than a droplet of fat (liposome), and drugs/targeting agents can be attached. Schematic image of a lipid nanoparticle is shown in Fig. 5.3.

Advantages of the lipid nanoparticle technology are:
Characteristics of lipid nanoparticles are:

- Unlike conventional bilayer liposomes, they do not randomly fuse with themselves or other membranes.
- The surface character, in terms of charge and molecular makeup, is easily modified.
- The spherical assemblies are easy to synthesize and very stable.
- Multivalent presentation of small-molecule ligands on nanoparticles offers vastly improved performance over the ligand alone.
- In addition to the multivalent attachment of ligands/antibodies to the surface, nanoparticles can carry a high payload of radiation, chemotherapeutic, or imaging agent to the target cell.
- Since the nanoparticle is larger than either antibodies or small-molecule ligands, the complex does not quickly leak from the blood vessel. As a result, the biodistribution and safety profile of the drug can be significantly improved.

Applications of Lipid Nanoparticles

The nanoparticle technology has broad therapeutic and diagnostic applications. The multivalent presentation of ligands or antibodies on nanoparticles makes this new class of drug ideally suited to treat diseases, which involve proliferation of blood vessels such as cancer, atherosclerosis, apoptosis, inflammation, rheumatoid arthritis, macular degeneration, unstable plaque, stroke, heart disease, and psoriasis.

When nanoparticles are used in the treatment of cancer, their powerful targeting ability and potential for large cytotoxic payload dramatically enhance the efficacy of conventional pharmaceuticals as well as novel therapeutics such as gene therapy, radioimmunotherapy, and photodynamic therapy. Integrin-targeted nanoparticles can be used for site-specific delivery of a therapeutic payload by using an anticancer gene. These targeted nanoparticles can deliver radionuclides and chemotherapeutics to tumors. Further applications are discussed under drug delivery for cancer.

Polymerized Liposomal Nanoparticle

Polymerized liposomal nanoparticle (PLN) is a nonviral nanoparticle incorporating a customizable drug delivery system for chemotherapeutic applications. PLNs are created using the self-assembling ability of a unique class of diacetylenic lipids that can be polymerized into stable, bimolecular membrane structures capable of delivering a drug payload. The PLN technology lends itself especially well to the display of multiple functions. These particles are comprised of individual lipid monomers, part of which are functionalized for the purpose of targeting and may include additional moieties to control other physical properties such as surface charge, polarity, and fluidity. Different functionalized lipids can be rapidly mixed and matched in an infinite number of combinations and relative concentrations to create tailor-made particles with desirable targeting and circulation properties. The nanoparticles are nonimmunogenic, display no acute toxicity, and can be highly concentrated. Intracellular degradation and excretion rates of the particles can be modulated by controlling the degree of polymerization.

Solid Lipid Nanoparticles

Solid lipid nanoparticles (SLNs) have emerged as oral bioavailability enhancer vehicles for various drugs. The protective effect of SLNs, coupled with their sustained/controlled-release properties, prevents drugs/macromolecules from premature degradation and improves their stability in the gastrointestinal tract. Review of various publications reveals that direct oral administration of SLNs improves the bioavailability of drugs 2- to 25-fold (Harde et al. 2011).

Lipid Nanocapsules

Due to their small size, lipid nanocapsules (LNC) might be promising as an injectable as well as for an oral drug delivery system. LNC provides sufficient drug solubility to avoid embolization during intravenous injection and facilitates drug absorption after oral administration. Biocompatible ibuprofen LNC has been developed with a particle size of ~50 nm. Pain relief after intravenous administration of ibuprofen is prolonged by at least 2 h when administering LNC formulation. A drug delivery system for intravenous administration of ibuprofen is available, which exhibits sustained-release properties by either oral or intravenous route and could be useful in the treatment of postoperative pain. LNCs can also be used as a transdermal drug delivery system as described later in this chapter.

Lipid Emulsions with Nanoparticles

An artificial lipoprotein-like particle, lipid nanosphere (LNS), is 25–50 nm in diameter and is composed of soybean oil and egg lecithin. Because of the lower uptake of LNS particles containing dexamethasone palmitate by the liver, LNS shows good recovery from the liver and prolonged the plasma half-life of after intravenous injection. In addition, higher anti-inflammatory efficacy of LNS is observed in targeting of dexamethasone palmitate into sites of inflammation. LNS easily and selectively pass through the leaky capillary wall by passive diffusion depending on the plasma concentration. LNS seems to be a promising carrier system for passive targeting of lipophilic drugs.

LNS has also been studied as a low-dose therapeutic system for amphotericin B (AmB), a potent antifungal drug. As a small-particle lipid emulsion, LNS is taken up by the liver to a lesser extent than a conventional lipid emulsion. As a result, LNS yields higher plasma concentrations of a radiochemical tracer than does the conventional lipid emulsion. LNS incorporating AmB (LNS-AmB) is a homogeneous emulsion with mean particle diameters ranging from 25 to 50 nm and yields higher plasma concentrations of AmB than Fungizone, a conventional intravenous dosage form of AmB, following intravenous administration in laboratory animals. This difference between LNS-AmB and Fungizone is also observed for constant intravenous infusion. In contrast to Fungizone, LNS-AmB shows a linear relationship between dose and area under the curve (AUC). These pharmacokinetic characteristics of LNS-AmB make it a suitable candidate for an effective low-dose therapeutic system for AmB.

Nanolipispheres are colloidal systems of drugs in a solid lipid matrix. These systems possess a submicron mean diameter and a uniform size distribution. A microemulsion–solidification process for manufacture produces a suspension of solid nanoparticles, which is then dried to obtain physically stable nanolipispheres in a powder form. Nanolipispheres provide for:

- Carrier incorporation of lipophilic and hydrophilic drugs
- Oral delivery of macromolecules that can be absorbed as a whole or as fragments through the gastrointestinal tract
- Therapeutic efficacy of some drugs by preferential and consistent absorption and metabolism through the lymphatic system
- Modified drug release

Nanotechnology has been applied to improve the absorption of CoQ10, a lipid-soluble compound found in the mitochondria of all living cells. It is a powerful antioxidant that is essential in the production of cellular energy and has been clinically shown to support healthy heart function, regulate blood pressure, increase energy and vitality, scavenge free radicals, and enhance the immune system. Although endogenously produced in the liver, there are conditions in which adequate production of CoQ10 in the body is impaired. In such situations, supplementation of CoQ10 has been shown to be very beneficial. However, many currently available dosage forms of CoQ10 exhibit negligible dissolution properties indicating potentially poor bioavailability, thereby limiting the therapeutic effect.

CoQ10-loaded PLGA nanoparticles, produced by a scalable emulsion–diffusion–evaporation method and measuring <100 nm, have been shown to significantly quench reactive oxygen species (ROS) with nearly tenfold higher efficacy than free CoQ10 (Swarnakar et al. 2011). Further, positively charged CoQ10-NPs were localized in two major sources of ROS generation: mitochondria and lysosomes. CoQ10 nanoparticles showed improved oral bioavailability (4.28 times) as compared to free CoQ10. The higher anti-inflammatory activity of CoQ10 nanoparticles is attributed to significant accumulation of these in the inflamed tissues.

Nanostructured Organogels

Organic gel nanomaterials can be used to encapsulate pharmaceutical, food, and cosmetic products. Using olive oil and six other liquid solvents, a simple enzyme has been added to chemically activate a sugar that changes the liquids to organic gels, thus using building blocks provided by nature to create new nanomaterials that are completely reversible and environmentally benign (John et al. 2006). In this study, researchers activated a sugar using a simple enzyme, which generated a compound that self-assembles into 3D fibers measuring approximately 50 nm in diameter. As the fibers entangle, a large amount of solvent gets packed together, trapping some 10,000 molecules.

Limitations of Liposomes for Drug Delivery

The use of liposomes may be limited because of problems related to stability, the inability to deliver to the right site, and the inability to release the drug when it gets to the right site. However, liposome surfaces can be readily modified by attaching polyethylene glycol (PEG) units to the bilayer (producing what is known as stealth liposomes) to enhance their circulation time in the bloodstream.

Several attempts have been made to use liposomes, targeted by specific ligands, for the delivery of antithrombotic/thrombolytic agents in order to increase their efficacy and decrease side effects. Although liposomes loaded with various anti-thrombotic drugs have been the subject of a significant number of experimental studies, they are not considered as candidates for clinical application in the near future (Elbayoumi and Torchilin 2008).

Liposomes Incorporating Fullerenes

Buckysomes are a new generation of liposomes that incorporate fullerenes to deliver drugs that are not water soluble, or tend to have large molecules, and are very hard to get into the body. Buckysomes appear to have much more flexibility in incorporating a wider range of drugs, as well as large molecule drugs, and delivering and releasing them more effectively. Buckysomes are being investigated for delivery of cancer therapeutics and anesthesia.

A study has examined the antioxidant activity of fullerene-C60 incorporated in liposome with a diameter of 75.6 nm, which was shown to have an antioxidant action characterized by long-term persistence, and is attributed to fullerene-C60 but hardly to liposome in all the tests examined (Kato et al. 2011). The combination is expected to be effective as a skin-protecting agent against oxidative stress.

Arsonoliposomes

Arsonolipids are analogs of phosphonolipids, in which P has been replaced by arsenic (As). Although arsonolipids possess interesting biophysical and biochemical properties, their anticancer or antiparasitic activity is not considered adequate for therapeutic applications. But when arsonolipids are incorporated in liposomes, arsonoliposomes show increased toxicity against cancer cells (compared to that of arsenic trioxide) but at the same time were less toxic than arsenic trioxide for normal cells (Fatouros et al. 2006). Furthermore, arsonoliposomes also demonstrate antiparasitic activity in vitro. Nevertheless, As is rapidly cleared from blood after in vivo administration of arsonoliposomes, and this will highly limit possible therapeutic applications. In addition, the fact that arsonoliposomes were observed to aggregate and subsequently fuse into larger particles in the presence of cations may also be considered as a problem. Thereby, methods to modulate the stability of arsonoliposomes and, perhaps, their in vivo distribution (as surface property modification) are currently being investigated. It has been shown that arsonoliposome pegylation results in the formation of liposomes with very high membrane integrity. In addition, pegylation results in increased physical stability of arsonoliposomes and abolishment of cation-induced aggregation and fusion. Nevertheless, further in vivo studies are required in order to prove if pegylation alters arsonoliposome in vivo kinetics in a positive way, without affecting their activity. Further development of arsonoliposomes to develop therapeutic systems for cancer or parasitic diseases is justified. Toxicity issues would need to be resolved

Liposome–Nanoparticle Hybrids

Small iron nanoparticles, quantum dots, liposomes, silica, and polystyrene nanoparticles have been incorporated into liposomes for a variety of applications. Different techniques to achieve encapsulation of solid or semisolid nanoparticles within liposomes have been described. These offer improvements in nanoparticle aqueous solubilization and offer a viable platform (the liposome surface) for further bioconjugation. Moreover, these hybrids have increased survival time in blood circulation following systemic administration and accumulate at sites of leaky vasculature such as in tumors or inflammatory lesions providing opportunities for a combination of diagnostic imaging and therapeutics (Al-Jamal and Kostarelos 2007). Some examples of liposome–nanoparticle hybrids and their applications are shown in Table 5.3.

Table 5.3 Liposome–nanoparticle hybrid systems

Nanoparticle	Method of encapsulation	Rationale	Applications
Superparamagnetic iron oxide (SPIO) particles/magnetite	Lipid film hydration and sonication	Cationic magnetoliposome provides selective intracellular hyperthermia and immune response induction	Cancer therapy
Quantum dots PEG-coated QD	QDs–liposome electrostatic complexation	High intracellular QD delivery	Intracellular trafficking
Phospholipid vesicles	Glass bead method	Preparation of double liposomes, which retard the drug release	Vaccines Drug delivery
Silicon-based nanoparticles	Adsorption and rupture of small unilamellar vesicle on nanoparticles surface forming a lipid monolayer or bilayer	Combining the intrinsic properties of silica and the bilayer	Design of biosensors
Polystyrene nanospheres	Adsorption and rupture of small unilamellar vesicle on nanoparticles surface forming a lipid monolayer or bilayer	Production of monodispersed and smooth bilayer nanosphere	Design of biosensors

Nanogels

Nanogels are colloidal microgel carriers in which cross-linked protonated polymer network binds oppositely charged drug molecules, encapsulating them into nanoparticles with a core–shell structure. The nanogel network also provides a suitable template for chemical engineering, surface modification, and vectorization.

Attempts are being made to develop novel drug formulations of nanogels with antiviral and antiproliferative nucleoside analogs in the active form of 5′-triphosphates. Notably, nanogels can improve the CNS penetration of nucleoside analogs that are otherwise restricted from passing through the blood–brain barrier. An efficient intracellular release of nucleoside analogs has been demonstrated that encourages applications of nanogel carriers for targeted drug delivery (Vinogradov 2007).

Nanogel–Liposome Combination

An appropriate assemblage of spherical nanogel particles and liposomes, termed "lipobead," combines the properties of both classes of materials and may find a variety of biomedical applications. Biocompatibility and stability, ability to deliver a broad range of bioactive molecules, environmental responsiveness of both inner nanogel core and external lipid bilayer, and individual specificity of both compartments make the liposome–nanogel design a versatile drug delivery system relevant for all known drug administration routes and suitable for different diseases with possibility of efficient targeting to different organs. New findings on reversible and irreversible aggregation of lipobeads can lead to novel combined drug delivery systems regarding lipobeads as multipurpose containers (Kazakov and Levon 2006).

Nanospheres

Hollow ceramic nanospheres (50–500 nm), created by using high-intensity ultrasound, were the first hollow nanocrystals that could be used in drug delivery (Dhas and Suslick 2005). A hollow nanocrystal of molybdenum oxide is prepared using high-intensity ultrasound to form a layer of amorphous material around a silica nanosphere. The nanosphere is then dissolved away with hydrofluoric acid and upon heating the shell crystallizes into a single hollow nanocrystal. TEM studies on the hollow ceramic materials indicate the formation of dispersed free spheres with a hollow core.

Nanotubes

Micro- and nanotubes or microstructures that resemble tiny drinking straws are alternatives that might offer advantages over spherical nanoparticles for some applications. Tubular structure of nanoparticles is highly attractive due to their structural attributes, such as the distinctive inner and outer surfaces, over conventional spherical nanoparticles.

When a PEG-silane is attached to the silica nanotubes, adsorption of IgG immunoglobulins is strongly suppressed relative to nanotubes that do not contain the attached PEG. This has potential usefulness for the delivery of biopharmaceuticals. A payload can be incorporated into the nanotubes by either covalent bonding or other chemical interactions between the payload and the inside walls of the nanotubes. For some applications, it might be useful to fill the nanotubes with the payload and then to apply caps to the nanotubes to keep the payload encapsulated. Uncapping and release of payload can be triggered by a biochemical signal.

Inner voids of nanotubes can be used for capturing, concentrating, and releasing species ranging in size from large proteins to small molecules. Distinctive outer surfaces can be differentially functionalized with environment-friendly and/or probe molecules to a specific target. By combining the attractive tubular structure with magnetic property, the magnetic nanotube can be an ideal candidate for the multi-functional nanomaterial toward biomedical applications, such as targeting drug delivery with MRI capability. Magnetic silica-iron oxide composite nanotubes have been successfully synthesized and shown to be useful for magnetic-field-assisted chemical and biochemical separations, immunobinding, and drug delivery.

Carbon Nanotubes for Drug Delivery

Carbon nanotubes (CNTs) are ready-made, strong, electrically useful microscopic tubes that form naturally in soot from sheets of carbon atoms. Various proteins adsorb spontaneously on the sidewalls of CNTs, enabling protein–nanotube conjugates. Proteins can be readily transported inside various mammalian cells via the endocytosis pathway with CNTs acting as the transporter. CNTs are useful for future in vitro and in vivo protein delivery applications. CNTs, used as liquid-filled nanoparticles, act as absorption enhancers and improve the bioavailability of erythropoietin to 11.5 % following administration into the small intestinal in experimental animals (Venkatesan et al. 2005).

CNTs can pierce cell membranes like tiny needles without damaging the cell. If proteins or nucleic acids are attached to the nanotubes, they can also go right through the cell membrane. CNTs can also carry small pharmaceutical molecules such as antibiotics or cancer drugs directly into cells and have been successfully used to inject antifungal agents into cells (Wu et al. 2005). It is also possible to attach two agents to nanotubes enabling combination therapies or to trace the uptake of a drug by adding a marker.

CNTs can be formed into nanopipettes by tapering the diameter from 700 nm to only a few nanometers with central channels that could sense chemicals at very specific locations and eventually deliver tiny amounts of fluids under the skin. Dense arrays of nanopipettes could be used for drug delivery.

CNT–Liposome Conjugates for Drug Delivery into Cells

CNTs can be used as carriers for drug delivery due to their facile transport through cellular membranes. However, the amount of loaded drug on a CNT is rather small; therefore, liposomes are employed as a carrier of a large amount of drug. In a CNT–liposome conjugate (CLC) drug delivery system, drug-loaded liposomes are covalently attached to CNT so that a high dose of the drug can be delivered into cells without potential adverse systemic effects when administered with CNTs without liposomes (Karchemski et al. 2012). This system is expected to provide versatile and controlled means for enhanced delivery of one or more agents that can be stably conjugated with liposomes.

Lipid–Protein Nanotubes for Drug Delivery

Nanotubes for drug or gene delivery applications can be developed with open or closed ends. The nanotubes could be designed to encapsulate and then open up to deliver a drug or gene in a particular location in the body. This can be achieved by manipulating the electrical charges of lipid bilayer membranes and microtubules (MT) from cells. Synchrotron X-ray scattering and electron microscopy studies of self-assembly of cationic liposome-MT complexes show that vesicles either adsorb onto MTs, forming a "beads on a rod" structure, or undergo a wetting transition and coat the MT. Tubulin oligomers then coat the external lipid layer, forming a tunable lipid–protein nanotube. The beads on a rod structure are a kinetically trapped state. The energy barrier between the states depends on the membrane bending rigidity and charge density. The inner space of the nanotube in these experiments measures about 16 nm in diameter, and the whole capsule is about 40 nm in diameter (Raviv et al. 2007; Safinya et al. 2011). By controlling the cationic lipid/tubulin stoichiometry, it is possible to switch between two states of nanotubes with either open ends or closed ends with lipid caps, a process that forms the basis for controlled chemical and drug encapsulation and release. Taxol is one type of drug that can be delivered with these nanotubes.

Halloysite Nanotubes for Drug Delivery

Halloysite is a natural clay material typically used in ceramics. Some clay reserves contain halloysite in the form of naturally occurring nanotubes that are approximately 10–100 nm in internal diameter and vary in length from a few hundred nanometers to several micrometers. Halloysite nanotubes can be loaded with drugs for sustained release, extending the effective life of drugs as they migrate out of the tubes over time. Once loaded, these tubes can also be encapsulated to further influence the rate of elution. This enables alteration of the drug release profile and extends the effectiveness of drugs without increasing potency. Compared to CNTs, halloysite nanotubes are far less expensive and have an extraordinarily large surface area. This feature promises significant advantages for drug delivery applications since surface area contact allows for greater control of drug loading and elution profiles.

Loaded nanotubes can also be combined with other technologies for noninvasive activation. Nanotubes can be coated with nanomagnetic material that can subsequently be heated selectively and noninvasively using specific electromagnetic energy. Heating can thus provide elution on demand. The benefits of using naturally occurring halloysite material for specific drug delivery applications are longer delivery times, more control of the drug release profile, and improved safety profiles. This technology can be applied to several product platforms, including transdermal drug delivery and drug-loaded wound care products. Transdermal delivery with halloysite nanotubes can enable a more controlled elution profile with several potential benefits:

- Eliminates the high initial delivery rate and improves the safety profile, particularly with drugs such as stimulants or hormones.
- More uniform delivery can result in better maintenance of the effective clinical dose.
- Less drug loading is required per patch. Since much of the drug is discarded when the patch is removed, this can lead to reduced costs.

Wound care products range from simple bandages to long-term treatments to promote healing and reduce the chances of infection and scarring. Drugs loaded into halloysite tubes and embedded into the base layer of a bandage can be released over an extended time period. This increases the duration of drug effectiveness and reduces the frequency with which a bandage needs to be changed. This novel delivery form can provide new dosage formulations with several advantages:

- Linear release ensures maintenance of clinically effective doses.
- Compliance and ease of use; longer elution times mean fewer bandage changes.
- Uniformity of drug delivery: elution from halloysite.

Nanocochleates

Cochleate, a lipid-based delivery system, is formed as a result of interaction between cations, e.g., Ca^{2+} and negatively charged phospholipids such as phosphatidylserine. Cochleates are stable precipitates with a unique structure consisting of a large, continuous, solid, lipid bilayer sheet rolled up in spiral, with no internal aqueous space. They are nontoxic and noninflammatory and have been used as vehicles for oral and parenteral delivery of protein and peptide antigens.

Smart Pharmaceuticals™ (BioDelivery Sciences International, BDSI) is based on Bioral™ technology, which uses nanocochleates and allows biopharmaceutical manufacturers to offer biologically active compounds with unparalleled convenience, shelf-life, and reduced side effects. This drug delivery technology has been applied to generic, off-patent injectable drugs to make them patent-protected oral drugs. The company has developed food processing technology to encochleate sensitive and easily degraded nutrients (beta-carotene, antioxidants, and others) for addition to processed food and beverages. Nanoencochleation's all-natural process encochleates and preserves essential nutrients like antioxidants into a protected "shell" for in high-temperature/pressure canning and bottling applications.

Bioral® technology was used to encapsulate and deliver a siRNA therapeutic in a mouse model of influenza. The siRNA targets critical gene segments shared by avian influenza (H5N1). A single intranasal dose of encochleated siRNA administered 4 h after influenza exposure reduced virus titers in the lung by 200 times. It was 25 times more effective in reducing the virus than intravenous delivery as siRNA is destroyed rapidly when introduced into the blood directly.

Nanobiotechnology and Drug Delivery Devices

There is a need to improve devices introduced into the human body. Some drug delivery devices are implanted in the body for release of therapeutic substances. The lining of these devices can be improved by nanotechnology. For example, implants can be coated by nanolayers of polymers.

Nanoencapsulation

The Nanocoat™ process (Nanotherapeutics Inc) is a patented, solvent-free encapsulation system for coating micron and submicron size powders. Adapted from commercial fluidization systems, a core nanoparticle/microparticle is excapsulated with a thin layer of a coating material, such as a surfactant or a biodegradable polymer. The coating may be applied to slow the rate of release of an active component, improve the dispersion/flow properties, or increase the absorption into the systemic circulation. The Nanocoat™ process has been shown to provide a method of producing complete, or continuous, coated drug particles of high encapsulation efficiency while requiring minimal processing. The process also has several advantages over current techniques including:

- It is a fast process with run-times on the order of minutes.
- A variety of coating materials can be used, making it possible to produce films from materials with proven biocompatibility.
- It is a dry, solvent-free technique that can be conducted under a sterile cGMP conditions.
- Particle agglomeration/adhesion can be minimized by applying coatings that affect the bonding nature and electrostatic charge on the surface.

Formation of microcapsules by depositing coatings onto the nanoparticle surface will make it possible to control drug release kinetics by (a) diffusion of the drug though a polymeric coating and (b) degradation of a biodegradable polymer coating on the drug particles, releasing the core drug material.

GeneSegues' nanocapsules are designed using a flexible formulation process and have the following characteristics:

- The drug is condensed to a small molecule <50 nm in diameter.
- The drug is fully encapsulated in a stable, controlled-release capsule.
- Ability to carry large or small molecules.
- Capsule coating can be made of ligands for receptor-mediated targeted delivery to different organs, tissues, and cells.
- Choices of route of administration include topical application, tablets, intravenous, or via devices.

Nanotechnology-Based Device for Insulin Delivery

One of the main aims of insulin therapy for diabetes is to appropriately mimic physiological insulin secretion levels and their correlation with glucose concentration in healthy individuals. A nanoscale device with channels and insulin monomers/dimers enclosed was proposed that can sense the increase in glucose levels and release monomeric insulin through channels in the nanocapsule (Koch et al. 2006). Ideally, insulin dimers would be blocked from passage, which will provide physiologically relevant insulin monomers to bind to the insulin receptor. Upon return of glucose levels to basal levels, the channels will close and insulin passage is stopped. Developments in functionalized nanocontainers will enable glucose-sensitive receptacles to be engineered. Such devices could be used to provide new therapeutic approaches in insulin treatment.

Nanoporous Materials for Drug Delivery Devices

Nanoporous materials with ordered and controlled pore structures, high surface area, and pore volume are particularly suited for implantable drug delivery systems. Considerable progress has been made in electrochemically engineered nanopores/nanotube materials such as nanoporous alumina and nanotubular titanium (Losic and Simovic 2009). Nanoporous devices can be used for cell encapsulation in hormonal therapy. Biosensors mounted on these devices can be used for noninvasive signal detection.

Nanopore Membrane in Implantable Titanium Drug Delivery Device

Implanted titanium drug delivery devices using silicon nanopore membranes can control the release by diffusion of an encapsulated drug at a nearly constant rate. Nearly zero-order drug kinetics can be achieved over long periods of time. Such nanodevices are suitable for the delivery of protein and peptide drugs and avoid the poor pharmacokinetics associated with injections, providing an optimized method of delivery for these compounds. The drug can be formulated as a dry powder or a concentrated suspension and maintains its stability. The drug is protected from the immunological reaction of the body by the nanopore membrane, which releases the drug but excludes entry of unwanted cells.

Measuring the Permeability of Nanomembranes

In order to design molecular transport systems effectively, one needs to know how big the pores in the vehicle's membranes are and how easily the contents can pass through them. This has proved quite difficult. A method for determining the permeability of thin films has been developed. A molecular beacon immobilized inside a porous silica particle that is subsequently encapsulated within a thin film can be used to determine the size of DNA that can permeate through the film (Johnston and Caruso 2005). Using this technique, it has been determined that over 3 h, molecules larger than 4.7 nm do not permeate 15-nm thick polyelectrolyte multilayers and after 75 h molecules larger than 6 nm were excluded. This technique has applications for determining the permeability of films used for controlled drug and gene delivery. A molecular beacon made from single DNA strands has been used to measure how easily DNA or genes can pass through the wall of drug delivery particles.

The beacons used are single DNA strands which have a light-emitting molecule (a fluorophore) at one end and a quencher at the other. The DNA strand self-assembles so that the two end segments pair up, forming a loop in the center – much like the shape of a round-bottomed flask. This is the closed molecular beacon. When the beacon is closed, the fluorophore on one end of the DNA strand is close to the "quencher" on the other end, which stops the fluorophore from giving off light. To determine the permeability of the capsule, the molecular beacons are placed inside the delivery vehicle. If DNA passes through the capsule wall, the beacon opens and the fluorophore emits light. So when DNA passes through the capsule, the beacon is switched "on." If no DNA passes through the capsule, the beacon remains switched off. This technique can be used in the design of intelligent drug delivery systems which can transport medicine to target locations and release the contents in a controlled way.

Nanovalves for Drug Delivery

A nanovalve that can be opened and closed at will to trap and release molecules could be used as a drug delivery system (Nguyen et al. 2005). This nanovalve consists of moving parts – switchable rotaxane molecules that resemble linear motors – attached to a tiny piece of glass (porous silica), which measures about 500 nm. It is big enough to let molecules in and out, but small enough so that the switchable rotaxane molecules can block the hole. The valve is uniquely designed so one end attaches to the opening of the hole that will be blocked and unblocked, and the other end has the switchable rotaxanes whose movable component blocks the hole in the down position and leaves it open in the up position. The researchers used chemical energy involving a single electron as the power supply to open and shut the valve, and a luminescent molecule that allows them to tell from emitted light whether a molecule is trapped or has been released. The nanovalve is much smaller than living cells. A nanovalve combined with biomolecules could be inserted into a cell and activated by light to release a drug inside a cell.

Switchable rotaxanes are molecules composed of a dumbbell component with two stations between which a ring component can be made to move back and forth in a linear fashion. Switchable rotaxanes have been used in molecular electronics and are now being adapted for use in the construction of artificial molecular machinery. Further research will test the size hole that can be blocked to see whether larger molecules such as enzymes can be transported inside the container.

Nanochips for Drug Delivery

MicroCHIPS Inc (Bedford, MA) is working on a silver dollar-size device to implant under a patient's skin or in the abdomen that would provide tiny, precise doses of hormones, pain medication, or other pharmaceuticals. The chips, made of silicon or polymer, feature hundreds of tiny micromachined wells that can be loaded with a mixture of medicines. A microcontroller could release small amounts of different chemicals on a customizable schedule. Or biosensors could trigger releases by detecting blood sugar levels or other biochemical conditions. If approved, such a device could provide diabetics with doses of insulin so that they could forgo daily injections for as much as a year. Or it could help liberate AIDS patients from following complicated daily regimens of multiple medications. To more closely imitate how the body releases hormones, the device could dispense compounds such as estrogen in periodic bursts.

Products currently in development include external and implantable microchips for the delivery of proteins, hormones, pain medications, and other pharmaceutical compounds. A clinical trial of an implantable microchip for delivery of human parathyroid hormone fragment ($hPTH_{1-34}$) has been conducted successfully in osteoporotic postmenopausal women. Potential advantages of these microchips include small size, low power consumption, absence of moving parts, and the ability to store and release multiple drugs or chemicals from a single device.

Nanobiotechnology-Based Transdermal Drug Delivery

Introduction

Transdermal drug delivery is an approach used to deliver drugs through the skin for therapeutic use as an alternative to oral, intravascular, subcutaneous, and transmucosal routes. Technical details are described in a special report on transdermal drug delivery (Jain 2012d). Nanoparticles and nanoemulsions have better skin penetration than larger particles.

There is experimental evidence for the potential of nanoparticles as delivery vectors for antigens and DNA for the purpose of transdermal vaccination protocols. Fluorescent particles ranging in size and charge were applied to the surface of full thickness pig skin in a diffusion chamber, and the receptor fluid was assayed to determine permeation (Kohli and Alpar 2004). Fluorescence microscopy was used to visualize the skin after experiments. The results showed that only 50- and 500-nm particles that were negatively charged were able to permeate the skin, indicating that negative particles with sufficient charge may be ideal carriers for this purpose.

Delivering genes and drugs within cells with devices approaching the nanoscale allows for new levels of precision and minimal damage to cells. Nanopatches can be used to target immunologically sensitive cells for DNA vaccination of malaria and allergies. This technology will also enable pain-free and needle-free immunotherapy of asthma.

The focus in this section is on the use of transdermal route for systemic delivery of therapeutics. Use of nanobiotechnology to improve skin penetration of drugs used for treatment of skin disorders will be described separately later in this book.

Delivery of Nanostructured Drugs from Transdermal Patches

Nanobiotechnology has been applied for the painless transdermal delivery of vaccines, peptide hormones, and other drugs. The patches are structured on the skin side with microprotrusions, which hold the drugs to be delivered. The protrusion face of the patch is applied to the skin where they cross the outer surface layer of the skin only reaching as far as the interstitial space avoiding nerves and blood vessels. In this interstitial space, the nanostructured drugs are released from the surface of the protrusions, and as the biocompatible polymer biodegrades, the drugs are released continuously from the body of the protrusions. The nanostructured drugs are either taken up by the cells of the immune system (for vaccination applications) or flow through the interstitial fluid to other compartments in the body.

Effect of Mechanical Flexion on Penetration of Buckyballs Through the Skin

Normally bucky amino clusters form spherical clusters that are up to 12 times larger than the width of the intercellular gaps in skin. In one study, confocal microscopy depicted dermal penetration of fullerene-substituted phenylalanine (Baa) derivative of a nuclear localization peptide at 8 h in skin flexed for 60 and 90 min, whereas Baa-Lys(FITC)-NLS, but did not penetrate into the dermis of unflexed skin until 24 h (Rouse et al. 2007). Transmission electron microscopy analysis revealed fullerene–peptide localization within the intercellular spaces of the stratum granulosum. This study shows that repetitive movement can speed the passage of nanoparticles through the skin.

Ethosomes for Transdermal Drug Delivery

Ethosomes – soft, malleable vesicles with size ranging from 30 nm to a few microns – form the basis of Ethosome Delivery System (Novel therapeutic Technologies). Ethosomal systems were found to be significantly superior at delivering drugs through the skin in terms of both quantity and depth when compared to liposomes and to many commercial transdermal and dermal delivery systems. Visualization by dynamic light scattering showed that ethosomes could be unilamellar or multilamellar through to the core. These novel delivery systems contain soft phospholipid vesicles in the presence of high concentrations of ethanol. Ethosomal systems are sophisticated conceptually but characterized by simplicity in their preparation, safety, and efficiency – a rare combination that can expand their applications.

Because of their unique structure, ethosomes are able to encapsulate and deliver through the skin highly lipophilic molecules such as cannabinoids, testosterone, and minoxidil, as well as cationic drugs such as propranolol and trihexyphenidyl. Results obtained in a double-blind two-armed randomized clinical study showed that treatment with the ethosomal acyclovir formulation significantly improved all the evaluated parameters (Godin and Touitou 2003).

Ethosomes penetrate cellular membrane releasing the entrapped molecule within cells. Studies focused on skin permeation behavior of fluorescently labeled bacitracin from ethosomal systems through human cadaver and rat skin demonstrated that the antibiotic peptide was delivered into deep skin layers through intercorneocyte lipid domain of stratum corneum (Godin and Touitou 2004). Ethosomal delivery systems could be considered for the treatment of a number of dermal infections, requiring intracellular delivery of antibiotics, whereby the drug must bypass two barriers: the stratum corneum and the cell membrane. Ethosomal formulation of testosterone could enhance testosterone systemic absorption and also be used for designing new products that could solve the weaknesses of the current testosterone replacement therapies (Ainbinder and Touitou 2005).

Advantages of ethosomes over other transdermal delivery systems are:

- Enhanced permeation
- Platform for the delivery of large and diverse group of drugs including peptides and very lipophilic molecules
- Safe and approved components
- Passive, noninvasive delivery system
- Available for immediate commercialization
- High patient compliance
- High cost to benefit ratio

NanoCyte Transdermal Drug Delivery System

NanoCyte drug delivery system (NanoCyte Inc) is based on a sophisticated injection system developed by the sea anemone during million years of evolution. The NanoCyte natural substance is extracted from aquatic invertebrates. Each microcapsule contains a coiled microscopic nanotube, which unfolds on activation – a process whereby high pressure of 200 atmospheres is developed within the microcapsule. The long thin nanotube evaginates out of the microcapsule and penetrates the skin at an acceleration of 40,000 g to deliver the drug efficiently in a fraction of a second into the epidermis skin layer. NanoCyte can be formulated as a suspension, lotion, cream, or a stick. NanoCyte can also be activated after attaching to an adhesive patch. Advantages of this system include:

- Immediate intradermal delivery
- Device-less active delivery
- Painless administration
- Avoidance of large dosages and side effects
- Treatment of large skin areas
- Multipoint penetration using nanoinjectors
- Ease of use

Safety Issues of Applications of Nanomaterial Carriers on the Skin

The European Commission has requested the Scientific Committee on Consumer Products (SCCP) to prepare an opinion on "Safety of Nanomaterials in Cosmetic Products." The preliminary version of the opinion can be found online (http://ec.europa.eu/health/ph_risk/committees/04_sccp/docs/sccp_o_099.pdf). The results obtained with nanosized delivery systems were not consistent. The following list of potential properties was considered that are relevant to transdermal drug delivery:

- Nanomaterial constituents (such as lipids or surfactants) may act as penetration enhancers by penetrating individually into the stratum corneum (after particle disruption on skin surface) and subsequently altering the intercellular lipid lamellae within this skin layer.
- Nanomaterials may serve as a depot for sustained release of dermally active compounds.
- Nanomaterials may serve as a rate-limiting membrane barrier for the modulation of systemic absorption, hence providing a controlled transdermal delivery system.

Transdermal Administration of Lipid Nanocapsules

Due to their small size, lipid nanocapsules (LNC) are a promising as a drug delivery system – as injectable or an oral or by transdermal route. Biocompatible ibuprofen LNC has been developed with a particle size of ~50 nm. Pain relief after intravenous administration of ibuprofen is prolonged by at least 2 h when administering LNC formulation. LNCs can also be used as a transdermal drug delivery system using ibuprofen as a model drug. A comparison to other lipid nanocarriers such as solid lipid nanoparticles (SLN) and nanostructured lipid carriers (NLC) and polymeric nanocarriers has been made. Polymeric carriers had fourfold higher accumulation in the skin compared to that of the LNC and twice the accumulation of SLN and NLC (Abdel-Mottaleb et al. 2011). These results would suggest that the LNC can be considered as efficient as SLN and NLC for the transdermal drug delivery whereas polymeric nanoparticles are more suitable for localized drug delivery to the skin.

Transdermal Nanoparticle Preparations for Systemic Effect

Nanoparticles have been used to facilitate transdermal delivery of several systemic drugs. Some examples are as follows:

$CaCO_3$–nanoparticle system successfully delivered insulin transdermally, as evidenced by a significant sustained decrease in blood glucose in normal mice and those with diabetes (Higaki et al. 2006). These results support the feasibility of developing transdermal nanoinsulin for human applications. A novel MicroArray Patch (Interstitial NS) is in preclinical development for insulin delivery using passive (diffusion) patches.

A multicenter, randomized, double-blind, placebo-controlled study showed that once-daily application of micellar nanoparticle estradiol emulsion was safe and effective in providing significant relief of vasomotor symptom frequency and severity in postmenopausal women (Simon et al. 2006). This is the basis of Estrasorb® (Novavax), a transdermal lotion containing estrogen for relief of menopausal symptoms, and Androsorb® (Novavax), a transdermal lotion containing androgen.

Indomethacin-loaded poly(*n*-butylcyanoacrylate) nanocapsules can improve the transdermal delivery of indomethacin compared to a conventional gel formulation. This might be due to their ultrafine particle size and their hydrophilic and hydrophobic surface characteristics.

Lipid nanoparticles have been used for transdermal delivery of flurbiprofen (Bhaskar et al. 2009). The bioavailability of flurbiprofen with reference to oral administration was found to increase by 4.4 times when gel formulations were applied and sustained drug release was demonstrated over a period of 24 h in studies on experimental animals.

Celecoxib, a selective cyclooxygenase-2 inhibitor, has been recommended orally for the treatment of arthritis and osteoarthritis. Long-term oral administration of celecoxib produces serious gastrointestinal side effects. It is a highly lipophilic, poorly soluble drug with oral bioavailability of around 40 % (capsule). The skin permeation mechanism and bioavailability of celecoxib by transdermally applied nanoemulsion formulation have been investigated (Shakeel et al. 2008). Optimized oil-in-water nanoemulsion of celecoxib was prepared by the aqueous phase titration method. Fourier transform infrared spectra and differential scanning calorimeter thermogram of skin treated with nanoemulsion indicated that permeation occurred due to the disruption of epidermal lipid bilayers by nanoemulsion, which was demonstrated by photomicrograph of skin sample. The absorption of celecoxib through transdermally applied nanoemulsion gel resulted in approximately threefold increase in bioavailability as compared to oral capsule formulation. Thus, nanoemulsions are potential vehicles for enhancement of skin permeation and bioavailability of poorly soluble drugs.

Nasal Drug Delivery Using Nanoparticles

The nasal cavity is an ideal site for delivery of both locally and systemically acting drugs. Topical administration includes agents for the treatment of nasal congestion, rhinitis, sinusitis, and related allergic and other chronic conditions. Various medications include corticosteroids, antihistaminics, anticholinergics, and vasoconstrictors. The focus in recent years has been on the use of nasal route for systemic drug delivery. Intranasal route is considered for drugs which are ineffective orally, are used chronically, and require small doses and where rapid entry into the circulation is desired. The rate of diffusion of the compounds through the nasal mucous membranes, like other biological membranes, is influenced by the physicochemical properties of the compound. Impressive improvements in bioavailability have been achieved with a range of compounds.

Chitosan, a naturally occurring polysaccharide derived from chitin, is used as an absorption enhancer for transnasal drug delivery. Chitosan is bioadhesive and binds to the mucosal membrane, prolonging retention time of the formulation on the nasal mucosa. It may also facilitate absorption through promoting paracellular transport. The chitosan nasal technology can be exploited as solution, dry powders, or nanoparticle formulations to further optimize the delivery system for individual compounds. For compounds requiring rapid onset of action, the nasal chitosan technology can provide a fast peak concentration compared with oral or subcutaneous administration. Density and size of PEG coating of poly(lactic acid)-poly(ethylene glycol) (PLA-PEG) nano- and microparticles have an important effect on their transport across the nasal mucosa. PLA-PEG particles with a high PEG coating density and a small size are more significantly transported than noncoated PLA nanoparticles or PLA-PEG nanoparticles with a lower coating density.

Nanoparticles made of low molecular weight chitosan are promising carriers for nasal vaccine delivery. Compacted DNA nanoparticles, encoding cystic fibrosis transmembrane regulator gene, can be safely administered by perfusion to the nose of cystic fibrosis subjects. A double-blind, dose escalation gene therapy trial with this technique showed evidence of vector gene transfer and partial correction of nasal potential difference that is typical for subjects with classic cystic fibrosis.

Mucosal Drug Delivery with Nanoparticles

The layers of mucus that protect sensitive tissue throughout can also prevent the entry of drugs into the body. The role of nanoparticles as drug delivery vehicles has been explored to overcome this hurdle. Cervicovaginal mucus was used for these investigations because its viscoelastic properties and mucin concentration are similar to those in many other human mucus secretions. Large nanoparticles, 200- to 500-nm in diameter, if coated with polyethylene glycol, diffused through mucus with an effective diffusion coefficient (Deff) only four- and sixfold lower than that for the same particles in water (Lai et al. 2007). In contrast, for smaller but otherwise identical 100-nm coated particles, Deff was 200-fold lower in mucus than in water. For uncoated particles 100–500 nm in diameter, Deff was 2,400- to 40,000-fold lower in mucus than in water. Much larger fractions of the 100-nm particles were immobilized or otherwise hindered by mucus than the large 200- to 500-nm particles. Thus, in contrast to the prevailing belief, these results demonstrate that large nanoparticles, if properly coated, can rapidly penetrate physiological human mucus, and they offer the prospect that large nanoparticles can be used for mucosal drug delivery.

Future Prospects of Nanotechnology-Based Drug Delivery

A desirable situation in drug delivery is to have smart drug delivery systems that can integrate with the human body. This is an area where nanotechnology will play an extremely important role. Even time-release tablets, which have a relatively simple coating that dissolves in specific locations, involve the use of nanoparticles. Pharmaceutical companies are already involved in using nanotechnology to create intelligent drug release devices. For example, control of the interface between the drug/particle and the human body can be programmed so that when the drug reaches its target, it can then become active. The use of nanotechnology for drug release devices requires autonomous device operation. For example, in contrast to converting a biochemical signal into a mechanical signal and being able to control and communicate with the device, autonomous device operation would require biochemical recognition to generate forces to stimulate various valves and channels in the drug delivery systems, so that it does not require any external control.

Subcellular or organelle-specific targeting has emerged as a new frontier in drug delivery. Nanocarriers will create the next generation of “magic bullets” that are capable of delivering a drug payload to a molecular target at a subcellular location (D’Souza and Weissig 2009). It now appears that we are on the verge of bioengineering molecular motors for specialized applications on nanoscale. These systems might be the key to yet unsolved biomedical applications that include nonviral gene therapy and interneuron drug delivery. Examples of some potential nanotechnology-based drug delivery systems are given in the following paragraphs.

Nanomolecular Valves for Controlled Drug Release

A macroscopic valve is a device with a movable control element that regulates the flow of gases or liquids by blocking and opening passageways. Construction of such a device on the nanoscale level requires (1) suitably proportioned movable control elements, (2) a method for operating them on demand, and (3) appropriately sized passageways. These three conditions can be fulfilled by attaching organic, mechanically interlocked, linear motor molecules that can be operated under chemical, electrical, or optical stimuli to stable inorganic porous frameworks (i.e., by self-assembling organic machinery on top of an inorganic chassis). A reversibly operating nanovalve has been demonstrated that can be turned on and off by redox chemistry (Nguyen et al. 2005). It traps and releases molecules from a maze of nanoscopic passageways in silica by controlling the operation of redox-activated bistable rotaxane molecules tethered to the openings of nanopores leading out of a nanoscale reservoir. Future applications could include nanofluidic systems and the controlled release of drugs from implants with nanoscopic properties.

Nanosponge for Drug Delivery

Nanosponges are hyper-cross-linked cyclodextrin polymers nanostructured to form 3D networks and are obtained by complexing cyclodextrin with a cross-linker such as carbonyldiimidazole. They have been used to increase the solubility and stability of poorly soluble drugs. β-cyclodextrin nanosponges loaded with anticancer agent tamoxifen have been used for oral drug delivery (Torne et al. 2012). In experimental studies, tamoxifen nanosponge complex with particle size of 400–600 nm was shown to be more cytotoxic than plain tamoxifen after 24 and 48 h of incubation.

Another method for making a nanosponge uses extensive internal cross-linking to scrunch a long, linear molecule into a sphere about 10 nm in diameter. Instead of trying to encapsulate drugs in nanoscale containers, this approach creates a nanoparticle with a large number of surface sites where drug molecules can be attached. A molecular transporter attached to the nanosponge can carry it and its cargo across biological barriers into specific intracellular compartments. The transporter can deliver large molecules – specifically peptides and proteins – into specific subcellular locations. A targeting unit attached to this delivery system can deliver drugs to the surface of tumors in the lungs, brain, and spinal cord. This delivery system can be adapted to carry the chemotherapy agents for targeted delivery to tumors.

Nanomotors for Drug Delivery

Basics of nanomotors – nanometer-scale machines, which are powered by chemical reactions – have been described in Chap. 3 as molecular motors. A technique to create catalytic nanomotors uses "dynamic shadowing growth," which involves a simple modification of existing methods to allow for greater flexibility in designing desired nanomotor structures (He et al. 2007). These could be used as tools to open constricted or clogged blood vessels too small for conventional stents, or they could deliver drugs by drilling through the cell wall of an organism. The researchers looked at the hundreds of moving parts in an automobile for designing each part of a nanomotor so as to achieve a controlled, flexible range of motion for the parts to work together. After successfully using the new technique to design nanorods to rotate, they broke the symmetry of the rods to form L-shaped rods which could then be aggregated to form larger particles. Then they transformed the rod into a spiral shape so that its rotation would mimic the turning of a drill. The team used the new technique to deposit a platinum or silver catalyst on different portions of the L-shaped rods and then designed different experiments to test their ability to control the motion. In a solution of hydrogen peroxide, they captured images of the nanorods turning precisely in the directions proscribed by the catalyst depositions.

Chapter 6
Role of Nanotechnology in Biological Therapies

Introduction

Biological therapies are playing an increasing role in modern medicine. This term include recombinant human proteins, monoclonal antibodies (MAbs), vaccines, cell therapy, gene therapy, antisense, and RNA interference (RNAi). Some technologies for cell and gene therapy are in themselves sophisticated methods of therapeutic delivery, whereas others require special methods of delivery. The role of nanobiotechnology in delivery of biologicals will be discussed in this chapter. MAbs are considered along with drug delivery for cancer in Chap. 7.

Nanotechnology for Delivery of Proteins and Peptides

Industrial production of therapeutic peptides and proteins is now possible, but delivery of these molecules by the oral route, which is the most desirable route, remains a challenge. Injection has been the traditional method of administration of most peptides and proteins for systemic effect. Orally ingested proteins are rapidly converted to constituent amino acids before absorption. The aim is absorption of proteins and peptides in an intact state. Three main obstacles to oral delivery of proteins are:

1. Destruction of proteins by acid and proteolytic enzymes in the stomach.
2. Difficulties of transporting molecules across the epithelial layer lining the intestine.
3. Bioavailability is low.

Nanotechnology-based methods have been explored for oral delivery of proteins. Advantages of delivery systems based on nanoparticles and cyclodextrins (CDs) are the protection of proteins from degradation, enhancement of absorption, and targeting and controlling the release of the drug. Physicochemical characteristics of nanoparticles influence the ability to successfully entrap the intended drug. Biodistribution

K.K. Jain, *The Handbook of Nanomedicine*, DOI 10.1007/978-1-61779-983-9_6,

and safety issues need also to be considered once material from the delivery system is absorbed by the body and thus interacts with biological components (Soares et al. 2007). Microemulsions with flexible formulations have been used for the solubilization of peptides and proteins to design encapsulation processes for oral delivery of insulin nanoparticles 320–350 nm in size (Graf et al. 2008).

Nanobiotechnology for Vaccine Delivery

Bacterial Spores for Delivery of Vaccines

Bacterial spores as described in Chap. 2 can be used for vaccine delivery. The spore coat can act as a vehicle for heterologous antigen presentation and protective immunization. Use of bacterial spores will solve the problem of stability and transport in the developing countries as they do not require refrigeration and the administration is needle-free.

Nanoparticles for DNA Vaccines

DNA vaccines, also referred to as genetic vaccines, are generating significant preclinical and clinical interest. It has been proven that the expression of an antigen or antigens from plasmid DNA (pDNA) may elicit both humoral and cellular immune responses. Therefore, DNA vaccines may have potential as new vaccines for important pathogens such as HIV, hepatitis C, tuberculosis, and malaria.

However, DNA vaccine delivery is not satisfactory. Relatively high doses of plasmid DNA (pDNA) are needed to elicit a response. The clinical results using "naked" pDNA have been disappointing in the breadth and depth of the immune response. Clinical trials with the gene gun have been promising, but it is unclear whether this technology will be commercially viable. Therefore, there is a need for new vaccine delivery systems that can be administered at low doses to elicit strong humoral and cellular immune responses. One promising approach is the development of microparticles and nanoparticles as delivery systems for DNA vaccines. Nanoemulsions or nanoparticle aerosol vaccines are also under development.

Nanoparticle-Based Adjuvants for Vaccines

For subunit vaccines, adjuvants play a key role in shaping immunological memory. NP delivery systems for antigens are promising adjuvants capable of promoting both cellular and humoral immune responses, but in most cases, the mechanisms of action of these materials are poorly understood. A report has documented the immune response elicited by NPs composed of multilamellar "stapled" lipid vesicles carrying a recombinant *Plasmodium vivax* circumsporozoite antigen, VMP001, both entrapped in the aqueous core and anchored to the lipid bilayer surfaces (Moon et al. 2012). Immunization with NPs and monophosphoryl lipid A (MPLA), an FDA-approved immunostimulatory agonist for Toll-like receptor-4, promoted high-titer, high-avidity antibody responses against VMP001, lasting more than 1 y in mice at tenfold lower doses than conventional adjuvants. NPs promoted robust germinal center (GC) formation at low doses of antigen where no GC induction occurred with soluble protein immunization, and GCs nucleated near depots of NPs accumulating in the draining lymph nodes over time. In parallel, NP vaccination enhanced the expansion of antigen-specific follicular helper T cells compared to vaccinations with soluble VMP001 or alum. Thus, NP vaccines may be a promising strategy to enhance the durability, breadth, and potency of humoral immunity by enhancing key elements of the B-cell response.

Although a number of adjuvants are currently approved for use in veterinary species, only alum has been widely used in humans. While it induces strong antibody responses, cell-mediated responses are often low and inflammatory reactions at the site of injection are common. Immunological properties of a novel nanobead adjuvant have been investigated in a large-animal sheep model (Scheerlinck et al. 2006). In contrast to alum, antigen covalently coupled to nanobeads induced substantial cell-mediated responses along with moderate humoral responses. No adverse reactions were seen at the site of immunization in the sheep. Thus, nanobead adjuvants in veterinary species may be useful for the induction of immunity to viral pathogens, where a cell-mediated response is required. These findings also highlight the potential usefulness of nanobead vaccines for intracellular pathogens in humans. Most adjuvants only stimulate antibodies against a particular disease. The nanobead technology gave the immune system a further boost, also producing T cells which are needed to eliminate viruses or cancer. Nanobeads measure 40 nm, and this size is critical as it is similar size to many viruses, where the nanobeads are taken up abundantly by the immune system and tricked into producing high levels of many types of T cells.

Nanospheres for Controlled Release of Viral Antigens

Aphios Corporation has developed an enhanced controlled-release technology for nanoencapsulating potent viral antigens in biodegradable polymer nanospheres by utilizing SuperFluids, which are supercritical, critical, or near-critical fluids with or without polar cosolvents. The use of SuperFluids reduces exposure of viral antigens such as HIV and influenza to potentially denaturing organic solvents such as methylene chloride and ethyl acetate and improves the stability of protein antigens in the body at ambient temperature for long periods, thereby enhancing the capability of nanoencapsulated vaccine antigens to induce the production of protective and neutralizing antibodies. This controlled-release vaccine delivery technology has the capability to deliver different types and combinations of HIV or influenza vaccine candidates including whole inactivated virus particles, DNA plasmids, and/or subunit protein antigens. SuperFluids polymer nanoencapsulation technology will reduce cost by eliminating unnecessary processing steps while improving the manufacturing environment. Unlike currently available technologies, this technology is portable, inexpensive, and amenable to large-scale processing.

Proteasomes™ as Vaccine Delivery Vehicles

Components of the immune system recognize particles more efficiently than soluble proteins. Proteasomes™ (GlaxoSmithKline) serve as vaccine delivery vehicles by virtue of their nanoparticulate nature, forming vesicles and vesicle clusters comparable to the size of small viruses. Proteosome™ vaccine vesicles and vesicle clusters may range in size from 20 to 800 nm, depending on the type and amount of antigen formulated with the Proteasomes™. The hydrophobic nature of the Proteosome™ porin proteins also contributes to vaccine delivery capabilities by facilitating interaction of the vaccine particles with, and uptake of the vaccine by, cells that initiate immune responses. The fact that Proteasomes™ are effective nasal vaccines is considered to be particularly related to this enhanced recognition and uptake afforded by Proteasomes' particulate and hydrophobic nature. This technology is being applied to develop vaccines for influenza, allergy, plague, respiratory syncytial virus, and Alzheimer's disease.

Targeted Synthetic Vaccine Particle (tSVP™) Technology

Selecta Biosciences' tSVP™ platform creates fully integrated synthetic nanoparticle vaccines engineered to mimic the properties of natural pathogens to elicit a maximal immune response. The tSVP vaccines are rationally designed to optimize the presentation of antigens to the immune system and ensure a focused and undistracted response. This is accomplished by delivering antigens and adjuvants, within the same biodegradable nanoparticle, directly to antigen-presenting cells. This approach maximizes the immune response while minimizing undesirable off-target effects. The tSVP platform includes a self-assembling nanoparticle platform that is synthetic, modular, and engineered for highly effective targeting to immune cells. The tSVP vaccines incorporate only the essential elements required for a specific, robust immune response based on precise engineering that is only possible with the proprietary nanoparticle self-assembly process.

Nanobiotechnology for Gene Therapy

Gene therapy is defined as the transfer of defined genetic material to specific target cells of a patient for the ultimate purpose of preventing or altering a particular disease state (Jain 2012c). It has three components: (1) identification of the gene that is mutated in the disease and obtaining a healthy copy of that gene, (2) carrier or delivery vehicle called vectors to deliver the healthy gene to a patient's cells, and (3) additional DNA elements that turn on the healthy gene in the right cells and at the right levels. As a higher level term, gene therapy covers other biological therapies such as antisense, RNAi, and use of genetically modified cells.

Vectors used in the past were mostly viral, but currently several nonviral techniques are being used as well. Genes and DNA are now being introduced without the use of vectors, and various techniques are being used to modify the function of genes in vivo without gene transfer. Nanoparticles and other nanostructures can be used for gene delivery.

Nanoparticle-Mediated Gene Therapy

The success of the gene therapy for clinical applications, in part, would depend on the efficiency of the expression vector as determined by the level as well as the duration of gene expression. Although various cationic polymers and lipid-based systems are being investigated, most of these systems exhibit higher level but transient gene expression. Most often, the emphasis is on the level of gene expression rather than on the duration of gene expression. In certain disease conditions, a relatively low level of gene expression (therapeutic level) but for a sustained duration may be more effective than higher level but transient gene expression. Therefore, a gene expression system that can modulate the level as well as the duration of gene expression in the target tissue is desirable. Polymer-based sustained-release formulations such as nanoparticles have the potential of developing into such a system.

Nanoparticles escape rapidly (within 10 min) from the endolysosomal compartment to the cytoplasmic compartment following their intracellular uptake via an endocytic process. The escape of nanoparticles is attributed to the reversal of their surface charge from anionic to cationic in the acidic pH of the endolysosomal compartment, causing nanoparticles to interact with the endolysosomal membrane and then escape into the cytoplasmic compartment. The rapid escape of nanoparticles from the endolysosomal compartment could protect nanoparticles as well as the encapsulated DNA from the degradative environment of the endolysosomes. Nanoparticles localized in the cytoplasmic compartment would release the encapsulated DNA slowly, thus resulting in sustained gene expression. Sustained gene expression could be advantageous, especially if the half-life of the expressed protein is very low and/or a chronic gene delivery is required for therapeutic efficacy.

Degradable nanoparticles are the only nonviral vectors that can provide a targeted intracellular delivery with controlled-release properties. Furthermore, the potential advantage of degradable nanoparticles over their nondegradable counterparts is the reduced toxicity and the avoidance of accumulation within the target tissue after repeated administration. Examples of application of nanoparticles for gene therapy are shown in Table 6.1, and some of these are described in the following text.

Table 6.1 Examples of application of nanoparticles for gene therapy

Nanoparticle	Application
Poly(D,L-lactide-co-glycolide) nanoparticles loaded with wild type p53 DNA	Inhibition of cell proliferation in cancer due to sustained gene expression following intracellular release of p53
Intravenous liposomal DOTAP:Chol-FUS1 complex of tumor suppressor gene FUS1	Suppresses tumor growth and has led to tumor regression in mouse models of metastatic lung cancer
Fluorescently labeled organically modified silica nanoparticles as a nonviral vector	For gene delivery and optical monitoring of intracellular trafficking and gene transfection
Cationized gelatin nanoparticles	Nonviral and nontoxic vectors for gene therapy
Calcium phosphate nanoparticles	Nonviral vectors for targeted gene therapy of liver
Nanotube spearing is based on the penetration of nickel-embedded nanotubes into cell membranes by magnetic field driving	This technique may provide a powerful tool for highly efficient gene transfer into a variety of cells, especially the hard-to-transfect cells
L-tyrosine-based polyphosphate nanoparticle	Degradable nonviral gene delivery systems (Ditto et al. 2009)
PAMAM dendrimers can hold DNA in cavities	Non-immunogenic vector for in vivo gene transfer
Nanoparticles: EGF-PEG-biotin-streptavidin-PEI-DNA complexes	Exhibit high transfection efficiency with no particle aggregation
Compacted DNA nanoparticles (20–25 nm): each DNA molecule is wrapped in a coat of positively charged peptides	Nanoparticles pass through a nuclear pore with 1,000-fold enhancement of gene expression compared to naked DNA. Used for transnasal gene therapy in cystic fibrosis
Cochleate delivery system is formed as a result of interaction between cations and negatively charged phospholipids	In vivo lipid-based delivery of DNA plasmids and antisense DNA
Nanorod binds plasmid DNA as well as proteins in spatially defined regions	A versatile gene delivery system that increases the plasmid's cellular internalization and cytoplasmic release
Combination of a gene, nanoparticle, and surfactant	Facilitation of gene transfer in the brain across the blood-brain barrier
Nanoneedles with a tip diameter of 1 nm for delivery of genetic material into cells	This method is the next stage in the refinement of microneedles for injecting genetic material into cells
Integrin-targeted nanoparticles	Site-specific delivery of anticancer genes
Nanocomposites: nanoparticles of titanium dioxide combined with oligonucleotide DNA that can be activated by light or radiation	Antisense genes can be delivered to a particular intracellular site and in combination with radiotherapy with the purpose of killing the cell in cancer patients
Nonionic polymeric micelles of poly(ethylene oxide)-poly(propylene oxide)-poly(ethylene oxide)	Stable gene transfer to the gastrointestinal tract can be achieved in mice by oral delivery

EGF epidermal growth factor, *PEI* polyethylenimine, *PEG* polyethylene glycol

Calcium Phosphate Nanoparticles as Nonviral Vectors

Calcium phosphate nanoparticles present a unique class of nonviral vectors, which can serve as efficient and alternative DNA carriers for targeted delivery of genes. DNA-doped calcium phosphate nanoparticles approximately 80 nm in diameter has been synthesized. DNA encapsulated inside the nanoparticle is protected from the external DNase environment and could be transferred safely under in vitro or in vivo conditions. Moreover, the surface of these nanoparticles can be suitably modified by adsorbing a highly adhesive polymer like polyacrylic acid followed by conjugating the carboxylic groups of the polymer with a ligand such as p-amino-1-thio-beta-galactopyranoside using 1-ethyl-3-(3-dimethylaminopropyl)-carbodiimide hydrochloride as a coupling agent. These surface-modified calcium phosphate nanoparticles can be used to target genes specifically to the liver.

Carbonate Apatite Nanoparticles for Gene Delivery

Biocompatible, inorganic nanoparticles of carbonate apatite have the unique features essentially required for smart delivery, as well as for the expression of genetic material in mammalian cells (Chowdhury 2007). The newly developed carbonate apatite, as with hydroxyapatite, adsorbs DNA, but, unlike the latter, it can prevent the growth of its crystals to a significant extent, enabling the synthesis of nanosized crystals to effectively carry the associated DNA across the cell membrane. It also possesses a high dissolution rate in endosomal acidic pH, leading to the rapid release of the bound DNA for a subsequent high level of protein expression. Carbonate apatite is a natural component of the body and is usually found in the hard tissues, such as bone and teeth. Moreover, because of their nanosize dimensions and sensitivity to low pH, particles of carbonate apatite are quickly degraded when taken up by cells in their acidic vesicles, without any indication of toxicity. Apatite nanoparticles are promising candidates for nonviral gene delivery and are superior to polymer- or lipid-based systems that are generally nonbiodegradable and inefficient.

Gelatin Nanoparticles for Gene Delivery

Construction of gelatin nanoparticles as biodegradable and low cell toxic alternative carrier to existing DNA delivery system has been described (Zwiorek et al. 2005). In order to bind DNA by electrostatic interactions onto the surface of the gelatin nanoparticles, the quaternary amine cholamine was covalently coupled to the particles. The modified nanoparticles were loaded with different amounts of plasmid in varying buffers and compared to polyethylenimine-DNA complexes (PEI polyplexes) as gold standard. In contrast to PEI polyplexes, cationized gelatin nanoparticles almost did not show any significant cytotoxic effects. Cationized gelatin nanoparticles have the potential of being a new effective carrier for nonviral gene delivery. The major benefit of gelatin nanoparticles is not only the very low cell toxicity but also their simple production combined with low costs and multiple modification opportunities offered by the matrix molecule.

The potential of engineered gelatin-based nanoparticles, nanovectors, has been investigated to deliver therapeutic genes to human breast cancer tumors implanted in mice (Kommareddy and Amiji 2007). Plasmid DNA encoding for the soluble form of the extracellular domain of VEGF-R1 or sFlt-1 was encapsulated in the control and PEG-modified gelatin-based nanoparticles. Following intravenous administration in female Nu/Nu mice bearing orthotopic MDA-MB-435 breast adenocarcinoma xenografts, 15 % of the dose found its way into the tumor. In vivo expression of sFlt-1 plasmid DNA was therapeutically active as shown by suppression of tumor growth and microvessel density measurements. The results of this study show that PEG-modified gelatin-based nanovectors can serve as a safe and effective systemically administered gene delivery vehicle for solid tumor. Clinical trials of this method are expected in the near future.

Immunolipoplex for Delivery of p53 Gene

A sterically stabilized immunolipoplex, containing an antitransferrin receptor single-chain antibody fragment-PEG molecule, has been developed to specifically and efficiently deliver a therapeutic gene to tumor cells (Yu et al. 2004). Lipoplex nanoparticles resembling virus particles can penetrate deeply into the tumor and move efficiently into cells. The immunolipoplex is spiked on the outside with antibody molecules that will seek out, bind to, and then enter cancer cells including metastases wherever they hide in the body. These molecules bind to the receptor for transferrin that is present in large numbers on cancer cells. Once inside, the immunolipoplex will deliver its payload, the p53 gene, whose protein helps to signal cells to self-destruct when they have the kind of genetic damage characterized by cancer and by cancer therapies. Immunolipoplex has shown promising results in animal tumor models, and a phase I clinical trial in patients with advanced solid cancers has been approved. The trial is supported by the NIH. Among the solid tumors approved for testing in the clinical trial are head and neck, prostate, pancreatic, breast, bladder, colon, cervical, brain, melanoma, liver, and lung cancers.

Immunolipoplex-based gene transfer represents an advance over the viral vectors that have been used to deliver gene therapy because these liposomes do not produce the immunologic response seen when disabled viruses are used to carry the payload. Immunolipoplex also substantially improves the anticancer effects of both chemotherapy and radiation therapy. These agents work synergistically with traditional therapies because the restoration of p53 protein helps push cancer cells that are now damaged to self-destruct. This approach will make it difficult for the cancer cells to become resistant to therapy and will be less likely to recur after therapy is complete.

Lipid Nanoparticles for Targeted Delivery of Nucleic Acids

Tekmira's delivery technology platform uses lipid nanoparticle (LNP), which fully encapsulates and systemically delivers a variety of nucleic acid molecules such as short interfering RNAs (siRNAs). Preclinical studies have shown them to be effective in delivering the drug to target organs and into cells where the nucleic acid–based drug

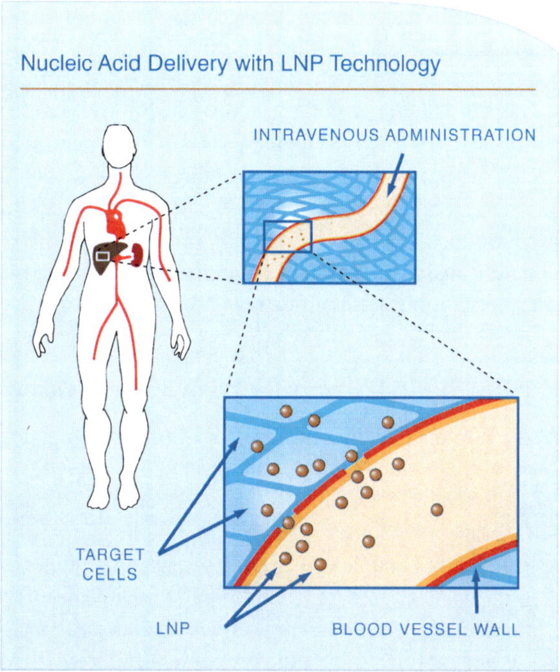

Fig. 6.1 Nucleic acid delivery with lipid nanoparticle (LNP) technology (Courtesy of Tekmira Pharmaceutical Corporation)

can carry out its desired effect while minimizing systemic toxicity. LNP technology relies on enhanced permeability and retention effect, which occurs because these nucleic acid-containing particles have a long circulation time in the blood, resulting in increased accumulation at sites of vascular leak such as those found at sites of tumor cell growth, infection, or inflammation. Once at the target site, cells take up the LNP through endocytosis, and the nucleic acid payload is delivered inside the cell resulting in a high degree of potency. LNP technology is shown schematically in Fig. 6.1.

Nanoparticles for Imaging and Intracellular Delivery of Nucleic Acids

Although materials have been developed and studied for polynucleotide transfer, the biological mechanisms and fate of the synthetic vehicle has remained elusive due to the limitations with current labeling technologies. Polymer beacons have been developed that enable the delivery of nucleic acids to be visualized at nanoscale (Bryson et al. 2009). The polycations have been designed to contain repeated oligoethyleneamines, for binding and compacting nucleic acids into nanoparticles, and lanthanide chelates (either luminescent europium Eu^{3+} or paramagnetic gadolinium Gd^{3+}). The chelated Lns allow the visualization of the delivery vehicle both on the nm/μm scale via microscopy and on the sub-mm scale via MRI. These delivery beacons effectively bind and compact plasmid DNA (pDNA) into nanoparticles and protect nucleic acids from nuclease damage. These delivery beacons efficiently deliver pDNA into cultured cells and do not exhibit toxicity. Micrographs of cultured cells exposed to the nanoparticle complexes formed with fluorescein-labeled pDNA and the europium-chelated polymers reveal effective intracellular imaging of the delivery process. MRI of bulk cells exposed to the complexes formulated with pDNA and the gadolinium-chelated structures show bright image contrast, allowing visualization of effective intracellular delivery. Because of their versatility, these delivery beacons possess remarkable potential for tracking and understanding nucleic acid transfer in vitro and have promise as in vivo vectors for gene therapy and agents for combining diagnostics and therapeutics.

Nanoparticles Linked to Viral Vectors for Photothermal Therapy

Hyperthermia can be produced by near-infrared laser irradiation of gold nanoparticles present in tumors and thus induce tumor cell killing via a bystander effect. However, selective delivery and physical targeting of gold nanoparticles to tumor cells are necessary to improve therapeutic selectivity. Covalent coupling of gold nanoparticles to retargeted adenoviral vectors enable selective delivery of the nanoparticles to tumor cells, thus facilitating hyperthermia and gene therapy as a combinatorial therapeutic approach. For this, sulfo-N-hydroxysuccinimide-labeled gold nanoparticles were linked to adenoviral vectors encoding a luciferase reporter gene driven by the cytomegalovirus promoter (Everts et al. 2006). The covalent coupling retains virus infectivity and ability to retarget tumor-associated antigens. These results show the feasibility of using adenoviral vectors as carriers for gold nanoparticles.

Nanoparticles for p53 Gene Therapy of Cancer

One of the important considerations in p53 gene delivery for tumor growth inhibition would be the sustained expression of the p53 protein in the target cells. A single-dose regimen results in only a weak and transient inhibition of cell proliferation. Multiple doses are required to obtain inhibition of cell proliferation comparable to that with viral vectors. Several mechanisms have been attributed to wt-p53 gene-mediated cancer therapy such as apoptosis of cancer cells, cell cycle arrest, and/or the antiangiogenic effect of the protein. Gene delivery with nanoparticles would require direct intratumoral injection in the case of a solid tumor or delivery via a catheter to an accessible diseased tissue. However, tumor targeting via intravascular administration would be possible if nanoparticle surface is modified to avoid extravasation by the reticuloendothelial system.

Nanoparticles with Viruslike Function as Gene Therapy Vectors

Novel multifunctional DNA carriers (MDCs) have been described, which self-assemble with DNA to form structured nanoparticles that possess viruslike functions for cellular trafficking (Glover et al. 2009). The new gene therapy vectors use the same machinery that viruses use to transport their cargo into cells. To create the new gene therapy vector, the scientists used pieces of different genes to create a protein called a "modular DNA carrier," which can be produced by bacteria. This protein carries therapeutic DNA and delivers it to a cell's nucleus, where it reprograms a cell to function properly. In the laboratory, these carrier proteins were combined with therapeutic DNA and attached to cell membrane receptors and the nuclear import machinery of target cells. In turn, the packaged DNA moved into the cell through the cytoplasm and into the nucleus. The nanoparticles were internalized in cell-specific fashion and subsequently exited the endosome into the cytoplasm. The nanoparticles interact with cellular nuclear transport proteins and are actively trafficked into the cell nucleus of nondividing cells, resulting in three- to fourfold higher reporter gene expression in growth-arrested human embryonic kidney cells, as well as lower cytotoxicity, than lipid and polyethylenimine vectors. MDCs that utilize cellular signaling pathways have enormous potential to safely and efficiently deliver therapeutic transgenes into the nucleus of nondividing cells.

Nanobiolistics for Nucleic Acid Delivery

Biolistic transfection using a gene gun is a method of incorporating DNA or RNA into cells that are difficult to transfect using traditional methods. Microparticles used in this technique are efficient at delivering DNA into cells but cannot transfect small cells and may cause significant tissue damage, thus limiting their potential usefulness. Nanobiolistic by use of 40 nm diameter nanoparticles results in ~30 % fewer damaged HEK293 cells following transfection and considerably enhances details of small cellular structures (O'Brien and Lummis 2011). The discovery that smaller projectiles are equally effective but cause less tissue damage could therefore have a significant impact on the feasibility of nanobiolistic transfection as a therapeutic technique. It may also be possible to modify the nanoparticles, e.g., with polyethylenimine to create cationic gold particles, which have been shown to deliver increased amounts of DNA. The use of nanoparticles as efficient carriers of genetic material also enhances the prospects of efficiently transfecting smaller organisms or specific regions of cells such as dendritic spines.

Silica Nanoparticles for Gene Delivery

Core shell silica particles with a diameter of 28 nm have been synthesized. The role of freeze-drying for the conservation of zwitterionic nanoparticles and the usefulness of different lyoprotective agents (LPA), DNA-binding capacity, and transfection efficiency have been investigated. Of the various LPAs screened in the investigations, trehalose and glycerol were found to be well suited for conservation of cationically modified silica nanoparticles with simultaneous preservation of their DNA-binding and transfection activity.

Fluorescently labeled organically modified silica nanoparticles are useful as nonviral vectors for gene delivery as well as for optically monitoring intracellular trafficking and gene transfection. Highly monodispersed, stable aqueous suspensions of organically modified silica nanoparticles, encapsulating fluorescent dyes and surface functionalized by cationic-amino groups, are produced by micellar nanochemistry. Gel electrophoresis studies reveal that the particles efficiently complex with DNA and protect it from enzymatic digestion of DNase 1. The electrostatic binding of DNA onto the surface of the nanoparticles, due to positively charged amino groups, is also shown by intercalating an appropriate dye into the DNA and observing the fluorescence resonance energy transfer between the dye (energy donor) intercalated in DNA on the surface of nanoparticles and a second dye (energy acceptor) inside the nanoparticles. Imaging by fluorescence confocal microscopy shows that cells efficiently take up the nanoparticles in vitro in the cytoplasm, and the nanoparticles deliver DNA to the nucleus. This work shows that the nanomedicine approach, with nanoparticles acting as a drug delivery platform combining multiple optical and other types of probes, provides a promising direction for targeted therapy with enhanced efficacy as well as for real-time monitoring of drug action.

Capped mesoporous silica nanoparticle (MSN) materials have been designed as efficient stimuli-responsive controlled-release systems with the advantageous "zero premature release" property. A variety of internal and external stimuli for controlled release of different cargos, as well as the biocompatibility of MSN both in vitro and in vivo, indicate that these multifunctional materials will find a wide variety of applications for gene delivery (Zhao et al. 2010).

Targeted Nanoparticle-DNA Delivery to the Cardiovascular System

Targeting gene therapy to the cardiovascular system is a challenge. Biodegradable polymeric superparamagnetic nanoparticle formulations have been formulated using a modified emulsification-solvent evaporation methodology with both the incorporation of oleate-coated iron oxide and a polyethylenimine oleate ion-pair surface modification for DNA binding (Chorny et al. 2007). The DNA was in the form of a plasmid, a circular molecule that carried a gene that coded for a growth-inhibiting protein adiponectin.

Magnetically driven nanoparticle-mediated gene transfer was studied using a green fluorescent protein reporter plasmid in cultured arterial smooth muscle cells and endothelial cells. Nanoparticle-DNA internalization and trafficking were examined by confocal microscopy. Cell growth inhibition after nanoparticle-mediated adiponectin plasmid transfection was studied as an example of a therapeutic end point. Nanoparticle-DNA complexes protected DNA from degradation and efficiently transfected quiescent cells under both low and high serum conditions after a 15-min exposure to a magnetic field. There was negligible transfection with nanoparticle in the absence of a magnetic field. Larger-sized nanoparticles (375 nm diameter) exhibited higher transfection rates compared with 185- and 240-nm-sized nanoparticles. Internalized larger-sized nanoparticles escaped lysosomal localization and released DNA in the perinuclear zone. Adiponectin plasmid DNA delivery using nanoparticles resulted in a dose-dependent growth inhibition of cultured arterial smooth muscle cells. It is concluded that magnetically driven plasmid DNA delivery can be achieved using biodegradable nanoparticles containing oleate-coated magnetite and surface modified with PEI oleate ion-pair complexes that enable DNA binding. The materials composing the nanoparticles are biodegradable, so they break down into simpler, nontoxic chemicals that can be carried away in the blood. This addresses the safety concerns of the use of nonbiodegradable nanoparticles in vivo. As a nonviral method, it avoids the unwanted immune system responses that have occurred when viruses are used to deliver genes.

Although the research, done in cell cultures, is in early stages, it may represent a new method for delivering gene therapy to benefit blood vessels damaged by arterial disease. Such nanoparticles could be magnetically directed into stents inserted into a patient's partially blocked vessels to improve blood flow. Delivering antigrowth genes to stents could help prevent restenosis. The magnetically driven delivery system also may find broader use as a vehicle for delivering drugs, genes, or cells to a target organ. After preloading genetically engineered cells with nanoparticles, researchers could use magnetic forces to direct the cells to a target organ. Furthermore, researchers might deliver nanoparticles to magnetically responsive, removable stents in sites other than blood vessels, such as airways or parts of the gastrointestinal tract. After the nanoparticles have delivered a sufficient number of genes, cells, or other agents to have a long-lasting benefit, the stent could be removed.

Dendrimers for Gene Transfer

Dendrimers are nanoparticles ranging in size from 1 to 20 nm and are capable of holding therapeutic substances such as DNA in their cavities. They are made up of precise three-dimensional branches called dendrons, with structure that mimics the bifurcation of tree branches. The dendrimers are so close in shape and size to a histone cluster that DNA wraps around them as it does around the natural protein complex. Dendrimers show great promise as DNA and drug delivery systems. The following factors, which drive the interest in use of dendrimers as gene, transfer vectors:

1. They can be produced as precise macromolecular structures.
2. They are composed of nanoscopic building blocks or modules.
3. They are non-immunogenic.

Activated polyamidoamine (PAMAM)-dendrimers provide a new technology for gene transfer that offers significant advantages over classical methods. QIAGEN reagents based on this technology provide high gene transfer efficiencies, minimal cytotoxicity, and can be used with a broad range of cell types. This technology could also be useful for in vivo gene transfer in gene therapy applications. Multifunctionalization of dendritic polymers provides gene vectors of low toxicity, significant transfection efficiency, specificity to certain biological cells, and transport ability through their membranes (Paleos et al. 2009).

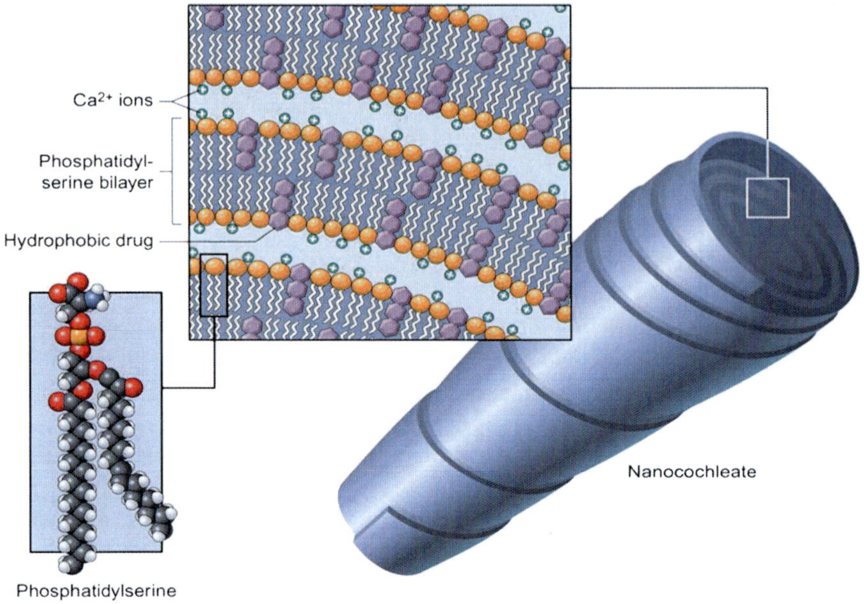

Fig. 6.2 Nanocochleate-mediated drug delivery. Addition of calcium ions to small phosphatidylserine vesicles induces their collapse into discs, which fuse into large sheets of lipid. These lipid sheets rolled up into nano-crystalline structures called nanocochleates (Courtesy of BioDelivery Sciences International Inc.)

Cochleate-Mediated DNA Delivery

Cochleate, a lipid-based delivery system, is formed as a result of interaction between cations, e.g., Ca^{2+} and negatively charged phospholipids such as phosphatidylserine. Cochleates are stable precipitates with a unique structure consisting of a large, continuous, solid lipid bilayer sheet rolled up in spiral, with no internal aqueous space. They are nontoxic and noninflammatory and have been used as vehicles for oral and parenteral delivery of protein and peptide antigens. Cochleate-mediated in vivo delivery of DNA plasmids and antisense DNA is under investigation. Protein and DNA cochleates are highly effective vaccines when given via mucosal or parenteral routes, including oral, intranasal, intramuscular, or subcutaneous. BioDelivery Sciences International Inc. is developing cochleate-based gene transfer as shown in Fig. 6.2.

Nanorod Gene Therapy

Gene therapy success has been limited in both viral and synthetic methods. In light of current gene therapy challenges, synthetic transfection systems provide several advantages over viral methods including ease of production, as well as reduced risk of cytotoxicity and immune responses. The drawbacks of synthetic vectors greatly stem from difficulty of controlling the vectors' properties at the nanoscale. The nanorod greatly overcomes those drawbacks by binding plasmid DNA as well as a potential variety of proteins in spatially defined regions. The proteins the nanorod binds can increase the plasmid's cellular internalization, cytoplasmic release, and/or nuclear internalization. The nanorods can be further guided via their magnetic properties. The potential of this versatile gene delivery system with precise composition and size has been demonstrated in cell transfection studies.

Nanomagnets for Targeted Cell-Based Cancer Gene Therapy

Using human cells as delivery vehicles for anticancer gene therapy is a promising approach for treating cancer. Monocytes naturally migrate from the bloodstream into tumors, and attempts have been made to use them to deliver therapeutic genes to these sites. However, transfected monocytes injected systemically fail to infiltrate tumors in large numbers. Therefore, nanoscale magnets have been developed to target cancer tumor cells more effectively (Muthana et al. 2008). The impact of gene therapy on cancer cells can be enhanced by "magnetic targeting," i.e., inserting nanomagnets into cells carrying genes so that the number of cells successfully reaching and invading cancer can be increased. Systemic administration of such "magnetic" monocytes to mice bearing solid tumors led to a marked increase in their extravasation into the tumor in the presence of an external magnet. Further studies are exploring the effectiveness of magnetic targeting in delivering a variety of cancer-fighting genes, including ones that could stop the spread of tumors. This technique could also be used to help deliver therapeutic genes in other diseases like arthritic joints or ischemic heart tissue.

Nanoneedles for Delivery of Genetic Material into Cells

Historically, plasmid molecules have been introduced into the nuclei by pricking cells on the nuclei with microneedles. Nanoneedles for delivery of genetic material into cells have been developed by Hayashi Wang at Challentech International Corp. (Taiwan). Needles with diameter of 2 μm and height of 60 μm have a tip 1 nm in size that can be used for injecting cells. A nanoneedle chip has been used to deliver pCMV EGFP (enhanced green fluorescent protein) plasmids into primary cells or stem cells. These can be coated with tungsten for electroporation.

Application of Pulsed Magnetic Field and Superparamagnetic Nanoparticles

A simple approach has been reported that enhances gene delivery using permanent and pulsating magnetic fields (Kamau et al. 2006). DNA plasmids and novel DNA fragments (PCR products) containing sequence encoding for green fluorescent protein were coupled to polyethylenimine-coated superparamagnetic nanoparticles (SPIONs). The complexes were added to cells that were subsequently exposed to permanent and pulsating magnetic fields. Presence of these magnetic fields significantly increased the transfection efficiency 40 times more than in cells not exposed to the magnetic field. The transfection efficiency was highest when the nanoparticles were sedimented on the permanent magnet before the application of the pulsating field, both for small (50 nm) and large (200–250 nm) nanoparticles. The highly efficient gene transfer already within 5 min shows that this technique is a powerful tool for future in vivo studies, where rapid gene delivery is required before systemic clearance or filtration of the gene vectors occurs.

Nanobiotechnology for Antisense Drug Delivery

Antisense molecules are synthetic segments of DNA or RNA, designed to mirror specific mRNA sequences and block protein production. One way to target the genetic material is to block the messenger RNA (mRNA) by using "antisense DNA," which prevents the message from ever becoming a protein. The use of antisense drugs to block abnormal disease-related proteins is referred to as antisense therapeutics. Synthetic short segments of DNA or RNA are referred to as oligonucleotides. Whereas typical drugs target the proteins, it is possible through antisense gene therapy to target the genetic material itself before it is ever made into copies of harmful proteins. Antisense drugs have the promise to be more effective than conventional drugs, but one of the problems with antisense therapy is delivery. The efficacy of antisense oligonucleotides is limited by the poor stability of the natural oligomers and the low efficacy of their cellular uptake. Nanotechnology has been used to improve this situation.

Antisense Nanoparticles

Scientists at Northwestern University have described the use of gold nanoparticle-oligonucleotide complexes, antisense nanoparticles, as intracellular gene regulation agents for the control of protein expression in cells (Rosi et al. 2006). Once inside cells, the DNA-modified nanoparticles act as mRNA "sponges" that bind to their targets and prevent them from being converted into proteins. By chemically tailoring the density of DNA bound to the surface of gold nanoparticles, they have demonstrated a tunable gene knockdown. In the future, this exciting new class of antisense material could be used for the treatment of cancer and other diseases that have a genetic basis. Advantages of attaching multiple strands of antisense DNA to the surface of a gold nanoparticle over conventional antisense oligonucleotides are:

- The DNA becomes more stable and can bind to the target mRNA more effectively than DNA that is not attached to a nanoparticle surface as in commercial agents such as Lipofectamine and Cytofectin.

- They are less susceptible to degradation by nuclease activity.
- They exhibit greater than 99 % cellular uptake.
- They can introduce oligonucleotides at a higher effective concentration than conventional transfection agents and are nontoxic to the cells under the conditions studied.

Dendrimers for Antisense Drug Delivery

Polypropylenimine dendrimers have been used for delivering a 31 nt triplex-forming oligonucleotide (ODN) in breast, prostate, and ovarian cancer cell lines, using 32P-labeled ODN (Santhakumaran et al. 2004). Dendrimers enhanced the uptake of ODN by 14-fold compared with control ODN uptake. Dendrimers exerted their effect in a concentration- and molecular weight-dependent manner, with generation 4 (G-4) dendrimer having maximum efficacy. The dendrimers had no significant effect on cell viability at concentrations at which maximum ODN uptake occurred. Gel electrophoretic analysis showed that ODN remained intact in cells even after 48 h of treatment. The hydrodynamic radii of nanoparticles formed from ODN in the presence of the dendrimers were in the range of 130–280 nm, as determined by dynamic laser light scattering. Taken together, these results indicate that polypropylenimine dendrimers might be useful vehicles for delivering therapeutic oligonucleotides in cancer cells.

Polymer Nanoparticles for Antisense Delivery System

ODNs have been shown to induce dystrophin expression in muscle cells of patients with Duchenne muscular dystrophy (DMD) and in the mdx mouse, the murine model of DMD. However, ineffective delivery of ODNs limits their therapeutic potential. Copolymers of cationic polyethylenimine (PEI) and nonionic polyethylene glycol (PEG) form stable nanoparticles when complexed with AOs, but the positive surface charge on the resultant PEG-PEI-ODN nanoparticles limits their biodistribution. A modified double emulsion procedure for encapsulating PEG-PEI-ODN polyplexes into degradable PLGA nanospheres has been described (Sirsi et al. 2009). Formulation parameters were varied including PLGA molecular weight, ester end-capping, and sonication energy/volume. The results showed successful encapsulation of PEG-PEI-ODN within PLGA nanospheres with average diameters ranging from 215 to 240 nm. Encapsulation efficiency ranged from 60 % to 100 %, and zeta potential measurements confirmed shielding of the PEG-PEI-ODN cationic charge. PLGA showed a rapid burst release of about 20 % of the PEG-PEI-ODN, followed by sustained release of up to 65 % over 3 weeks. PEG-PEI-AO polyplexes were loaded into PLGA nanospheres using an ODN that is known to induce dystrophin expression in dystrophic mdx mice. Intramuscular injections of this compound into mdx mice resulted in over 300 dystrophin-positive muscle fibers distributed throughout the muscle cross-sections, approximately 3.4

times greater than for injections of ODN alone. It is concluded that PLGA nanospheres are effective compounds for the sustained release of PEG-PEI-ODN polyplexes in skeletal muscle and concomitant expression of dystrophin and may have potential in treating DMD.

Nanoparticle-Mediated siRNA Delivery

Delivery of DNA and siRNA into mammalian cells is a powerful technique in treating various diseases caused by single-gene defects. Potent sequence selective gene inhibition by short interfering RNA (siRNA)-"targeted" therapeutics promises the ultimate level of specificity, but siRNA therapeutics is hindered by poor intracellular uptake, limited blood stability, and nonspecific immune stimulation. Use of viral vectors for siRNA delivery has also some problems. Nonviral carrier systems, especially nanoparticles, have been investigated extensively for siRNA delivery and may be utilized in clinical applications in the future. So far, a few preliminary clinical trials of nanoparticles have produced promising results. However, further research is still required to pave the way to successful clinical applications. The most important issues that need to be focused on include encapsulation efficiency, formulation stability of siRNA, degradation in circulation, endosomal escape and delivery efficiency, targeting, toxicity, and off-target effects (Yuan et al. 2011).

Chitosan-Coated Nanoparticles for siRNA Delivery

Overexpression of RhoA in cancer indicates a poor prognosis because of increased tumor cell proliferation and invasion and tumor angiogenesis. Anti-RhoA siRNA inhibits aggressive breast cancer more effectively than conventional blockers of Rho-mediated signaling pathways. A study reports the efficacy and lack of toxicity of intravenously administered encapsulated anti-RhoA siRNA in chitosan-coated polyisohexylcyanoacrylate (PIHCA) nanoparticles in xenografted aggressive breast cancers (Pille et al. 2006). The siRNA treatment inhibited the growth of tumors by 90 %, and necrotic areas were observed in tumors, resulting from angiogenesis inhibition. In addition, this therapy was found to be devoid of toxic effects. Because of its efficacy and the absence of toxicity, it is suggested that this strategy of anti-RhoA siRNA holds significant promise for the treatment of aggressive cancers.

Delivery of Gold Nanorod-siRNA Nanoplex to Dopaminergic Neurons

A nanotechnology approach that uses gold nanorod-DARPP-32 siRNA complexes (nanoplexes) can target this dopaminergic signaling pathway in the brain (Bonoiu et al. 2009). The shift in the localized longitudinal plasmon resonance peak of gold nanorods was used to show their interaction with siRNA. Plasmon-enhanced dark-field imaging was used to visualize the uptake of these nanoplexes in dopaminergic neurons in vitro. Gene silencing of the nanoplexes in these cells was evidenced by the reduction in the expression of key proteins (DARPP-32, ERK, and PP-1) belonging to this pathway, with no observed cytotoxicity. Moreover, these nanoplexes were shown to transmigrate across an in vitro model of the BBB. Therefore, these nanoplexes appear to be suited for brain-specific delivery of appropriate siRNA for therapy of drug addiction and other brain diseases.

Polymer-Based Nanoparticles for siRNA Delivery

Polyethylenimine Nanoparticles for siRNA Delivery

A highly efficient delivery system has been described using 1,4-butanediol diglycidyl ether (bisepoxide) cross-linked polyethylenimine (PEI) nanoparticles (Swami et al. 2007). The nanoparticle-DNA complexes (nanoplexes) exhibited approximately 2.5- to 5.0-fold gene transfer efficacy and decreased cytotoxicity in cultured cell lines compared to the native PEI used as gold standard and commercially available transfection agents such as Lipofectamine 2000. The bisepoxide cross-linking results in change in amine ratio in PEI; however, it retains the net charge on nanoparticle unaltered. A series of nanoparticles obtained by varying the degree of cross-linking was found to be in the size range of 69–77 nm and the zeta potential varying from +35 to 40 mV. The proposed system was also found to deliver siRNA efficiently into HEK cells, resulting in approximately 70 % suppression of the targeted gene for green fluorescent protein.

siRNA-PEG Nanoparticle-Based Delivery

Bioneer's patented siRNA-PEG nanoparticle-based delivery system has been developed to overcome many of the obstacles of siRNA delivery. In this novel and efficient delivery system, the synthetic siRNA is conjugated to polyethylene glycol (PEG) via a disulfide linkage. siRNA-PEG nanoparticles are then generated by adding a cationic core forming agent which interacts with negative-charged siRNA to form the core of a nanoparticle. The outer layer of PEG protects siRNA from degradation by ribonuclease in serum and thus significantly extends the circulation time of siRNA in blood.

The disulfide linkage between siRNA and PEG was introduced to be cleaved specifically in a reductive condition of the cytoplasm. Because the redox-potential of an intracellular environment is about two orders of magnitude lower than that of an extracellular environment, the intact form of siRNA is released in the cytoplasm after cellular uptake. Release of intact siRNA in the cytoplasm is essential for efficient gene silencing.

The efficiency of the siRNA-PEG nanoparticle formulation has been demonstrated by evaluating anticancer property of VEGF siRNA in human prostate carcinoma.

The results showed that VEGF siRNA-PEG nanoparticles almost completely inhibited the expression of secreted VEGF in a human prostate cancer cell line. More importantly, the tail vein injection of VEGF siRNA-PEG nanoparticles dramatically inhibited tumor growth in a PC3 tumor xenograft model, which demonstrated that siRNA-PEG nanoparticles are effective as a delivery vehicle for in vivo experiments.

The interaction between PEG-conjugated VEGF siRNA and PEI was shown to lead to the spontaneous formation of nanoscale polyelectrolyte complex (PEC) micelles (VEGF siRNA-PEG/PEI PEC micelles), having a characteristic siRNA/PEI PEC inner core with a surrounding PEG shell layer (Kim et al. 2008). Intravenous as well as intratumoral administration of the PEC micelles significantly inhibited VEGF expression at the tumor tissue and suppressed tumor growth in an animal tumor model without showing any detectable inflammatory responses in mice. Following intravenous injection, enhanced accumulation of the PEC micelles was also observed in the tumor region. This study demonstrates the feasibility of using PEC micelles as a potential carrier for therapeutic siRNAs in local and systemic treatment of cancer.

Polycation-Based Nanoparticles for siRNA Delivery

Polycation-based nanoparticles (polyplexes) formed by self-assembly with RNA can be used to modulate pharmacokinetics and intracellular trafficking to improve the therapeutic efficacy of RNAi-based therapeutics (Howard 2009). Polyplexes can be used for extracellular and intracellular delivery of synthetic RNA molecules. Flexibility in design and a capability to introduce different functional groups into a wide range of polymer types add multifunctional properties needed to fulfill delivery requirements. Surface modification with PEG during polyplex self-assembly has been used for steric stabilization and results in "stealth-like" nanoparticles that reduce serum protein interactions and capture by the mononuclear phagocyte system.

Endosomolytic and cleavable polymers can be built into the design of siRNA polyplexes to facilitate cytosolic release of siRNA needed to permit siRNA interaction with the intracellular target. Endosomal buffering, pH-activated polymers and membrane interactive peptides, which allow endosomal escape and transport to the cytosol, can be used to overcome capture within the endosomal-lysosomal pathway associated with cellular endocytosis of nanoparticles. The advent of bioresponsive nanoparticles whose function is triggered by biological conditions is a method used to control spatial delivery. Nanoparticles composed of reducible disulfide-linked polycations cleaved in response to intracellular redox conditions is an exciting strategy to install extracellular stability while allowing for intracellular breakdown and maximal release of the siRNA cargo.

Calando's Technology for Targeted Delivery of Anticancer siRNA

Calando Pharmaceuticals combines its proprietary technologies in targeted polymeric delivery systems and siRNA design to create effective therapeutics. Cyclodextrin-containing polymers form the foundation for a two-part siRNA delivery system.

The first component is a linear, cyclodextrin-containing polycation that, when mixed with siRNA, binds to the anionic "backbone" of the siRNA. The polymer and siRNA self-assemble into nanoparticles of approximately 50 nm diameter that fully protect the siRNA from nuclease degradation in serum. The cyclodextrin in the polymer enables the surface of the particles to be decorated by stabilizing agents and targeting ligands.

CALAA-01 employs a novel nanoparticle delivery system containing non-chemically modified siRNA and a transferrin (Tf) protein targeting agent formulated with Calando's RONDEL™ (RNA/oligonucleotide nanoparticle delivery). The effects of administering escalating, intravenous (IV) doses of targeted nanoparticles CALAA-01 targeting the M2 subunit of ribonucleotide reductase to nonhuman primates have been studied (Heidel et al. 2007a). The data show that multiple, systemic doses of targeted nanoparticles containing non-chemically modified siRNA can safely be administered to nonhuman primates. Further studies have shown that CALAA-01 exhibits significant antiproliferative activity in cancer cells of varying human type and species (mouse, rat, monkey); these findings suggest that this duplex is a promising candidate for therapeutic development (Heidel et al. 2007b). CALAA-01 is in clinical trials.

Delivery of siRNA by Nanosize Liposomes

siRNA incorporated into the neutral nanosize liposome 1,2-dioleoyl-sn-glycero-3-phosphatidylcholine (DOPC) has been used for efficient in vivo siRNA delivery. Getting the siRNA to the targeted protein tumor cells, focal adhesion kinase (FAK), is difficult as it is located inside the cell, rather than on the cell surface where most proteins targeted by cancer drugs are found. FAK, which is difficult to target with a drug, can be attacked with the liposomal siRNA approach, which penetrates deeply into the tumor (Halder et al. 2006). Mice infected with three human ovarian cancer cell lines derived from women with advanced cancer were treated with liposomes that contained either the FAK siRNA, a control siRNA, or were empty. Some mice received siRNA liposomes plus the chemotherapy docetaxel. Mice receiving the FAK-silencing liposome had reductions in mean tumor weight ranging from 44 % to 72 % compared with mice in the control groups. Combining the FAK-silencing liposome with docetaxel boosted tumor weight reduction to the 94–98 % range In addition to its anticancer effect, the therapeutic liposome also had an anti-angiogenic effect when combined with chemotherapy. By inducing apoptosis among blood vessel cells, the treatment steeply reduced the number of small blood vessels feeding the tumor, cut the percentage of proliferating tumor cells, and increased cell suicide among cancer cells. Two advantages of this approach are:

1. The FAK-targeting liposome ranges between 65 and 125 nm in diameter. Blood vessels that serve tumors are more porous than normal blood vessels, with pores of 100–780 nm wide. The liposomes do not enter the normal blood vessel whose pores are 2 nm or less in diameter.

2. The liposome DOPC has no electrical charge. Its neutrality provides an advantage over positively or negatively charged liposomes when it comes to binding with and penetrating cells.

These studies show the feasibility of siRNA as a clinically applicable therapeutic modality. The next step for the FAK siRNA-DOPC liposome is toxicity testing. In addition to ovarian cancer, FAK is overexpressed in colon, breast, thyroid, and head and neck cancers.

Quantum Dots to Monitor RNAi Delivery

A critical issue in using RNA interference (RNAi) for identifying genotype-phenotype correlations is the uniformity of gene silencing within a cell population. Variations in transfection efficiency, delivery-induced cytotoxicity, and "off-target" effects at high siRNA concentrations can confound the interpretation of functional studies. To address this problem, a novel method of monitoring siRNA delivery has been developed that combines unmodified siRNA with semiconductor quantum dots (QDs) as multicolor biological probes (Chen et al. 2005). siRNA was co-transfected with QDs using standard transfection techniques, thereby leveraging the photostable fluorescent nanoparticles to track delivery of nucleic acid, sort cells by degree of transfection, and purify homogenously silenced subpopulations. Compared to alternative RNAi tracking methods (co-delivery of reporter plasmids and end-labeling the siRNA), QDs exhibit superior photostability and tunable optical properties for an extensive selection of nonoverlapping colors. This simple, modular system can be used for multiplexed gene knockdown studies, as demonstrated in a two-color proof-of-principle study with two biological targets. When the method was applied to investigate the functional role of T-cadherin (T-cad) in cell-cell communication, a subpopulation of highly silenced cells obtained by QD labeling was required to observe significant downstream effects of gene knockdown. QDs are compatible with a variety of transfection techniques (other reagents, electroporation, and microinjection) and therefore amenable to nucleic acid monitoring in cells that are susceptible to liposome-triggered cytotoxicity. Primary cells may be particularly suited to benefit from this method, as nonviral delivery of siRNA has to date been technically difficult.

Chapter 7
Nanodevices and Techniques for Clinical Applications

Introduction

This chapter contains use of technologies for clinical applications in general. More detailed description of these technologies will be given in chapters dealing with special therapeutic areas.

Clinical Nanodiagnostics

Role of nanotechnology in molecular diagnostics was discussed in Chap. 4. This will have a tremendous impact on the practice of medicine. Biosensor systems based on nanotechnology could detect emerging disease in the body at a stage that may be curable. This is extremely important in management of infections and cancer. Some of the body functions and responses to treatment will be monitored without cumbersome laboratory equipment. Some examples are a radio transmitter small enough to put into a cell and acoustical devices to measure and record the noise a heart makes.

Nanoendoscopy

Endoscopic microcapsules are being developed that can be ingested and precisely positioned in the gastrointestinal system by nanorobotic techniques. A control system allows a capsule to attach to the walls of the digestive tract and move within its lumen. Several different methods are being researched for the attachment of microcapsules including both dry and wet adhesion as well as mechanical methods such as a set of tripod legs with adhesive on feet. Simple models with surface characteristics similar to that of the digestive tract are being constructed to test these methods.

Precisely positioned microcapsules would enable physicians to view any part of the inside lining of the digestive tract in detail resulting in more efficient, accurate, and less

K.K. Jain, *The Handbook of Nanomedicine*, DOI 10.1007/978-1-61779-983-9_7,

invasive diagnosis. In addition, these capsules could be modified to include treatment mechanisms as well, such as the release of a drug or chemical near an abnormal area.

Given Imaging Inc. (Duluth, GA) pioneered capsule endoscopy, which is now the gold standard for small bowel visualization. Its PillCam® capsule endoscope, a tool to visualize small intestine abnormalities, was approved in 2001 and more than one million patients worldwide have benefited from PillCam capsule endoscopy. Other companies are now producing ingestible capsules for this purpose. The patient ingests the capsule, which contains a tiny camera, and intestinal peristalsis propels the capsule for approximately 8 h. During this time, the camera snaps the pictures and images that are transmitted to a data recorder worn by the patient. The physicians can review the images later on to make the diagnosis. Controlling the positioning and movement on a nanoscale greatly improves the accuracy of this method. Video capsule endoscopy is a major innovation that provides high-resolution imaging of the entire small intestine in its entirety. Capsule endoscopy is the first-line investigation in patients with obscure gastrointestinal bleeding after a negative esophagogastroduodenoscopy and colonoscopy, and it has a positive impact on the outcome. Video capsule endoscopy is also useful in the evaluation of inflammatory and neoplastic disorders of the small bowel.

Colon capsule endoscopy represents a new diagnostic technology for colonic exploration, and there are still some controversies. A conference has critically evaluated the available results obtained by colon capsule endoscopy in clinical studies, in order to identify the proper test indications and proposed a shared preparation protocol and colon capsule endoscopy procedure (Spada et al. 2011).

The so-called gutbots are based on nanotechnology including nanosensors and sticking devices. If such devices are successful, their use may be extended to the large intestine. Although colon is currently examined by colonoscopy, physicians might be interested in introducing a pill-sized camera through the anus to visualize the suspicious area. Similar nanorobots are under development for other parts of the body.

Application of Nanotechnology in Radiology

X-ray radiation is widely used in medical diagnosis. The basic design of the X-ray tube has not changed significantly in the last century. Now medical diagnostic X-ray radiation can be generated using CNT-based field emission cathode. The device can readily produce both continuous and pulsed X-rays with programmable wave form and repetition rate. The X-ray intensity is sufficient to image a human extremity. The CNT-based cold-cathode X-ray technology can potentially lead to portable and miniature X-ray sources for industrial and medical applications

Xintek Inc. (Research Triangle Park, NC) has invented a new X-ray device based on CNT that emits a scanning X-ray beam composed of multiple smaller beams while also remaining stationary. As a result, the device can create images of objects from numerous angles and without mechanical motion, which is a distinct advantage for any machine since it increases imaging speed, can reduce the size of the device, and requires less maintenance. This technology can lead to smaller and faster X-ray imaging systems for tomographic medical imaging such as CT scanners. Other advantages will be that scanners will be cheaper, use less electricity, and produce higher-resolution images. This new technology enables the generation of digitized X-ray radiation with fine control of the spatial distribution of the X-ray pixels and temporal modulation of the radiation. The additional degrees of freedom in the source configuration will enable system vendors to design imaging systems with enhanced performance and new capabilities. In particular, the technology enables the design of stationary tomography imaging systems with faster scanning speed and potentially better imaging quality compared to currently available commercial scanners. Xintek's field emission X-ray technology is used for diagnostic medical imaging and in vivo imaging of small animal models for preclinical cancer studies.

High-Resolution Ultrasound Imaging Using Nanoparticles

Early diagnoses would involve using nanotechnology to improve the quality of images produced by one of the most common diagnostic tools used in physicians' offices – the ultrasound machine. In an experimental study, mice were injected intravenously with silica nanospheres (100 nm) in a suspension dispersed in agarose and imaged by a high-resolution ultrasound imaging system (Liu et al. 2006). B-mode images of the livers were acquired at different time points after particle injection. An automated computer program was used to quantify the grayscale changes. Ultrasonic reflections were observed from nanoparticle suspensions in agarose gels. The image brightness, i.e., mean gray scale level, increased with particle size and concentration. The mean gray scale of mouse livers also increased following particle administration. These results indicated that it is feasible to use solid nanoparticles as contrast-enhancing agents for ultrasonic imaging. The long-term goal is to use this technology to improve the ability to identify very early cancers and other diseases. The ultimate aim is to identify disease at its cellular level, at its very earliest stage.

Another study has demonstrated improvement of a pulsed magneto-motive ultrasound (pMMUS) image quality by using large size superparamagnetic nanoclusters characterized by strong magnetization per particle (Mehrmohammadi et al. 2011). Water-soluble magnetic nanoclusters of two sizes (15 and 55 nm) were synthesized from 3 nm iron precursors in the presence of citrate capping ligand. The size distribution of synthesized nanoclusters and individual nanoparticles was characterized using dynamic light scattering analysis and TEM. Tissue-mimicking phantoms containing single nanoparticles and two sizes of nanoclusters were imaged using a custom-built pMMUS imaging system. While the magnetic properties of citrate-coated nanoclusters are identical to those of superparamagnetic nanoparticles, the magneto-motive signal detected from nanoclusters was larger, i.e., the same magnetic field produced larger magnetically induced displacement. Therefore, this study demonstrated that clusters of superparamagnetic nanoparticles result in pMMUS images with higher contrast and SNR.

Nanobiotechnology in Tissue Engineering

Tissue engineering is an interdisciplinary field, which applies the principles of engineering and the life sciences to the development of biological substitutes that restore, maintain, or improve tissue function. Tissue engineering is an emerging field between traditional medical devices and regular pharmaceuticals. It faces many challenges, and it is also a field that is extremely interdisciplinary requiring the efforts of physicians, cell biologists, material scientists, chemical engineers, and chemists. Apart from the use of nanoparticles for diagnostic and therapeutic purposes, nanotechnology has applications for the development of tissue engineering as indicated by some of the studies in life sciences.

The response of cell motility and metabolism to changes in substrates has been thoroughly studied in the past decade. Size, structure, geometry, integrin binding, and other factors have all been investigated. Various techniques have been employed to create micropatterned surfaces of different materials to study cell behavior. In the presence of patterned stripes of bovine serum albumin and laminin, Schwann cells aggregate preferentially on the laminin regions. Osteogenic cells have been cultured in 3D nanohydroxyapatite/collagen matrix, which is precipitated such a manner that hydroxyapatite crystals are uniformly distributed in a matrix of collagen, seemingly ideal for bone construction.

Microfluidic devices enable the study of methods for patterning cells, topographical control over cells and tissues, and bioreactors. They have not been used extensively in tissue engineering but major contributions are expected in two areas. The first is growth of complex tissue, where microfluidic structures ensure a steady blood supply, thereby circumventing the well-known problem of providing larger tissue structures with a continuous flow of oxygen as well as nutrition and removal of waste products. The second and probably more important function of microfluidics, combined with micro/nanotechnology, lies in the development of in vitro physiological systems for studying fundamental biological phenomena.

Nanoscale Surfaces for Stem Cell Culture

Adult stem cells spontaneously differentiate in culture, resulting in a rapid diminution of the multipotent cell population and their regenerative capacity. There is currently an unmet need for the supply of autologous, patient-specific stem cells for regenerative therapies in the clinic. MSC differentiation can be driven by the material/cell interface suggesting a unique opportunity for manipulating stem cells in the absence of complex soluble chemistries or cellular reprogramming. A nanostructured surface has been identified that retains stem cell phenotype and maintains stem cell growth over 8 weeks (McMurray et al. 2011). The authors had previously designed and produced a polycaprolactone-based support with nanoscale features that enabled osteogenesis from stem and progenitor mesenchymal populations cultured in osteogenic media. In the current study, by reduction of the level of offset to as close to zero as possible, the resulting nanotopography induced a switch from osteogenic stimulation to a surface conducive to MSC growth while allowing prolonged retention of MSC biomarkers and multipotency. The authors postulate that nanoscale modifications to surface topography alter the interaction of integrin receptors within cell adhesions, resulting in changes in intracellular tension. The demonstrable sensitivity of MSCs to materials, with <50 nm alterations in feature placement and the role of such defined topographies on cell fate and function, offers nanoscale patterning as a powerful tool for the noninvasive manipulation of stem cells. By incorporating nanoscale features into tissue-engineering scaffolds that support reservoirs of progenitor cells for a range of tissue-specific stem cell types, this approach can improve the regenerative capacity of in vitro-fabricated tissue and organs.

3D Nanofilament-Based Scaffolds

Ideally the tissue-engineering scaffolds should be analogous to native extracellular matrix (ECM) in terms of both chemical composition and physical structure. Polymeric nanofiber matrix is similar, with its nanoscaled nonwoven fibrous ECM proteins, and thus is a candidate ECM-mimetic material (Ma et al. 2005). Scaffolds for tissue engineering are typically solid or porous materials with isotropic characteristics and present regenerative cues such as growth factors or extracellular matrix proteins, but these do not explicitly guide tissue regeneration. Novel 3D nanofilament-based scaffolds have been developed for tissue regeneration. This mimics the strategy used by collagen and other fibrillar structures to guide cell migration or tissue development and/or regeneration in a guided, direction-sensitive manner. The critical advantage of this technology is that it provides directional cues for cell and tissue regeneration. This strategy can be used to guide the migration of endogenous or transplanted cells and tissues to damaged tissues of the peripheral and central nervous systems to restore function but could also be applied to tissue engineering.

The development of effective biological scaffold materials for tissue engineering and regenerative medicine applications hinges on the ability to present precise environmental cues to specific cell populations to guide their position and function. Natural extracellular matrices have an ordered nanoscale structure that can modulate cell behaviors critical for developmental control, including directional cell motility. A method has been described for fabricating fibrin gels with defined architecture on the nanometer scale in which magnetic forces are used to position thrombin-coated magnetic microbeads in a defined 2D array and thereby guide the self-assembly of fibrin fibrils through catalytic cleavage of soluble fibrinogen substrate (Alsberg et al. 2006). Time-lapse and confocal microscopy confirmed that fibrin fibrils nucleate near the surface of the thrombin-coated beads and extend out in a radial direction to form these gels. When controlled magnetic fields were used to position the beads in hexagonal arrays, the fibrin nanofibrils that polymerized from the beads oriented preferentially along the bead-bead axes in a geodesic (minimal path) pattern. These biocompatible scaffolds supported adhesion and spreading of human microvascular endothelial cells, which exhibited coalignment of internal actin stress fibers with underlying fibrin nanofibrils within some membrane extensions at the cell periphery. This magnetically guided, biologically inspired microfabrication system is unique in that large scaffolds may be formed with little starting material, and thus it may be useful for in vivo tissue-engineering applications in the future.

In addition to fabricating 3D microfabricated scaffolds as templates for cell aggregate formation, nanoscale technologies can be used for controlling the features such as shape and pore architecture, as templates for microtissue formation or as improved bioreactors (Khademhosseini et al. 2006). The nanoscale control of cellular environments can also be used to probe the influence of the spatial and temporal effects of specific cell-cell, cell-extracellular matrix, and cell-soluble factor interactions.

Electrospinning Technology for Nanobiofabrication

Jet-based technologies are increasingly being explored as potential high-throughput and high-resolution methods for the manipulation of biological materials. Previously shown to be of use in generating scaffolds from biocompatible materials, electrospinning technology has been used to deposit active biological threads and scaffolds comprised of living cells (Townsend-Nicholson and Jayasinghe 2006). This has been achieved by use of a coaxial needle arrangement where a concentrated living biosuspension flows through the inner needle, and a medical-grade polydimethylsiloxane medium with high viscosity and low electrical conductivity flows through the outer needle. Cells cultured after electrospinning were shown to be viable with no evidence of having incurred any cellular damage during the nanobiofabrication process. This demonstrates the feasibility of using coaxial electrospinning technology for biological and biomedical applications requiring the deposition of living cells as composite nanothreads for forming active biological scaffolds. The process could enable significant advances to be made in technologies ranging from tissue engineering to regenerative medicine. Perhaps in the future such living nanothreads might be spun directly into wounds.

Organogenesis Inc is developing electrospinning technology, which makes it possible to mimic the 3D architectural structure that is essential for the body's natural growth and repair processes to develop designer scaffolds for the purposes of regenerative medicine. An implantable scaffold with the correct nanofiber diameter, orientation, and architecture is virtually indistinguishable from native tissue and is recognized as "self" by the body and facilitates regeneration. Electrospinning techniques have also been used to produce a variety of material, including vascular grafts, nerve guides, tendon, and skin.

Nanomaterials for Tissue Engineering

Several nanomaterials, particularly nanofibers, have been investigated for use as scaffolds in tissue engineering. A naturally occurring nanofiber- and nanoparticle-based nanocomposite from the adhesive of Sundew can be used for tissue engineering and opens the possibility for further examination of natural plant adhesives for biomedical applications (Zhang et al. 2010). Tissue engineering of skeletal muscle is a promising method for the treatment of soft tissue defects in reconstructive surgery. Nanomaterials are useful for this purpose.

Carbon Nanotubes for Artificial Muscles

Research on carbon multiwalled nanotubes (MWCNTs) could lead to new materials that will mimic biological tissues and artificial muscles. MWCNTs can withstand repeated stress and still be able to retain their structural and mechanical integrity, similar to the behavior of soft tissue. A study has shown that under repeated high compressive strains, long, vertically aligned MWCNTs exhibit viscoelastic behavior similar to that observed in soft tissue membranes (Suhr et al. 2006). Under compressive cyclic loading, the mechanical response of the nanotube arrays shows preconditioning, characteristic viscoelasticity-induced hysteresis, nonlinear elasticity and stress relaxation, and large deformations. Furthermore, no fatigue failure is observed at high strain amplitudes up to half a million cycles. This combination of soft tissue-like behavior and outstanding fatigue resistance suggests that properly engineered nanotube structures could mimic artificial tissues, and that their good electrical conductivity could lead to their use as compliant electrical contacts in a variety of applications. The springiness is similar to real muscles' ability to return to their original shapes over a lifetime of perpetual extension and contraction. Because real muscles create a smoother motion than jerky electric motors or pneumatic devices, some of the new materials would be used to power robots and prosthetic limbs, as well as artificial tissue for implantation. MWCNTs are being combined with different polymers, which control when an artificial muscle gets stretched, to improve their resistance to fatigue.

Nanofibers for Tissue Engineering of Skeletal Muscle

Rapid lysis and contraction of pure collagen I or fibrin matrices have been a problem in the past efforts for tissue engineering of skeletal muscle. This problem could be overcome by combining both materials as significant proliferation of cultivated myoblasts has been detected in collagen I-fibrin matrices and collagen nanofibers. Seeding cells on parallel-orientated nanofibers results in strongly aligned myoblasts. Collagen I-fibrin mixtures as well as collagen nanofibers yield good proliferation rates and myogenic differentiation of primary rat myoblasts in vitro (Beier et al. 2009). In addition, parallel-orientated electrospun nanofibers enable the generation of aligned cell layers and therefore represent the most promising step toward successful engineering of skeletal muscle tissue.

Nanobiotechnology Combined with Stem Cell-Based Therapies

Nanobiotechnology is used for tracking stem cells introduced into the human body and can also be applied for delivery of gene therapy using genetically modified stem cells. In their natural environment in the body, stem cells transform into other cell types based on chemical triggers they receive from their surroundings. The nature and the location of these triggers are not known for most stem cells. The current ability to introduce specific chemicals at select locations on a cell is also very limited as one must bathe the entire surface of stem cells in various chemicals to search for a response. A nano lab has been used to experiment with individual adult stem cells. Each lab essentially consists of a capsule on a silicon chip, around which up to 1,000 nanoreservoirs hold roughly a millionth of a billionth of a milliliter of liquid, comparable to the size of secretions cells use to communicate. This is an artificial cell-interface unit for a stem cell to establish chemical communication in much the same way real cells do. Nanotechnology is essential for this approach. Larger systems cannot provide the number of different reservoirs and chemicals within a space small enough to select different areas on a cell.

Nanofibrous scaffolds are being developed for stem cells, which mimic the nanometer-scale fibers normally found in that matrix (Kang et al. 2005). These biodegradable scaffolds can nurture stem cells derived from adipose tissue. Preadipocytes grown on 3D matrices acquire morphology and biological features of mature adipocytes. This culture model has significant utility for in vitro studies of adipocyte cell biology and development.

One study has assessed bone formation from mesenchymal stem cells (MSCs) on a novel nanofibrous scaffold in a rat model (Shin 2004). A highly porous, degradable polycaprolactone scaffold with an extracellular matrix-like topography was produced by electrostatic fiber spinning. MSCs derived from the bone marrow of neonatal rats were cultured, expanded, and seeded on the scaffolds. The cell-polymer constructs were cultured with osteogenic supplements and maintained the size and shape of the original scaffolds when explanted. Morphologically, the constructs were rigid and had a bone-like appearance. Cells and extracellular matrix formation were observed throughout the constructs. In addition, mineralization and type I collagen were also detected. This study establishes the ability to develop bone grafts on electrospun nanofibrous scaffolds in a well-vascularized site using MSCs.

A challenge in vascular tissue engineering is to develop optimal scaffolds and establish expandable cell sources for the construction of tissue-engineered vascular grafts that are nonthrombogenic and have long-term patency. Tissue-engineered vascular grafts have been used as a model to demonstrate the potential of combining nanofibrous scaffolds and bone marrow MSCs for vascular tissue engineering (Hashi et al. 2007). Biodegradable nanofibrous scaffolds with aligned nanofibers were used to mimic native collagen fibrils to guide cell organization in vascular grafts. The results from artery bypass experiments showed that nanofibrous scaffolds allowed efficient infiltration of vascular cells and matrix remodeling. Acellular grafts (without MSCs) resulted in significant intimal thickening, whereas cellular grafts (with MSCs) had excellent long-term patency and exhibited well-organized layers of endothelial cells (ECs) and smooth muscle cells (SMCs), as in native arteries. Short-term experiments showed that nanofibrous scaffolds alone induced platelet adhesion and thrombus formation, which was suppressed by MSC seeding. In addition, MSCs, as ECs, resisted platelet adhesion in vitro, which depended on cell-surface heparan sulfate proteoglycans. These data, together with the observation on the short-term engraftment of MSCs, suggest that the long-term patency of cellular grafts may be attributed to the antithrombogenic property of MSCs. These results demonstrate several favorable characteristics of nanofibrous scaffolds, the excellent patency of small-diameter nanofibrous vascular grafts, and the unique antithrombogenic property of MSCs.

Nanomaterials for Combining Tissue Engineering and Drug Delivery

A variety of organic and inorganic nanostructures have been developed for scaffolds in tissue regeneration as well as drug delivery. These nanostructures provide favorable biological integration of implants and have applications in many areas, including orthopedics, cardiovascular medicine, and ophthalmology. Additionally, these nanostructures are capable of delivering drugs in a localized and controlled manner, accounting for the short biological half-life, lack of long-term stability and tissue-selectivity, and potential toxicity of many therapeutic compounds.

Elastin-like polypeptides (ELPs) are artificial polypeptides, derived from Val-Pro-Gly-Xaa-Gly (VPGXG) pentapeptide repeats found in human tropoelastin. The potential of ELPs to self-assemble into nanostructures in response to environmental triggers is another interesting feature of these polypeptides that promises to lead to a host of new applications (Chilkoti et al. 2006). Genetically encodable ELPs are monodisperse, stimuli responsive, and biocompatible, properties that make them attractive for combining drug delivery and tissue engineering.

In sophisticated tissue-engineering strategies, the biodegradable scaffold is preferred to serve as both a 3D substrate and a growth factor delivery vehicle to promote cellular activity and enhance tissue neogenesis. A novel approach has been described for fabrication of tissue-engineering scaffolds capable of controlled growth factor delivery whereby growth factor containing microspheres are incorporated into 3D scaffolds with good mechanical properties, well-interconnected macroporous and nanofibrous structures (Wei et al. 2006). The microspheres were uniformly distributed throughout the nanofibrous scaffold, and their incorporation did not interfere the macro-, micro-, and nanostructures of the scaffold. The release kinetics of platelet-derived growth factor-BB (PDGF-BB) from microspheres and scaffolds was investigated using PLGA50 microspheres. Incorporation of microspheres into scaffolds significantly reduced the initial burst release. Sustained release from several days to months was achieved through different microspheres in scaffolds. Released PDGF-BB was demonstrated to possess biological activity as evidenced by stimulation of human gingival fibroblast DNA synthesis in vitro. The successful generation of 3D nanofibrous scaffold incorporating controlled-release factors indicates significant potential for more complex tissue regeneration.

Nanobiotechnology for Organ Replacement and Assisted Function

Several devices are used to repair, replace, or assist the function of damaged organs such as kidneys. The technologies range from those for tissue repair to those for device to take over or assist the function of the damaged organs. The following sections include some examples of these applications.

Exosomes for Drug-Free Organ Transplants

Exosomes are nanovesicles shed by dendritic cells. They may hold the key to achieving transplant tolerance, i.e., the long-term acceptance of transplanted organs without the need for drugs. Exosomes are no larger than 65–100 nm, yet each contains a potent reserve of major histocompatibility complex (MHC) molecules – gene products that cells use to determine self from nonself. Millions of exosomes scurry about within the bloodstream, and while their function has been somewhat of a mystery, researchers are beginning to surmise that they play an important role in immune regulation and response.

Because certain dendritic cells have tolerance-enhancing qualities, several approaches under study involve giving recipients donor dendritic cells that have been modified in some way. The idea is that the modified donor cells would convince recipient cells that a transplanted organ from the same donor is not foreign. MHC-rich vesicles, siphoned from donor dendritic cells, are captured by recipient dendritic cells and processed in a manner important for cell-surface recognition. Thus one can efficiently deliver donor antigen using the exosomes as a magic bullet. The exosomes are caught by the dendritic cells of the spleen, the site where dendritic cells typically present antigens as bounty to T cells. However, these dendritic cells internalize the exosomes instead of displaying them to T cells despite the exosomes' rich endowment of donor MHC molecules. Once internalized, the exosomes are ushered inside larger vesicles, special endosomes called MHC-II enriched compartments, where they are processed with the dendritic cell's own MHC molecules. This hybrid MHC-II molecule, now loaded with a peptide of donor MHC, is then expressed on the cell's surface. As one family of MHC molecules, MHC-II serves as a beacon for a specific population of T cells called CD4+ T cells. Such cells are activated during chronic rejection in a process associated with the indirect pathway of immune recognition.

This finding is significant because current immunosuppression therapies used in the clinical setting are not able to efficiently prevent T cell activation via the indirect pathway. Perhaps the CD4+ T cells normally involved in this pathway would retreat from attack if they encountered a cell-surface marker that is of both donor and recipient origin, such as that which is observed following the dendritic cell's internalization of the donor-derived exosomes. The process of internalizing the donor exosomes does not affect maturation of the dendritic cell. Only immature dendritic cells can capture antigens efficiently and are believed to participate in the induction of transplant tolerance. By contrast, once mature, dendritic cells are capable of triggering the T cell activation that leads to transplant rejection. Additional research will be required to determine whether donor-derived exosomes will enhance the likelihood that an organ transplant from the same donor will be accepted. Only a few research groups are engaged in active study of exosomes with most of the research taking place in Europe.

Nanobiotechnology and Organ-Assisting Devices

Organ-assisting devices (OAD) are an emerging area for application of nanobiotechnology. This includes implants and other devices to assist or replace the impaired function of various organs. One example of this is restoration of function of the tympanic membrane of the ear by magnetically responsive nanoparticles. Other examples would be given in the later sections of this chapter.

Superparamagnetic iron oxide nanoparticles (SNP) composed of magnetite (Fe(3)O(4)) were studied preliminarily as vehicles for therapeutic molecule delivery to the inner ear and as a middle ear implant capable of producing biomechanically relevant forces for auditory function (Kopke et al. 2006). Magnetite SNP were synthesized, then encapsulated in either silica or poly (D,L-lactide-co-glycolide) or obtained commercially with coatings of oleic acid or dextran. Permanent magnetic fields generated forces sufficient to pull them across tissue in several round window membrane models (in vitro cell culture, in vivo rat and guinea pig, and human temporal bone) or to embed them in middle ear epithelia. Biocompatibility was investigated by light and electron microscopy, cell culture kinetics, and hair cell survival in organotypic cell culture, and no measurable toxicity was found. A sinusoidal magnetic field applied to guinea pigs with SNP implanted in the middle ear resulted in displacements of the middle ear comparable to 90 dB SPL.

Nanosurgery

Miniaturization in Surgery

Historically, surgery was macrosurgery and most of general surgery still involves gross manipulation of organs and tissues by human hands and handheld instruments. Some branches of surgery such as ophthalmology and otorhinolaryngology started to miniaturize early and start using microsurgery. In the last quarter of twentieth century, miniaturization started to develop most branches of surgery. The basic feature was minimization of trauma to the body tissues during surgery. Trends were small incisions, laparoscopic surgery by fiberoptic visualization through tubular devices, vascular surgery by catheters and microsurgery under operating microscopes to refine the procedures and reduce trauma. Many of the devices such as robotics and implants will be a part of this miniaturization process.

Nanotechnology for Hemostasis During Surgery

There are few effective methods to stop bleeding during surgery without causing tissue damage. More than 57 million Americans undergo nonelective surgery each year, and as much as 50 %of surgical time is spent working to control bleeding. Current tools used to stop bleeding include clamps, pressure, cauterization, vasoconstriction, and sponges. Some simple liquids composed of peptides, when applied to open wounds in rodents, self-assemble into a nanoscale protective barrier gel that seals the wound and stop bleeding in less than 15 s. Once the injury heals, the nontoxic gel is broken down into molecules that cells can use as building blocks for tissue repair. The exact mechanism of the action of such solutions is still unknown, but one explanation is that the peptides interact with the extracellular matrix surrounding the cells. The hemostatic action has been demonstrated in open wounds in several different types of tissue: brain, liver, skin, spinal cord, and intestine.

Minimally Invasive Surgery Using Catheters

Surgery is continuously moving toward more minimally invasive methods. The main driver of this technical evolution is patient recovery: the lesser the trauma inflicted on the patient, the shorter is the recovery period. Minimally invasive surgery, often performed by use of catheters navigating the vascular system, implies that the operator has little to no tactile or physical information about the environment near or at the surgical site. This information can be provided by biosensors implanted in the catheters. Verimetra Inc. is developing such devices. Minimally invasive and in vivo surgery is limited by the ability to provide controllable and powerful motion at scales appropriate for navigation within the human body. Nanotechnology will play an important role in the construction of miniaturized biosensing devices. These sensors improve outcomes, lower risk, and help control costs by providing the surgeon with real-time data about:

- Instrument force and performance
- Tissue density, temperature, or chemistry
- Better or faster methods of preparing tissue or cutting tissue
- Extracting tissue and fluids

Examples of procedures and applications where such an approach would be useful are:

- Cardiovascular surgery
- Stent insertion
- Percutaneous transluminal coronary angioplasty
- Coronary artery bypass graft (CABG)
- Atrial fibrillation
- Cardiac surgery in utero
- Cerebrovascular surgery
- Surgery of intracranial aneurysms
- Embolization of intracranial vascular malformations

Nanorobotics

Robotics is already developing for applications in life sciences and medicine. Robots can be programmed to perform routine surgical procedures. Nanobiotechnology introduces another dimension in robotics leading to the development of nanorobots also referred to as nanobots. Instead of performing procedures from outside the body, nanobots will be miniaturized for introduction into the body through the vascular system or at the end of catheters into various vessels and other cavities in the human body. Surgical nanobot, programmed by a human surgeon, could act as an autonomous on site surgeon inside the human body. Various functions such as searching for pathology, diagnosis, and removal or correction of the lesion by nanomanipulation can be performed and coordinated by an on-board computer. Such concepts, once science fiction, are now considered to be within the realm of possibility. Nanorobots will have the capability to perform precise and refined intracellular surgery which is beyond the capability of manipulations by the human hand.

A device is being developed for facilitating minimally invasive intrapericardial interventions on the beating heart (Riviere et al. 2004). This is based on the concept of an endoscopic robotic device that adheres to the epicardium by suction and navigates by crawling like an inchworm to any position on the surface under the control of a surgeon. This approach obviates cardiac stabilization, lung deflation, differential lung ventilation, and reinsertion of laparoscopic tools for accessing different treatment sites, thus offering the possibility of reduced trauma to the patient. The device has a working channel through which various tools can be introduced for treatment. The current prototype demonstrated successful prehension, turning, and locomotion on beating hearts in a limited number of trials in a porcine model.

A motor for in vivo microbot propulsion has been constructed with a diameter of 250 μm, demonstrating the potential to directly drive a flagellum for swimming at up to 1,295 rpm with a torque of 13 nN m (Watson et al. 2009). The motor uses coupled axial-torsional vibration at 652–682 kHz in a helically cut structure excited by a thickness-polarized piezoelectric element. The motor has been named the

"Proteus motor" after the miniature submarine that traveled through the human body in the science fiction movie, "Fantastic Voyage." The output power is 4.25 μW, on the order of what is necessary to navigate small human arteries. This micromotor, small enough to be injected into the human bloodstream, could be used for a range of complex surgical operations necessary to treat stroke victims, confront hardened arteries, or address blockages in the bloodstream.

Nanoscale Laser Surgery

Scalpel and needle may remain adequate instruments for most surgery work, and biological compounds may still be needed to prod cells to certain actions. Introduction of lasers in surgery more than a quarter of century ago has already refined surgery and experimental biological procedures to enable manipulations beyond the capacity of the human handheld instruments. Laser microsurgery was used both for ablation and repair of tissues (Jain 1983). Mechanical devices such as microneedles are too large for the cellular scale, while biological and chemical tools can only act on the cell as a whole rather than on any one specific mitochondrion or other structure. Further developments are leading to manipulation of cellular structures at the micrometer and nanometer scale. This is opening up the field of nanoscale laser surgery.

Femtosecond (one millionth of a billionth of a second) laser pulses can selectively cut a single strand in a single cell in the worm and selectively knock out the sense of smell. One can target a specific organelle inside a single cell (e.g., a mitochondrion or a strand on the cytoskeleton) and zap it out of existence without disrupting the rest of the cell. The lasers can neatly zap specific structures without harming the cell or hitting other mitochondria only a few hundred nanometers away. It is possible to carve channels slightly less than 1 μm wide, well within a cell's diameter of 10–20 μm. By firing a pulse for only 10–15 fs in beams only one micron wide, the amount of photons crammed into each burst becomes incredibly intense: 100 quadrillion watts per square meter, 14 orders of magnitude greater than outdoor sunlight. That searing intensity creates an electric field strong enough to disrupt electrons on the target and create a microexplosion. But because the pulse is so brief, the actual energy delivered into the cell is only a few nanojoules. To achieve that same intensity with nanosecond or millisecond, pulses would require so much more energy that the cell would be destroyed.

That opens the door to researching how cytoskeletons give a cell its shape, or how organelles function independently from each other rather than a whole system. The technology might be scaled up to do surgery without scarring or perhaps to deliver drugs through the skin. Near-infrared femtosecond laser pulses have been applied in a combination of microscopy and nanosurgery on fluorescently labeled structures within living cells (Sacconi et al. 2005). Femtolasers are already in use in corneal surgery.

Chapter 8
Nanooncology

Introduction

Application of nanotechnology in cancer can be termed nanooncology and includes both diagnostics and therapeutics (Jain 2008). Various applications in diagnosis and drug delivery for cancer are discussed in this chapter. Two nanotechnology-based products are already approved for the treatment of cancer – Doxil (a liposome preparation of doxorubicin) and Abraxane (paclitaxel in nanoparticle formulation). Approximately 150 drugs in development for cancer are based on nanotechnology. Some of the nanotechnologies and their applications in developing cancer therapies are described in this section. The most important factor in the fight against cancer, besides prevention, is early detection.

Nanobiotechnology for Detection of Cancer

Nanobiotechnology offers a novel set of tools for detection of cancer. It will contribute to early detection of cancer as follows:

- It can complement existing technologies and make significant contributions to cancer detection, prevention, diagnosis, and treatment.
- It would be extremely useful in the area of biomarker research and provide additional sensitivity in assays with relatively small sample volumes.
- Examples of applications of nanobiotechnology in cancer diagnostics include quantum dots and use of nanoparticles for tumor imaging.

Dendrimers for Sensing Cancer Cell Apoptosis

Poly(amidoamine) (PAMAM) dendrimers have been used as a platform for the targeted delivery of chemotherapeutic drugs in cancer. A PAMAM nanodevice can be used to monitor the rate and extent of cell-killing or apoptosis caused by the

K.K. Jain, *The Handbook of Nanomedicine*, DOI 10.1007/978-1-61779-983-9_8,

delivered chemotherapeutic drug, which is important for predicting clinical efficacy (Myc et al. 2007). Whereas other approaches to detect apoptosis rely on the human protein annexin V, which binds to a hidden cell membrane component revealed in the initial stages of apoptosis, this method detects caspase-3, an enzyme activated early in the apoptosis process. This enzyme cleaves the bond between two specific amino acids, and this specificity has been exploited to design fluorescence resonance energy transfer (FRET)-based assays for caspase-3. The fluorescence appears only when caspase-3 breaks a valine–aspartic acid bond in a specially designed substrate for this enzyme. To create a tumor-specific apoptosis detector, folic acid and the caspase-3 substrate were attached to a PAMAM dendrimer. Folic acid acts as a tumor-targeting agent, binding to folic acid that many types of tumor cells produce in abundance. Apoptotic tumor cells bearing this folic acid receptor take up the dendrimer and fluoresce brightly. In contrast, apoptotic cells lacking the folic acid receptor do not fluoresce. An optical fiber device, capable of detecting FRET emissions in tumors, has been used to quantify apoptosis in live mice with tumors bearing the folic acid receptor.

Detection of Circulating Cancer Cells

A method has been described for magnetically capturing circulating tumor cells (CTCs) in the bloodstream of mice followed by rapid photoacoustic detection (Galanzha et al. 2009). Magnetic nanoparticles, which were functionalized to target a receptor commonly found in breast cancer cells, bound and captured circulating tumor cells under a magnet. To improve detection sensitivity and specificity, gold-plated carbon nanotubes conjugated with folic acid were used as a second contrast agent for photoacoustic imaging. By integrating in vivo multiplex targeting, magnetic enrichment, signal amplification, and multicolor recognition, this approach enables circulating tumor cells to be concentrated from a large volume of blood in the vessels of tumor-bearing mice and has potential applications for the early diagnosis of cancer and the prevention of metastasis in humans.

A nano-Velcro technology, engineered into a 2.5 × 5-cm microfluidic chip is a second-generation CTC-capture technology, which is capable of highly efficient enrichment of rare CTCs captured in blood samples collected from prostate cancer patients (Wang et al. 2011). It is based on the research team's earlier development of "flypaper" technology that involves a nanopillar-covered silicon chip whose stickiness resulted from the interaction between the nanopillars and nanostructures on CTCs known as microvilli, creating an effect much like the top and bottom of Velcro. The new device adds an overlaid microfluidic channel to create a fluid flow path that increases mixing. In addition to the Velcro-like effect from the nanopillars, the mixing produced by the microfluidic channel's architecture causes the CTCs to have greater contact with the nanopillar-covered floor, further enhancing the device's efficiency. The device features high flow of the blood samples, which travel at increased speed, bouncing up and down inside the channel; get slammed against the surface; and get caught.

Differentiation Between Normal and Cancer Cells by Nanosensors

Rapid and effective differentiation between normal and cancer cells is an important challenge for the diagnosis and treatment of tumors. A nanoparticle array-based system has been described for identification of normal and cancer cells based on a "chemical nose/tongue" approach that exploits subtle changes in the physicochemical nature of different cell surfaces (Bajaj et al. 2009). Differential interactions with functionalized nanoparticles are transduced through displacement of a multivalent polymer fluorophore that is quenched when bound to the particle and fluorescent after release. This sensing method can rapidly (minutes/seconds) and effectively distinguish (1) different cell types; (2) normal, cancerous, and metastatic human breast cells; and (3) isogenic normal, cancerous, and metastatic murine epithelial cell lines.

Gold Nanoparticles for Cancer Diagnosis

Gold nanoparticles conjugated to anti-epidermal growth factor receptor (anti-EGFR) monoclonal antibodies (MAbs) specifically and homogeneously bind to the surface of the cancer cells with 600% greater affinity than to the noncancerous cells. This specific and homogeneous binding is found to give a relatively sharper SPR absorption band with a red-shifted maximum compared to that observed when added to the noncancerous cells (El-Sayed et al. 2005). The particles that worked the best were 35 nm in size. These results suggest that SPR scattering imaging or SPR absorption spectroscopy generated from antibody-conjugated gold nanoparticles can be useful in molecular biosensor techniques for the diagnosis and investigation of living oral epithelial cancer cells in vivo and in vitro. Advantages of this technique are:

- It is not toxic to human cells. A similar technique with QDs uses semiconductor crystals to mark cancer cells, but the semiconductor material is potentially toxic to the cells and humans.
- It does not require expensive high-powered microscopes or lasers to view the results. All it takes is a simple, inexpensive microscope and white light.
- The results are instantaneous. If a cancerous tissue is sprayed with gold nanoparticles containing the antibody, the results can be seen immediately. The scattering is so strong that a single particle can be detected.

An animal study has successfully demonstrated the safety of diagnostic use of Raman-silica-gold-nanoparticles (R-Si-Au-NPs), which overcome the inherently weak nature of Raman effect by producing larger Raman signals through surface-enhanced Raman scattering (Thakor et al. 2011). R-Si-Au-NPs were bound to PEG molecules to improve biological tolerance. Molecules that home in on cancer cells can be attached to PEG-R-Si-Au-NPs, and the overall diameter is 100 nm. Photoimaging with these nanoparticles holds the promise of very early disease detection in colorectal cancer (CRC), even before any gross anatomical changes show up, without physically removing any tissue from the patient. Both rectal and intravenous administration of the particles did not show any systemic toxicity in experimental animals. Furthermore, the nanoparticles were quickly excreted. The intravenously administered nanoparticles were rapidly sequestered by scavenger cells resident in organs such as the liver and spleen. This opens the door to human tests of intravenous injections of these nanoparticles to search for tumors throughout the body. Molecules targeting breast, lung, or prostate cancer can be attached to these nanoparticles. The investigators are now filing for FDA approval to proceed to clinical studies of the nanoparticles for the diagnosis of CRC.

Chemiluminescence resonance energy transfer (CRET) with gold nanoparticles (AuNPs) has been used as an efficient long-range energy acceptor in sandwich immunoassays. A CRET-based sandwich immunoassay has been developed for alpha-fetoprotein (AFP) cancer biomarker (Huang and Ren 2011). In immunoassay, two antibodies (anti-AFP-1 and anti-AFP-2) are conjugated to AuNPs and horseradish peroxidase, respectively. The sandwich-type immunoreactions between the AFP (antigen) and the two different antibodies bridge the donors (luminol) and acceptors (AuNPs), which leads to the occurrence of CRET from luminol to AuNPs upon chemiluminescent reaction. We observed that the quenching of chemiluminescence signal depended linearly on the AFP concentration within a range of concentration from 5 to 70 ng mL^{-1} and the detection limit of AFP was 2.5 ng mL^{-1}. This method was successfully applied for determination of AFP levels in sera from cancer patients, and the results were in good agreement with ELISA assays. This approach is expected to be extended to other assay designs, i.e., using other antibodies, analytes, and chemiluminescent substance.

Gold nanoparticles can be heated rapidly whenever exposed to infrared light of the right wavelength. Heating of gold nanoparticles results in variations in pressure surrounding them, which in turn is expressed in the generation of ultrasound – a phenomenon called plasmon resonance. The shape of the particles determines the wavelength at which this happens. In this way, light from a laser results in sound. By attaching MAbs to gold nanoparticles or nanorods, which can recognize a specific cancer cell, the heating phenomenon can be used in cancer detection. This acoustic signal gives valuable information about the presence of cancer cells. Scientists at the University of Twente (UT) in the Netherlands expect better results with this approach than is currently possible with imaging techniques. The temperature rise can be up to 100 °C. Photothermal therapy would use the heated gold to destroy the tumor. Another option would be to include gold particles in capsules filled with cancer medication: the capsule attaches to the cancer cell, is heated and the medicine is released locally. Both diagnostic and therapeutic applications will be investigated by the UT scientists together with colleagues from the Erasmus Medical Center in Rotterdam and two companies: Esoate Europe and Luminostix.

Gold Nanorods for Detection of Metastatic Tumor Cells

Scientists at Purdue University have developed a technique for producing biocompatible gold nanorods of various sizes to which antibodies can be attached.

Gold nanorods interact with light to produce plasmons, a wavelike motion of electrons on the surface of the nanorods. Depending on the ratio of a nanorod's length to its diameter, these plasmons trigger light emission at a specific frequency that is easily detected using SPR spectroscopy. An antibody that recognizes one specific cancer cell surface biomarker is attached to each nanorod of a given length and diameter. A gold nanorod-antibody construct that recognizes a biomarker found on all cell surfaces serves as an internal reference control that enables calculation of relative amounts of the various tumor biomarkers on a given cancer cell. Using a panel of three different antibody-labeled gold nanorods, the investigators were able to characterize breast tumors according to their cellular composition and correlate their findings to the metastatic potential of each given cell type. These results were validated using flow cytometry, the standard technique used to classify cells according to surface biomarkers. Gold nanorods enabled monitoring of as many as 15 different antibody-nanorod constructs simultaneously.

Implanted Biosensor for Cancer

An implant for biosensing of cancer, developed at the Massachusetts Institute of Technology (Cambridge, MA), contains nanoparticles that can be designed to test for different substances, including metabolites such as glucose and oxygen that are associated with tumor growth. It can also be used to test the effects of anticancer drugs in individual patients; the implant could reveal how much of a drug has reached the tumor. The nanoparticles are encased in a silicone delivery device, enabling their retention in patients' bodies for an extended period of time. The device can be implanted directly into a tumor, allowing a more direct look at what is happening in the tumor over a period of time. The technique makes use of detection nanoparticles composed of iron oxide and coated with dextran. Antibodies specific to the target molecules are attached to the surface of the nanoparticles. When the target molecules are present, they bind to the particles and cause them to clump together. That clumping can be detected by MRI. The nanoparticles are trapped inside the silicone device, which is sealed off by a porous membrane. The membrane allows molecules smaller than 30 nm to get in, but the detection particles are too large to get out. In addition to monitoring the presence of chemotherapy drugs, the device could also be used to check whether a tumor is growing or shrinking, or whether it has spread to other locations, by sensing the amount and location of tumor biomarkers. Preclinical testing is being done for this device for human chorionic gonadotropin that can be considered a biomarker for cancer because it is produced by tumors but not normally found in healthy individuals except pregnant women.

Nanotubes for Detection of Cancer Proteins

Single-wall carbon nanotubes (SWCNTs) are being developed for monitoring cancer-specific proteins. These are hundreds of times smaller than nanocantilevers, highly sensitive to single-protein binding events, and can be massively multiplexed with millions of tubes per chip for proteomic profiling. The tubes have extraordinary strength, unique electronic properties, and the ability to tag cancer-specific proteins to their surface. These tubes can be fabricated by decomposition of carbon-based gas in a furnace, using iron nanoparticles as catalyst material. With diameter of 1 nm and length of 1 μm, these tubes are smaller than a single strand of DNA. In other words, such a tube is an atomic arrangement of one layer of carbon atoms, which are on the surface. Protein binding events occurring on the surface of these tubes produce a measurable change in the mechanical and electrical properties.

By coating the surfaces of SWCNTs MAbs, it is possible to detect circulating tumor cells (CTCs) in the blood. SWCNTs covered with MAbs, particularly those for insulin-like growth factor-1 receptor (IGF-1), which is commonly found at high levels on cancer cells, home in on target protein "antigens" on the surface of CTCs. This method can be used for detection of recurring CTCs or residual micrometastases from the originally treated tumor. The technique could be cost-effective and could diagnose whether cells are cancerous or not in seconds versus hours or days required for conventional histology examination. It will enable large-scale production methods to make thousands of biosensors and have microarrays of these to detect the fingerprints of specific kinds of CTCs. Eventually, it may be possible to design an assay that can detect CTCs on a handheld device no bigger than a cell phone. Limitation of the technique is that it may not detect more than one antigen at a time on a single cell.

Nanobiochip Sensor Technique for Analysis of Oral Cancer Biomarkers

A pilot study has described a nanobiochip sensor technique for analysis of oral cancer biomarkers in exfoliative cytology specimens, targeting both biochemical and morphologic changes associated with early oral tumorigenesis (Weigum et al. 2010). Oral lesions from dental patients, along with normal epithelium from healthy volunteers, were sampled using a noninvasive brush biopsy technique. Specimens were enriched, immunolabeled, and imaged in the nanobiochip sensor according to previously established assays for the epidermal growth factor receptor (EGFR) biomarker and cytomorphometry. Four key parameters were significantly elevated in both dysplastic and malignant lesions relative to healthy oral epithelium, including the nuclear area and diameter, the nuclear-to-cytoplasmic ratio, and EGFR biomarker expression. Further examination using logistic regression and receiver-operating characteristic curve analyses identified morphologic features as the best predictors of disease individually, whereas a combination of all features further enhanced discrimination of oral cancer and precancerous conditions with high sensitivity and specificity. Further clinical trials are necessary to validate the regression model and evaluate other potential biomarkers. Nanobiochip sensor technique is a promising tool for early detection of oral cancer, which could enhance patient survival.

Nanodots for Tracking Apoptosis in Cancer

Apoptosis is a hallmark effect triggered by anticancer drugs. Researchers at Seoul National University (Korea) have developed a biocompatible, fluorescent nanoparticle that could provide an early sign that apoptosis is occurring as a result of anticancer therapy (Yu et al. 2007). The team created their fluorescent surface-enhanced Raman spectroscopic (F-SERS) nanodots to boost the optical signal generated by typical, biocompatible fluorescent dyes. The nanodots consist of silver nanoparticles embedded in a silica sphere. Attached to the silica core are fluorescent dye molecules and molecules known as Raman labels that enhance the electronic interactions between the silver nanoparticles and the dye molecules. The researchers also linked annexin-V, a molecule that binds specifically to a chemical that appears on cells undergoing apoptosis, to the silica–silver nanoparticle construct. Toxicity tests showed that the silica–silver nanodots were not toxic to various human cells growing in culture. The investigators then added the nanodots to cells triggered to undergo apoptosis and were able to image those cells as they went through programmed cell death. Based on these results, the researchers prepared other nanodots containing antibodies that bind to other molecules involved in apoptosis. They then added these antibody-linked nanodots and the annexin-V-linked nanodots to cultured human lung cancer cells. The investigators were able to track the appearance of all three molecules simultaneously, which has been difficult to do using conventional cell-staining techniques.

Nanolaser Spectroscopy for Detection of Cancer in Single Cells

Nanolaser scanning confocal spectroscopy can be used to identify a previously unknown property of certain cancer cells that distinguishes them with single-cell resolution from closely related normal cells (Gourlay et al. 2005). This property is the correlation of light scattering and spatial organization of mitochondria; normally, it is well scattered, but in cancer cells, the mitochondria are disorganized and scatter light poorly. These optical methods are promising powerful tools for detecting cancer at an early stage.

Nanoparticles Designed for Dual-Mode Imaging of Cancer

Scientists at Yonsei University, Korea, have combined the best characteristics of QDs and magnetic iron oxide nanoparticles to create a single nanoparticle probe that can yield clinically useful images of both tumors and the molecules involved in cancer (Choi et al. 2006). They start by synthesizing 30-nm-diameter silica nanoparticles impregnated with Rhodamine, a bright fluorescent dye, and 9-nm-diameter water-soluble iron oxide nanoparticles. They then mix these two nanoparticles with a chemical linker, yielding the dual-mode nanoparticle. On average, ten magnetic iron oxide particles link to a single-dye-containing silica nanoparticle, and the resulting construct is approximately 45 nm in diameter. The combination nanoparticle performed better in both MRI and fluorescent-imaging tests than did the individual components. In MRI experiments, the combination nanoparticle generated an MRI signal that was over threefold more intense than did the same number of iron oxide nanoparticles. Similarly, the fluorescent signal from the dual-mode nanoparticle was almost twice as bright as that produced by dye molecules linked directly to iron oxide nanoparticles. Next, the researchers labeled the dual-mode nanoparticles with an antibody that binds to molecules known as polysialic acids, which are found on the surface of certain nerve cell and lung tumors. These targeted nanoparticles were quickly taken up by cultured tumor cells and were readily visible using fluorescence microscopy.

Nanotechnology-Based Single-Molecule Assays for Cancer

Information about the biological processes in living cells is required for the detection and diagnosis of cancer for the following reasons:

- To recognize the important changes, which occur when cells undergo malignant transformation.
- There are situations when primary cells from a surgical procedure cannot be propagated due to the type of cell or the low number of cells available.
- Detection of cancer at an early stage is a critical step for improving cancer treatment.

Early detection will require sensitive methods for isolating and interrogating individual cells with high spatial and temporal resolution without disrupting their cellular biochemistry. Probes designed to penetrate a cell and report on the conditions within that cell must be sufficiently small, exceedingly bright, and stable for a long time in the intracellular environment without disrupting the cell's normal biochemical functioning. A series of silver nanoparticles have been prepared that meet many of the criteria listed above. Although smaller than 100 nm in diameter, these particles are bright enough to be seen by eye using optical microscopy. Unlike fluorophores, fluorescent proteins, or quantum dots, silver nanoparticles do not photodecompose during extended illumination. Therefore, they can be used as a probe to continuously monitor dynamic events in living cells during studies that last for weeks or even months. Because the color of the scattered light from nanoparticles depends upon their size, they have been used to measure the change in single-membrane pores in real time using dark-field optical microscopy. Intracellular and extracellular nanoparticles can also be differentiated by the intensity of light scattering. Next challenge is to develop methods for modifying the surface of the nanoparticles to make them more biocompatible, so that biological processes can be observed without disturbing or destroying the cell's intrinsic biochemical machinery. Ultimately, these probes may be combined to produce highly sensitive assays with high spatial and temporal resolution. This advance will enable researchers to study the interactions of multiple genes in the same cell simultaneously by using different colored reporter molecules. In addition to transcription and translation, similar live-cell single-molecule assays will offer the prospect of studying more complex cellular processes, such as cell signaling. Continuous advances and evolution along these research fronts is necessary to unravel biochemical processes in vivo and to develop tools that can be used to detect and diagnose cancer using only a single cell from the patient.

QDs for Detection of Tumors

QD bioconjugates that are highly luminescent and stable can be used for studying gene and enable visualization of cancer cells in living animals. QDs can be combined with fluorescence microscopy to follow cells at high resolution in living animals. These offer considerable advantages over organic fluorophores for this purpose. QDs and emission spectrum scanning multiphoton microscopy have been used to develop a means to study extravasation of tumor cells in vivo.

QD-Based Test for DNA Methylation

DNA methylation contributes to carcinogenesis by silencing key tumor suppressor genes. An ultrasensitive and reliable nanotechnology assay, MS-qFRET (fluorescence resonance energy transfer), can detect and quantify DNA methylation (Bailey et al. 2009). Bisulfite-modified DNA is subjected to PCR amplification with primers that would differentiate between methylated and unmethylated DNA. QDs are then used to capture PCR amplicons and determine the methylation status via FRET. The specific target of the test is DNA methylation which occurs when methyl attaches to cytosine, a DNA building block. When this happens at specific gene locations, it can stop the release of tumor-suppressing proteins; cancer cells then more easily form and multiply. The method involves singling out the DNA strands with methyl attachments through bisulfite conversion, whereby all nonmethyl segments are converted into another nucleotide. Copies of the remaining DNA strands are made, two molecules (a biotin protein and a fluorescent dye) are attached at either end, and the strands are mixed with QDs that are coated with a biotin-attractive chemical. Up to 60 DNA strands are attracted to a single QD. An UV light or blue laser activates the QDs, which pass the energy to the fluorescent molecules on the DNA strands which then light up and are identifiable via a spectrophotometer, which both identifies and can count the DNA methylation.

Key features of MS-qFRET include its low-intrinsic background noise, high resolution, and high sensitivity. This approach detects as little as 15 pg of methylated DNA in the presence of a 10,000-fold excess of unmethylated alleles, enables reduced use of PCR (as low as eight cycles), and allows for multiplexed analyses. The high sensitivity of MS-qFRET enables one-step detection of methylation at PYCARD, CDKN2B, and CDKN2A genes in patient sputum samples that contain low concentrations of methylated DNA, which normally would require a nested PCR approach.

The direct application of MS-qFRET on clinical samples offers great promise for its translational use in early cancer diagnosis and prognostic assessment of tumor behavior, as well as monitoring response to therapeutic agents. Gene DNA methylation indicates a higher risk of developing cancer and is also seen as a warning sign of genetic mutations that lead to development of cancer. Moreover, since different cancer types possess different genetic markers, e.g., lung cancer biomarkers differ from leukemia, the test should identify which cancer a patient is at risk of developing. This test could be used for frequent screening for cancer and replacing traditionally invasive methods with a simple blood test. It could also help determine whether a cancer treatment is effective and thus enable personalized chemotherapy.

Nanobiotechnology for Early Detection of Cancer to Improve Treatment

Cancer cells themselves may be difficult to detect at an early stage, but they leave a fingerprint, i.e., a pattern of change in biomarker proteins that circulate in the blood. There may be 20–25 biomarkers, which may require as many as 500 measurements, all of which should be made from a drop of blood obtained by pinprick. Thus, nanoscale diagnostics will play an important role in this effort. Nanowire sensors are in development at California Institute of Technology (Pasadena, CA) for very early diagnosis of cancer, when there are just a few thousand cells. Nanowires can electronically detect a few protein molecules along with other biochemical markers that are early signs of cancer. Nanowires in a set are coated with different compounds, each of which binds to a particular biomarker and changes the conductivity of the nanowire that can be measured. Thousands of such nanowires are combined on a single chip that enables identification of the type of cancer. Currently, such a chip can detect between 20 and 30 biomarkers and is being used for the early diagnosis of brain cancer.

Cancer is easier to treat and less likely to develop drug resistance when treatment is started very early. Cancer cells in very early stages are less likely to have mutations that make them resistant to treatment.

An automated gold nanoparticle biobarcode assay probe has been described for the detection of prostate-specific antigen (PSA) at 330 fg/mL, along with the results of a clinical pilot study designed to assess the ability of the assay to detect PSA in the serum of 18 men who have undergone radical prostatectomy for prostate cancer. Available PSA immunoassays are often not capable of detecting PSA in the serum of men after radical prostatectomy. This new bio-barcode PSA assay is approximately 300 times more sensitive than commercial immunoassays, and all patients in this study had a measurable serum PSA level after radical prostatectomy. Because the patient outcome depends on the level of PSA, this ultrasensitive assay enables (1) informing patients, who have undetectable PSA levels with conventional assays, but detectable and nonrising levels with the barcode assay, that their cancer will not recur; (2) earlier detection of recurrence, earlier because of the ability to measure increasing levels of PSA before conventional tools can make such assignments; and (3) use of PSA levels, which would otherwise not be detectable with conventional assays, to follow the response of patients to treatment.

Nanobiotechnology-Based Drug Delivery in Cancer

Drug delivery in cancer is important for optimizing the effect of drugs and reducing toxic side effects. Several nanobiotechnologies, mostly based on nanoparticles, have been used to facilitate drug delivery in cancer. A classification of the nanotechnologies for drug delivery in cancer is shown in Table 8.1.

Approximately 150 drugs in development for cancer are based on nanotechnology. Those approved are listed in Table 8.2, and several more are in clinical trials.

Table 8.1 Classification of nanobiotechnology approaches to drug delivery in cancer

Nanoparticles
Nanoparticle formulations of anticancer drugs, e.g., paclitaxel
Exosomes for cancer drug delivery
Nanoencapsulation and enclosure of anticancer drugs
Enclosing drugs in lipid nanocapsules
Encapsulating drugs in hydrogel nanoparticles
Micelles for drug delivery in cancer
Targeted delivery of anticancer therapy
Targeted drug delivery with nanoparticles
PEGylated nanoliposomal formulation
Folate-linked nanoparticles
Carbon magnetic nanoparticles for targeted drug delivery in cancer
Targeted drug delivery with nanoparticle–aptamer bioconjugates
Nanodroplets for site-specific cancer treatment
Lipid-based nanocarriers
Targeted antiangiogenic therapy using nanoparticles
Nanoparticles for delivery of drugs to brain tumors
Combination of nanoparticles with radiotherapy
Combination with boron neutron capture therapy
Nanoengineered silicon for brachytherapy
Combination with physical modalities of cancer therapy
Combination with laser ablation of tumors
Combination with photodynamic therapy
Combination with thermal ablation
Combination with ultrasound
Nanoparticle-mediated gene therapy
p53 Gene therapy of cancer
Immunolipoplex for delivery of p53 gene
Intravenous delivery of FUS1 gene
Strategies combining diagnostics and therapeutics
Nanoshells as adjuncts to thermal tumor ablation
Perfluorocarbon nanoparticles
Nanocomposite devices

Table 8.2 Approved anticancer drugs using nanocarriers

Trade name/compound	Manufacturer	Nanocarrier
Abraxane/paclitaxel	Abraxis Biosciences	Albumin-bound paclitaxel
Bexxar/anti-CD20 conjugated to iodine-131	Corixa/GlaxoSmithKline	Radio-immunoconjugate
DaunoXome/daunorubicin	Diatos, available in France	Liposome
Doxil/Caelyx/doxorubicin	Ortho Biotech	Liposome
Myoset/doxorubicin	Cephalon, available in Europe	Non-PEGylated liposome
Oncaspar/PEG-L-asparaginase	Enzon	Polymer–protein conjugate
Ontak/IL2 fused to diphtheria toxin	Eisai Inc.	Immunotoxic fusion protein
SMANCS/zinostatin	Yamanouchi Pharma	Polymer–protein conjugate
Zevalin/anti-CD20 conjugated to yttrium-90	Cell Therapeutics Inc	Radio-immunoconjugate
Zoladex/goserelin acetate	AstraZeneca	Polymer rods

SMANCS styrene maleic anhydride neocarzinostatin

Nanoparticle Formulations for Drug Delivery in Cancer

Anticancer Drug Particles Incorporated in Liposomes

Several injectable and biodegradable systems have been synthesized based on incorporation of antiestrogens (AEs) in nanoparticles and liposomes. Both nanospheres and nanocapsules (polymers with an oily core in which AEs were solubilized) incorporated high amounts of 4-hydroxytamoxifen (4-HT) or RU 58668. Liposomes containing various ratios of lipids enhanced the apoptotic activity of RU 58668 in several multiple myeloma cell lines tested by flow cytometry. These cell lines expressed both estrogen receptor alpha and beta subtypes. RU-loaded liposomes, administered intravenously in an animal model, induce the arrest of tumor growth. Thus, the drug delivery of antiestrogens enhances their ability to arrest the growth of tumors which express estrogen receptors and are of particular interest for estrogen-dependent breast cancer treatment. In addition, it represents a new potent therapeutic approach for multiple myeloma.

SuperFluids™ technology (Aphios Corporation) involves biodegradable polymer nanospheres utilizing supercritical, critical, or near-critical fluids with or without polar cosolvents. These nanospheres are utilized to encapsulate proteins with controlled-release characteristics without usage of toxic organic solvent. The patented technology can be utilized to form stable biocompatible aqueous formulations of poorly soluble anticancer drugs such as paclitaxel and camptothecin. An improved process utilizing SuperFluids™ results in the formation of small, uniform liposomes (nanosomes) to improve the delivery and therapeutic efficacy of poorly water-soluble drugs while reducing their toxicities. The process has been used for the nanoencapsulation of paclitaxel in a formulation called Taxosomes™, which has been tested in nude mice with breast cancer xenografts. Taxosomes™ will lead to (1) enhanced therapeutic efficacy, (2) elimination of premedication to counteract castor oil, (3) reduction of drug toxicity side effects, (4) prolonged circulation time and therapeutic effect, and (5) improved quality of life.

The process has also been used for the nanoencapsulation of camptothecin, a potent and exciting anticancer agent, in a stable aqueous liposomal formulation called Camposomes™. Water-soluble derivatives of camptothecin, a unique topoisomerase 1 inhibitor, have recently been approved by the FDA for use in colorectal cancer. Camposomes™ have been shown to be very effective against lymphomas in nude mice.

Nanocapsules, which are small aggregates of cisplatin covered by a single-lipid bilayer, have an unprecedented drug-to-lipid ratio and an in vitro cytotoxicity up to 1,000-fold higher than the free drug. Analysis of the mechanism of nanocapsule formation suggests that the method may be generalized to other drugs showing low water solubility and lipophilicity.

In Protein Stabilized Liposome (PSL™) nanotechnology of Azaya Therapeutics, the liposome product is prepared in a single step that encapsulates the active drug docetaxel (ATI-1123) in the lipid layer of the liposome while forming active nanoparticles in situ (100–130 nm). This process is geared toward the formulation of hydrophobic molecules that would otherwise have limited success as developmental drugs using traditional formulation methodologies. Azaya intends to use its PSL nanotechnology to improve the performance and reduce the nonspecific cytotoxicity of leading marketed chemotherapeutics such as Taxotere (docetaxel) and Camptosar®, as well as several experimental drugs that have been withdrawn from development due to their nonspecific cytotoxicity and formulation difficulties.

Cerasomes

Ceramide is a lipid molecule in plasma membrane of the cell and controls cell functions such as cell aging. Ceramide selectively kills cancer cells but is not toxic to normal cells. However, as a lipid, ceramide cannot be delivered effectively as a drug. To solve this limitation, cerasome is created to turn the insoluble lipid into a soluble form. Cerasomes are molecular-sized bubbles (size range from 60 to 200 nm) filled with C6-ceramide for use as anticancer agents. Paclitaxel-loaded cerasomes exhibit sophisticated controlled-release behavior for drug delivery in cancer (Cao et al. 2010). Cerasomes have already been shown to effectively treat cellular and animal models of breast cancer and melanoma. Systemic administration of nanoliposomal C6-ceramide to mice engrafted with SK-HEP-1 tumors reduced tumor vascularization and proliferation, induced tumor cell apoptosis, decreased phosphorylation of AKT, and ultimately blocked tumor growth (Tagaram et al. 2011). These studies show that nanoliposomal ceramide is an efficacious antineoplastic agent for the treatment of in vitro and in vivo models of human hepatocellular carcinoma.

Encapsulating Drugs in Polymeric Nanoparticles

Curcumin, an element found in the cooking spice turmeric, has long been known to have potent anticancer properties as demonstrated in several human cancer cell line and animal carcinogenesis models. Nevertheless, widespread clinical application of this relatively efficacious agent in cancer and other diseases has been limited due to poor aqueous solubility and, consequently, minimal systemic bioavailability. This problem has been overcome by encapsulating free curcumin with a polymeric nanoparticle, creating nanocurcumin (Bisht et al. 2007). Further, nanocurcumin's mechanisms of action on pancreatic cancer cells mirror that of free curcumin, including induction of cellular apoptosis, blockade of nuclear factor kappa-B (NF-kappaB) activation, and downregulation of steady-state levels of multiple pro-inflammatory cytokines (IL-6, IL-8, and TNF-α). No evidence of toxicity was found in tests with empty versions of the polymeric nanoparticle. Their findings show no evidence of weight loss, organ changes, or behavioral changes in live mice after administering a relatively large dosage of the empty nanoparticles. Nanocurcumin provides an opportunity to expand the clinical repertoire of this efficacious agent by enabling ready aqueous dispersion. Future studies utilizing nanocurcumin are warranted in preclinical in vivo models of cancer and other diseases that might benefit from the effects of curcumin.

Encapsulating Drugs in Hydrogel Nanoparticles

A versatile chemical technique has been developed for creating ultrafine nanosized hydrogels, essentially a network of polymer chains that absorb as much as 99% of their weight in water (Gao et al. 2007). Polyacrylamide was used to create nanoparticles of 2 nm diameter that have no charge on their surfaces. This lack of charge prevents blood proteins from sticking to the surface of the nanoparticles. Combined with the fact that these nanoparticles are too small to be recognized by the immune system, the result is a nanoscale drug-delivery vehicle with the ability to remain in circulation long enough to reach and permeate tumors before being excreted through the kidneys. These nanoscale hydrogels were first tested as a drug-delivery vehicle for a water-insoluble photosensitizer called meta-tetra(hydroxyphenyl) chlorin (mTHPC), which is approved in the European Union for use in treating head and neck cancer. mTHPC produces cell-killing reactive oxygen when irradiated with red light, but not without serious side effects resulting from the method now used to deliver this drug to tumors. When added to the chemical mixture used to create the nanoparticles, mTHPC becomes trapped within the polymer framework. Characterization experiments showed that this photosensitizer does not escape from the nanoparticles, but is still capable of producing the same amount of reactive oxygen as if it were free in solution. When added to human brain cancer cells growing in culture and irradiated with red light, this formulation kills the cells rapidly. Empty nanoparticles had no effect on the cells. Neither did drug-loaded nanoparticles added to the cells that were kept in the dark.

Exosomes

Exosomes are small (50–100 nm), spherical vesicles produced and released by most cells to facilitate intercellular communication. These vesicles are of endosomal origin and are secreted in the extracellular milieu following fusion of late endosomal multivesicular bodies with the plasma membrane. Exosomes have a defined protein composition, which confers specific biological activities contingent on the nature of the producing cell. Although exosomes express tumor antigens, leading to their proposed utility as tumor vaccines, they also can suppress and induce T-cell signaling molecules.

Exosomes produced by dendritic cells are called dexosomes and contain essential components to activate both adaptive and innate immune responses. Anosys is developing dexosome vaccines that use patient-specific dexosomes loaded with tumor antigen-derived peptides to treat cancer. Exosome research continues to reveal unique properties which broaden their fields of application. Anosys' Exosome Display Technology provides the ability to manipulate exosome composition and tailor exosomes with new desirable properties opening up opportunities in the field of recombinant vaccine and MAb preparation. This is achieved by generating genes coding for chimeric proteins linking an exosome addressing sequence to antigens or biologically active proteins. The resulting proteins are targeted to exosomal compartment and released in the extracellular milieu bound to exosomes.

Exosomes are emerging as novel approaches for cancer vaccine development. Safety of exosomes has been established in clinical trials that can be administered, but their potency for eliciting appropriate immune responses to kill cancer cells leaves much to be desired (Tan et al. 2010). Most of the investigational evidence is about solid tumors, and it has not been demonstrated that nonsolid tumors (e.g., hematological malignancies) can be treated using exosome technology. Moreover, exosomal immunotherapy relies on the immune system, and cancer patients, who are immunocompromised and/or immunosuppressed due to chemotherapy and radiotherapy, might not be able to overcome cancer with their immune system alone.

Folate-Linked Nanoparticles

PEG-coated biodegradable nanoparticles can be coupled to folic acid to target the folate-binding protein; this molecule is the soluble form of the folate receptor that is overexpressed on the surface of many tumor cells. The specific interaction between the conjugate folate nanoparticles and the folate-binding protein has been evaluated by surface plasmon resonance and confirmed a specific binding of the folate nanoparticles to the folate-binding protein. Thus, folate-linked nanoparticles represent a potential new drug carrier for tumor cell-selective targeting.

Iron Oxide Nanoparticles

A novel water-dispersible oleic acid (OA)–Pluronic-coated iron oxide magnetic nanoparticle formulation can be loaded easily with high doses of water-insoluble anticancer agents (Jain et al. 2005). Drug partitions into the OA shell surrounding iron oxide nanoparticles and the Pluronic that anchors at the OA–water interface confers aqueous dispersity to the formulation. Neither the formulation components nor the drug loading affects the magnetic properties of the core iron oxide nanoparticles. Sustained release of the incorporated drug is observed over 2 weeks under in vitro conditions. The nanoparticles have further demonstrated sustained intracellular drug retention relative to drug in solution and a dose-dependent antiproliferative effect in breast and prostate cancer cell lines. This nanoparticle formulation can be used as a universal drug carrier system for systemic administration of water-insoluble drugs while simultaneously enabling magnetic targeting and/or imaging.

Lipid-Based Nanocarriers

LiPlasome Pharma's proprietary prodrug and drug-delivery technology is based on smart lipid-based nanocarriers (LiPlasomes) that can be applied for targeted transport of anticancer drugs (Andresen et al. 2005). The targeted drug-delivery principle consists of long-circulating nanoparticles such as liposomes or micelles that accumulate in porous cancer tissue with a high PLA2 activity. The carrier nanoparticles are composed of special prodrug lipids whose degradation products, after exposure to PLA2, are converted to active drugs such as anticancer lysolipids and/or fatty acid drug derivatives. The PLA2 hydrolysis products will furthermore act as locally generated permeability enhancers that promote the absorption of the released drugs across the cancer cell membranes into putative intracellular target sites. This innovative prodrug and drug-delivery concept allows for intravenous transport of high concentrations of anticancer drugs directly to the tumor target. It enables, without any prior knowledge of the position and size of the tumor, to release the anticancer drugs specifically at the tumor target site. The delivery system is formulated with PEG to prolong the serum half-life of the drugs and prodrugs and avoid the nanocarriers being removed by the reticuloendothelial system.

Micelles for Drug Delivery in Cancer

Block copolymer micelles are spherical supramolecular assemblies of amphiphilic copolymers that have a core-shell-type architecture. The core is a loading space that can accommodate hydrophobic drugs, and the shell is a hydrophilic brushlike corona that makes the micelle water soluble, thereby allowing delivery of the poorly soluble contents (Fig. 8.1).

However, a key issue with the contained cytotoxic drugs is an understanding of how the micelle and the micelle-incorporated agent are distributed. By using fluorescently labeled polymer and organelle-specific dyes in combination with confocal microscopy, it has been shown that the micelles localize in several cytoplasmic organelles, including the mitochondria, but not the nucleus. Furthermore, the micelles increase the amount of a drug delivered to the cells and have the potential to deliver drugs to particular subcellular targets. Antibodies can be attached to

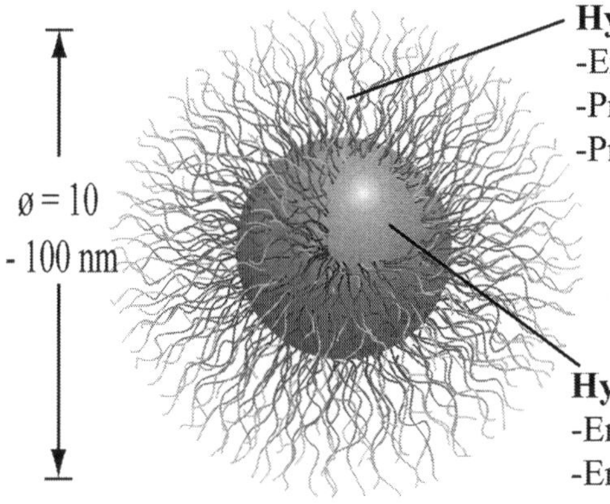

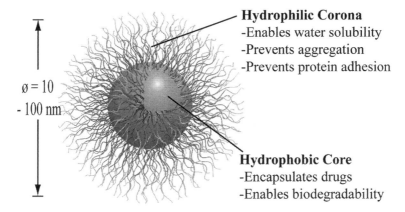

Fig. 8.1 Use of micelles for drug delivery (Source: Sutton et al. 2007)

the polymers that make up the micelles. Administering immunomicelles loaded with the sparingly soluble anticancer drugs. Paclitaxel micelle (NanoCarrier Ltd), a tumor-targeted drug-delivery system, has been investigated in clinical trials by Nippon Kayaku Co in Japan and is entering phase III in 2012. These micelles have a size of ~28 nm and exhibit a sustained drug release accompanied with the decay of the carrier itself in physiological saline. They show remarkably prolonged blood circulation and effectively accumulate in solid tumors indicating a potential for the targeted therapy of solid tumors.

DACH-platin-PEG-polyglutamic acid (DACH Platin Medicelle™) from NanoCarrier, based on Medicelle™ technology, has demonstrated enhanced permeability and retention of the compound in the tumor, leading to improved efficacy and toxicity profiles in animal experiments. The mechanism of action of Medicelle™ delivery system is based on the formation of micelles, including hydrophilic–hydrophobic block copolymers, with a hydrophobic inner core and hydrophilic outer shell. This allows the chemical entrapment of various drugs into the micelles. The drugs are then released slowly into the organism. This product is being developed for clinical application.

Camptothecin (CPT) is a topoisomerase I inhibitor that is effective against cancer, but clinical application of CPT is limited by insolubility, instability, and toxicity problems. Biocompatible, targeted sterically stabilized micelles (SSM) have been used as nanocarriers for CPT (CPT-SSM). CPT solubilization in SSM is reproducible and is attributed to avoidance of drug aggregate formation. Furthermore, SSM composed of polyethylene glycol (PEGylated) phospholipids are attractive nanocarriers for CPT delivery because they are sufficiently small (~14 nm) to extravasate through the leaky microvasculature of tumor and inflamed tissues for passive targeting of solid cancers in vivo, resulting in high drug concentration in tumors and reduced drug toxicity to the normal tissues (Koo et al. 2006).

Stealth micelle formulations have stabilizing PEG coronas to minimize opsonization of the micelles and maximize blood circulation times. Clinical data have been reported on three stealth micelle systems: SP1049C, NK911, and Genexol-PM (Sutton et al. 2007). SP1049C is formulated as doxorubicin (DOX)-encapsulated Pluronic micelles, NK911 is DOX-encapsulated micelles from a copolymer of PEG and DOX-conjugated poly(aspartic acid), and Genexol-PM is a paclitaxel-encapsulated PEG–PLA micelle formulation. Polymer micelles are becoming a powerful nanotherapeutics platform that affords several advantages for targeted drug delivery in cancer, including increased drug solubility, prolonged circulation half-life, selective accumulation at tumor sites, and a decrease in toxicity.

Minicells for Targeted Delivery of Nanoscale Anticancer Therapeutics

Indiscriminate drug distribution and severe toxicity of systemic administration of chemotherapeutic agents can be overcome through encapsulation and cancer cell-specific targeting of chemotherapeutics in 400-nm minicells (EnGeneIC Delivery Vehicle). Targeted minicells enter the cancer cells via receptor-mediated endocytosis, while the bacteria carrying nanoparticles enter the mammalian cells in a nonspecific manner, i.e., via phagocytosis. Scientists at EnGeneIC discovered that minicells can be packaged with therapeutically significant concentrations of chemotherapeutics of differing charge, hydrophobicity, and solubility (MacDiarmid et al. 2007). Targeting of minicells via bispecific antibodies to receptors on cancer cell membranes results in endocytosis, intracellular degradation, and drug release. Doses of drugs delivered via minicells are ~1,000 times less than the dose of the free drug required for equivalent or better tumor shrinkage. It produces significant tumor growth inhibition and regression in mouse xenografts and lymphoma in dogs despite administration of minute amounts of drug and antibody, a factor critical for limiting systemic toxicity that should allow the use of complex regimens of combination chemotherapy. Phase I clinical trials are in progress in Australia.

In a further study, minicells were shown to specifically and sequentially deliver to tumor xenografts siRNA- or shRNA-encoding plasmids to counteract drug resistance by knocking down a multidrug resistance protein (MacDiarmid et al. 2009). Subsequent administration of targeted minicells containing cytotoxic drugs eliminates formerly drug-resistant tumors. The dual sequential treatment, involving minicells loaded with both types of payload, enables complete survival without toxicity in mice with tumor xenografts, while involving several-thousand-fold less drug, siRNA, and antibody than needed for conventional systemic administration of cancer therapies.

Nanocarriers Enhance Doxorubicin Uptake in Drug-Resistant Cancer

Resistance to anthracyclines and other chemotherapeutics due to P-glycoprotein (pgp)-mediated export is a frequent problem in cancer treatment. Iron oxide–titanium dioxide (TiO_2) core-shell nanocomposites can serve as efficient carriers for doxorubicin to overcome this common mechanism of drug resistance in cancer cells (Arora et al. 2012). Doxorubicin nanocarriers (DNC) increased effective drug uptake in drug-resistant ovarian cells. Doxorubicin binds to the TiO_2 surface by a labile bond that is severed upon acidification within cell endosomes. Upon its release, doxorubicin traverses the intracellular milieu and enters the cell nucleus by a route that evades pgp-mediated drug export. Confocal and X-ray fluorescence microscopy and flow cytometry have been used to show the ability of DNCs to modulate transferrin uptake and distribution in cells. Increased transferrin uptake occurs through clathrin-mediated endocytosis, indicating that nanocomposites and DNCs may both interfere with removal of transferrin from cells. Together, these findings show that DNCs not only provide an alternative route of delivery of doxorubicin to pgp-overexpressing cancer cells but also may boost the uptake of transferrin-tagged therapeutic agents.

Nanoconjugates for Subcutaneous Delivery of Anticancer Drugs

Most of the anticancer drugs are administered intravenously. Nanoformulation of anticancer drugs is being developed for subcutaneous delivery of existing and new chemotherapeutics. Subcutaneous nanocarrier delivery of hyaluronan-conjugated doxorubicin or cisplatin has demonstrated significantly improved efficacy with decreased toxicity compared with standard agent combination therapy at all doses tested, achieving complete pathologic tumor response in mice implanted with human tumors (Cohen et al. 2011). Advantages of subcutaneous anticancer drug delivery are:

- Avoids complicated and expensive intravenous infusions
- Improves safety and efficacy for existing chemotherapy drugs
- Highly localized drug delivery to primary tumor sites to prevent recurrence
- Enables development of multidrug combinations to overcome drug resistance
- Incorporation of imaging agents to monitor penetration of the drug into tumor

Nanomaterials for Delivery of Poorly Soluble Anticancer Drugs

Nanomaterials have been successfully manipulated to create a new drug-delivery system that can solve the problem of poor water solubility of most promising currently available anticancer drugs and thereby increase their effectiveness. The poorly soluble anticancer drugs require the addition of solvents in order for them to be easily absorbed into cancer cells. Unfortunately, these solvents not only dilute the potency of the drugs but create toxicity as well. A novel approach has been devised using silica-based nanoparticles to deliver the anticancer drug camptothecin and other water-insoluble drugs into human cancer cells (Lu et al. 2007). The method incorporates a hydrophobic anticancer drug camptothecin into the pores of fluorescent mesoporous silica nanoparticles and delivers the particles into a variety of human cancer cells to induce cell death. The results suggest that the mesoporous silica nanoparticles might be used as a vehicle to overcome the insolubility problem of many anticancer drugs.

Nanoparticle Formulation for Enhancing Anticancer Efficacy of Cisplatin

Cisplatin is first-line chemotherapy for most types of cancer. However, its use is dose-limited due to severe nephrotoxicity. Rational engineering of a novel nanoplatinate has been reported, which self-assembles into a nanoparticle at unique platinum-to-polymer ratio and releases cisplatin in a pH-dependent manner (Paraskar et al. 2010). The nanoparticles are rapidly internalized into the endolysosomal compartment of cancer cells and exhibit an IC50 comparable to that of free cisplatin and superior to carboplatin. The nanoparticles showed significantly improved anticancer efficacy in terms of tumor growth delay in breast and lung cancers. Furthermore, the nanoparticle treatment resulted in reduced systemic and nephrotoxicity, validated by decreased biodistribution of platinum to the kidney. Given the need for a better platinate, this coupling of nanotechnology and structure-activity relationship to rationally reengineer cisplatin is anticipated to have a major impact on the treatment of cancer.

Nanoparticle Formulations of Paclitaxel

Paclitaxel is active and widely used to treat multiple types of solid tumors. The commercially available paclitaxel formulation uses Cremophor/ethanol (C/E) as the solubilizers. Other formulations including nanoparticles have been introduced. A study evaluated the effects of nanoparticle formulation of paclitaxel on its tissue distribution in experimental animals (Yeh et al. 2005). The nanoparticle and C/E formulations showed significant differences in paclitaxel disposition; the nanoparticles yielded 40% smaller area under the blood concentration-time curve and faster blood clearance of total paclitaxel concentrations (sum of free, protein-bound, and nanoparticle-entrapped drug). Tissue specificity of the two formulations was different. The nanoparticles showed longer retention and higher accumulation in organs and tissues, especially in the liver, small intestine, and kidney. The most striking difference was an eightfold greater drug accumulation and sustained retention in the kidney. These data indicate that nanoparticulate formulation of paclitaxel affects its clearance as well as distribution in tissues with preferential accumulation in the liver, spleen, small intestine, and kidney. Solid tumors have unique features, such as leaky tumor blood vessels and defective lymphatic drainage, that promote the delivery and retention of macromolecules or particles, a phenomenon recognized as the enhanced permeability and retention effect. Tissue specificity of the gelatin nanoparticles warrants further investigations before using nanoparticle formulations of anticancer drugs for tumors in various organs.

AI-850 (Acusphere Inc.) is a rapidly dissolving porous particle formulation of paclitaxel, created by using the company's Hydrophobic Drug Delivery Systems. The patented spray-drying technology embeds small drug particles inside hydrophobic water-soluble matrices so that the whole composition is a mixture of microparticles and nanoparticles. AI-850 was compared to Taxol following intravenous administration in a rat pharmacokinetic study, a rat tissue distribution study, and a human xenograft mammary tumor model in nude mice (Straub et al. 2005). The volume of distribution and clearance for paclitaxel following intravenous bolus administration of AI-850 were sevenfold and fourfold greater, respectively, than following intravenous bolus administration of Taxol. There were no significant differences between AI-850 and Taxol in tissue concentrations and area under the curve for the tissues examined. Nude mice implanted with mammary tumors showed improved tolerance of AI-850, enabling higher administrable dose of paclitaxel, which resulted in improved efficacy as compared to Taxol administered at its maximum-tolerated dose.

Gold nanoparticles (2 nm) have been covalently functionalized with paclitaxel (Gibson et al. 2007). The synthetic strategy involves the attachment of a flexible hexaethylene glycol linker at the C-7 position of paclitaxel followed by coupling of the resulting linear analogue to phenol-terminated gold nanocrystals. The reaction yields the product with a high molecular weight, while exhibiting an extremely low polydispersity index. The organic shell of hybrid nanoparticles contains 67% by weight of paclitaxel, which corresponds to ~70 molecules of the drug per one nanoparticle. High-resolution TEM was employed for direct visualization of the inorganic core of hybrid nanoparticles, which were found to retain their average size, shape, and high crystallinity after multiple synthetic steps and purifications. The interparticle distance substantially increases after the attachment of paclitaxel as revealed by low-magnification TEM, suggesting the presence of a larger organic shell. Thus organic molecules with exceedingly complex structures can be covalently attached to gold nanocrystals in a controlled manner and fully characterized by traditional analytical techniques. In addition, this approach gives a rare opportunity to prepare hybrid particles with a well-defined amount of drug and offers a new alternative for the design of nanosized drug-delivery systems. Follow-up studies will determine the potency of the paclitaxel-loaded nanoparticles. Since each ball is loaded with a uniform number of drug molecules, it will be relatively easy to compare the effectiveness of the nanoparticles with the effectiveness of generally administered paclitaxel. This technique could help to deliver more of the drug directly to the cancer cells and reduce the side effects of chemotherapy. The aim is to improve the effectiveness of the drug by increasing its ability to stay bound to microtubules within the cell.

Albumin nanoparticle technology enables the transportation of hydrophobic drugs such as paclitaxel without the need of potentially toxic solvents. Nab-paclitaxel can be administered without premedication, in a shorter infusion time and without the need for a special infusion set. Moreover, this technology allows the selective delivery of larger amounts of anticancer drug to tumors, by exploiting endogenous albumin pathways. Nab-paclitaxel is approved for the treatment of metastatic breast cancer, after the failure of first-line standard therapy, when anthracyclines are not indicated. Efficacy and safety data, along with a more convenient administration, confirm the potential for nab-paclitaxel to become a reference taxane in breast cancer treatment (Guarneri et al. 2012).

Nanoparticles Containing Albumin and Antisense Oligonucleotides

Nanoparticles consisting of human serum albumin (HSA) and containing different antisense oligonucleotides (ASO) have been used for drug delivery to tumors. The preparation process has been optimized regarding the amount of solving agent, stabilization conditions, as well as nanoparticle purification. The glutaraldehyde cross-linking procedure of the particle matrix is a crucial parameter for biodegradability and drug release of the nanoparticles. The drug-loading efficiency increases with longer chain length and employment of a phosphorothioate backbone. The resulting nanoparticles can be tested in cell cultures for cytotoxicity and cellular uptake. All cell lines show a significant cellular uptake of HSA nanoparticles. The entrapment of a fluorescent-labeled oligonucleotide within the particle matrix can be used for the detection of the intracellular drug release of the carrier systems. Confocal laser scanning microscopy reveals that nanoparticles cross-linked with low amounts of glutaraldehyde rapidly degrade intracellularly, leading to a significant accumulation of the ASO in cytosolic compartments of the tumor cells.

PEGylated Nanoliposomal Formulation

PEG-coated nanoparticles remain in the tumors and bloodstream longer compared to gelatin nanoparticles. The coating prevents the nanoparticles from removal by the reticuloendothelial system. This property has led to more effective nanoparticles with tumor-targeting properties.

Ceramide, an antimitogenic and proapoptotic sphingolipid, accumulates in cancer tissues and helps to kill cancer cells when patients undergo chemotherapy and radiation. Although the mechanism remains unknown, ceramide is inherently attracted to tumor cells. In vitro tumor cell culture models have shown the potential therapeutic utility of raising the intracellular concentration of ceramide. However, therapeutic use of systemically delivered ceramide is limited by its inherent insolubility in the blood as it is a lipid as well as its toxicity when injected directly into the bloodstream. Packaging ceramide in nanoliposome capsules allows them to travel through the bloodstream without causing toxicity and release the ceramide in the tumor. Systemic intravenous delivery of C6-ceramide (C6) in a PEGylated liposomal formulation significantly limited the growth of solid tumors in a syngeneic BALB/c mouse tumor model of breast adenocarcinoma (Stover et al. 2005). A pharmacokinetic analysis of systemic liposomal-C6 delivery showed that the PEGylated liposomal formulation follows first-order kinetics in the blood and achieves a steady-state concentration in tumor tissue. Intravenous liposomal-C6 administration was also shown to diminish solid tumor growth in a human xenograft model of breast cancer. In this study, in mice, the ceramide bundles targeted and destroyed only breast cancer cells, sparing the surrounding healthy tissue. Together, these results indicate that bioactive ceramide analogues can be incorporated into PEGylated liposomal vehicles for improved solubility, drug delivery, and antineoplastic efficacy. The next step is to explore how additional chemotherapeutic agents could be incorporated into the liposomes for a more lasting effect.

Peptide-Linked Nanoparticle Delivery

The coupling chemistry and surface charge effects of peptide labeling in nanoparticle drug-delivery strategies are more difficult to control than using folate. Chemical conjugation to peptides reduces colloidal stability, which is a limiting factor in the development of targeting nanoparticles. However, the successful peptide targeting of structural, hormonal cytokine and endocrine receptors in the delivery of therapeutic and diagnostic radionuclides provides justification for finding methods to synthesize peptide-targeted nanoparticles (Franzen 2011). Although most of the work so far has been done using gold nanoparticles, biological and polymer nanoparticles are more colloidally stable and present enormous opportunities for coupling to peptides. Further studies are needed to develop peptide targeting for nanoparticles to rival the selectivity that has been achieved with the small-molecule folate.

Poly-2-Hydroxyethyl Methacrylate Nanoparticles

Poly-2-hydroxyethyl methacrylate nanoparticles can potentially be used for the controlled release of the anticancer drug doxorubicin and reduction of its toxicity (Chouhan and Bajpai 2009). Suspension polymerization of 2-hydroxyethyl methacrylate (HEMA) results in the formation of swellable nanoparticles of defined composition. Release profiles of doxorubicin can be greatly modified by varying the experimental parameters such as percent loading of doxorubicin and concentrations of HEMA, cross-linker, and initiator. Swelling of nanoparticles and the release of doxorubicin increases with the increase in percentage loading of drug. Absorption spectra of doxorubicin do not change following its capture and release from the nanoparticles, indicating that chemical structure of the drug is likely to be unaffected by the procedure.

Polypeptide–Doxorubicin-Conjugated Nanoparticles

Artificial recombinant chimeric polypeptides (CPs), produced from genetically altered *Escherichia coli*, have been shown to spontaneously self-assemble into 50-nm nanoparticles on conjugation with various chemotherapeutics regardless of their water solubility (MacKay et al. 2009). CPs contain a biodegradable polypeptide that is attached to a short Cys-rich segment. Covalent modification of the Cys residues with a structurally diverse set of chemotherapeutics leads to spontaneous formation of nanoparticles over a range of CP compositions and molecular weights. Attachment to one of CPs induces characteristics that the drug alone does not possess. Most chemotherapeutics do not dissolve in water, which limits their ability to be taken in by cells, but attachment to a nanoparticle makes the drug soluble. When used to deliver chemotherapeutics to a murine cancer model, CP nanoparticles have a fourfold higher maximum-tolerated dose than free drug and induce nearly complete tumor regression after a single dose. After delivering the drug to the tumor, the delivery vehicle breaks down into harmless by-products, markedly decreasing the toxicity for the recipient. This simple as well as inexpensive strategy can promote co-assembly of drugs, imaging agents, and targeting moieties into multifunctional nanomedicines. Since blood vessels supplying tumors are more porous, or leaky, than normal vessels, the nanoformulation can more easily enter and accumulate within tumor cells. This means that higher doses of the drug can be delivered, increasing its anticancer effects while decreasing the side effects associated with systematic chemotherapy.

Protosphere Nanoparticle Technology

Protosphere™ nanoparticle technology (Abraxis Bioscience Inc), also referred to as nanoparticle albumin-bound or nab™ technology, was used to integrate biocompatible proteins with drugs to create the nanoparticle form of the drug having a size of about 100–200 nm. SPARC (secreted protein acidic and rich in cysteine), a protein overexpressed and secreted by cancer cells, binds albumin to concentrate albumin-bound cytotoxic drugs at the tumor.

The product Abraxane (ABI-007) is a patented albumin-stabilized nanoparticle formulation of paclitaxel (nab-paclitaxel) designed to overcome insolubility problems encountered with paclitaxel. The solvent Cremophor EL, used previously in formulations of paclitaxel, causes severe hypersensitivity reactions. To reduce the risk of allergic reactions when receiving Taxol, patients must undergo premedication using steroids and antihistamines and be given the drug using slow infusions. The active component (paclitaxel) can be delivered into the body at a 50% higher dose over 30 min. This contrasts with Taxol infusions, which can take up to 3 h. Because Abraxane is solvent-free, solvent-related toxicities are eliminated, premedication is not required, and administration can occur more rapidly. Abraxane also has a different toxicity profile than solvent-based paclitaxel, including a lower rate of severe neutropenia. In a randomized phase III trial, the response rate of Abraxane was almost twice that of the solvent-containing drug Taxol. Because Abraxane does not contain solvents, higher doses of paclitaxel could be given which may account in part for its increased anticancer activity. In addition, albumin is a protein that normally transports nutrients to cells and has been shown to accumulate in rapidly growing tumors. Therefore, Abraxane's increased effectiveness may also be due to preferential delivery of albumin-bound paclitaxel to cancer cells. In addition to the standard infusion formulation of Abraxane, oral and pulmonary delivery formulations are also being investigated.

A randomized controlled phase III clinical trial compared the safety and efficacy of 260 mg/m^2 of Abraxane to 175 mg/m^2 of Taxol administered every 3 weeks in patients with metastatic breast cancer (Gradishar et al. 2005). Abraxane was infused over 30 min without steroid pretreatment and at a higher dose than Taxol, which requires steroid therapy and infusion over 3 h. Abraxane was found to be superior to Taxol on lesion response rate as well as on tumor progression rate. In 2005, the FDA approved Abraxane for the treatment of metastatic breast cancer. Abraxane also is being evaluated in non-small-cell lung, ovarian, melanoma, and cervical cancers.

Zinc Oxide Nanoparticles for Drug Delivery in Cancer

Zinc oxide (ZnO) nanomaterials provide versatile platforms for biomedical applications and therapeutic intervention, and recent studies demonstrate that they hold considerable promise as anticancer agents. Several of these are under development at the experimental, preclinical, and clinical stages. Through a better understanding of the mechanisms of action and cellular consequences resulting from nanoparticle interactions with cells, the inherent toxicity and selectivity of ZnO nanoparticles against cancer may be improved further to make them attractive new anticancer agents (Rasmussen et al. 2010).

Nanoparticles for Targeted Delivery of Anticancer Therapeutics

Nanosystems are emerging that may be very useful for tumor-targeted drug delivery: novel nanoparticles are preprogrammed to alter their structure and properties during the drug-delivery process to make them most effective for the different extra- and intracellular delivery steps (Wagner 2007). This is achieved by the incorporation of molecular sensors that are able to respond to physical or biological stimuli, including changes in pH, redox potential, or enzymes. Tumor-targeting principles include systemic passive targeting and active receptor targeting. Physical forces (e.g., electric or magnetic fields, ultrasound, hyperthermia, or light) may contribute to focusing and triggered activation of nanosystems. Biological drugs delivered with programmed nanosystems also include plasmid DNA, siRNA, and other therapeutic nucleic acids.

Drug-delivery systems are being developed that attempt to destroy tumors more effectively by using synthesized smart nanoparticles that target and kill cancer cells while sparing healthy cells. Such particles can be injected intravenously into the blood circulation. Each particle, chemically programmed to have an affinity for the cell wall of tumor, can recognize the cancer cell, anchor itself to it, and diffuse inside the cell. Once inside, the particle disintegrates, causing a nearly instantaneous release of the drug precisely where it is needed. To be effective, the nanoparticles must evade the body's immune system, penetrate into the cancer cells, and discharge the drugs before being recognized by the cancer cells. Advantages of such systems are:

- They can fool cancer cells, which are very good at detecting and rejecting drugs.
- Provide rapid drug delivery at sufficiently high concentration that can overwhelm the cancer cell's resistance mechanisms.
- Reduction of side effects because the cancer cells are targeted selectively, sparing the normal cells.

Another approach is to use a nanoparticle made of a hydrogen and carbon polymer with anticancer drug bound up in its fabric and attached to a substance that targets cancer cells. Following intravenous injection, the polymer would gradually dissolve on reaching the target and gradually release the drug. One limitation of systemic introduction of such nanoparticles, unless properly targeted, is that they may end up in the liver and spleen. This is an unwanted side effect because once the nanoparticles dissolve in those organs, they release toxic levels of chemotherapy in healthy tissues.

Canine Parvovirus as a Nanocontainer for Targeted Drug Delivery

The canine parvovirus (CPV) utilizes transferrin receptors (TfRs) for binding and cell entry into canine as well as human cells. TfRs are overexpressed by a variety of tumor cells and are widely being investigated for tumor-targeted drug delivery. To explore the natural tropism of CPV to TfRs for targeting tumor cells, CPV viruslike particles (VLPs) produced by expression of the CPV-VP2 capsid protein in a baculovirus expression system were examined for attachment of small molecules and delivery to tumor cells (Singh et al. 2006a). Structural modeling suggested that six lysines per VP2 subunit are presumably addressable for bioconjugation on the CPV capsid exterior. Between 45 and 100 of the possible 360 lysines/particle could be routinely derivatized with dye molecules depending on the conjugation conditions. Dye conjugation also demonstrated that the CPV-VLPs could withstand conditions for chemical modification on lysines. Attachment of fluorescent dyes neither impaired binding to the TfRs nor affected internalization of the 26-nm-sized VLPs into several human tumor cell lines. CPV-VLPs therefore exhibit highly favorable characteristics for development as a novel nanomaterial for tumor targeting.

Carbon Magnetic Nanoparticles for Targeted Drug Delivery in Cancer

Using dense medium plasma technology, carbon magnetic nanoparticles (CMNP) have been synthesized at room temperature and atmospheric pressure. Results of X-ray photoelectron spectroscopy, Fourier transform infrared spectroscopy, and scanning electron microscopy show that these nanoparticles are composed of spherical particles, 40–50 nm in diameter, with iron/iron oxide particles dispersed in a carbon-based host structure (Ma et al. 2004). Thermal gravimetry/differential thermal gravimetry analysis shows these nanoparticles are stable to temperatures as high as 600 °C. The synthesized CMNP were treated by argon plasma, aminated with ethylene diamine and subsequently activated by generating aldehyde groups on them. Free doxorubicin (DOX) molecules are then immobilized onto the surfaces of activated CMNP particles to form CMNP–DOX conjugates. The in vitro antiproliferative activity of immobilized doxorubicin in the conjugates has been demonstrated in tumor cell cytotoxicity assays. It is suggested that this CMNP–DOX system can be used for targeted drug-delivery systems in cancer.

Carbon Nanotubes for Targeted Drug Delivery to Cancer Cells

CNTs with DNA and RNA wrapped around them to make them biocompatible can be targeted toward cancer cells by attaching additional molecules. Such CNT composites, in combination with laser treatment, may be used to destroy the cancer cells. An improved delivery scheme for intracellular tracking and anticancer therapy uses a novel double functionalization of a carbon nanotube delivery system containing antisense oligodeoxynucleotides as a therapeutic gene and CdTe QDs as fluorescent-labeling probes via electrostatically layer-by-layer assembly (Jia et al. 2007).

Chemically functionalized SWCNTs have shown promise in tumor-targeted accumulation in mice and exhibit biocompatibility, excretion, and little toxicity. The anticancer drug paclitaxel (PTX) has been conjugated to branched PEG chains on SWCNTs via a cleavable ester bond to obtain a water-soluble SWCNT–PTX conjugate (Liu et al. 2008a). SWCNT–PTX is more efficient in suppressing tumor growth than Taxol in a murine 4 T1 breast cancer model, owing to prolonged blood circulation and tenfold higher tumor PTX uptake by SWCNT delivery, likely through enhanced permeability and retention. Drug molecules carried into the reticuloendothelial system are released from SWCNTs and excreted via biliary pathway without toxic effects on normal organs. Thus, CNT drug delivery is promising for enhancing treatment efficacy and minimizing side effects of cancer therapy by use of low drug doses. Water-dispersed carbon nanohorns, prepared by adsorption of polyethylene glycol–doxorubicin conjugate (PEG–DXR) onto oxidized single-wall carbon nanohorns, have been shown to be effective anticancer drug-delivery carriers when administered intratumorally to human NSCLC-bearing mice (Murakami et al. 2008). There was significant retardation of tumor growth associated with prolonged DXR retention in the tumor.

Although considerable further work is required before any new drugs based on CNTs are developed, it is hoped that it will eventually lead to more effective treatments for cancer. However, it is too early to claim whether carbon-based nanomaterials will become clinically viable tools to combat cancer, although there is definitely room for them to complement existing technologies.

Cyclosert System for Targeted Delivery of Anticancer Therapeutics

Cyclosert™ (Calando/Insert Therapeutics) is the first nanoparticle drug transport platform to be designed de novo and synthesized specifically to overcome limitations in existing technologies used for the systemic transport of therapeutics to targeted sites within the body. Based on small cyclic repeating molecules of glucose called cyclodextrins, Cyclosert promotes the ability of cytotoxic drugs to inhibit the growth of human cancer cells while reducing toxicity and remaining nonimmunogenic at therapeutic doses. In particular, the system is designed to reduce the toxicity of the drugs until they actually reach the targeted tumor cells where the active drug is released in a controlled fashion. Animal studies have shown that the Cyclosert system can safely deliver tubulysin A, a potent, but highly toxic, antitumor agent. In vitro studies have shown the tubulysin–Cyclosert conjugate to be effective against multiple human cancer cell lines. The conjugate is stable and 100 times more water soluble than the free drug. Calando is developing CALAA01, a siRNA, for anticancer use using Cyclosert as a delivery system.

IT-101 (Calando) is a de novo-designed experimental therapeutic comprised of linear, cyclodextrin(CD)-containing polymer conjugates of camptothecin (CPT) that assemble into 40-nm-diameter nanoparticles via polymer–polymer interactions that involve inclusion complex formation between the CPT and the CD. Particle size, near-neutral surface charge, and CPT release rate were specifically designed into IT-101. Cyclosert platform forms nanoscale constructs with hydrodynamic diameters between 30 and 60 nm. This makes Cyclosert-based drugs ideal for effective delivery to solid tumors. Preclinical animal studies show extended circulation times, tumor accumulation, slow release of the CPT, and anticancer efficacy that directly correlate to the properties of the nanoparticle. Release of CPT can disassemble the nanoparticle into individual polymer chains ~10 nm in size that are capable of renal clearance. IT-101 has been evaluated in patients with relapsed or refractory cancer following two cycles of therapy by intravenous infusion. Interim analysis shows that IT-101 is well tolerated and pancytopenia is the dose-limiting toxicity (Yen et al. 2007). Pharmacokinetics data were favorable and consistent with results from preclinical animal studies. In the patients studied, IT-101 showed longer half-life, lower clearance, and lower volume of distribution than seen in patients treated with other camptothecin-based drugs. It is in phase II clinical trials.

DNA Aptamer–Micelle for Targeted Drug Delivery in Cancer

Design of a self-assembled aptamer–micelle nanostructure has been reported that achieves selective and strong binding of otherwise low-affinity aptamers at physiological conditions (Wu et al. 2009). Specific recognition ability is directly built into the nanostructures. The attachment of a lipid tail onto the end of nucleic acid aptamers provides these unique nanostructures with an internalization pathway. Other merits include extremely low off rate once bound with target cells, rapid recognition ability with enhanced sensitivity, low critical micelle concentration values, and dual-drug-delivery pathways. To prove the potential detection/delivery application of this aptamer–micelle in biological living systems, the authors mimicked a tumor site in the blood stream by immobilizing tumor cells onto the surface of a flow channel device. Flushing the aptamer–micelles through the channel demonstrated their selective recognition ability under flow circulation in human whole-blood sample. The aptamer–micelles show great dynamic specificity in flow channel systems that mimic drug delivery in the blood system. Therefore, DNA aptamer–micelle assembly has shown high potential for cancer cell recognition and for targeted in vivo drug-delivery applications.

Fullerenes for Enhancing Tumor Targeting by Antibodies

Although it was previously possible to attach drug molecules directly to antibodies, scientists have not been able to attach more than a handful of drug molecules to an antibody without significantly changing its targeting ability. That happens, in large part, because the chemical bonds that are used to attach the drugs – strong, covalent bonds – tend to block the targeting centers on the antibody's surface. If an antibody is modified with too many covalent bonds, the chemical changes will destroy its ability to recognize the cancer it was intended to attack.

In order to overcome this limitation, a new class of anticancer compounds have been created that contain both tumor-targeting antibodies and nanoparticles called fullerenes (C60), which can be loaded with several molecules of anticancer drugs like Taxol® (Ashcroft et al. 2006). It is possible to load as many as 40 buckyballs into a single skin-cancer antibody called ZME-018, which can be used to deliver drugs directly into melanoma tumors. Certain binding sites on the antibody are hydrophobic (water repelling) and attract the hydrophobic fullerenes in large numbers, so multiple drugs can be loaded into a single antibody in a spontaneous manner. No covalent bonds are required, so the increased payload does not significantly change the targeting ability of the antibody. The real advantage of fullerene immunotherapy over other targeted therapeutic agents is likely to be the fullerene's potential to carry multiple drug payloads, such as Taxol plus other chemotherapeutic drugs. Cancer cells can become drug resistant, and one can cut down on the possibility of their escaping treatment by attacking them with more than one kind of drug at a time. The first fullerene immunoconjugates have been prepared and characterized as an initial step toward the development of fullerene immunotherapy.

Gold Nanoparticles for Targeted Drug Delivery in Cancer

Gold and silica composite nanoparticles have been investigated as nanobullets for cancer. Gold atoms bind to silicon atoms with dangling bonds and serve as seeds for the growth of Au islands. The large electron affinity of gold causes a significant change in the electronic structure of silica resulting in a substantial reduction in the highest occupied and the lowest unoccupied molecular orbital and the optical gap, thus allowing it to absorb near-infrared radiation. This suggests that a small cluster can have a similar effect in the treatment of cancer as the large-size nanoshell, but with a different mechanism.

The unique chemical properties of colloidal gold make it a promising targeted delivery approach for drugs or genes to specific cells. The physical chemical properties of colloidal gold permit more than one protein molecule to bind to a single particle of colloidal gold. CytImmune Sciences Inc has shown that tumor necrosis factor (TNF) can be bound to gold nanocrystals and delivered safely and effectively to tumor-burdened mice and dogs. CytImmune scientists have characterized and modified the colloidal gold (cAu) particles to optimize binding of TNF to the nanocrystals and also the targeting of the particles to the tumor. The therapeutic compounds that CytImmune is developing are new formulations of the TNF-α, which causes the death of tumors but is toxic to healthy organs. Coupling TNF-α to colloidal gold is expected to improve the safety and effectiveness of anticancer therapy. Specifically, two drugs are in development: Aurimune-T and AuriTax. Aurimune-T is manufactured by covalently linking molecules of TNF-α and thiol-derivatized polyethylene glycol (PEG-THIOL) onto the surface of 25-nm colloidal gold. Intravenously administered Aurimune-T rapidly accumulates in solid tumors implanted in mice and shows little to no accumulation in the reticuloendothelial system or in other healthy organs. Coincident with the sequestration of gold is a tenfold accumulation of TNF-α in the tumor when compared to animals treated with native TNF-α. By getting more TNF-α to the tumor, Aurimune-T improves the safety and efficacy of TNF-α treatment since maximal tumor responses were achieved at lower doses of the drug. The second nanoparticle drug, AuriTax, consists of TNF-α, a chemotherapeutic (paclitaxel), and PEG-THIOL, which are bound to the same cAu nanoparticle. Like Aurimune-T, AuriTax delivers tenfold more TNF-α and paclitaxel to the solid tumor when compared to each drug alone. These results support the continued development of the colloidal gold platform for cancer therapy and TNF-α as a tumor-targeting ligand.

Biocompatible and nontoxic PEGylated gold nanoparticles with surface-enhanced Raman scattering have been used for in vivo tumor targeting and detection (Qian et al. 2008). Colloidal gold has been safely used to treat rheumatoid arthritis for 50 years and has recently been found to amplify the efficiency of Raman scattering by 14–15 orders of magnitude. It has been shown that large optical enhancements can be achieved under in vivo conditions for tumor detection in live animals. An important finding is that small-molecule Raman reporters such as organic dyes are not displaced but were stabilized by thiol-modified polyethylene glycols. These PEGylated SERS nanoparticles are considerably brighter than semiconductor QDs with light emission in the near-infrared window. When conjugated to tumor-targeting ligands such as single-chain variable fragment antibodies (ScFv), the conjugated nanoparticles are able to target tumor biomarkers such as EGFRs on human cancer cells and in xenograft tumor models. ScFv peptides bind cancer cells and the gold particles latch onto tumors after their injection into a mouse. When illuminated with a laser beam, the tumor-bound particles emit a signal that is specific to the dye. The signal from the dye tags is very bright and the distinct peaks in the dye signal mean several different probes could be used at the same time. The tags' rich spectroscopic signatures provide the capability of using several probes at once, but that will require more sophisticated computational tools. The authors are developing data processing tools and making them available to the NCI's caBIG (cancer biomedical informatics grid) so that the research community can use them. Compared with QDs, the gold particles are more than 200 times brighter on a particle-to-particle basis, although they are about 60 times larger by volume. Covered with a nontoxic polymer, the gold particles are about 60–80 nm in diameter, that is, 150 times smaller than a typical human cell and thousands of times smaller than a human hair. The researchers were able to detect human cancer cells injected into a mouse at a depth of 1–2 cm. That makes the gold particles especially appropriate tools for gathering information about head or neck tumors, which tend to be more accessible. The technology will need further adaptation for use with abdominal or lung cancers deep within the body.

PEGylated gold nanoparticles are decorated with various amounts of human transferrin (Tf) to give a series of Tf-targeted particles with near-constant size and electrokinetic potential. Studies in experimental animals with tumors show that quantitative biodistribution of the nanoparticles 24 h after intravenous injections results in their accumulations in the tumors and other organs independent of Tf (Choi et al. 2010). However, the nanoparticle localization within a particular organ is influenced by the Tf content. In tumor tissue, the content of targeting ligands significantly influences the number of nanoparticles localized within the cancer cells. In liver tissue, high Tf content leads to small amounts of the nanoparticles residing in hepatocytes, whereas most nanoparticles remain in nonparenchymal cells. These results suggest that targeted nanoparticles can provide greater intracellular delivery of therapeutic agents to the cancer cells within solid tumors than their nontargeted analogues.

Selective transport of gold nanoparticles to the nuclei of cancer cells has been achieved by properly conjugating them with specific peptides (Kang et al. 2010). Localization of gold nanoparticles at the nucleus of a cancer cell damages the DNA resulting in double-strand breaks. Dark-field imaging of live cells in real time revealed that the nuclear targeting of gold nanoparticles specifically induces cytokinesis arrest in cancer cells leading apoptosis.

Magnetic Nanoparticles for Remote-Controlled Drug Delivery to Tumors

Remotely controlled nanoparticles, when pulsed with an electromagnetic field, can release anticancer drugs into tumors (Derfus et al. 2007). The innovation could lead to the improved diagnosis and targeted treatment of cancer. In an earlier work, injectable multifunctional nanoparticles were designed to flow through the bloodstream, home to tumors, and clump together. Clumped particles help visualization of tumors by MRI. The system that makes it possible consists of particles that are superparamagnetic, a property that causes them to give off heat when they are exposed to a magnetic field. Tethered to these particles are active molecules, such as anticancer drugs. Exposing the particles to a low-frequency electromagnetic field causes the particles to radiate heat, which melts the tethers and releases the drugs. The waves in this magnetic field have frequencies between 350 and 400 kHz – the same range as radio waves. These waves pass harmlessly through the body and heat only the nanoparticles. The tethers in the system consist of strands of DNA. Two strands of DNA link together through hydrogen bonds that break when heated. In the presence of the magnetic field, heat generated by the nanoparticles breaks these, leaving one strand attached to the particle and allowing the other to float away with its cargo. One advantage of a DNA tether is that its melting point is tunable. Longer strands and differently coded strands require different amounts of heat to break. This heat-sensitive tunability makes it possible for a single particle to simultaneously carry many different types of cargo, each of which can be released at different times or in various combinations by applying different frequencies or durations of electromagnetic pulses. To test the particles, the researchers implanted mice with a tumorlike gel saturated with nanoparticles. They placed the implanted mouse into the well of a cup-shaped electrical coil and activated the magnetic pulse. The results confirm that without the pulse, the tethers remain unbroken. With the pulse, the tethers break and release the drugs into the surrounding tissue. The experiment is a proof of principle demonstrating a safe and effective means of tunable remote activation. However, work remains to be done before such therapies become viable in the clinic.

Mesoporous Silica Nanoparticles

Mesoporous silica nanoparticles (MSNs) have a considerable potential for drug-delivery applications due to their flexibility and high drug load potential. MSNs are biodegradable and mostly eliminated through renal clearance. Although numerous reports demonstrate sophisticated drug-delivery mechanisms in vitro, the therapeutic benefit of these systems for in vivo applications has been uncertain in the past. Recent preclinical data has demonstrated that MSNs are safe and a biocompatible technology platform for targeted drug delivery, passive as well as folate-directed, in cancer. MSN medical applicability in vivo has been demonstrated for both drug delivery and diagnostics. Therapeutic efficacy of MSNs has been shown in vivo following oral, local, subcutaneous, or intravenous administration, and MSNs have been approved for clinical trials. Incorporation of multiple therapeutic and diagnostic agents into MSNs is feasible, and this will enable diagnostic-guided therapy. There are still several issues to be considered before MSNs can be used in clinical practice (Rosenholm et al. 2012).

Nanobees for Targeted Delivery of Cytolytic Peptide Melittin

The in vivo application of cytolytic peptides for cancer therapeutics is hampered by toxicity, nonspecificity, and degradation. A specific strategy was developed to synthesize a nanoscale delivery vehicle for cytolytic peptides by incorporating the synthetic version of a toxin called melittin that is found in bees into the outer lipid monolayer of a perfluorocarbon (PFC) nanoparticle. The composite structures, called nanobees, are engineered to travel directly to tumor cells without harming any others. They spare the healthy cells but attach to tumor blood vessels, which express a particular protein to which a substance on the nanobees has a chemical affinity. Melittin, which would destroy red blood cells and other normal tissues if it is delivered intravenously, is completely safe when carried on a nanoparticle. Favorable pharmacokinetics of this nanocarrier have been demonstrated, which allow accumulation of melittin in murine tumors in vivo and a dramatic reduction in tumor growth without any apparent signs of toxicity (Soman et al. 2009). Furthermore, direct assays demonstrated that molecularly targeted nanocarriers selectively delivered melittin to multiple tumor targets, including endothelial and cancer cells, through a hemifusion mechanism. In cells, this hemifusion and transfer process did not disrupt the surface membrane but did trigger apoptosis and in animals caused regression of precancerous dysplastic lesions. Collectively, these data suggest that the ability to restrain the wide-spectrum lytic potential of a potent cytolytic peptide in a nanovehicle, combined with the flexibility of passive or active molecular targeting, represents an innovative molecular design for chemotherapy with broad-spectrum cytolytic peptides for the treatment of cancer at multiple stages. So far, nanobees have been tested only on mice, with promising results, but what works in mice does not always work in humans. If proven to be effective in humans, this therapy could become widely available in about 5–10 years.

Nanovehicles for Targeted Delivery of Paclitaxel

Nanoparticles have been used to deliver paclitaxel, an antitumor drug, directly to tumors for targeted anticancer treatment. The nanoparticles were loaded with paclitaxel and then mixed with lipids to form nanoparticle-like clusters, which were then coated with a glycosaminoglycan (GAG). The nanovehicle was formed of clusters of loaded nanoparticles then coated with a GAG. When these clusters come into contact with cancerous cells, the paclitaxel is released from the individual nanoparticles directly into the cancerous cell. This targeted release allows the treatment to be focused to the cancerous cells, reducing the negative effects of the chemotherapy treatment.

The ability of the nanovehicles to specifically target the cancerous cells is due to the specific GAG used in the coating of the clusters. The researchers use hyaluronan, a sugar that is recognized by receptors on many different types of cancer cell. When the nanovehicle interacts with the cancer cell, the sugar is recognized by the receptors, triggering a structural change and the subsequent release of the paclitaxel directly into the cell. The release of the paclitaxel directly into the target cell also overcomes the issues associated with the solubility of the drug (Rivkin et al. 2010). The vehicle is similar to a cluster bomb; when the delivery vehicle, comprising of multiple nanoparticles, comes into contact with cancer cells, it releases the chemotherapeutic payload directly into the cell. Peer also mentions that the device can be used to treat a variety of cancers, such as blood, colon, breast, pancreatic, ovarian, and several types of brain cancer. The ability of the nanovehicle to be recognized by the cancer cells means that it increases the effectiveness of the treatment while healthy cells are unaffected by the treatment. Tests on tumor-bearing mice have demonstrated that the nanoclusters are more effective when compared with free paclitaxel and other albumin nanoparticles containing paclitaxel. The nanoclusters also demonstrated a high safety profile and a high level of accumulation within tumors.

The fabrication of the nanovehicles means that the treatment may potentially be safer than alternative current therapies. The nanocluster is formed of the naturally occurring lipid hyaluronan, a lipid that decomposes in the body once the nanoparticles have delivered the paclitaxel. The researchers hope that GAGs may provide vehicles for other chemotherapy drugs, such as taxanes, and may also be used as carriers for other therapeutic applications.

Nanocell for Targeted Drug Delivery to Tumor

Simultaneous delivery of chemotherapeutic and antiangiogenic drugs is clearly beneficial, but because chemotherapy is blood-borne, shutting down the tumor's blood supply with antiangiogenic drugs may decrease the delivery of drugs designed to kill the tumor cells. A more effective strategy would be to use a delivery vehicle that became concentrated in tumors before the vasculature shuts down and allows the staged release of the two drugs (Sengupta et al. 2005). The delivery of the antiangiogenic factor could lead to a collapse of the vascular network and imprison the vehicle, which would still be carrying its second payload of chemotherapeutic drug, inside the tumor. The subsequent release of the latter drug within the tumor would kill the cancer cells. For example, a composite nanocell can be constructed with a solid biodegradable polymer core surrounded by a lipid membrane in which the outer membrane is loaded with the antiangiogenic drug combretastatin and the inner membrane with the chemotherapy drug doxorubicin. The nanocells are small enough to pass through tumor blood vessels, but they are too large to pass through the pores of normal vessels. Once inside the tumor, the nanocell's outer membrane can disintegrate, releasing the antiangiogenic drug and causing the collapse of the blood vessels feeding the tumor. The collapsed blood vessels trap the nanocell inside the tumor. The nanocell can then slowly release the chemotherapy drug. Although the effect of the sequential delivery of these two drugs on tumor growth is dramatic, these results cannot be quickly translated into therapy for humans. There is a concern that antiangiogenic drugs may promote the spread of tumors to other tissues. Also, in contrast to combretastatin, many antiangiogenic drugs require prolonged tissue exposure to shut down the vasculature and so may not be amenable to this particular approach with a short exposure time.

Nanodiamonds for Local Delivery of Chemotherapy at Site of Cancer

Nanodiamonds (NDs) with diameter of 2–8 nm, physically bound with doxorubicin, and sandwiched between a base and thin layer of polymer parylene enable extended targeted and controlled release at diseased areas in cancer, viral infection, and inflammation (Lam et al. 2008). A substantial amount of drug can be loaded onto clusters of nanodiamonds, which have a high surface area. Nanodiamonds have many other advantages for drug delivery. They can be functionalized with nearly any type of therapeutic. They can be suspended easily in water, which is important for biomedical applications. They are very scalable and can be produced in large quantities. In control experiments, where the drug was administered without the nanodiamonds, virtually the whole drug was released within 1 day. By adding the drug-laden nanodiamonds to the device, drug release was instantly lengthened to the month-long timescale. The FDA-approved polymer parylene displayed the stable and continuous slow release of drug for at least 1 month due to the powerful sequestration abilities of the DOX–ND complex. The device also avoids the massive initial release of the drug, which is a disadvantage of conventional therapy.

The nanodiamonds are quite economical and have already been mass-produced as lubrication components for automobiles and for use in electronics. Because the fabrication process is devoid of any destructive steps, the DOX–ND conjugates are unaffected and unaltered. The flexible microfilm device resembles a piece of plastic wrap and can be customized easily into different shapes. It can transform conventional treatment strategies and reduce patients' unnecessary exposure to toxic drugs. The biocompatible and minimally invasive device could be used to deliver chemotherapy drugs locally to sites where malignant tumors have been surgically removed.

Nanoimmunoliposome-Based System for Targeted Delivery of siRNA

Low transfection efficiency, poor tissue penetration, and nonspecific immune stimulation by in vivo administered siRNAs have delayed their therapeutic applications. Their potential as anticancer therapeutics hinges on the availability of a vehicle that can be systemically administered, safely and repeatedly, and will deliver the siRNA specifically and efficiently to both primary tumors and metastases. A nanosized immunoliposome-based delivery complex (scL) has been developed that will preferentially target and deliver molecules, including plasmid DNA and antisense oligonucleotides, to tumor cells following systemic administration (Pirollo et al. 2007). This tumor-targeting nanoparticle delivery vehicle can also deliver siRNA to both primary and metastatic disease. The efficiency of this complex has been enhanced by the inclusion of a pH-sensitive histidine–lysine peptide in the complex (scL–HoKC) and by delivery of a modified hybrid (DNA–RNA) anti-HER-2 siRNA molecule. Scanning probe microscopy confirms that this modified complex maintains its nanoscale size. More importantly, this nanoimmunoliposome anti-HER-2 siRNA complex can sensitize human tumor cells to chemotherapeutics, silence the target gene, affect its downstream pathway components in vivo, and significantly inhibit tumor growth in a pancreatic cancer model. This complex has the potential to help translate the potent effects of siRNA into a clinically viable anticancer therapeutic.

Nanoparticle-Mediated Targeting of MAPK Signaling Pathway

The MAPK signal transduction cascade is dysregulated in a majority of human tumors, and nanoparticle-mediated targeting of this pathway can optimize cancer chemotherapy. Nanoparticles engineered from a polymer that is chemically conjugated to a selective MAPK inhibitor, PD98059, are taken up by cancer cells through endocytosis and demonstrate sustained release of the active agent, resulting in the inhibition of phosphorylation of downstream extracellular signal-regulated kinase (Basu et al. 2009). Modification of the polymer, which is biocompatible as well as biodegradable and approved by the FDA, leads to a 20-fold increase in drug-loading capacity. Nanoparticle-mediated targeting of MAPK has been shown to inhibit the proliferation of melanoma and lung carcinoma cells and induce apoptosis in vitro. Administration of the PD98059-nanoparticles in melanoma-bearing mice inhibits tumor growth and enhances the antitumor efficacy of cisplatin chemotherapy. This study shows the nanoparticle-mediated delivery of signal transduction inhibitors is a potentially effective method of cancer chemotherapy

Nanoparticles for Targeted Antisense Therapy of Cancer

Antisense oligonucleotides (ASO) against specific molecular targets (e.g., Bcl-2 and Raf-1) are important reagents in cancer biology and therapy. Phosphorothioate modification of the ASO backbone has resulted in an increased stability of ASO in vivo without compromising, in general, their target selectivity. Although the power of antisense technology remains unsurpassed, dose-limiting side effects of modified ASO and inadequate penetration into the tumor tissue have necessitated further improvements in ASO chemistry and delivery systems. Oligonucleotide delivery systems may increase stability of the unmodified or minimally modified ASO in plasma, enhance uptake of ASO by tumor tissue, and offer an improved therapy response. An overview of ASO design and in vivo delivery systems with focus on preclinical validation of a liposomal nanoparticle containing minimally modified raf antisense oligodeoxynucleotide (LErafAON) has been published (Zhang et al. 2007). Intact rafAON (15-mer) is present in plasma and in normal and tumor tissues of athymic mice systemically treated with LErafAON. Raf-1 expression is decreased in normal and tumor tissues of LErafAON-treated mice. Therapeutic benefit of a combination of LErafAON and radiation or an anticancer drug exceeds radiation or drug alone against human prostate, breast, and pancreatic tumors grown in athymic mice. Further improvements in ASO chemistry and nanoparticles are promising avenues in antisense therapy of cancer.

Nanoparticles for Delivery of Suicide DNA to Prostate Tumors

A prostate-specific, locally delivered gene therapy has been developed for the targeted killing of prostate cells using C32/DT-A, a degradable polymer a nanoparticulate system, to deliver a diphtheria toxin suicide gene (DT-A) driven by a prostate-specific promoter to cells (Peng et al. 2007b). These nanoparticles were directly injected to the normal prostate and to prostate tumors in mice. Nearly 50% of normal prostates showed a significant reduction in size, attributable to cellular apoptosis, whereas injection with naked DT-A-encoding DNA had little effect. A single injection of C32/DT-A nanoparticles triggered apoptosis in 80% of tumor cells present in the tissue. It is expected that multiple nanoparticle injections would trigger a greater percentage of prostate tumor cells to undergo apoptosis. These results suggest that local delivery of polymer/DT-A nanoparticles may have application in the treatment of benign prostatic hypertrophy and prostate cancer.

Nanoparticles for Targeted Delivery of Concurrent Chemoradiation

The development of chemoradiation – the concurrent administration of chemotherapy and radiotherapy – has led to significant improvements in local tumor control and survival. However, it is limited by its high toxicity. A novel NP therapeutic, ChemoRad NP, has been developed, which can deliver biologically targeted chemoradiation (Wang et al. 2010a). This is a biodegradable and biocompatible lipid–polymer hybrid NP that is capable of delivering both chemotherapy and radiotherapy. Using docetaxel, indium^{111}, and yttrium^{90} as model drugs, ChemoRad NP was shown to encapsulate chemotherapeutics (up to 9% of NP weight) and radiotherapeutics (100 mCi of radioisotope per gram of NP) efficiently and deliver both effectively. Targeted delivery of ChemoRad NPs and high therapeutic efficacy of ChemoRad NPs was demonstrated using prostate cancer as a disease model.

ChemoRad NP represents a new class of therapeutics that holds great potential to improve cancer treatment.

Nanoparticle-Based Therapy Targeted to Cancer Metastases

Early detection of metastases plays an important role in the management of metastatic cancer. In patients with prostate cancer who undergo surgical lymph node resection or biopsy, MRI with lymphotropic superparamagnetic nanoparticles can correctly identify all patients with nodal metastases. This diagnosis is not possible with conventional MRI alone and has implications for the management of men with metastatic prostate cancer, in whom adjuvant androgen-deprivation therapy with radiation is the mainstay of management.

Nanoparticle formulations of anticancer drugs may be more effective against cancer metastases. Nanoparticles have the ability to transport complex molecular cargoes to the major sites of metastasis, such as the lungs, liver, and lymph nodes, as well as targeting to specific cell populations within these organs (Schroeder et al. 2012). Oral administration of alpha-TEA formulated in liposome or biodegradable poly(D, L-lactide-co-glycolide) nanoparticle has been shown to significantly reduce tumor burden in a mammary cancer mouse model (Wang et al. 2007a). Both formulations reduced lymph node and lung micrometastatic tumor foci, but nanoparticle formulation was more effective in reducing metastases. Tumor targeting with nanoparticles facilitates systemic delivery of immunomodulatory cytokine genes to remote sites of cancer metastasis. Targeted delivery and localized expression of the intravenously administered nanoparticles bearing the gene encoding granulocyte/macrophage colony-stimulating factor was confirmed in a patient with metastatic cancer, as was the recruitment of significant tumor-infiltrating lymphocytes (Gordon et al. 2008).

Nanostructured Hyaluronic Acid for Targeted Drug Delivery in Cancer

Active targeting of bioactive molecules by nanoparticulate delivery systems that include hyaluronic acid (HA) in their structures is an attractive approach to drug delivery because HA is biocompatible, nontoxic, and noninflammatory. To make HA useful as an intravenous targeting carrier, strategies have to be devised to reduce its clearance from the blood, suppress its uptake by liver and spleen, and provide tumor-triggered mechanisms of release of an active drug from the HA carrier (Ossipov 2010).

HA nanoparticles (HA-NPs), which are formed by the self-assembly of hydrophobically modified HA derivatives, have been tested for their physicochemical characteristics and fates in tumor-bearing mice after systemic administration (Choi et al. 2010a). Irrespective of the particle size, significant amounts of HA-NPs circulated for 2 days in the bloodstream and selectively accumulated into the tumor. The smaller HA-NPs were able to reach the tumor more effectively than larger HA-NPs. The concentration of HA-NPs in the tumor site was dramatically reduced when mice were pretreated with an excess of free HA. These results indicate that HA-NPs can accumulate into the tumor site by a combination of passive and active-targeting mechanisms.

Perfluorocarbon Emulsion for Targeted Chemotherapeutic Delivery

Kereos Inc's emulsion particles consist of a perfluorocarbon core surrounded by a lipid monolayer, which stabilizes the particle in addition to providing a virtually unlimited number of anchoring sites for targeting ligands and payload molecules. The result is an oil-in-water emulsion of particles with an average size of approximately 250 nm, referred to as "targeted nanoparticles." Delivered by injection, this approach offers the following advantages:

- High molecular specificity of MAbs, small-molecule ligands, and other targeting ligands for disease biomarkers translates directly into high specificity of the emulsion particles for disease sites.
- Although only 10–100 targeting ligand molecules are needed to direct and securely bind an individual emulsion particle to the disease site, each particle can carry 100,000 or more payload molecules. This "signal amplification" opens up opportunities not otherwise possible.
- Both in terms of size and composition, the emulsion particles are designed to be both safe and effective and to avoid potential problems with distribution, metabolism, or excretion.

Polymer Nanoparticles for Targeted Drug Delivery in Cancer

Cerulean Nanopharmaceuticals' CRLX101 (formerly IT-101), a cyclodextrin polymer-based nanoparticle containing camptothecin, is in phase IIa clinical development for the treatment of cancer. PET data from ^{64}Cu-labeled CRLX101 to quantify the in vivo biodistribution in mice-bearing tumors shows that ~8% of the injected dose is rapidly cleared as a low molecular weight fraction through the kidneys, and the remaining material circulates in plasma with a terminal half-life of 13.3 h (Schluep et al. 2009). A three-compartment model is used to determine vascular permeability and nanoparticle retention in tumors and is able to accurately represent the experimental data. The calculated tumor vascular permeability indicates that the majority of nanoparticles stay intact in circulation and do not disassemble into individual polymer strands. A key assumption to modeling the tumor dynamics is that there is a sink for the nanoparticles within the tumor. Histological measurements using confocal microscopy show that CRLX101 localizes within tumor cells and provides the sink in the tumor for the nanoparticles.

Several mechanisms have been proposed to explain nanoparticle retention in tumors:

1. Dextran-coated iron oxide nanoparticles accumulate in the interstitial fluid and are taken up by tumor vascular endothelial cells, which observed mostly in areas of neovascularization, whereas intracellular concentrations are highest in tumor cells.
2. Long-circulating liposomes accumulate predominantly in tumor stroma, either in the extracellular space or in tumor-associated macrophages in a breast cancer tumor model. A Her2-targeted version of the same liposomes achieves the same

over all tumor concentration, but more internalization by cancer cells through endocytosis is observed.

3. Cyclodextrin-based polymer (CDP) conjugates have been shown to be avidly taken up by cancer cells. This result may be a function of the unique surface characteristics of CDP nanoparticles, which contain hydrophobic pockets within the cyclodextrin molecules that have been shown to interact with lipid rafts of cell membranes.

Scientists at the MIT-Harvard Center for Cancer Nanotechnology Excellence (Cambridge, MA) have studied the effects of altering nanoparticle polymer composition, drug loading, and solvents on the ability of the resulting nanoparticles to target and deliver drugs to tumors. As a targeting agent for all the polymer nanoparticles studied, they used a molecule that recognizes the prostate-specific membrane antigen. The aim of this study was to develop formulation parameters that would control the size of the resulting polymer nanoparticles, which the investigators believe play a major role in optimizing tumor targeting. Nanoparticles were prepared from a biocompatible material. Experimenting with a variety of polymer concentrations and solvent mixtures, they found that they could systematically control the size of the resulting polymers. The results were so consistent that the investigators believe that they may have developed a broadly applicable approach to reproducibly tuning the size of polymer nanoparticles during their formulation. In a final experiment, the researchers added the targeting agent to their optimized nanoparticles. The targeted nanoparticles were able to significantly increase drug delivery to human prostate tumors growing in mice.

Polymersomes for Targeted Cancer Drug Delivery

Polymersomes, hollow shell nanoparticles, have unique properties of that allow them to deliver two distinct drugs, paclitaxel and doxorubicin, directly to tumors implanted in mice (Ahmed et al. 2006). Loading, delivery, and cytosolic uptake of drug mixtures from degradable polymersomes are shown to exploit both the thick membrane of these block copolymer vesicles and their aqueous lumen as well as pH-triggered release within endolysosomes. Drug-delivering polymersomes break down in the acidic environment of the cancer cells resulting in targeted release of these drugs within tumor cells. While cell membranes and liposomes (vesicles often used for drug delivery) are created from a double layer of fatty molecules called phospholipids, a polymersome is comprised of two layers of synthetic polymers. The individual polymers are degradable and considerably larger than individual phospholipids but have many of the same chemical features. The large polymers making up the shell allow paclitaxel, which is water insoluble, to embed within the shell. Doxorubicin, which is water soluble, stays within the interior of the polymersome until it degrades. The polymersome and drug combination is self-assembling and the structure spontaneously forms when all of the components are suitably mixed together. Recent studies have shown that cocktails of paclitaxel and doxorubicin lead to better tumor regression than either drug alone, but previously, there was no carrier system that could carry both drugs as efficiently to a tumor. Polymersomes get around those limitations.

Another approach is by assembling diverse bioactive agents, such as DNA, proteins, and drug molecules into core-shell multifunctional polymeric nanoparticles (PNPs) that can be internalized in human breast cancer cells (Bertin et al. 2006). Using ring-opening metathesis polymerization, block copolymers containing small-molecule drug segments (>50% w/w) and tosylated hexaethylene glycol segments were prepared and assembled into PNPs that allowed for the surface conjugation of single-stranded DNA sequences and/or tumor-targeting antibodies. The resulting antibody-functionalized particles were readily uptaken by breast cancer cells that overexpressed the corresponding antigens.

Quantum Dots and Quantum Rods for Targeted Drug Delivery in Cancer

A single-particle QD conjugated with a tumor-targeting MAb (anti-HER2) has been tracked in tumors of live mice (Tada et al. 2007). The researchers used a dorsal skinfold chamber and a high-speed confocal microscope with a high-sensitivity camera to track the antibody-labeled QDs and made 30-frame-per-second movies of these nanoparticles (NPs) as they traveled through the bloodstream. The HER2 MAb binds to a protein found on the surface of certain breast and other tumors. This was injected, conjugated to the QDs, into mice with HER2-overexpressing breast cancer to analyze the molecular processes of its mechanistic delivery to the tumor. The investigators identified six distinct "stop-and-go" steps in the process involved in the antibody-labeled QDs traveling from the injection site to the cell where they bind HER2: within a blood vessel in the circulation, during extravasation, in the extracellular region, binding HER2 on the cell membrane, moving into the perinuclear region, and within the perinuclear region. The image analysis of the delivery processes of single particles in vivo thus provides valuable information on antibody-conjugated therapeutic nanoparticles, which will be useful in increasing therapeutic efficacy.

Water-soluble CdSe/CdS/ZnS quantum rods (QRs) have been developed as targeted probes for imaging cancer cell lines using two-photon fluorescence imaging. The researchers first developed a new method of creating QRs that would remain well dispersed in water and then refined the technique to allow the attachment of targeting molecules (in this case, transferrin, which binds to a receptor that is overexpressed in many types of cancer cells) to the QR surface. QRs, similar to the spherical QDs, fluoresce and can be made to fluoresce in a range of colors. However, since QRs have larger dimensions than QDs, they are easier to excite with incoming light than QDs. This research showed that the QRs were only taken up by targeted transferrin-positive cells and accumulated within these cells, being easily visible using low-intensity near-infrared light, which helps to protect cell integrity. If future research can further our understanding of QDs and QRs following these studies, it is hoped that we could then improve the ability of NPs to deliver drugs specifically to tumors, thus resulting in improved cancer diagnostics and therapeutics.

Remote-Controlled Drug Delivery from Magnetic Nanocrystals

Combination of magnetic nanocrystals with ability to exhibit hyperthermic effects when placed in an oscillating magnetic field and mesoporous silica nanoparticles that can contain and release drug cargos could provide a unique drug-delivery system for cancer, which is a nanosystem that incorporates zinc-doped iron oxide nanocrystals within a mesoporous silica framework that has been surface-modified with pseudorotaxanes (Thomas et al. 2010). Upon application of an AC magnetic field, the nanocrystals generate local internal heating, causing the molecular machines to disassemble and allowing the drug cargos to be released. Breast cancer cell (MDA-MB-231) death was achieved in vitro when doxorubicin-loaded particles were exposed to an AC field. This material has potential as a noninvasive, externally controlled drug-delivery system with cancer-killing properties.

Targeted Delivery of Nanoparticulate Drugs into Lymphatic System

The lymphatic system plays a major role in the defense cancer and is one of the main pathways for the metastasis of tumors. The regional lymph nodes, when invaded by cancer cells, act as reservoirs from where these cells spread to other parts of the body. The lymphatic system is not easily accessible by conventional intravenous infusion of chemotherapeutics, thus limiting the amount of drug that reaches lymphatic tissues including lymph node metastases. The lymphatics, however, can be exploited as a route for drug delivery as these channels can transport certain lipophilic compounds and chemotherapeutics.

Nanoparticles can be effectively taken up into lymphatics as well as retained in lymph nodes for several days, and without using any specific targeting ligand, they are internalized exclusively by nodal resident dendritic cells (DCs) and other antigen-presenting cells. Animal studies have demonstrated that nanoparticles made of natural or synthetic polymers and liposomal carriers have higher accumulation in the lymph nodes and surrounding lymphatics compared to conventional intravenous therapies (Xie et al. 2009). In vivo studies have shown that up to 40–50% of resident lymph node DCs internalize nanoparticles, further supporting the feasibility of this delivery strategy. Bioavailability and biodistribution can be controlled easily by varying the size of nanoparticles. Biodegradable nanoparticles of 20–45 nm have shown the potential for immunotherapeutic applications that specifically target DCs in lymph nodes, e.g., targeted delivery of immunomodulating formulations and vaccines (Reddy et al. 2006a). This can diminish toxicity of highly toxic active drugs.

Targeted Drug Delivery with Nanoparticle–Aptamer Bioconjugates

Nucleic acid ligands (aptamers) are potentially well suited for the therapeutic targeting of drug encapsulated controlled-release polymer particles in a cell- or tissue-specific manner. Scientists at the Massachusetts Institute of Technology (Cambridge, MA) have synthesized poly(lactic acid)-block-polyethylene glycol (PLA-PEG) copolymer with a terminal carboxylic acid functional group (PLA-PEG-COOH) and encapsulated Rhodamine-labeled dextran (as a model drug) within PLA-PEG-COOH nanoparticles (Farokhzad et al. 2004). These nanoparticles have the following desirable characteristics:

- Negative surface charge, which may minimize nonspecific interaction with the negatively charged nucleic acid aptamers
- Carboxylic acid groups on the particle surface for potential modification and covalent conjugation to amine-modified aptamers
- Presence of PEG on particle surface, which enhances circulating half-life while contributing to decreased uptake in nontargeted cells

Nanoparticle–aptamer bioconjugates were generated with RNA aptamers that bind to the prostate-specific membrane antigen (PSMA), a well-known prostate cancer tumor marker that is overexpressed on prostate acinar epithelial cells. These bioconjugates could efficiently target and get taken up by the prostate epithelial cells, which express the PSMA protein. The uptake of these particles was not enhanced in cells that do not express the prostate-specific membrane antigen protein. This represents the first report of targeted drug delivery with nanoparticle–aptamer bioconjugates.

Numerous investigators have used aptamers as replacements for antibodies in various therapeutic and diagnostic applications, and now, a team at McMaster University has found a third use for these versatile molecules – as the heart of a DNA-protein nanoengine that can be programmed to release therapeutically useful molecules in response to a programmed molecular signal (Nutiu and Li 2005). To construct their nanoengine, the researchers first create an aptamer that binds to a molecule that would signal "release cargo here." The researchers call this signaling molecule the "input." The researchers then prepare a complementary piece of DNA that binds to the aptamer according to the Watson–Crick rules. A drug molecule, or even a therapeutic gene, can be linked to this piece of DNA, and the combination is called the "output." When the output piece of DNA is then mixed with the aptamer, the two bind to one another until the aptamer comes in contact with the input signal. The aptamer folds around the input signal, causing it to release its cargo, the output DNA-drug molecule combination. As an example, the researchers used this construct to carry and release an enzyme.

Use of T Cells for Delivery of Gold Nanoparticles to Tumors

Gold nanoparticles (AuNPs) are injected intravenously and are allowed to accumulate within the tumor via the enhanced permeability and retention (EPR) effect. Although reliance on the EPR effect for tumor targeting has proven adequate for vascularized tumors in small animal models, the efficiency and specificity of tumor delivery in vivo, particularly in tumors with poor blood supply, may not be adequate. Human T cells, loaded with 45-nm gold colloid nanoparticles, can be used as cellular delivery vehicles for AuNP transport into tumors, without affecting viability or function (e.g., migration and cytokine production). Using a human tumor xenograft mouse model, it was demonstrated that AuNP-loaded T cells retain their capacity to migrate to tumor sites in vivo (Kennedy et al. 2011). In addition, the efficiency of AuNP delivery to tumors in vivo is increased by more than four-fold compared to injection of free PEGylated AuNPs, and the use of the T cell delivery system also dramatically alters the overall nanoparticle biodistribution. Thus, the use of T cell chaperones for AuNP delivery could enhance the efficacy of nanoparticle-based therapies and imaging applications by increasing AuNP tumor accumulation. This could also be used for thermal destruction of tumor by application of NIR laser.

Dendrimers for Anticancer Drug Delivery

Earlier studies of dendrimers in drug-delivery systems focused on their use for encapsulating drug molecules. However, it was difficult to control the release of the drug. One solution to this problem involves the use of dendrimers with pH-sensitive hydrophobic acetal groups on the dendrimer periphery. Loss of acetal group at mildly acidic pH triggers the disruption of micelles and release of the drug. Another approach is to attach the drug to the periphery of the dendrimer so that the release of the drug can be controlled by incorporating a degradable linkage between the drug and the dendrimer. Dendrimers have been used to facilitate boron neutron capture therapy as well as photodynamic therapy of cancer.

Developments in polymer and dendrimer chemistry have provided a new class of molecules called "dendronized polymers," i.e., linear polymers that bear dendrons at each repeat unit. Their behavior differs from that of linear polymers and provides drug-delivery advantages because of their longer circulation time and numerous possibilities for peripheral attachments of drugs.

Another approach is to attach the drug to the periphery of the dendrimer so that the release of the drug can be controlled by incorporating a degradable linkage between the drug and the dendrimer. New developments in polymer and dendrimer chemistry have provided a new class of molecules called "dendronized polymer," i.e., linear polymer that bear dendrons at each repeat unit. Their behavior differs from that of linear polymers and provides drug-delivery advantages because of their longer circulation time and numerous possibilities peripheral attachments of drugs. Modified PAMAM dendritic polymers <5 nm in diameter have been used as drug carriers (Kukowska-Latallo et al. 2005). They are conjugated to folic acid as a targeting agent and then coupled to methotrexate and injected intravenously into animals bearing tumor that overexpress the folate receptor. Folate molecules bind to receptors on tumor cell membranes and facilitate the transport of methotrexate to inside of the tumor cell.

Doxorubicin (DOX) has been conjugated to a biodegradable dendrimer with optimized blood circulation time through size and molecular architecture, drug loading through multiple attachment sites, solubility through PEGylation, and drug release through the use of pH-sensitive hydrazone linkages (Lee et al. 2006). Dendrimer–DOX is >10 times less toxic than free DOX toward colon carcinoma cells in culture. Upon intravenous administration to tumor-bearing mice, tumor uptake of dendrimer–DOX was ninefold higher than intravenous free DOX and caused complete tumor regression as well as 100% survival of the mice over the 60-day experiment. No cures were achieved in tumor-implanted mice treated with free DOX, drug-free dendrimer, or dendrimer–DOX in which the DOX was attached by means of a stable carbamate bond. The antitumor effect of dendrimer–DOX was similar to that of an equimolar dose of liposomal DOX (Doxil). The remarkable antitumor activity of dendrimer–DOX results from the ability of the dendrimer to favorably modulate the pharmacokinetics of attached DOX.

Application of Dendrimers in Boron Neutron Capture Therapy

Boron neutron capture therapy (BNCT) offers a potential method for localized destruction of tumor cells. The technology is based on the nuclear reaction between thermal neutrons and boron-10 (10B) to yield alpha particles and lithium-7 nuclei. The destructive effect of this reaction is limited to a range of about the diameter of a single cell. In order for BNCT to be effective in cancer therapy, there must be selective delivery of an adequate concentration of 10B to tumors. Various types of antibodies as well as epidermal growth factor have been utilized to investigate receptor-mediated boron delivery; however, in vivo studies have demonstrated only a small percentage of the total administered dose actually accumulates in tumors, while high concentrations end up in the liver.

In normal as well as cancer cells, the low molecular weight vitamin, folic acid, is required for a number of enzymatic pathways. Cell membrane receptors mediating endocytic transport of folic acid into cells are expressed in elevated levels in a variety of human tumors. Folic acid conjugates with macromolecules such as toxins, enzymes, antibodies, genes, and liposomes have been shown to be internalized into tumor cells overexpressing folate receptors. These strategies have been employed to enhance the effect of BNCT. The use of dendrimers as boron carriers for antibody conjugation is based on their well-defined structure and multivalency.

The use of dendrimers as boron carriers for antibody conjugation is based on their well-defined structure and multivalency. Boronated PAMAM dendrimers have been designed to target the epidermal growth factor receptor, a cell surface receptor that is frequently overexpressed in brain tumor cells.

Preclinical evaluation has been described of a multipurpose STARBURST PAMAM (polyamidoamine) dendrimer prototype (Dendritic Nanotechnologies Inc) that exhibits properties suitable for use as (1) targeted, diagnostic MRI/NIR (near-IR) contrast agents, and/or (2) for controlled delivery of cancer therapies (Tomalia et al. 2007). The lead candidate is 1,4-diaminobutane, a dendritic nanostructure ~5 nm diameter, which was selected on the basis of a very favorable biocompatibility profile on in vitro studies, i.e., benign and nonimmunogenic.

The expectation is that it will exhibit desirable mammalian kidney excretion properties and demonstrated targeting features.

Application of Dendrimers in Photodynamic Therapy

Photodynamic therapy (PDT) uses light-activated drugs called photosensitizers to treat a range of diseases characterized by rapidly growing tissue, including the formation of abnormal blood vessels, such as cancer and age-related macular degeneration. The more traditional name for this therapy is photoradiation therapy. Treatment with PDT consists of a two-step process that starts with administration of the drug, or photosensitizer, by intravenous injection. Once the drug enters the bloodstream, it attaches itself to low-density lipoproteins already circulating. As cells undergoing rapid growth require an above-average supply of lipoproteins, the drug reaches these types of cells more quickly and in higher concentrations. Once the necessary level of concentration is attained, the second step is to activate the drug with a specific dose of light of a particular wavelength. This causes the conversion of normal oxygen found in tissue to a highly energized form called singlet oxygen, which in turn disrupts normal cellular functions. Neither the drug nor the light exerts any effect until combined.

Numerous studies have used liposomes, oils, and polymeric micelles as encapsulation methods, with some success. However, all of these techniques suffer from one unpleasant side effect: after controlled release and photosensitization, the drug is free to circulate the body, accumulating in the eyes and skin. This leads to phototoxic side effects, rendering the patient highly sensitive to light. A further disadvantage is that liposomes can be engulfed and destroyed by cells of the reticuloendothelial system. Such problems have limited the emerging field of PDT, but combination of this technique with nanotechnology is promising.

The possibility of improving dendrimers through appropriate functionalization of their periphery makes them promising carriers of PDT. The use of 5-aminolevulinic acid (ALA) is one approach to PDT based on dendrimers. ALA is a natural precursor of the photosensitizer protoporphyrin IX (PIX), and its administration increases the cellular concentrations of PIX. Cellular uptake of the dendrimer occurs through endocytic routes predominantly via a macropinocytosis pathway. A dendrimer conjugate, which incorporated 18 aminolevulinic acid residues attached via ester linkages to a multipodent aromatic core, has been investigated (Battah et al. 2007). The ability of the dendrimer to deliver and release 5-ALA intracellularly for metabolism to the photosensitizer, protoporphyrin IX, was studied in the transformed PAM 212 murine keratinocyte and A431 human epidermoid carcinoma cell lines. The macromolecular dendritic derivatives were shown to be capable of delivering 5-ALA efficiently to cells for sustained porphyrin synthesis.

Another approach to deep tissue penetration is based on two-photon excitation with near-infrared lasers. Multivalent aspects of dendrimer scaffold can be used to conjugate several two-photon-absorbing chromophores to the porphyrin core. Such a system can generate singlet oxygen efficiently on light irradiation at 780-nm wavelength.

Dendrimer-Based Synthetic Vector for Targeted Cancer Gene Therapy

A synthetic vector system based on polypropylenimine dendrimers has the desired properties of a systemic delivery vehicle and mediates efficient transgene expression in tumors after intravenous administration (Dufes et al. 2005). Specifically, the systemic injection of dendrimer nanoparticles containing a TNF-α expression plasmid regulated by telomerase gene promoters (hTR and hTERT) leads to transgene expression, regression of remote xenograft murine tumors, and long-term survival of up to 100% of the animals. The combination of pharmacologically active synthetic transfection agent and transcriptionally targeted antitumor gene creates an efficacious gene medicine for the systemic treatment of experimental solid tumors. The promising results of these experiments could make it possible to treat inaccessible tumors in humans using gene therapy in the future. This new treatment can selectively target cancer cells, without causing damage to surrounding healthy cells.

Poly-L-Lysine Dendrimer as Antiangiogenic Agent

Poly-L-lysine (PLL) sixth-generation (G6) dendrimer molecules exhibit systemic antiangiogenic activity that could lead to arrest of growth of solid tumors. Intravenous administration of the PLL-dendrimer molecules into C57BL/6 mice inhibits vascularization of tumors grown within dorsal skinfold window chambers as demonstrated by intravital microscopy (Al-Jamal et al. 2010). The in vivo toxicological profile of the PLL-dendrimer molecules shows that it is safe at the dose regime studied. The antiangiogenic activity of the PLL dendrimer is further shown to be associated with significant suppression of B16F10 solid tumor volume and delayed tumor growth. Enhanced apoptosis/necrosis within tumors of PLL-dendrimer-treated animals only and reduction in the number of CD31 positive cells are observed in comparison to protamine treatment. This study suggests that PLL-dendrimer molecules can exhibit a systemic antiangiogenic activity that may be used for therapy of solid tumors and, in combination with their capacity to carry other therapeutic or diagnostic agents, may potentially offer capabilities combining diagnosis with therapy.

RNA Nanotechnology for Delivery of Cancer Therapeutics

RNA has immense promise as a therapeutic agent against cancer, but the problem has been to have an efficient system to bring multiple therapeutic agents directly into specific cancer cells where they can perform different tasks. The 25-nm RNA nanoparticles enable repeated long-term administration and avoid the problems of short retention time of small molecules and the difficulties in the delivery of particles larger than 100 nm. Nanoparticles, which are assembled from three short pieces of RNA and resemble miniature triangles, possess both the right size to gain entry into cells and also the right structure to carry other therapeutic strands of RNA inside with them, where they are able to halt viral growth or cancer's progress.

RNA molecules come in many variant forms, and the one mimicked from the phi29 virus – called pRNA – also can be linked to other types of RNA to form longer hybrid strands with properties that could be assigned. Incubation of cancer with the pRNA dimer, one subunit of which harbored the receptor-binding moiety and the other harboring the gene-silencing molecule, resulted in their binding and entry into the cells and subsequent silencing of anti-/proapoptotic genes. The chimeric pRNA complex was found to be processed into functional double-stranded siRNA by Dicer (RNA-specific endonuclease). Animal studies have confirmed the suppression of tumorigenicity of cancer cells by ex vivo delivery.

RNA nanotechnology has been used to engineer both therapeutic siRNA and a receptor-binding RNA aptamer into individual pRNAs of phi29's motor. The RNA building block harboring siRNA or other therapeutic molecules is subsequently incorporated in a trimer through the interaction of engineered right and left interlocking RNA loops. The incubation of the protein-free nanoscale particles containing the receptor-binding aptamer or other ligands results in the binding and co-entry of the trivalent therapeutic particles into cells, which can modulate the apoptosis of cancer cells as shown in animal studies. The use of such antigenicity-free 20–40-nm particles holds promise for the repeated long-term treatment of cancer and other chronic diseases.

Delivery of siRNAs for Cancer

Targeted delivery of siRNAs is considered to be safer and more effective therapeutics for oncology applications. Although macromolecules accumulate nonspecifically in tumors through the enhanced permeability and retention (EPR) effect, previous studies using nanoparticles to deliver siRNA demonstrated that attachment of cell-specific targeting ligands to the surface of nanoparticles leads to enhanced potency relative to nontargeted formulations. Although both nontargeted and transferrin-targeted siRNA nanoparticles exhibit similar biodistribution and tumor localization by PET, transferrin-targeted siRNA nanoparticles reduce tumor luciferase activity by ~50% relative to nontargeted siRNA nanoparticles one day after injection (Bartlett et al. 2007). Compartmental modeling is used to show that the primary advantage of targeted nanoparticles is associated with processes involved in cellular uptake in tumor cells rather than overall tumor localization. Optimization of internalization may, therefore, be a key to the development of effective nanoparticle-based targeted siRNA therapeutics.

Tumor Priming for Improving Delivery of Nanomedicines to Solid Tumors

Effectiveness of nanomedicines in cancer therapy is limited in part by inadequate delivery and transport in tumor interstitium. Tumor priming to overcome these limitations includes measures for extravasation and interstitial transport (Wang et al. 2011). Experimental approaches to improve delivery and transport of nanomedicines in solid tumors include tumor vasculature normalization, interstitial fluid pressure modulation, enzymatic extracellular matrix degradation, and apoptosis-inducing tumor priming technology, which is exemplified by enhancement of delivery and efficacy of liposomal doxorubicin by paclitaxel.

Nanotechnology-Based Cancer Therapy

Devices for Nanotechnology-Based Cancer Therapy

Convection-Enhanced Delivery with Nanoliposomal CPT-11

Combination of convection-enhanced delivery (CED) with a novel, highly stable nanoparticle/liposome containing CPT-11 (nanoliposomal CPT-11) is a potential dual-drug-delivery strategy for brain tumor treatment. Following CED in rat brains, tissue retention of nanoliposomal CPT-11 was shown to be greatly prolonged, with >20% injected dose remaining at 12 days (Noble et al. 2006). In contrast, CED of free CPT-11 resulted in rapid drug clearance. At equivalent CED doses, nanoliposomal CPT-11 increased area under the time-concentration curve by 25-fold and tissue t1/2 by 22-fold over free CPT-11; CED in intracranial U87 glioma xenografts showed even longer tumor retention. Plasma levels were undetectable following CED of nanoliposomal CPT-11. Importantly, prolonged exposure to nanoliposomal CPT-11 resulted in no measurable CNS toxicity at any dose tested, whereas CED of free CPT-11 induced severe CNS toxicity. In the intracranial U87 glioma xenograft model, a single CED infusion of nanoliposomal CPT-11 resulted in significantly improved median survival compared with CED of control liposomes. The study concluded that CED of nanoliposomal CPT-11 greatly prolonged tissue residence while also substantially reducing toxicity, resulting in a highly effective treatment strategy in preclinical brain tumor models.

Nanoengineered Silicon for Brachytherapy

BrachySil™ (^{32}P BioSilicon, pSivida Corporation) is a nanoparticle in which the isotope 32-phosphorus is immobilized. It demonstrates a very high degree of isotope retention following injection into the liver, thus reducing the risk of soluble radioactive material affecting healthy hepatic tissue, or entering the circulation and causing systemic toxicity. Unlike titanium seeds, which remain forever in the body, phosphorus seeds degrade over time and enable repetition of treatment if necessary. Other treatments for primary liver cancer include a variety of embolization and radiofrequency ablation techniques. BrachySil offers a more versatile and safer product for the treatment of such tumors. The procedure is undertaken without surgery under local anesthetic, and patients can be discharged the following day. A phase IIa trial in primary liver cancer has shown that it is safe and effective in tumor regression with increased efficacy. An efficacy/safety study for the treatment of pancreatic cancer was completed in 2008 and showed that the treatment was well tolerated with disease control in 82% of patients and an overall median survival of 309 days.

Anticancer Effect of Nanoparticles

Antiangiogenic Therapy Using Nanoparticles

Integrin-targeted nanoparticles can be used for site-specific delivery of a therapeutic payload. Selective targeting of upregulated $\alpha_v\beta_3$ and Flk-1 on the neovasculature of tumors is a novel antiangiogenesis strategy for treating a wide variety of solid tumors. A study provides proof of principle that targeted radiotherapy works using different targeting agents on a NP, to target both the integrin $\alpha_v\beta_3$ and the vascular endothelial growth factor receptor (Li et al. 2004). These encouraging results demonstrate the potential therapeutic efficacy of the IA-NP-90Y and anti-Flk-1 MAb-NP-90Y complexes as novel therapeutic agents for the treatment of a variety of tumor types.

The mechanism of inhibition of the function of pro-angiogenic heparin-binding growth factors (HB-GFs), such as vascular endothelial growth factor 165 (VEGF165) and basic fibroblast growth factor (bFGF) by gold nanoparticles (GNPs), has been investigated (Arvizo et al. 2011). It was shown that a naked GNP surface is required and core size plays an important role to inhibit the function of HB-GFs and subsequent intracellular signaling events. The authors also demonstrated that the inhibitory effect of GNPs is due to the change in HB-GFs conformation/configuration (denaturation) by the NPs, whereas the conformations of non-HB-GFs remain unaffected. This study will help structure-based design of therapeutic NPs.

Cytotoxic Effects of Cancer Nanoparticles

Nanoparticles may have a direct cytotoxic effect on cancer cells by various mechanisms. DNA degradation and anticancer activity of copper nanoparticles of 4–5 nm size have been reported, e.g., dose-dependent degradation of isolated DNA molecules by copper nanoparticles through generation of singlet oxygen. Singlet oxygen scavengers such as sodium azide and tris[hydroxylmethyl]aminomethane were able to prevent the DNA degradation (Jose et al. 2011). Additionally, it was observed that the copper nanoparticles are able to exert cytotoxic effect toward U937 and Hela cells of human histiocytic lymphoma and human cervical cancer origins, respectively, by inducing apoptosis.

Nanoshell-Based Cancer Therapy

Nanoshells may be combined with targeting proteins and used to ablate target cells. This procedure can result in the destruction of solid tumors or possibly metastases not otherwise observable by the oncologist. In addition, Nanoshells can be utilized to reduce angiogenesis present in cancer. Experiments in animals, in vitro and in tissue, demonstrate that specific cells (e.g., cancer cells) can be targeted and destroyed by an amount of infrared light that is otherwise not harmful to surrounding tissue. This procedure may be performed using an external (outside the body) infrared laser. Prior research has indicated the ability to deliver the appropriate levels of infrared light at depths of up to 15 cm, depending upon the tissue. Photothermal tumor ablation in mice has been achieved by using near-infrared-absorbing nanoparticles. The advantages of nanoshell-based tumor cell ablation include:

- Targeting to specific cells and tissues to avoid damage to surrounding tissue.
- Superior side-effect profile than targeted chemotherapeutic agents or photodynamic therapy.
- Repeatability because of:
  - No "tissue memory" as in radiation therapy
  - Biocompatibility
  - Ability to treat metastases and inoperable tumors
- Nanoshells enable a seamless integration of cancer detection and therapy.

Nanobody-Based Cancer Therapy

A nanobody with subnanomolar affinity for the human tumor-associated carcinoembryonic antigen (CEA) has been identified (Cortez-Retamozo et al. 2004). This nanobody was conjugated to *Enterobacter cloacae* beta-lactamase, and its site-selective anticancer prodrug activation capacity was evaluated. The conjugate was readily purified in high yields without aggregation or loss of functionality of the constituents. In vitro experiments showed that the nanobody–enzyme conjugate effectively activated the release of phenylenediamine mustard from the cephalosporin nitrogen mustard prodrug 7-(4-carboxybutanamido) cephalosporin mustard at the surface of CEA-expressing LS174T cancer cells. In vivo studies demonstrated that the conjugate had an excellent biodistribution profile and induced regressions and cures of established tumor xenografts. The easy generation and manufacturing yield of nanobody-based conjugates together with their potent antitumor activity makes nanobodies promising vehicles for new-generation cancer therapeutics.

Nanoparticles Combined with Physical Agents for Tumor Ablation

Several physical agents have been used for ablation of tumors. Nanoparticles can be combined with these techniques, and some examples are shown here.

Boron Neutron Capture Therapy Using Nanoparticles

Boron carbide nanoparticles are proposed as a system for T cell-guided boron neutron capture therapy (Mortensen et al. 2006). Nanoparticles were produced by ball milling in various atmospheres of commercially available boron carbide. The physical and chemical properties of the particles were investigated using transmission electron microscopy, photon correlation spectroscopy, X-ray photoelectron spectroscopy, X-ray diffraction, vibrational spectroscopy, gel electrophoresis, and chemical assays and revealed profound changes in surface chemistry and structural characteristics. In vitro thermal neutron irradiation of B16 melanoma cells incubated with sub-100-nm nanoparticles induced complete cell death. The nanoparticles alone induced no toxicity.

A cancer therapeutic plus diagnostic has been developed that is a variation of BNCT using radioactivate boron-nitride (BN) nanotubes. BNs are covalently bound to tumor-cloned antibodies or immunoglobulins (IgGs) to deliver intense, short-lived, therapeutic doses of radiation specifically to active tumor sites. The therapy involves activation of the BN nanotubes with a neutron beam (as in BNCT) once the IgG carrier molecules reach their target tissue. In contrast to conventional BNCT, instant BN nanotubes can deliver significant numbers of boron atoms (100–1,000 s) specifically to the tumor site while avoiding exposures to surrounding tissue. BNCT is a technique that relies on (nonradioactive) 10B delivery specifically to a tumor site and then activating it using an accurate beam of epithermal neutrons (low-energy neutrons with velocities adjusted to penetrate tissue to the specific tumor depth where the 10B has lodged). BN nanotube structure is similar to the "rolled-up-graphite" structure of a CNT, six member rings but with boron atoms bound to three surrounding nitrogen atoms, and the nitrogen atoms bound to surrounding boron atoms (no conjugation). Thus, each BN nanotube is composed of a substantial number of boron atoms, e.g., 50%, meaning hundreds to thousands for each nanotube. Boron has a relatively large radioactive cross section and can be easily made radioactive in a neutron flux.

Gold Nanoparticles Combined with Radiation Therapy

High-atomic number metals, such as gold, preferentially absorb much more X-ray energy than soft tissues and thus augment the effect of ionizing radiation when delivered to cells. Proteins that regulate poly-SUMO (small ubiquitin-like modifier)-chain conjugates play important roles in cellular response to DNA damage, such as those caused by cancer radiation therapy. A study has demonstrated that conjugation of a weak SUMO-2/3 ligand to gold nanoparticles (AuNPs) facilitates selective multivalent interactions with poly-SUMO-2/3 chains leading to efficient inhibition of poly-SUMO-chain-mediated protein–protein interactions (Li et al. 2012). The ligand–gold particle conjugate significantly sensitized cancer cells to radiation but was not toxic to normal cells. This study demonstrates a viable approach for selective targeting of poly-Ubl chains through multivalent interactions created by nanoparticles that can be chosen based on their properties, such as abilities to augment radiation effects.

Laser-Induced Cancer Destruction Using Nanoparticles

Biological systems are known to be highly transparent to 700–1,100-nm NIR light. It is shown here that the strong optical absorbance of SWCNT in this special spectral window, an intrinsic property of carbon nanotubes, can be used for optical stimulation of nanotubes inside living cells to afford multifunctional nanotube biological transporters. For oligonucleotides transported inside living cells by nanotubes, the oligos can translocate into cell nucleus upon endosomal rupture triggered by NIR laser pulses. Continuous NIR radiation can cause cell death because of excessive local heating of carbon nanotubes in vitro. Selective cancer cell destruction can be achieved by functionalization of carbon nanotubes with a folate moiety, selective internalization of carbon nanotubes inside cells labeled with folate receptor tumor markers, and NIR-triggered cell death, without harming receptor-free normal cells. Thus, the transporting capabilities of carbon nanotubes combined with suitable functionalization chemistry and their intrinsic optical properties can lead to new classes of novel nanomaterials for drug delivery and cancer therapy (Kam et al. 2005). One example for application is lymphoma as lymphoma cells have well-defined surface receptors that recognize unique antibodies. When attached to a carbon nanotube, the antibody would play the role of a Trojan horse. This approach is being tested in laboratory mice with lymphoma. The researchers want to determine if shining NIR on the animal's skin will destroy lymphatic tumors while leaving normal cells intact. Carbon nanotubes also can be delivered to diseased cells by direct injection. The idea is to use the nanotube to deliver therapeutic molecules of DNA, RNA, or protein directly into the cell nucleus to fight various infections and diseases.

Plasmon-resonant gold nanorods, which have large absorption cross sections at near-infrared frequencies, are excellent candidates as multifunctional agents for image-guided therapies based on localized hyperthermia. The controlled modification of the surface chemistry of the nanorods is of critical importance, as issues of cell-specific targeting and nonspecific uptake must be addressed prior to clinical evaluation. Nanorods coated with cetyltrimethylammonium bromide (a cationic surfactant used in nanorod synthesis) are internalized within hours into cancer cells by a nonspecific uptake pathway, whereas the careful removal of cetyltrimethylammonium bromide from nanorods functionalized with folate results in their accumulation on the cell surface over the same time interval. Thus, the nanorods render the tumor cells highly susceptible to photothermal damage when irradiated at the nanorods' longitudinal plasmon resonance, generating extensive blebbing of the cell membrane at laser fluences as low as 30 J/cm^2 (Huff et al. 2007).

A light-controlled delivery system that can be tailored to release nonbiological molecules into living cells can be remotely controlled and can release quantifiable amounts on demand. The technique utilizes gold nanoparticles, in the form of nanoshells, to transport the target molecule into the cell, where it can subsequently be released. dsDNA nanoshells can be loaded with molecules, which are associated with the DNA; these molecules can be released inside the cell when triggered by light (Huschka et al. 2010). The research describes how the nanoshell complexes were loaded with 4′,6-diamino-2-phenylindole (DAPI), a fluorescent blue dye that is able to reversibly bind to DNA. The nanoshells were then introduced to cancer cells, and once uptake of the nanoparticles by the cells was confirmed, the cells were illuminated using a continuous wave laser at a specified wavelength. The wavelength of the laser excitation is tailored to the specific DNA, and the plasmon resonance wavelength dehybridizes the DNA, causing the release of the DAPI molecule. The DAPI molecule is released from the nanoshell and diffuses through the cytoplasm into the cell nucleus. The diffusion of the DAPI molecule into the cell nucleus was confirmed by the staining of the nuclear DNA. The ability of DAPI to reversibly stain DNA fluorescent blue allowed the intracellular release process to be easily visualized. The research concluded that the light-triggered release of DAPI using a laser did not have an adverse effect on the cells, due to the low power of the laser and the minimal irradiation times required to stimulate the release of the molecule. The researchers also discerned that the uptake of the nanoshells had no adverse effects on the living cells.

Poly(lactic-co-glycolic acid) nanoparticles have been produced, which encapsulate the photosensitizer meso-tetraphenylporpholactol and are stable and nonphototoxic upon systemic administration (McCarthy et al. 2005). Upon cellular internalization, the photosensitizer is released from the nanoparticle and becomes highly phototoxic. Irradiation with visible light results in cell-specific killing of several cancer cell lines. In vivo experiments have shown complete eradication of cancers in mouse models. The concept of photosensitizers with selective phototoxicity should have widespread applications in cancer therapy.

A nanocarrier consisting of polymeric micelles of diacylphospholipid-poly(ethylene glycol) (PE-PEG) coloaded with the photosensitizer drug 2-[1-hexyloxyethyl]-2-devinyl pyropheophorbide-a (HPPH) and magnetic Fe3O4 nanoparticles has been used for guided drug delivery together with light-activated photodynamic therapy for cancer (Cinteza et al. 2006). The nanocarrier shows excellent stability and activity over several weeks. The loading efficiency of HPPH is practically unaffected upon coloading with the magnetic nanoparticles, and its phototoxicity is retained. The magnetic response of the nanocarriers was demonstrated by their magnetically directed delivery to tumor cells in vitro. The magnetophoretic control on the cellular uptake provides enhanced imaging and phototoxicity. These multifunctional nanocarriers demonstrate the exciting prospect offered by nanochemistry for targeting photodynamic therapy.

In a novel nanoformulation of for PDT of cancer, the photosensitizer molecules are covalently incorporated into organically modified silica (ORMOSIL) nanoparticles (Ohulchanskyy et al. 2007). The incorporated photosensitizer molecules retain their spectroscopic and functional properties and can robustly generate cytotoxic singlet oxygen molecules upon photoirradiation. The synthesized nanoparticles are of ultralow size (approximately 20 nm) and are highly monodispersed and stable in aqueous suspension. The advantage offered by this covalently linked nanofabrication is that the drug is not released during systemic circulation, which is often a problem with physical encapsulation. These nanoparticles are also avidly taken up by tumor cells and demonstrate phototoxic action, thereby improving the diagnosis as well as PDT of cancer.

Magnetic Nanoparticles for Thermal Ablation of Cancer

An experimental procedure for the treatment of breast cancer is called magnetic thermal ablation. Magnetic nanoparticles are promising tools for the minimal invasive elimination of small tumors in the breast using magnetically induced heating. The approach complies with the increasing demand for breast conserving therapies and has the advantage of offering a selective and refined tuning of the degree of energy deposition allowing an adequate temperature control at the target.

Anti-HER2 antibody can induce antitumor responses and can be used in delivering drugs to HER2-overexpressing cancer. Anti-HER2 immunoliposomes containing magnetite nanoparticles, which act as tumor-targeting vehicles, have been used to combine anti-HER2 antibody therapy with hyperthermia in experimental studies. SWCNTs emit heat when they absorb energy from NIR light. Tissue is relatively transparent to NIR, which suggests that targeting SWCNTs to tumor cells, followed by noninvasive exposure to NIR light, will ablate tumors within the range of NIR. One study has demonstrated the specific binding of MAb-coupled SWCNTs to tumor cells in vitro, followed by their highly specific ablation with NIR light (Chakravarty et al. 2008). Only the specifically targeted cells were killed after exposure to NIR light.

Targeted nanotherapeutics (TNT) system is an innovation of thermal ablation of cancer that bonds iron nanoparticles and MAbs into bioprobes. The magnetic field energy is converted to lethal heat by the particles causing a rapid temperature increase to more than 170 C at the surface of the cancer cells, killing them and their blood supply with negligible damage to surrounding healthy tissues. To evaluate the potential of TNT for in vivo tumor targeting, efficacy and predictive radionuclide-based heat dosimetry were studied using ^{111}In-ChL6 bioprobes (ChL6 is chimeric L6) in a human breast cancer xenograft model (Denardo et al. 2007). Mice in the study received a series of alternating magnetic field (AMF) bursts in a single 20-min treatment. Dosing was calculated using an equation that included tumor concentration of bioprobes, heating rate of particles at different amplitudes, and the spacing of AMF bursts. MAb-guided bioprobes (iron oxide nanoparticles) effectively targeted the tumors without causing particle-related toxicity. Tumor total heat dose, calculated using empirically observed ^{111}In-bioprobe tumor concentration and in vitro nanoparticle heat induction by AMF, correlated with tumor growth delay. The biggest problem of thermotherapy of cancer has been how to apply it to the tumor alone, how to predict the amount needed, and how to determine its effectiveness. By combining nanotechnology, focused AMF therapy, and quantitative molecular-imaging techniques, a safe technique has been developed that could be considered for clinical use as a treatment for breast and other cancers.

Nanoshells for Thermal Ablation of Cancer

Metal nanoshells belong to a class of nanoparticles with tunable optical resonances that have been used for thermal ablative therapy for cancer. Nanoshells can be tuned to strongly absorb light in the NIR, where optical transmission through tissue is optimal. Nanoshells placed at depth in tissues can be used to deliver a therapeutic dose of heat by using moderately low exposures of extracorporeally applied NIR. In vivo studies under MRI guidance have revealed that exposure to low doses of NIR in solid tumors treated with metal nanoshells reach temperatures capable of inducing irreversible tumor destruction within minutes. Gold nanoshells are ~120 nm in diameter and a cancer cell is 170 times bigger. Therefore, nanoshells can penetrate the tumor capillaries and lodge in the tumor. Application of NIR light, which passes through the skin harmlessly, heats the nanoshells and kills the tumor cells. Since no drug is used, the cancer cells are unlikely to develop drug resistance.

The ability to control both wavelength-dependent scattering and absorption of nanoshells offers the opportunity to design nanoshells which provide both diagnostic and therapeutic capabilities in a single nanoparticle. A nanoshell-based all-optical platform technology can integrate cancer imaging and therapy applications. Immunotargeted nanoshells, engineered to both scatter light in the near-infrared range enabling optical molecular cancer imaging and to absorb light, enable selective destruction of targeted carcinoma cells through photothermal therapy. In a proof of principle experiment, dual-imaging/therapy-immunotargeted nanoshells were used to detect and destroy breast carcinoma cells that overexpress HER2, a clinically relevant cancer biomarker. This approach has some significant advantages over alternatives that are under development. For example, optical imaging is much faster and less expensive than other medical imaging techniques. Gold nanoparticles are also more biocompatible than other types of optically active nanoparticles, such as QDs.

Nanospectra Biosciences Inc is already developing nanoshells for the targeted destruction of various cancers using nanoshells (AuroShell™). AuroLase™ cancer therapy combines the unique physical and optical properties of AuroShell™ microparticles with a near-infrared laser source to thermally destroy cancer cells without significant damage to surrounding tissue. AuroShell™ microparticles are injected intravenously and specifically collect in the tumor through the associated leaky vasculature (the Enhanced Permeability and Retention effect, or EPR). After the particles accumulate in a tumor, the area is illuminated with a near-infrared laser at wavelengths chosen to allow the maximum penetration of light through tissue. Unlike solid metals and other materials, AuroShell™ microparticles are designed to specifically absorb this wavelength, converting the laser light into heat. This results in the rapid destruction of the tumor along its irregular boundaries. The basics of this approach have been tested experimentally.

The blood vessels inside tumors develop poorly, allowing small particles like nanoshells to leak out and accumulate inside tumors. An animal trial involved 25 mice with tumors ranging in size from 3 to 5.5 mm. The mice were divided into three groups. The first group was given no treatment. The second received saline injections, followed by three-minute exposure to near-infrared laser light. The final group received nanoshell injections and laser treatments. In the test, researchers injected nanoshells into the mice, waited 6 h to give the nanoshells time to accumulate in the tumors, and then applied a 5-mm laser beam on the skin above each tumor. Surface temperature measurements taken during the laser treatments showed a marked increase that averaged about 46 °F (7.7 °C) for the nanoshells group. There was no measurable temperature increase at the site of laser treatments in the saline group. Likewise, sections of laser-treated skin located apart from the tumor sites in the nanoshells group also showed no increase in temperature, indicating that the nanoshells had accumulated as expected within the tumors. All signs of tumors disappeared in the nanoshells group within 10 days. These mice remained cancer-free after treatment. Tumors in the other two test groups continued to grow rapidly. All mice in these groups were euthanized when the tumors reached 10 mm in size. The mean survival time of the mice receiving no treatment was 10.1 days; the mean survival time for the group receiving saline injections and laser treatments was 12.5 days. The advantages of nanoshell-based tumor cell ablation include:

- Targeting to specific cells and tissues to avoid damage to surrounding tissue
- Less adverse effects than targeted chemotherapeutic agents or photodynamic therapy
- Repeatability because of lack of "tissue memory" as in radiation therapy and biocompatibility
- Ability to treat malignancies such as glioblastoma multiforme, metastases, and inoperable tumors

Thermosensitive Affibody-Conjugated Liposomes

Thermosensitive liposomes have been used as vehicles for the delivery and release of drugs to tumors. To improve the targeting efficacy for breast cancer treatment, a HER2-specific affibody molecule was conjugated to the surface of thermosensitive small unilamellar liposomes measuring 80–100 nm, referred to as "affisomes," to study effects of this modification on physical characteristics and stability of the resulting preparation (Puri et al. 2008). Affisomes released calcein, a water-soluble fluorescent probe, in a temperature-dependent manner, with optimal leakage (90–100%) at 41 °C. Affisomes, when stored at room temperature, retained >90% entrapped calcein up to 7 days. Affisomes are promising candidates for targeted thermotherapy of breast cancer.

Ultrasound Radiation of Tumors Combined with Nanoparticles

Nanoparticles have been introduced in tumors followed by ultrasound-induced cavitation for safe and efficient drug and gene delivery. In a study on athymic nude mice-bearing human colon KM20 tumors, polystyrene nanoparticles (100 and 280 nm in diameter) were injected intravenously in combination with ultrasound to enhance delivery of chemotherapeutic agent 5-fluorouracil (Larina et al. 2005). This combination significantly decreased tumor volume and resulted in complete tumor regression at optimal irradiation conditions.

Impact of Nanotechnology-Based Imaging in Management of Cancer

The role of nanotechnology in diagnostic imaging of cancer, particularly MRI, has already been described earlier in this chapter. Nanotechnology-based cancer imaging will lead to sensitive and accurate detection of early-stage cancer. Nanoparticle-enabled imaging can help accurate delivery of cancer therapy.

Cornell Dots for Cancer Imaging

Cornell dots (C dots) are ultrasmall, cancer-targeted, multimodal silica nanoparticle <7 nm in diameter, which has been surface functionalized with cyclic arginine–glycine–aspartic acid peptide ligands and radioiodine. C dots exhibit high-affinity binding, favorable tumor-to-blood residence time ratios, and enhanced tumor-selective accumulation in $\alpha v\beta 3$ integrin-expressing melanoma xenografts in mice (Benezra et al. 2011). The silica shell, essentially glass, is chemically inert and small enough to pass through the body and out in the urine. Coating the dots by PEGylation further protects them from being recognized by the body as foreign substances, giving them more time to find targeted tumors. The outside of the shell can also be coated with organic molecules that can attach to desired targets on tumor surfaces or within tumors. The cluster of dye molecules in a single dot fluoresces under near-infrared light much more brightly than single-dye molecules, and the fluorescence identifies malignant cells, showing a surgeon exactly what needs to be cut out and helping ensure that all malignant cells are found. C dots can reveal the extent of a tumor's blood vessels, cell death, treatment response, and invasive or metastatic spread to lymph nodes and distant organs. The FDA has approved the first clinical trial in humans of C dots that can light up cancer cells in PET–optical imaging. The technology aims to safely show surgeons extent of tumors in human organs. The trial is being conducted at Memorial Sloan-Kettering Center, New York. Commercial development will be in collaboration with Hybrid Silica Technologies Inc.

Nanoparticle MRI for Tracking Dendritic Cells in Cancer Therapy

Several techniques have been developed that allow an effective cellular internalization of clinical SPIO formulations without affecting cell proliferation, differentiation, and function, with "magnetoelectroporation" being the most recent labeling paradigm. Animal studies have shown that the MR distribution pattern is reliable when cells have limited cell division, as validated by conventional histological techniques. Magnetically labeled stem cells are not yet in clinical use due to safety concerns about the in vivo behavior of stem cells. A phase I trial has shown the feasibility and safety of imaging autologous dendritic cells that were labeled with a clinical superparamagnetic iron oxide formulation or ^{111}In-oxine and were co-injected intranodally in melanoma patients under ultrasound guidance. In contrast to scintigraphic imaging, MRI allowed assessment of the accuracy of dendritic cell delivery and of inter- and intranodal cell migration patterns of MRI cell tracking using iron oxides that appear clinically safe and well suited to monitor cellular therapy in humans. It is believed that MRI cell tracking will become an important technique that someday may become routine in standard radiological practice once stem cell therapy enters clinical practice.

Nanoparticle CT Scan

Use of nanomaterials for one of the most common imaging techniques, computed tomography (CT), has remained unexplored. Current CT contrast agents are based on small iodinated molecules. They are effective in absorbing X-rays, but nonspecific distribution and rapid pharmacokinetics have rather limited their microvascular and targeting performance. While most of the nanoparticles are designed to be used in conjunction with MRI, bismuth sulfide (Bi_2S_3) nanoparticles naturally accumulate in lymph nodes containing metastases and show up as bright white spots in CT images (Rabin et al. 2006). A polymer-coated Bi_2S_3 nanoparticle preparation has been proposed as an injectable CT imaging agent. This preparation demonstrates excellent stability at high concentrations, high X-ray absorption (fivefold better than iodine), very long circulation times (>2 h) in vivo, and an efficacy/safety profile comparable to or better than iodinated imaging agents. The utility of these polymer-coated Bi_2S_3 nanoparticles for enhanced in vivo imaging of the vasculature, the liver, and lymph nodes has been demonstrated in mice. These nanoparticles and their bioconjugates are expected to become an important adjunct to in vivo imaging of molecular targets and pathological conditions. Tumor-targeting agents are now being added to the surfaces of these polymer-coated Bi_2S_3 nanoparticles.

QDs Aid Lymph Node Mapping in Cancer

An improved method for performing sentinel lymph node (SLN) biopsy, which depends on illuminating lymph nodes during cancer surgery, has been developed using QDs that emit NIR light, a part of the spectrum that is transmitted through biological tissue with minimal scattering. SLN mapping is a common procedure used to identify the presence of cancer in a single "sentinel" lymph node, thus avoiding the removal of a patient's entire lymph system. SLN mapping relies on a combination of radioactivity and organic dyes, but the technique is inexact during surgery, often leading to removal of much more of the lymph system than necessary, causing unwanted trauma. QD technique is a significant improvement over the dye/radioactivity method currently used to perform SLN mapping. The imaging system and QDs allowed the pathologist to focus on specific parts of the SLN that would be most likely to contain malignant cells, if cancer were present.

Different varieties of PEG-coated QDs have been injected directly into tumors in mouse models of human cancer and their course tracked through the skin using NIR fluorescence microscopy to image and map SLNs (Ballou et al. 2007). In tumors that drained almost immediately to the SLNs, the QDs were confined to the lymphatic system, mapping out the connected string of lymph nodes. This provided easy tagging of the SLNs for pathology, and there was little difference in results among the different QD types used. Examination of the SLNs identified by QD localization showed that at least some contained metastatic tumor foci. The animals used in this study were followed for >2 years, with no evidence of toxicity, even though QDs could still be observed within the animals. SLN mapping has already revolutionized cancer surgery. NIR QDs have the potential to improve this important technique even further. Because the QDs in the study are composed of heavy metals, which can be toxic, they have not yet been approved for clinical use until safety has been established.

Nanosensor Device as an Aid to Cancer Surgery

Scientists at the University of Nebraska-Lincoln have developed a high-resolution touch sensor, one that uses a self-assembling nanoparticle device and acts much like a human finger. The self-assembly process developed by the research team involves no complex lithography, thus proving to be cost-effective and would be relatively easy to reproduce. This device has the ability to sense texture by touch, which is vital for surgeons who need the "touch sensation" in order to operate with precision and accuracy, such as when it comes to detecting and removing cancer cells from the body. One of the most important applications of this newly created sensor is the potential it holds for cancer surgeons, who are faced with the difficult task of knowing where to stop cutting when removing cancer cells in the body. In the development of artificial skin, the nanodevice structure can attain resolution of ~20 μm or even less. As this dimension is comparable to single-cell dimension, one can hope to "see" a single cancer cell in a tissue. The next goal is to make a high-resolution thermal imaging device and develop an ultrasound detector with a much better image resolution to enable detection of malignant tumors at early stages.

Role of Nanoparticle-Based Imaging in Oncology Clinical Trials

Currently, CT scans are used as surrogate end points in cancer clinical trials. The size of the tumor gives only limited information about the effectiveness of therapy. New imaging agents could speed the clinical trials process in two ways: (1) better imaging data could help oncologists better select which therapies to use on a particular patient and (2) and increasingly sensitive and specific imaging agents will be able to provide real-time information about whether a therapy is working. Currently, oncologists and their patients must wait months to determine if a given therapy is working. Shorter clinical trials would mean that effective new drugs would reach patients quicker and ineffective drugs would be dropped from clinical trials sooner, allowing drug discoverers to better focus their efforts on more promising therapies.

Nanoparticle-Based Anticancer Drug Delivery to Overcome MDR

Although multidrug resistance (MDR) is known to develop through a variety of molecular mechanisms within the tumor cell, many tend to converge toward the alteration of apoptotic signaling. The enzyme glucosylceramide synthase (GCS), responsible for bioactivation of the proapoptotic mediator ceramide to a nonfunctional moiety glucosylceramide, is overexpressed in many MDR tumor types and has been implicated in cell survival in the presence of chemotherapy.

A study has to investigate the therapeutic strategy of coadministering ceramide with paclitaxel in an attempt to restore apoptotic signaling and overcome MDR in the human ovarian cancer cell line using modified poly(epsilon-caprolactone) (PEO-PCL) nanoparticles to encapsulate and deliver the therapeutic agents for enhanced efficacy (van Vlerken et al. 2007). Results show that indeed the complete population of MDR cancer cells can be eradicated by this approach. Moreover, with nanoparticle drug delivery, the MDR cells can be resensitized to a dose of paclitaxel near the IC50 of non-MDR (drug sensitive) cells, indicating a 100-fold increase in chemosensitization via this approach. Molecular analysis of activity verified the hypothesis that the efficacy of this therapeutic approach is due to a restoration in apoptotic signaling, although the beneficial properties of PEO-PCL nanoparticle delivery enhanced the therapeutic success even further, showing the promising potential for the clinical use of this therapeutic strategy to overcome MDR. Besides MDR, this novel paclitaxel–ceramide nanoparticle therapy also shows great potential for use in the treatment of non-MDR cancer types, in which therapeutic efficacy of paclitaxel is also enhanced.

Nanotechnology, used in conjunction with existing therapies, such as gene therapy and P-glycoprotein inhibition, has been shown to improve the reversal of drug resistance. The mechanisms involved include specific targeting of drugs, enhanced cellular uptake of drugs, and improved bioavailability of drugs. Important strategies in the reversal of drug resistance include (Palakurthi et al. 2012):

- A multifunctional nanoparticulate system
- Therapeutics to kill resistant cancer cells and cancer stem cells
- Release of encapsulated cytotoxic therapeutics in a stimuli-responsive tumor microenvironment

Nanoparticles for Targeting Tumors

Nanoparticles can deliver chemotherapy drugs directly to tumor cells and then give off a signal after the cells are destroyed. Drugs delivered this way are 100 times more potent than standard therapies. Gold nanoparticles can help X-rays kill cancerous cells more effectively in experiments on mice. Combination of nanoparticles followed by X-ray treatment reduced the size of the tumors, or completely eradicated them, whereas tumors that had received only X-ray therapy continued to grow. The gold nanoparticles had no therapeutic effect on their own. The technique works because gold, which strongly absorbs X-rays, selectively accumulates in tumors. This increases the amount of energy that is deposited in the tumor compared with nearby normal tissue.

Efficient conversion of strongly absorbed light by plasmonic gold nanoparticles to heat energy and their easy bioconjugation suggest their use as selective photothermal agents in molecular cancer cell targeting (El-Sayed et al. 2006). Two oral squamous carcinoma cell lines and one benign epithelial cell line were incubated with anti-epithelial growth factor receptor (EGFR) antibody-conjugated gold nanoparticles and then exposed to continuous visible argon ion laser at 514 nm. Malignant cells required less than half the laser energy to be killed than the benign cells after incubation with anti-EGFR antibody-conjugated Au nanoparticles. No photothermal destruction was observed for all types of cells in the absence of nanoparticles at four times energy required to kill the malignant cells with anti-EGFR/Au conjugates bonded. Au nanoparticles thus offer a novel class of selective photothermal agents using a CW laser at low powers. The ability of gold nanoparticles to detect cancer was demonstrated previously. Now, it will be possible to design an "all-in-one" active agent that can be used to noninvasively find the cancer and then destroy it. This selective technique has a potential in molecularly targeted photothermal therapy in vivo.

Nanocarriers with TGF-β Inhibitors for Targeting Cancer

TGF-β inhibitors can prevent the growth and metastasis of certain cancers. However, there may be adverse effects caused by TGF-β signaling inhibition, including the induction of cancers by the repression of TGF-β-mediated growth inhibition. Application of a short-acting, small-molecule TGF-β type I receptor (TR-I) inhibitor at a low dose has been shown to be effective in treating several experimental intractable solid tumors, including pancreatic adenocarcinoma and diffuse-type gastric cancer, characterized by hypovascularity and thick fibrosis in tumor microenvironments. Low-dose TR-I inhibitor alters neither TGF-β signaling in cancer cells nor the amount of fibrotic components (Kano et al. 2007). However, it decreases pericyte coverage of the endothelium without reducing endothelial area specifically in tumor neovasculature and promotes accumulation of macromolecules, including anticancer nanocarriers, in the tumors. Compared with the absence of TR-I inhibitor, anticancer nanocarriers exhibit potent growth-inhibitory effects on these cancers in the presence of TR-I inhibitor. The use of TR-I inhibitor combined with nanocarriers may thus be of significant clinical and practical importance in treating intractable solid cancers.

Nanobombs for Cancer

Nanobombs are nanoscale bombs, which infiltrate into tumors in a minimally invasive manner and then explode on exposure to physical or chemical triggers. Various nanomaterials have been used for the construction of nanobombs including gold and silica nanoparticles as well as carbon nanotubes. Nanobombs are effective anticancer agents as the shock waves that are generated after local explosion inside the tumor kill cancer cells and also disrupt cancer pathways so that the effect spreads beyond the area of explosion.

Temperature change can be used to trigger explosion. Nanogels fabricated by light cross-linking exhibit abrupt volume expansion upon exposure to sudden temperature change, causing cell death (Lee et al. 2009). In another approach, nanoclusters (gold nanobombs) can be activated in cancer cells only by confining near-infrared laser pulse energy within the critical mass of the nanoparticles in the nanocluster (Zharov et al. 2005). Once the nanobombs are exploded and kill cancer cells, macrophages can effectively clear the cell debris and the exploded nanotube along with it.

Blending of supramolecular chemistry and mechanostereochemistry with mesoporous silica nanoparticles has led to a new class of materials that are biological nanoscale bombs with the potential to infiltrate cells and explode upon the pulling of a chemical trigger (Cotí et al. 2009). The triggers are initiated by changes in pH, light, and redox potentials, in addition to enzymatic catalysis. This approach has been tried in in vitro experiments where loaded mechanized silica nanoparticles are endocytosed selectively by cancer cells and an intracellular trigger causes release of a cytotoxin, effectively leading to apoptosis.

Combination of Diagnostics and Therapeutics for Cancer

Aptamer-Conjugated Magnetic Nanoparticles

Magnetic nanoparticles have shown promise for targeted drug delivery, hyperthermia, and MRI imaging in cancer. Aptamer-conjugated magnetic nanoparticles controlled by an externally applied 3-D rotational magnetic field have been developed as a nanosurgical approach for the removal of cancerous cells selectively from the interior of an organ or tissue without any collateral damage (Nair et al. 2010). This system could be upgraded for the selective removal of complex cancers from diverse tissues by incorporating various target-specific ligands on magnetic nanoparticles.

Biomimetic Nanoparticles Targeted to Tumors

Nanoparticle-based diagnostics and therapeutics hold great promise because multiple functions can be built into the particles. One such function is an ability to home to specific sites in the body. Biomimetic particles that not only home to tumors but also amplify their own homing have been described (Simberg et al. 2007). The system is based on a peptide that recognizes clotted plasma proteins and selectively homes to tumors, where it binds to vessel walls and tumor stroma. Iron oxide nanoparticles and liposomes coated with this tumor-homing peptide accumulate in tumor vessels, where they induce additional local clotting, thereby producing new binding sites for more particles. The system mimics platelets, which also circulate freely but accumulate at a diseased site and amplify their own accumulation at that site. The self-amplifying homing is a novel function for nanoparticles. The clotting-based amplification greatly enhances tumor imaging, and the addition of a drug carrier function to the particles is envisioned.

Dendrimer Nanoparticles for Targeting and Imaging Tumors

Dendrimer nanoparticles have been used to entrap metal nanoparticles, a combination that could serve as a potent imaging and thermal therapy agent for tumors if it were not for associated toxicity issues. To eliminate the toxicity associated with dendrimer–metal nanoparticle combinations, methods have been developed for modifying the surface of dendrimers laden with gold nanoparticles. This chemical treatment greatly reduces the toxicity of the hybrid nanoparticle without changing its size. Construction of novel dendrimers with biocompatible components and the surface modification of commercially available dendrimers by PEGylation, acetylation, glycosylation, and amino acid functionalization have been proposed to solve the safety problem of dendrimer-based nanotherapeutics (Cheng et al. 2011). There are several opportunities and challenges for the development of dendrimer-based nanoplatforms for targeted cancer diagnosis and therapy.

Gold Nanoparticle Plus Bombesin for Imaging and Therapy of Cancer

Bombesin (BBN) peptides have demonstrated high affinity toward gastrin-releasing peptide (GRP) receptors in vivo that are overexpressed in prostate, breast, and small-cell lung carcinoma. In vivo studies using gold nanoparticles (AuNPs)–BBN and its radiolabeled surrogate ^{198}AuNP–BBN constructs are GRP-receptor specific showing accumulation with high selectivity in GRP-receptor-rich prostate tumors implanted in severe combined immunodeficient mice (Chanda et al. 2010). The intraperitoneal mode of delivery was found to be efficient as AuNP–BBN conjugates showed reduced RES organ uptake with concomitant increase in uptake at tumor targets. The selective uptake of this new generation of GRP-receptor-specific AuNP–BBN peptide analogues have clinical potential in molecular imaging using CT techniques as the contrast numbers in prostate tumor sites are severalfold higher as compared to the pretreatment group. They also provide synergistic advantages by combining molecular imaging with therapy of cancer.

Gold Nanorods for Diagnosis Plus Photothermal Therapy of Cancer

Photothermal therapy is based on the enhancement of electromagnetic radiation by noble metal nanoparticles due to strong electric fields at the surface. The nanoparticles also absorb laser light more easily, so that the coated malignant cells only require half the laser energy to be killed compared to the benign cells. This makes it relatively easy to ensure that only the malignant cells are being destroyed. These unique properties provide the potential of designing novel optically active reagents for simultaneous molecular imaging and photothermal cancer therapy. Gold nanorods with suitable aspect ratios (length divided by width) can absorb and scatter strongly in the NIR region (650–900 nm). Changing the spheres into rods lowers the frequency to which the nanoparticles respond from the visible light spectrum used by the nanospheres to the NIR spectrum. Since these lasers can penetrate deeper under the skin than lasers in the visible spectrum, they can reach tumors that are inaccessible to visible lasers.

In vitro studies have demonstrated that gold nanorods are novel contrast agents for both molecular imaging and photothermal cancer therapy (Huang et al. 2006). Nanorods are synthesized and conjugated to anti-epidermal growth factor receptor (anti-EGFR) monoclonal antibodies (MAbs) and incubated in cancer cell cultures. The anti-EGFR antibody-conjugated nanorods bind specifically to the surface of the malignant-type cells with a much higher affinity due to the overexpressed EGFR on the cytoplasmic membrane of the malignant cells. As a result of the strongly scattered red light from gold nanorods in dark field, observed using a laboratory microscope, the malignant cells are clearly visualized and diagnosed from the nonmalignant cells. It is found that, after exposure to continuous red laser at 800 nm, malignant cells require about half the laser energy to be photothermally destroyed than the nonmalignant cells. Thus, both efficient cancer cell diagnostics and selective photothermal therapy are realized at the same time.

Magnetic Nanoparticles for Imaging as well as Therapy of Cancer

Several multifunctional nanoparticles are being developed for simultaneous imaging and therapeutic applications in cancer. Tumor-targeting dendrimers can contain an imaging as well as a delivery agent for drugs and genetic materials. A dendrimer linked to a fluorescent-imaging agent and paclitaxel can identify tumor cells and kill them simultaneously.

In ovarian cancer, metastasis occurs when cells slough off the primary tumor and float free in the abdominal cavity. If one could use the magnetic nanoparticles to trap drifting cancer cells and pull them out of the abdominal fluid, it may be possible to predict and perhaps prevent metastasis. With this aim, magnetic cobalt-spinel ferrite nanoparticles, which have cobalt-spiked magnetite at their core, were coated with biocompatible polygalacturonic acid and functionalized with ligands specific for targeting expressed EphA2 receptors on ovarian cancer cells (Scarberry et al. 2008). By using such magnetic nanoparticle-peptide conjugates, targeting and extraction of malignant cells were achieved with a magnetic field. The particles, which are just 10 nm or less in diameter, are not magnetic most of the time, but when a magnet is present, they become strongly attracted to it. Targeting ovarian cancer cells with receptor-specific peptide-modified magnetic nanoparticles resulted in cell capture from a flow stream in vitro and from the peritoneal cavity of mice in vivo. Successful removal of metastatic cancer cells from the abdominal cavity and from circulation using magnetic nanoparticle conjugates indicate the feasibility of a dialysis-like treatment and may improve long-term survival rates of ovarian cancer patients. This approach can be applied for treating other cancers, such as leukemia, once the receptors on malignant cells are identified and the efficacy of targeting ligands is established. This technique will provide a way to test for and even treat metastatic ovarian cancer. Although the nanoparticles were tested inside the bodies of mice, it is possible to construct an external device that would remove a patient's abdominal fluid, magnetically filter out the cancer cells, and then return the fluid to the body. After surgery for removal of the primary tumor, a patient would undergo such a treatment to remove any residual cancer cells. The researchers are currently developing such a filter and testing it on abdominal fluid from human ovarian cancer patients.

Micelles for Targeted Drug Delivery and PET Imaging in Cancer

H40-DOX-cRGD, a multifunctional unimolecular micelle made of a hyperbranched amphiphilic block copolymer with attached doxorubicin (DOX), was tested for targeted anticancer drug delivery and PET imaging in tumor-bearing mice (Xiao et al. 2012). A uniform size distribution and pH-sensitive drug release behavior was observed. There was a much higher cellular uptake in U87MG human glioblastoma cells due to integrin $\alpha v\beta 3$-mediated endocytosis than nontargeted unimolecular micelles (i.e., H40-DOX), thereby leading to a significantly higher cytotoxicity. Thus, unimolecular micelles formed by hyperbranched amphiphilic block copolymers integrate passive and active tumor-targeting abilities with pH-controlled drug release. Simultaneous PET imaging for diagnosis provides the basis for personalized cancer therapy.

Nanobialys for Combining MRI with Delivery of Anticancer Agents

Although gadolinium has been the dominant paramagnetic metal for MRI contrast, the recent association of this lanthanide with nephrogenic systemic fibrosis, an untreatable disease, has spawned renewed interest in alternative metals for molecular MRI. Manganese was one of the first examples of a paramagnetic contrast material studied in cardiac and hepatic MRI because of efficient site-specific MR T1-weighted molecular imaging. Similar to Ca^{2+} and unlike the lanthanides, manganese is a natural cellular constituent and often a cofactor for enzymes and receptors. Mangafodipir trisodium, a manganese blood pool agent, has been approved as a hepatocyte-specific contrast agent with transient side effects due to dechelation of manganese from the linear chelate. A self-assembled, manganese(III)-labeled nanobialys MRI nanoparticle has been developed for combined diagnosis and delivery of a chemotherapeutic agent (Pan et al. 2008). The "bialy" shape affords increased stability. Nanobialys nanoparticles have been characterized for targeted detection of fibrin, a major biochemical feature of thrombus. A complementary ability of nanobialys to incorporate anticancer compounds with greater than 98% efficiency and to retain more than 80% of these drugs after infinite sink dissolution point to the potential of this platform technology to combine a therapeutic agent with a diagnostic agent.

Nanoparticles: MRI and Thermal Ablation of Tumors

Nanostructures with surface-bound ligands can be used for the targeted delivery and ablation of colorectal cancer (CRC), the third most common malignancy and the second most common cause of cancer-related mortality in the USA. Normal colonic epithelial cells as well as primary CRC and metastatic tumors all express a unique surface-bound guanylyl cyclase C (GCC), which binds the bacterial heat-stable enterotoxin (ST) – a peptide. This makes GCC a potential target for metastatic tumor ablation using ST-bound nanoparticles in combination with thermal ablation with near-infrared or radio-frequency energy absorption (Fortina et al. 2007). Furthermore, the incorporation of iron or iron oxide nanoparticles into such structures would provide advantages for MRI.

Gold nanoshell-based, targeted, multimodal contrast agents in the near-IR are fabricated and utilized as a diagnostic and therapeutic probe for MRI, fluorescence optical imaging, and photothermal cancer therapy of breast carcinoma cells in vitro (Bardhan et al. 2009). This may enable diagnosis as well as treatment of cancer during one hospital visit.

In the future, it may be possible for a patient to be screened for breast cancer using MRI techniques with engineered enhanced ferrites as the MRI contrast agent. Enhanced ferrites are a class of ferrites that are specially engineered to have enhanced magnetic or electrical properties and are created through the use of core-shell morphology. Magnetic nanoparticles are coupled to the radio frequency of the MRI, which converts the radio frequency into heat. If a tumor is detected, the physician could increase the power to the MRI coils and localized heating would destroy the tumor without damage to the surrounding healthy cells. The only hindrance to the development of enhanced ferrites for 100-MHz applications is a lack of understanding of the growth mechanisms and synthesis–property relationships of these nanoparticles. By studying the mechanism for the growth of the enhanced ferrites, it will be possible to create shells that help protect the metallic core from oxidation in biologically capable media.

pHLIP Nanotechnology for Detection and Targeted Therapy of Cancer

The pH-selective insertion and folding of a membrane peptide, pHLIP (pH low insertion peptide), can be used to target acidic tissue in vivo, including acidic foci in tumors. pHLIP nanotechnology is considered to be a promising approach for mapping areas of elevated acidity in the body. The peptide has three states: soluble in water, bound to the surface of a membrane, and inserted across the membrane. At physiological pH, the equilibrium is toward water, which explains its low affinity for cells in healthy tissue; at acidic pH, the equilibrium shifts toward membrane insertion and tissue accumulation. This peptide acts like a nanosyringe to deliver tags or therapy to cells. Tumors can be detected by labeling pHLIP peptide with Cy5.5 and imaging by use of NIR fluorescence with wavelengths in the range of 700–900 nm. In a mouse breast adenocarcinoma model, fluorescently labeled pHLIP detects solid acidic tumors with high accuracy and accumulates in them even at a very early stage of tumor development (Andreev et al. 2007). The fluorescence signal is stable and is approximately five times higher in tumors than in healthy counterpart tissue. Tumor targeting is based on the fact that most tumors, even very small ones, are acidic as a result of the way they grow, known as the Warburg effect (Nobel Prize 1931). Tumors may be treated by attaching and delivering anticancer agents with pHLIP.

QD Conjugates Combine Cancer Imaging, Therapy, and Sensing

The specificity and sensitivity of a QD–aptamer–doxorubicin (QD-Apt(Dox)) conjugate as a targeted cancer imaging, therapy, and sensing system has been demonstrated in vitro (Bagalkot et al. 2007). By functionalizing the surface of fluorescent QD with a RNA aptamer, which recognizes the extracellular domain of the prostate-specific membrane antigen (PSMA), the system is capable of differential uptake and imaging of prostate cancer cells that express the PSMA. The intercalation of Dox, an anticancer drug with fluorescent properties, in the double-stranded stem of the aptamer results in a targeted conjugate with reversible self-quenching properties based on a Bi-FRET mechanism. A donor–acceptor model FRET between QD and Dox and a donor–quencher model FRET between Dox and aptamer result when Dox is intercalated within the aptamer. This simple multifunctional nanoparticle system can deliver Dox to the targeted prostate cancer cells and sense the delivery of Dox by activating the fluorescence of QD, which concurrently images the cancer cells.

Squalene-Based Nanocomposites for Tumor Imaging and Therapy

Nanocomposites, constructed of magnetite nanocrystals into NPs by self-assembling molecules of the squalenoyl gemcitabine (SQgem) bioconjugated, are characterized by an unusually high drug loading, a significant magnetic susceptibility, and a low burst release. When injected into a subcutaneous mice tumor model, these magnetite/SQgem NPs were magnetically guided and displayed considerably greater anticancer activity than other anticancer treatments including nonmagnetically guided magnetite/SQgem NPs (Arias et al. 2011). The histology and immunohistochemistry investigation of the tumor biopsies clearly evidenced the therapeutic superiority of the magnetically guided nanocomposites, while Prussian blue staining confirmed their accumulation at the tumor periphery. The superior therapeutic activity and enhanced tumor accumulation has been successfully visualized using T2-weighted MRI imaging. This concept was further enlarged by (1) the design of squalene-based NPs containing the T1 Gd^{3+} contrast agent instead of magnetite and (2) the application to other anticancer squalenoyls, such as, cisplatin, doxorubicin, and paclitaxel. This nanotechnology platform is expected to have important applications in imaging-guided cancer therapy.

Radiolabeled Carbon Nanotubes for Tumor Imaging and Targeting

Single-walled carbon nanotubes (SWCNTs) with covalently attached multiple copies of tumor-specific MAbs, radiometal-ion chelates, and fluorescent probes can target lymphomas and deliver both imaging and therapeutic molecules to these tumors (McDevitt et al. 2007). Each nanotube, which contained approximately six antibody molecules and 114 radioactive atoms, proved to be stable in human plasma for at least 96 h and was able to bind to targeted tumor cells. Most importantly, the chemical linkages binding the radioactive element indium-111 were completely stable in human plasma for the entire 4-day experiment. Tests using a mouse model of human lymphoma showed that the nanotube construct successfully targeted tumors while avoiding healthy cells. The ability to specifically target tumor with prototype-radiolabeled or fluorescent-labeled, antibody-appended SWCNT constructs was encouraging and suggested further investigation of these as diagnostic combined with drug delivery for cancer.

Ultrasonic Tumor Imaging and Targeted Chemotherapy by Nanobubbles

Drug delivery in polymeric micelles combined with tumor irradiation by ultrasound results in effective drug targeting, but this technique requires prior tumor imaging. A new targeted drug-delivery method uses ultrasound to image tumors, while also releasing the drug from nanobubbles into the tumor (Rapoport et al. 2007). Mixtures of drug-loaded polymeric micelles and perfluoropentane (PFP) nanobubbles stabilized by the same biodegradable block copolymer were prepared. Size distribution of nanoparticles was measured by dynamic light scattering. Cavitation activity (oscillation, growth, and collapse of microbubbles) under ultrasound was assessed based on the changes in micelle-to-nanobubble volume ratios. The effect of the nanobubbles on the ultrasound-mediated cellular uptake of doxorubicin (Dox) in MDA MB231 breast tumors in vitro and in vivo (in mice-bearing xenograft tumors) was determined by flow cytometry. Phase state and nanoparticle sizes were sensitive to the copolymer-to-perfluorocarbon volume ratio. At physiologic temperatures, nanodroplets converted into nanobubbles. Doxorubicin was localized in the nanobubble walls formed by the block copolymer. Upon intravenous injection into mice, Dox-loaded micelles and nanobubbles extravasated selectively into the tumor interstitium, where the nanobubbles coalesced to produce microbubbles. When exposed to ultrasound, the bubbles generated echoes, which made it possible to image the tumor. The sound energy from the ultrasound popped the bubbles, releasing Dox, which enhanced intracellular uptake by tumor cells in vitro to a statistically significant extent relative to that observed with unsonicated nanobubbles and unsonicated micelles and resulted in tumor regression in the mouse model. In conclusion, multifunctional nanoparticles that are tumor-targeted drug carriers, long-lasting ultrasound contrast agents, and enhancers of ultrasound-mediated drug delivery have been developed and deserve further exploration as cancer therapeutics.

Nanorobotics for Management of Cancer

It is within the realm of possibility to use molecular tools to design a miniature device, e.g., a nanobot that can be introduced in the body, locate and identify cancer cells, and finally destroy them. The device would have a biosensor to identify cancer cells and a supply of anticancer substance that could be released on encountering cancer cells. A small computer could be incorporated to program and integrate the combination of diagnosis and therapy and provide the possibility to monitor the in vivo activities by an external device. Since there is no universal anticancer agent, the computer program could match the type of cancer to the most appropriate agent. Such a device could be implanted as a prophylactic measure in persons who do not have any obvious manifestations of cancer. It would circulate freely and could detect and treat cancer at the earliest stage. Such a device could be reprogrammed through remote control and enable change of strategy if the lesion encountered is other than cancer.

Bacterial Nanorobots for Targeting Cancer

Flagellated nanomotors combined with the nanometersized magnetosomes of a single magnetotactic bacterium (MTB) can be used as an effective integrated propulsion and steering system for devices such as nanorobots designed for targeting locations only accessible through the smallest capillaries in humans while being visible for tracking and monitoring purposes using modern medical imaging modalities such as MRI (Martel et al. 2009). Through directional and magnetic field intensities, the displacement speeds, directions, and behaviors of swarms of these bacterial actuators can be controlled from an external computer. Such a device can be used for diagnosis as well as therapy of cancer.

DNA Robots for Targeting Cancer

DNA nanotechnology is widely investigated for its potential to deliver drugs and molecular signals to cells in the body because DNA is a biocompatible and biodegradable material. However, opinions differ as about the best nanorobot design, i.e., the ideal structure to load, transport, and deliver molecules. Various designs include a spiderlike robot that moves along a chemical track, a nanofactory with mobile robotic walkers and molecular forklifts, and DNA tweezers that open and close to grasp and release molecules.

An autonomous DNA nanorobot has been described that is capable of transporting molecular payloads to cells, sensing cell surface inputs for conditional, triggered activation, and reconfiguring its structure for payload delivery (Douglas et al. 2012). The nanorobot, constructed using a computer-aided design tool called DNA origami, is a hexagonal barrel, 35 nm in diameter, and opens like a clam shell. The device can be loaded with a variety of materials and is controlled by an aptamer-encoded logic gate, enabling it to respond to a wide array of cues that have demonstrated their efficacy in selective regulation of nanorobot function. This barrel-shaped DNA nanorobot seeks out cancer cells and delivers self-destruct instructions. It can successfully deliver antibody fragments to surfaces of cancer cells to kill them and bacterial proteins to activate T cells.

Fullerenes for Protection Against Chemotherapy-Induced Cardiotoxicity

Therapeutic use doxorubicin as an anticancer drug is limited due to its cardiotoxicity. Generation of free radicals plays an important role in the mechanism of doxorubicin-induced cardiotoxicity. There is significant evidence indicating that mitochondria are the principal targets in this pathological process. Efficacy of fullerenol ($C_{60}OH_{24}$) in preventing single, high-dose doxorubicin-induced cardiotoxicity has been investigated in rats with malignant neoplasm (Injac et al. 2008). Study was performed on adult female Sprague Dawley rats with chemically induced mammary carcinomas. The animals were sacrificed 2 days after the application of doxorubicin and/or fullerenol, and the serum activities of cardiac enzymes were determined. The results obtained showed that the administration of a single dose of 8 mg/kg in all treated groups induces statistically significant cardiotoxicity. There were significant changes in the enzymes' lactate dehydrogenase and creatine kinase and increase in level of tissue malondialdehyde (MDA), a product of lipid peroxidation, after intraperitoneal administration of doxorubicin. The results revealed that doxorubicin-induced oxidative damage and that the fullerenol antioxidant effect caused significant changes in the levels of biomarker MDA in the heart. Thus, fullerenol may have an important role as for cardioprotection in doxorubicin-treated individuals.

Concluding Remarks and Future Prospects of Nanooncology

The rationale for using nanobiotechnology in oncology is that nanoparticles have optical, magnetic, or structural properties that are not available from larger molecules or bulk solids. When linked with tumor-targeting ligands such as MAbs, peptides, or small molecules, nanoparticles can be used to target tumor antigens (biomarkers) as well as tumor vasculatures with high affinity and specificity. In the size range of 5–100 nm diameter, nanoparticles have large surface areas and functional groups for conjugating to multiple diagnostic and therapeutic anticancer agents. Recent advances have led to bioaffinity nanoparticle probes for molecular and cellular imaging, targeted nanoparticle drugs for cancer therapy, and integrated nanodevices for early cancer detection and screening. These developments have provided opportunities for personalized oncology in which biomarkers are used to diagnose and treat cancer based on the molecular profiles of individual patients.

Nanoparticles have shown promise for incorporating multiple functions including diagnosis and therapy of cancer. Most of the work done in this area is still experimental, and some challenges need to be resolved before clinical applications. These include the following:

- Preventing capture/removal by the reticuloendothelial system
- Difficulties in selective targeting as well as penetration of tumor by systemic administration of anticancer nanostructures, which requires identification of receptors unique to a particular cancer
- Investigation of long-term fate and toxicity concerns of nanoparticles

Efforts are being made to use nanostructures to develop anticancer treatment strategies based on various mitochondrial targets that play vital roles in cancer development and progression. Cancer mitochondria-targeted multifunctional compounds have been identified that could provide an alternative strategy for the development of novel solutions for cancer diagnosis and therapy (Zhang et al. 2011).

Chapter 9
Nanoneurology

Introduction

Neurology deals with the study and management of disorders of the nervous system. Considerable research is in progress in basic neurosciences and clinical neurology. The management is mostly medical. Many neurological disorders require surgical intervention, and the closely related specialty of surgical neurology or neurosurgery will also be considered in this chapter. There is a considerable scope for application of nanobiotechnology in neurology and hence the term nanoneurology (Jain 2009a). Nanobiotechnology has been applied for neurophysiological studies, diagnosis, neuropharmacology, and refinement of surgical tools (Jain 2012i). Neuroprotection is an important objective in treatment of diseases of the central nervous system (CNS).

Nanobiotechnology for Neurophysiological Studies

Use of Nanoelectrodes in Neurophysiology

Insulated microelectrodes are used in neurophysiological studies since 1950s with minor modifications. Single large neuron recordings are possible with electrodes in μm diameter range. For small neurons, it is worthwhile to have electrodes with nanoscale tips for recording. It is now possible to grind the bare tip of a tungsten microelectrode down to 100–1,000 nm and remove the insulation at the tip. A 700-nm tipped electrode was demonstrated to record well-isolated action potentials extracellularly from single visual neurons in vivo (Qiao et al. 2005). Experimental studies have shown that sub-100 nm silicon nanowires can be integrated into live cells without causing detrimental effects (Kim et al. 2007).

Transistor arrays of silicon nanowires with diameter of 30 nm and fabricated on transparent substrates can be reliably interfaced to acute brain slices (Qing et al. 2010). These can record across a wide range of length scales, whereas the transparent device chips only provide imaging of individual cell bodies. Combination of

K.K. Jain, *The Handbook of Nanomedicine*, DOI 10.1007/978-1-61779-983-9_9,

arrays with patch clamp studies enables identification of action potential signals. The result is a recording with high temporal and spatial resolution, as well as mapping of functional connectivity. This provides a powerful platform for studying neural circuits in the brain.

Nanowires for Monitoring Brain Activity

Electrical recording from spinal cord vascular capillary bed has been achieved demonstrating that the intravascular space may be utilized as a means to address brain activity without violating the brain parenchyma. Working with platinum nanowires and using blood vessels as conduits to guide the wires, researchers have successfully detected the activity of individual neurons lying adjacent to the blood vessels (Llinas et al. 2005). This can provide an understanding of the brain at the neuron-to-neuron interaction level with nonintrusive, biocompatible and biodegradable nanoprobes. This technique may one day enable monitoring of individual brain cells and perhaps provide new treatments for neurological diseases. Because the nanowires can deliver electrical impulses as well as receive them, the technique has potential as a treatment for Parkinson's disease (PD). It has already been shown that patients with PD can experience significant improvement from direct stimulation of the affected area of the brain. But the stimulation is currently carried out by inserting wires through the skull and into the brain, a process that can cause scarring of the brain tissue. By stimulating the brain with nanowires threaded through blood vessels, patients can receive benefits of the treatment without the damaging side effects. The challenge is to precisely guide the nanowire probes to a predetermined spot through the thousands of branches in the brain's vascular system. One solution is to replace the platinum nanowires with new conducting polymer nanowires. Not only do the polymers conduct electrical impulses, they change shape in response to electrical fields, which would allow the researchers to steer the nanowires through the brain's circulatory system. Polymer nanowires have the added benefit of being 20–30 times smaller than the platinum ones used in the reported laboratory experiments. They are biodegradable and therefore suitable for short-term brain implants.

Gold Nanoparticles for In Vivo Study of Neural Function

As a novel in vivo method to study interactions between gold nanoparticles AuNPs and the nervous system, negatively charged AuNPs, 50 nm in diameter, were injected into the CNS of a cockroach (Rocha et al. 2011). The charged nanoparticles affected the insect's locomotion and behavior but no significant effect on the life expectancy of the cockroach after 2 months of observation, apparently due to the encapsulation of AuNPs inside the insect's brain. This inexpensive method offers an opportunity to further understand how nanoparticles affect neural communication by monitoring insect activity and locomotion.

Nanodiagnosis and Nanoparticle-Based Brain Imaging

Nanodiagnostic technologies described in Chap. 4 are applicable to neurological disorders. Relation of various technologies to diagnosis of neurological disorders is shown in Fig. 9.1.

Applications of Nanotechnology in Molecular Imaging of the Brain

Some of the applications of nanobiotechnology in brain imaging are the following:

- Tracking of stem cells by tagging with nanoparticles such as SPIONs so they can be detected with MRI
- QDs for molecular imaging in cerebrovascular disorders
  - Early aneurysm detection and guide endovascular intervention
  - Imaging of vessels prone to spasm
  - To distinguish penumbra from infarction
  - To identify unstable arterial plaques for targeted intervention
- Early diagnosis of neurodegenerative disorders
- Early diagnosis of brain tumors

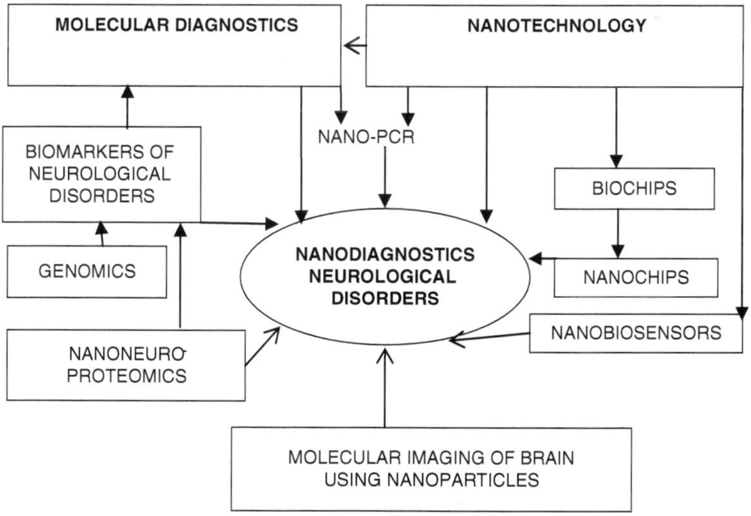

Fig. 9.1 Nanodiagnostics for neurological disorders (© Jain PharmaBiotech)

Nanoparticles and MRI for Macrophage Tracking in the CNS

Activated macrophages, acting in concert with other immune competent cells, are an index of inflammatory/immune reaction in CNS disorders such as multiple sclerosis, ischemic stroke lesions, and tumors. The MRI detection of brain macrophages defines precise spatial and temporal patterns of macrophage involvement that helps to characterize individual neurological disorders. Macrophage tracking by MRI with iron oxide nanoparticles has been developed during the last decade for numerous diseases of the CNS. Experimental studies on animal models were confirmed by clinical applications of MRI technology of brain macrophages. This approach is being explored as an in vivo biomarker for the clinical diagnosis of cerebral lesion activity, in experimental models for the prognosis of disease development, and to determine the efficacy of immunomodulatory treatments under clinical evaluation (Petry et al. 2007). Comparative brain imaging follow-up studies of blood–brain barrier leakage by MRI with gadolinium chelates, microglia activation by PET with radiotracer ligand PK11195, and MRI detection of macrophage infiltration provide more precise information about the pathophysiological cascade of inflammatory events in cerebral diseases. Such multimodal characterization of the inflammatory events should help in the monitoring of patients, in defining precise time intervals for therapeutic interventions, and in developing and evaluating new therapeutic strategies.

Nanoparticles for Tracking Stem Cells for Therapy of CNS Disorders

Cellular MRI using superparamagnetic iron oxide nanoparticles (SPION) can visualize and track cells in living organisms. MRI studies have been conducted in rat models of CNS injury and stroke to track stem cells that were either grafted intracerebrally, contralaterally to a cortical photochemical lesion, or injected intravenously (Sykova and Jendelova 2007). ESCs and MSCs were labeled with iron oxide nanoparticles (Endorem®) and human $CD34^{+}$ cells were labeled with magnetic MicroBeads (Miltenyi). During the first posttransplantation week, grafted MSCs or ESCs migrated to the lesion site in the cortex as well as in the spinal cord and were visible in the lesion on MRI as a hypointensive signal, persisting for more than 30 days. In rats with an SCI, an increase in functional recovery was noted after the implantation of MSCs or after an injection of granulocyte colony stimulating factor (G-CSF). Morphometric measurements in the center of the lesions showed an increase in white matter volume in cell-treated animals. Prussian blue staining confirmed a large number of iron-positive cells, and the lesions were considerably smaller than in control animals. To obtain better results with cell labeling, new polycation-bound SPIONs (PC-SPIONs) were developed. In comparison with Endorem, PC-SPIONs demonstrated a more efficient intracellular uptake into MSCs with no decrease in cell viability. These studies demonstrate that MRI of grafted adult as well as ESCs labeled with iron oxide nanoparticles is a useful method for evaluating cellular migration toward a lesion site.

Autologous bone marrow CD34+ cells labeled with magnetic nanoparticles have been delivered into the spinal cord via lumbar puncture in a study on patients with chronic SCI (Callera and de Melo 2007). One group received their own labeled CD34+ cells, whereas the others received an injection containing only magnetic nanoparticles without stem cells to serve as controls. CD34+ cells were labeled with magnetic nanoparticles coated with a monoclonal antibody specific for the CD34 cell membrane antigen. MRI showed that magnetically labeled CD34+ cells were visible at the lesion site as hypointense signals following transplantation, but these signals were not visible in any patient in the control group. This study shows that autologous bone marrow CD34+ cells labeled with magnetic nanoparticles, when delivered intrathecally, migrate into the site of injury in patients with chronic SCI and can be tracked by MRI. This shows the feasibility of treatment of SCI with intrathecal cell therapy.

Multifunctional NPs for Diagnosis and Treatment of Brain Disorders

Multifunctional NPs (MFNPs) are particularly suited for combining diagnostics with therapeutics of brain disorders. Tailoring the size, contents, and surface electronic properties through chemistry and physical methods within sub-200 nm nanoparticles will be key factors for using MFNPS (Suh et al. 2009). Functions such as directing neuronal growth and influencing stem cell differentiation for brain repair seem to be the next logical step in nanobiotechnology utilizing MFNPS. Studies involving stem cell differentiation and transplantation, neural implants, targeted drug delivery with real-time monitoring capabilities, and in vivo RNAi will be of great interest. Advances in neuroscience will arise from systematic investigations starting from synthesis to application where the efforts are focused on probing and understanding events occurring at the nano–bio interface.

Nanotechnology-Based Drug Delivery to the CNS

Delivery of drugs to CNS is a challenge, and the basics as well as various strategies are discussed in a special report on this topic (Jain 2012e). Molecular motors, operating at nanoscale, can deliver drugs to the CNS by peripheral muscle injection. An advantage is the use of nanomotors in native environment for intraneural drug delivery. The disadvantages are that this approach requires engineered molecular motors for use in cells and neurotoxicity may be a problem.

Nanoencapsulation for Delivery of Vitamin E for CNS Disorders

Vitamin E is used for the treatment of neurological disorders, particularly those where oxidative stress plays a role. Oxidative stress is an early hallmark of affected neurons in Alzheimer's disease (AD). The antioxidant vitamin E provided limited neuroprotection in AD, which may have derived from its lipophilic nature and resultant inability to quench cytosolic reactive oxygen species (ROS), including those generated from antecedent membrane oxidative damage. Encapsulation into polyethylene glycol (PEG)-based nanospheres, which can enter the cytosol, improved the efficacy of vitamin E against Aβ-induced ROS (Shea et al. 2005). These findings suggest that nanosphere-mediated delivery methods may be a useful adjunct for antioxidant therapy in AD.

Nanoparticle Technology for Drug Delivery Across BBB

Currently most of the strategies are directed at overcoming the blood–brain barrier (BBB). Role of nanobiotechnology in overcoming BBB is described elsewhere (Jain 2012f). Very small nanoparticles may just pass through the BBB but this uncontrolled passage is not desirable. Most of the strategies described in this report for passage of drugs across the BBB can be enhanced by nanotechnology and some examples are the following (Barbu et al. 2009):

- Nanoparticles open the tight junctions between endothelial cells and enable the drug to penetrate the BBB either in free form or together with the nanocarrier.
- Nanoparticles are transcytosed through the endothelial cell layer and allow the direct transport of their therapeutic cargo.
- Nanoparticles are endocytosed by endothelial cells and release the drug inside the cell, as a precursor step to the transport of active ingredients, which occurs by exocytosis at the abluminal side of the endothelium.
- Nanoparticles, which combine an increased retention at the brain capillaries with adsorption onto the capillary walls, improve delivery to the brain by creating a concentration gradient that promotes transport across the endothelial cell layer.
- Drug transport is enhanced by the solubilization of the endothelial cell membrane lipids by surfactant, which leads to membrane fluidization (surfactant effect).
- Coating agents (such as polysorbates) inhibit the transmembrane efflux systems, i.e., P-glycoprotein.
- Nanoparticles induce local toxic effects at the brain vasculature, which leads to a limited permeabilization of the brain endothelial cells.

BBB represents an insurmountable obstacle for a large number of drugs, including antibiotics, antineoplastic agents, and a variety of central nervous system (CNS)-active drugs, especially neuropeptides. One of the possibilities to overcome this barrier is a drug delivery to the brain using nanoparticles. Drugs that have successfully been transported into the brain using this carrier include the hexapeptide dalargin, the dipeptide kytorphin, loperamide, tubocurarine, the NMDA receptor antagonist MRZ 2/576, and doxorubicin.

The mechanism of the nanoparticle-mediated transport of the drugs across the BBB at present is not fully elucidated. The most likely mechanism is endocytosis by the endothelial cells lining the brain blood capillaries. Nanoparticle-mediated drug transport to the brain depends on the overcoating of these materials with polysorbates, especially polysorbate 80, which seems to lead to the adsorption of apolipoprotein E from blood plasma onto the nanoparticle surface. The particles then seem to mimic low-density lipoprotein (LDL) particles and could interact with the LDL receptor leading to their uptake by the endothelial cells. After this, the drug may be released in these cells and diffuse into the brain interior or the particles may be transcytosed. Other processes such as tight junction modulation or P-glycoprotein (Pgp) inhibition also may occur. Moreover, these mechanisms may run in parallel or may be cooperative thus enabling a drug delivery to the brain.

The use of NPs to deliver drugs to the brain across the BBB may provide a significant advantage to current strategies. The primary advantage of NP carrier technology is that NPs mask the BBB limiting characteristics of the therapeutic drug molecule. Furthermore, this system may slow drug release in the brain, decreasing peripheral toxicity. Various factors that influence the transport include the type of polymer or surfactant, NP size, and the drug molecule. Use of metallic NPs such as AuNPs is associated with risk of neurotoxicity, and special precautions such as coating of NPs are required to prevent this. Other nanomaterials used for delivery across BBB include dendrimers, lipid NPs, liposomes, micelles, nanogels, PLGA, poly-ε-caprolactone, and polymeric NPs.

Polymeric nanoparticles have been shown to be promising carriers for CNS drug delivery due to their potential both in encapsulating drugs, hence protecting them from excretion and metabolism, and in delivering active agents across the BBB without inflicting any damage to the barrier (Tosi et al. 2008). Polymeric NPs for delivery across BBB should have the following ideal properties: biocompatible, nontoxic, nonthrombogenic, and nonimmunogenic (Martin-Banderas et al. 2011).

By designing well-controlled and appropriate preclinical and clinical translational studies, use of nanotechnologies for safely, efficiently, and specifically delivering drugs and other molecules across the BBB may prove one of their highest impact contributions to clinical neuroscience (Silva 2010). Two strategies for transporting drugs across the BBB are in commercial development: G-technology and LipoBridge.

G-Technology®

G-Technology® (to-BBB) platform utilizes nanoliposomes coated with glutathione-conjugated PEG to mediate safe targeting and enhanced delivery of drugs to the brain. Glutathione, an endogenous tripepeptide transporter, is highly expressed on the BBB. Intravenous injections of PEGylated liposomes are already on the market (Doxil), and high dosages of glutathione in supportive therapy in cancer as well. Glutathione, a natural antioxidant, is found at high levels in the brain, and its receptor is abundantly expressed at the BBB. Therefore, glutathione minimizes adverse effects such as adverse immunological reactions or interference with essential physiological pathways. None of the other technologies for delivery of drugs to the brain have the favorable pharmacokinetic and safety profile of the G-Technology®. This technology utilizes an endogenous receptor-mediated endocytosis mechanism in combination with nanosized drug-loaded liposomes. This approach is unique in that it does not require drug modification and at the same time gives rise to metabolic protection during transport and increased bioavailability at the target site.

LipoBridge™ Technology

LipoBridge™ (Genzyme Pharmaceuticals) temporarily and reversibly opens tight junctions to facilitate transport of drugs across the BBB and into the CNS. LipoBridge itself forms a clear suspension of nanoparticles in water and can solubilize or stabilize some drugs, is nonimmunogenic, and is excreted unmetabolized. It has been demonstrated in several laboratories that intracarotid injections of a simple mixture of Lipobridge™ and model compounds or pharmaceutical actives can deliver these actives into one or both hemispheres of the brain allowing for increased concentration in a selected hemisphere. It can be administered orally as well as intravenously. LipoBridge has been used to administer anticancer drugs for brain cancer in animals. Safety clinical studies in humans are in progress.

Nanovesicles for Transport Across BBB

According to US patent application #20070160658, scientists at the Ben-Gurion University of the Negev, Israel, are developing a targeting moiety conjugated to the nanovesicle, which comprises a therapeutic composition. These nanovesicles are useful in treatment of a wide spectrum of disorders. This technology solves the problem of transport through the BBB by using nanovesicles that are able to cross the BBB and which carry the desired drugs by using a targeted delivery mechanism where the drug will be released from the vesicle in the brain. The drug to be delivered is encapsulated within stable nanosized vesicles (20–100 nm) possessing surface moieties that facilitate the release of the drug at target sites, such as the brain. The method of targeting is based on head groups that are selectively cleaved at the target site by enzymatic activity, thus releasing the encapsulated material primarily at the target organ. Injection of an encapsulated analgesic peptide, encephalin, into mice showed an analgesic effect comparable to morphine, while encephalin in its free form did not to penetrate the BBB and had no effect. Potential applications of this technology include cancer, pain, and neurodegenerative diseases such as Alzheimer's and Parkinson's. The advantages are as follows:

- Vesicles are stable and flexible, allowing penetration through biological barriers.
- Unique surface chemistry allows the incorporation of selective targeting proteins.
- V-Smart targeting mechanism allows better precision in drug delivery by unloading drug from the vesicle only in predetermined location characterized by unique enzyme which causes drug release from the vesicle.

Nanotechnology-Based Drug Delivery to Brain Tumors

The focus of this section is glioblastoma multiforme (GBM), a primary malignant tumor of the brain. Treatment of GBM is one of the most challenging problems. Surgery remains the basic treatment in which the bulk of the tumor is removed and the peripheral infiltrating part is the target of supplementary treatments. GBM is not easily targeted but advances in nanobiotechnology have improved the prospects of delivery of therapeutics to GBM (Jain 2007a).

Multifunctional Nanoparticles for Treating Brain Tumors

One approach combines two promising approaches for diagnosing and treating cancer, creating a targeted multifunctional polymer nanoparticle that successfully images and kills brain tumors in laboratory animals (Reddy et al. 2006). The team developed of a 40-nm-diameter polyacrylamide nanoparticle loaded with Photofrin, a photosensitizing agent, and iron oxide. When irradiated with laser light, Photofrin, which is used to treat several types of cancer, including esophageal, bladder, and skin cancers, triggers the production of reactive oxygen species that destroy a wide variety of molecules within a cell. The iron oxide nanoparticles function as an MRI contrast agent. As the targeting agent, the researchers used a 31-amino-acid-long peptide developed by members of the NCI-funded Center of Nanotechnology for Cancer at the University of California (San Diego, CA). This peptide targets an unknown receptor found on the surface of new blood vessels growing around tumors and also triggers cell uptake of nanoparticles attached to it. Researchers tested the nanoparticles in cell cultures and animal models. The studies showed that the nanoparticles traveled to the tumor, resulting in less Photofrin exposure throughout the body, and enhanced exposure within the tumor. This allowed a larger window for activating the drug with light, which was accomplished by threading a fiber optic laser into the brain. In humans, this approach could reduce or eliminate a common side effect of photodynamic therapy, in which healthy skin becomes sensitive to light.

Nanoparticles for Delivery of Drugs to Brain Tumors Across BBB

Nanoparticles may be especially helpful for the treatment of malignant brain tumors. Nanoparticles made of poly(butyl cyanoacrylate) (PBCA) or PLGA coated with polysorbate 80 or poloxamer 188 enable the transport of cytostatics such as doxorubicin across the BBB. Following intravenous injection to rats bearing intracranial glioblastoma, these particles loaded with doxorubicin significantly increased the survival times and led to a complete tumor remission in 20–40 % of the animals (Kreuter and Gelperina 2008). Moreover, these particles considerably reduced the dose-limiting cardiotoxicity and also the testicular toxicity of this drug. The drug transport across the BBB by nanoparticles appears to be due to a receptor-mediated interaction with the brain capillary endothelial cells, which is facilitated by certain plasma apolipoproteins adsorbed by nanoparticles in the blood.

SPION conjugates, prepared using a novel circulating system, have been used to locate brain tumors earlier and more accurately than current methods and to target the tumors (Zhang et al. 2004). A simple dialysis method was developed to immobilize nanoparticles with functional biopolymers and targeting agents, which avoids the use of the normal centrifugation process that may cause particle agglomeration during the coating process. To enhance the specific targeting capability of the nanoparticles, a new chemical scheme was introduced, in which folic acid (FA) was chosen as the targeting agent combined with PEG serving to improve biocompatibility of nanoparticles. AFM characterization showed that the nanoparticles produced are well dispersed with a narrow size distribution. The biological part of the study showed that coating nanoparticles with PEG-FA significantly enhanced the intracellular uptake of nanoparticles by target cells. The researchers plan to attach a variety of small molecules, such as tumor receptor target, and even chemotherapy agents, to the nanoparticles.

MRI can detect the incorporation into brain tumor vasculature of systemically administered bone marrow stem cells labeled with superparamagnetic iron oxide nanoparticles as part of ongoing angiogenesis and neovascularization (Anderson et al. 2005). This technique can be used to directly identify neovasculature in vivo and to facilitate gene therapy by noninvasively monitoring these cells as gene delivery vectors.

A polymeric nanobioconjugate based on biodegradable, nontoxic, and nonimmunogenic polymalic acid as a universal delivery nanoplatform is used for design of a nanomedicine for intravenous treatment of brain tumors (Ding et al. 2010). The polymeric drug passes through the BTB and tumor cell membrane using tandem monoclonal antibodies targeting the BTB and tumor cells. The next step for polymeric drug action is inhibition of tumor angiogenesis by specifically blocking the synthesis of a tumor neovascular trimer protein, laminin-411, by attached antisense oligonucleotides, which are released into the target cell cytoplasm via pH-activated trileucine, an endosomal escape moiety. Introduction of a trileucine endosome escape unit results in significantly increased antisense oligonucleotide delivery to tumor cells, inhibition of laminin-411 synthesis, specific accumulation in brain tumors, and suppression of intracranial glioma growth compared with pH-independent leucine ester. The availability of a systemically active polymeric drug delivery system that crosses BTB, targets tumor cells, and inhibits tumor growth is a promising strategy of glioma treatment.

NP Delivery Across the BBB for Imaging and Therapy of Brain Tumors

In vivo application of nanoparticle-based platforms in brain tumors is limited by insufficient accumulation and retention within tumors due to limited specificity for the target, and an inability to traverse the BBB. A nanoprobe has been designed that can cross the BBB and specifically target brain tumors in a genetically engineered mouse model, by using in vivo magnetic resonance and biophotonic imaging, as well as histologic and biodistribution analyses (Veiseh et al. 2009). The nanoprobe is made of an iron oxide nanoparticle coated with biocompatible PEG-grafted chitosan copolymer, to which a tumor-targeting agent, chlorotoxin (a small peptide isolated from scorpion venom), and a near-IR fluorophore are conjugated. The particle was about 33 nm in diameter when wet, i.e., about a third the size of similar particles used in other parts of the body. The nanoprobe shows an innocuous toxicity profile and sustained retention in tumors. The nanoparticles remained in mouse tumors for up to 5 days and did not show any evidence of damaging the BBB. With the versatile affinity of the targeting ligand and the flexible conjugation chemistry for alternative diagnostic and therapeutic agents, this nanoparticle platform can be potentially used for the diagnosis and treatment of a variety of brain tumors. The fluorescent nanoparticles improved the contrast between the tumor tissue and the normal tissue in both MRI and optical imaging, which are used during surgery to see the tumor boundary more precisely. Precise imaging of brain tumor margins is important because patient survival for brain tumors is directly related to the amount of tumor that can be resected.

Nano-imaging could also help with early detection of brain tumors. Current imaging techniques have a maximum resolution of 1 mm. Nanoparticles could improve the resolution by a factor of 10 or more, allowing detection of smaller tumors and earlier treatment. Future research will evaluate this nanoparticle's potential for treating tumors.

Intravenous Gene Delivery with Nanoparticles into Brain Tumors

Brain tumors may be amenable to gene therapy with cytotoxic genes, such as the proapoptotic Apo2 ligand/tumor necrosis factor-related apoptosis-inducing ligand (Apo2L/TRAIL). Gene therapy of gliomas ideally employs intravenously given vectors, thus excluding viral vectors as they cannot cross the BBB. Cationic albumin-conjugated pegylated nanoparticles (CBSA-NP) have been synthesized and shown to accumulate in mouse brain cells upon IV administration. Plasmid pORF-hTRAIL (pDNA) has been incorporated into CBSA-NP, and the resulting CBSA-NP-hTRAIL was evaluated as a nonviral vector for gene therapy of gliomas (Lu et al. 2006). Thirty minutes after IV administration of CBSA-NP-hTRAIL to BALB/c mice bearing intracranial C6 gliomas, CBSA-NP-hTRAIL colocalized with glycoproteins in brain and tumor microvasculature and, via absorptive-mediated transcytosis, accumulated in tumor cells. At 24 and 48 h after intravenous administration of CBSA-NP-hTRAIL, respectively, hTRAIL mRNA and protein were detected in normal brain and tumors. Furthermore, repeated IV injections of CBSA-NP-hTRAIL induced apoptosis in vivo and significantly delayed tumor growth. In conclusion, this study indicates that CBSA-NP-hTRAIL is a promising candidate for noninvasive gene therapy of malignant glioma.

PLA Nanoparticles for Controlled Delivery of BCNU to Brain Tumors

BCNU-loaded biodegradable PLA nanoparticles have been combined with transferrin, an iron-transporting serum glycoprotein, which binds to receptors expressed on surface of glioma cells (Kang et al. 2009). In vitro drug release studies have

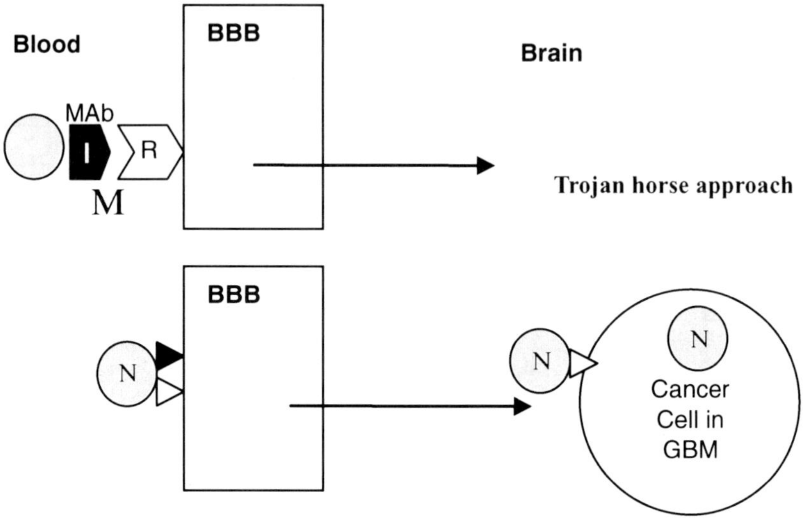

Fig. 9.2 A concept of targeted drug delivery to GBM across the BBB (Nanoparticle (N) combined with a monoclonal antibody (MAb) for receptor (R) crosses the blood brain barrier (BBB) into brain by Trojan horse approach. N with a ligand targeting BBB ▶ traverses the BBB by receptor-mediated transcytosis. Ligand ▷ docks on a cancer cell receptor and N delivers anticancer payload to the cancer cell in glioblastoma multiforme (GBM). © Jain PharmaBiotech)

demonstrated that BCNU-loaded PLA nanoparticles show certain sustained-release characteristics. The biodistribution of transferrin-coated nanoparticles, investigated by 99Tc-labeled SPECT, showed that the surface-containing transferrin PLA nanoparticles were concentrated in the brain and no radioactive foci could be found outside the brain. Inhibition of tumor growth in the C6 tumor-bearing animal model showed that BCNU-loaded PLA NPs had stronger cytotoxicity and prolonged the average survival time of rats. In contrast to the BCNU wafer approach, the stereotactic method of delivery used in this study may be useful in the development of a new method for delivery of chemotherapy to malignant brain tumors.

NP-Based Targeted Delivery of Chemotherapy Across the BBB

Some of techniques used for facilitating transport of therapeutic substances across the BBB involve damage to the BBB, which is not desirable. Technologies based on nanoparticles targeted delivery of anticancer drugs across the BBB. A concept of targeted drug delivery to GBM across the BBB is shown in Fig. 9.2.

Nanoparticles as Nonviral Vectors for CNS Gene Therapy

Viral vectors for gene delivery to neuronal cells can achieve high transfection efficiency, but problems, such as host immune responses and safety concerns, currently restrict their use in humans. Nonviral nanoparticles represent a good alternative to viral vectors, but transfection efficiency has to be increased to reach levels that would be relevant for therapeutic purposes.

Silica Nanoparticles for CNS Gene Therapy

Use of organically modified silica (ORMOSIL) nanoparticles ($\approx$30 nm) has been reported as a nonviral vector for efficient in vivo gene delivery without toxic effects and with efficacy equaling or exceeding that obtained in studies using a viral vector (Bharali et al. 2005). Highly monodispersed, stable aqueous suspensions of nanoparticles, surface-functionalized with amino groups for binding of DNA, were prepared and characterized. Intraventricular and intracerebral stereotaxic injections of nanoparticles, complexed with plasmid DNA encoding for EGFP (enhanced green fluorescent protein), were made into the mouse brain. Use of an optical fiber in vivo imaging technique enabled observation of the brain cells expressing genes without having to sacrifice the animal. The ORMOSIL-mediated transfections also were used to manipulate the biology of the neural stem/progenitor cells in vivo. Transfection of a plasmid expressing the nucleus-targeting fibroblast growth factor receptor type 1 resulted in significant inhibition of the in vivo incorporation of bromodeoxyuridine into the DNA of the cells in the subventricular zone and the adjacent rostral migratory stream. Targeted dopamine neurons, which degenerate in Parkinson's disease, take up and express a fluorescent marker gene, demonstrating the ability of nanoparticle technology to effectively deliver genes to specific types of cells in the brain. The gene–nanoparticle complexes were shown to activate adult brain stem/progenitor cells in vivo, which could be effective replacements for those destroyed by neurodegenerative diseases. Thus, ORMOSIL nanoparticles have a potential for effective therapeutic manipulation of the neural stem/progenitor cells as well as in vivo targeted brain therapy. The structure and composition of ORMOSIL allow for the development of an extensive library of tailored nanoparticles to target gene therapies for different tissues and cell types.

Cationic Lipids for CNS Gene Therapy

Cationic lipids show very low transfection efficiency in neurons. They are useful for single-cell studies but not for lack-of-function studies. Generally, cationic lipids are toxic to neurons, although different lipidic formulations can decrease toxicity. They are not effective for gene delivery to the brain when administered intravenously, although "Trojan horse" liposomes can be an exception.

Polyethylenimine-Based Nanoparticles for CNS Gene Therapy

Polyethylenimine (PEI) nanoparticles have higher transfection efficiency (20 %) than cationic lipids in neurons, but this is still very low for therapeutic purposes. At least 70–80 % transfection efficacy is required for removal of a protein.

These nanoparticles are toxic for neurons, and modifications of the molecule, such as PEGylation, are required to decrease neurotoxicity. Moreover, PEI-based nanoparticles are only effective for gene delivery when injected locally, which precludes their development for clinical use at present.

Dendrimers for CNS Gene Therapy

Dendrimers are capable of very efficient neuronal transfection in vitro (transfection efficiencies of 75 % have been achieved) with low toxicity when external amino groups are masked by surface functionalization. Further developments need to be carried out to enable efficient BBB crossing, in order to deliver genetic material to neurons and glial cells. Dendrimers are the most promising particles for genetic material delivery to the CNS either alone or in combination with carbon-based nanoparticles (nanotubes and nanohorns).

Carbon Nanotubes for CNS Gene Therapy

CNTs avoid endosomes and, once functionalized, their solubility is increased to make them biocompatible and capable of delivery of genetic material to different cells. Coupled to dendrimers, CNTs represent a new concept that can play a relevant role in gene therapy in the nervous system, if toxicological issues are solved (Posadas et al. 2010). Once the safety has been established, CNT-based vectors should be able to perform an "enhanced" gene transfer in target cells. Potential applications include cerebral ischemia and Rett syndrome.

Nanoparticle-Based Drug Delivery to the Inner Ear

Drug delivery to the inner ear is important for the treatment of inner ear disorders such as those involving hearing. Another disorder, tinnitus, is a problem in management and several innovative approaches are under investigation. An obstacle to effective treatment of inner ear diseases is the atraumatic delivery of therapeutics into inner ear perilymph. It is feasible to use SPIONs as drug delivery vehicles. As a minimally invasive approach, intratympanic delivery of multifunctional nanoparticles (MFNPs) carrying genes or drugs to the inner ear is a future therapy for treating inner ear diseases, including sensorineural hearing loss (SNHL) and Meniere's disease. Liposome nanoparticles encapsulating gadolinium-tetra-azacyclo-dodecane-tetra-acetic acid (LPS + Gd-DOTA) are visible by MRI in the inner ear in vivo after either intratympanic or intracochlear administration demonstrating transport from the middle ear to the inner ear and their distribution in the inner ear (Zou et al. 2010). Passive diffusion of fluorescent NPs through the round window membrane (RWM) within the freshly frozen human temporal bone has been demonstrated, and these NPs were subsequently found to be distributed in the sensory hair cells, nerve fibers, and to other cells of the cochlea (Roy et al. 2012). Nontoxic NPs have a great potential for controlled drug delivery to the human inner ear across the RWM.

Nanotechnology-Based Devices and Implants for CNS

Nanoparticle-mediated drug delivery to the brain, as described in previous sections, will minimize the need for use of invasive delivery devices, but there will still be need for implants and direct delivery of drugs to the brain and the cerebral ventricles. Nanomaterials, because of their action in preventing the formation of scar due to astrocyte proliferation, would improve the construction of nonreactive cerebroventricular catheters for administration of drugs into the cerebral ventricles. Nanoengineered probes can deliver drugs at the cellular level using nanofluidic channels.

Nanobiotechnology and Neuroprotection

Nanoparticles can improve drug delivery to the CNS and facilitate crossing of BBB and more precisely target a CNS injury site. These technologies were described in Chap. 6 and the topic of neuroprotection is dealt with in detail in a handbook on this topic (Jain 2011). Some nanoparticles have a neuroprotective effect.

QD technology has been used to gather information about how the CNS environment becomes inhospitable to neuronal regeneration following injury or degenerative events by studying the process of reactive gliosis. Other research is looking at how QDs might spur growth of neurites by adding bioactive molecules to the QDs, in a way to provide a medium that will encourage this growth in a directed way. PLGA nanoparticles loaded with superoxide dismutase have neuroprotective effect seen up to 6 h after H_2O_2-induced oxidative stress, which appears to be due to the stability of the encapsulated enzyme and its better neuronal uptake after encapsulation (Reddy et al. 2008).

Gold salts, known to have an immunosuppressive effect, have been considered for treatment of TBI, which results in loss of neurons caused not only by the initial injury but also by the resulting neuroinflammation as a secondary effect. The systemic use of gold salts is limited by nephrotoxicity. However, implants of pure metallic gold release gold ions, which do not spread in the body but are taken up by cells near the implant. This is a safer method of using to reduce local neuroinflammation. Release or dissolucytosis of gold ions from metallic gold surfaces requires the presence of disolycytes, i.e., macrophages, and the process is limited by their number and activity. In one study, the investigators injected 20–45-μm gold particles into the neocortex of mice before generating a cryo-injury (Larsen et al. 2008). Comparison of gold-treated and untreated cryolesions showed that the release of gold reduced microgliosis and neuronal apoptosis accompanied by a transient astrogliosis and an increased neural stem cell response indicating anti-inflammatory and neuroprotective effect. Intracerebral application of metallic gold as a pharmaceutical source of gold ions bypasses the BBB and enables direct drug delivery to inflamed brain tissue. The method of delivery is invasive and a gold implant could produce foreign body reaction leading to an epileptic focus. This can be refined by the use of gold nanoparticles.

Cadmium telluride (CdTe) nanoparticles (NPs) can efficiently prevent amyloid beta (Aβ) fibril formation, a pathological feature of Alzheimer's disease (AD), based on the multiple binding of Aβ oligomers to CdTe NPs (Yoo et al. 2011). By introducing tetrahedral CdTe NPs that were comparable in size with growing fibrils, the researchers discovered that the Aβ plaque readily bonded to them and CdTe NP geometry was strongly distorted resulting in complete inhibition of further growth of Aβ fibrils. CdTe NPs can inhibit the Aβ fibril formation in minute quantities with much greater efficiency; 1 CdTe NP can capture more than 100 amyloid peptides. This high efficiency of CdTe NPs is similar to some proteins that the human body uses to prevent formation of Aβ fibrils and protect itself against the progression of AD. These findings provide new opportunities for the development of drugs to prevent AD.

Nanobiotechnology for Regeneration and Repair of the CNS

Nanobiotechnology applications, aimed at the regeneration and neuroprotection of the CNS, will significantly benefit from basic nanotechnology research conducted in parallel with advances in cell biology, neurophysiology, and neuropathology. The aim is to help neuroscientists better understand the physiology of and develop treatments for disorders such as traumatic brain injury (TBI), spinal cord injury (SCI), degenerative retinal disorders, and neurodegenerative disorders such as Alzheimer's disease (AD) and Parkinson's disease (PD).

Nanowire Neuroprosthetics with Functional Membrane Proteins

Living organisms use a sophisticated arsenal of membrane receptors, channels, and pumps to control signal transduction to a degree that is unmatched by man-made devices. Electronic circuits that use such biological components could achieve drastically increased functionality; however, this approach requires nearly seamless integration of biological and man-made structures. A versatile hybrid platform for such integration has been constructed that uses shielded nanowires coated with a continuous lipid bilayer (Misra et al. 2009). When shielded silicon nanowire transistors incorporate transmembrane peptide pores gramicidin A and alamethicin in the lipid bilayer, they can achieve ionic to electronic signal transduction by using voltage-gated or chemically gated ion transport through the membrane pores. The membrane pore could be opened and closed by changing the gate voltage of the device to enable monitor specific transport and also to control the membrane protein. The work shows promise for enhancing biosensing and diagnostics tools, and neural prosthetics such as cochlear implants.

Nanotube–Neuron Electronic Interface

Thin films of CNTs deposited on transparent plastic can also serve as a surface on which cells can grow, and these nanotube films could potentially serve as an electrical interface between living tissue and prosthetic devices or biomedical instruments. University of Texas Medical Branch's scientists have shown that there is some kind of electrical communication between these two things, by stimulating cells through the transparent conductive layer (Liopo et al. 2006). The scientists employed two different types of cells in their experiments, neuroblastoma cells commonly used in test-tube experiments and neurons cultured from experimental rats. Both cell types were placed on ten-layer-thick "mats" of SWCNTs deposited on transparent plastic. This enabled the researchers to use a microscope to position a tiny electrode next to individual cells and record their responses to electrical pulses transmitted through the SWCNTs. In addition to their electrical stimulation experiments, the scientists also studied how different kinds of SWCNTs affected the growth and development of neuroblastoma cells. They compared cells placed on mats made of "functionalized" SWCNTs, carbon nanotubes with additional molecules attached to their surfaces that may be used to guide cell growth or customize nanotube electrical properties, to cells cultured on unmodified "native" carbon nanotubes and conventional tissue culture plastic. Native CNTs supported neuron attachment and growth well better than the two types of functionalized nanotubes tested. Next step in the research is to find a way to functionalize the nanotubes to make neuron attachment and communication better and make these surfaces more biocompatible. If nanotubes turn out to be sensitive enough to record ongoing electrical activity in cells, they could form the basis of a device that can both sense and deliver stimuli to cells for prosthetic control.

Role of Nanobiotechnology in Regeneration and Repair Following CNS Trauma

Repair and regeneration following CNS trauma requires a multifaceted approach (Jain 2012g). Role of nanobiotechnology in various strategies for regeneration and repair following CNS trauma are listed in Table 9.1.

Nanofibers as an Aid to CNS Regeneration by Neural Progenitor Cells

One approach to growing nerve cells in tissue cultures is to encapsulate neural progenitor cells in vitro within a 3D network of nanofibers formed by self-assembly of peptide amphiphile molecules. The self-assembly of nanofiber scaffold is initiated by mixing cell suspensions in media with dilute aqueous solutions of the molecules, and cells survive the growth of the nanofibers around them. These nanofibers are designed to present to cells the neurite-promoting laminin epitope. Relative to laminin

Table 9.1 Role of nanobiotechnology in regeneration and repair following CNS trauma

Neuroprotective nanoparticles to prevent further damage
Nanofibers to provide scaffolds for regeneration and reducing or eliminating scar formation
Nanofibers for providing cues to axons for regeneration
Nanoparticles for repair of injured neurons and nerve fibers by sealing them
Nanoparticles to track stem cells implanted to replace the loss and to promote growth of neural tissues
Nanoparticle-based delivery of drugs to promote growth of neural tissues

or soluble peptide, the artificial nanofiber scaffold induces very rapid differentiation of neural progenitor cells into neurons, while discouraging the development of astrocytes.

These new materials, because of their chemical structure, interact with cells of the CNS in ways that may help prevent the formation of scar due to astrocyte proliferation that is often linked to paralysis resulting from traumatic spinal cord injury (SCI). Silicon neural electrodes are engineered with a nanostructured form of silicon called porous silicon, which acts as a scaffold that reduces glial scarring from electrode implantation and enhances neural growth at the brain recording sites to create a superior interface with neurons. This would be useful in the procedure of electrode implantation in neurological disorders such as PD and epilepsy.

Peptide Nanostructures for Repair of the CNS

Peptide nanostructures containing bioactive signals offer novel therapies with potential impact on regenerative medicine. These nanostructures can be designed through self-assembly strategies and supramolecular chemistry and can combine bioactivity for multiple targets with biocompatibility. It is also possible to multiplex their functions by using them to deliver proteins, nucleic acids, drugs, and cells. Self-assembling peptide nanostructures can facilitate regeneration of the CNS. Other self-assembling oligopeptide technologies and the progress made with these materials toward the development of potential therapies have been reviewed elsewhere (Webber et al. 2010).

Nanobiotechnology for Repair and Regeneration Following TBI

Challenges of using a tissue engineering approach for regeneration in TBI include a complex environment and variables that are difficult to assess. For optimal benefit, the brain should be in a condition that minimizes immune response, inflammation, and rejection of the grafted material. Tissue engineering, using a bioactive scaffold, counters some of the hostile factors and facilitates integration of donor cells into the brain, but transplantation of a combination biologic construct to the brain has not yet been successfully translated into clinical use (Stabenfeldt et al. 2011).

The next generation of tissue engineering scaffolds for TBI may incorporate nanoscale surface feature dimensions, which mimic natural neural tissue. Nanomaterials can enhance desirable neural cell activity while minimizing unwanted astrocyte reactivity. Composite materials with zinc oxide nanoparticles embedded into a polymer matrix can provide an electrical stimulus when mechanically deformed through ultrasound, which can act as a cue for neural tissue regeneration (Seil and Webster 2010).

Nanoparticles for Repair Following SCI

SCI can lead to serious neurological disability, and the most serious form of it is paraplegia or quadriplegia. Currently, over 250,000 persons in the USA and several millions worldwide are living with permanent disability due to chronic SCI. There are approximately 12,000 new cases of acute SCI in the USA each year. Over 90 % of acute SCI victims now survive their injuries and go on to become part of the chronic SCI population, living paralyzed for an average of more than 40 years after injury.

Local spinal cord lesions are often greatly enlarged by secondary damage, which is accompanied by additional massive cell death that involves neurons, microglia, and macroglia and is virtually complete at 12 h. Immediate care involves stabilization of the patient's general condition by supportive measures. Surgery is carried out in some cases for removal of compressing lesions and stabilization of spinal fractures. A number of neuroprotective strategies are under investigation. Stem cell therapies are also under investigation for neuroregeneration, and nanoparticles can be used to track the course of stem cells. There is no therapeutic measure available currently that enhances functional recovery significantly.

Nanomaterials injected into the severed spinal cords of mice enable them to walk again after several weeks of therapy. The nanomaterials used in these studies were designed to self-assemble into nanofibers, which provide the framework for regeneration of nerve fibers. In a nanofiber network, progenitor cells develop into neurons and not astrocytes that form scar tissue and hinder regeneration. The research offers new insights into the near-term research potential of nanotechnology and offers hope for patients with severe neuron damage due to other causes as well.

Repair of SCI by Nanoscale Micelles

Another key approach for repairing injured spinal cord is to seal the damaged membranes at an early stage. Axonal membranes injured by compression can be effectively repaired using self-assembled monomethoxy PEG–PLA di-block copolymer micelles (Shi et al. 2010). The micelles might be used instead of conventional PEG. A critical feature of micelles is that they combine two types of polymers, one being hydrophobic and the other hydrophilic, meaning they are either unable or able to mix with water. Because of the nanoscale size and the PEG shell of the micelles, they are not quickly filtered by the kidney or captured by the liver, enabling them to remain in the bloodstream long enough to circulate to damaged tissues.

Injured spinal tissue incubated with micelles (60 nm diameter) showed rapid restoration of compound action potential and reduced calcium influx into axons for micelle concentrations much lower than the concentrations of PEG, approximately 1/100,000, for early-stage SCI. Intravenously injected micelles effectively recovered locomotor function and reduced the volume and inflammatory response of the lesion in injured rats, without any adverse effects. These results show that copolymer micelles can interrupt the spread of primary SCI damage with minimal toxicity. The research also showed that without the micelles treatment, about 18 % of axons recover in a segment of damaged spinal cord tested, whereas the micelles treatment boosted the axon recovery to about 60 %. The researchers used the chamber to study how well micelles repaired damaged nerve cells by measuring the "compound action potential," or the ability of a spinal cord to transmit signals.

The experiment mimics what happens during a traumatic SCI. Findings showed that micelles might be used to repair axon membranes damaged by compression injuries, a common type of spine injury. Dyed micelles were also tracked in rats, demonstrating that the nanoparticles were successfully delivered to injury sites. Findings also showed micelles-treated animals recovered the coordinated control of all four limbs, whereas animals treated with conventional PEG did not. Further research will include work to learn about the specific mechanisms that enable the micelles to restore function to damaged nerve cells.

Nanobiotechnology-Based Devices for Restoration of Neural Function

The remarkable optical and electrical properties of nanostructured materials are now considered to be a source for a variety of biomaterials, biosensing, and cell interface applications. Some of the characteristics of nanoparticles can be exploited to custom-build new materials from the bottom up with characteristics such as compatibility with living cells and the ability to turn light into tiny electrical currents that can produce responses in nerves. A study reports construction of a hybrid bionanodevice where absorption of light by thin films of quantum-confined semiconductor nanoparticles of HgTe produced by the layer-by-layer assembly stimulate adherent neural cells via a sequence of photochemical and charge-transfer reactions (Pappas et al. 2007). The development opens the door to applying the unique properties of nanoparticles to a wide variety of light-stimulated nerve-signaling devices including the possible development of a nanoparticle-based artificial retina.

Nanobiotechnology-Based Artificial Retina

Although light signals have previously been transmitted to nerve cells using silicon (whose ability to turn light into electricity is employed in solar cells and in the imaging sensors of video cameras), nanoengineered materials promise far greater efficiency and versatility. It should be possible to tune the electrical characteristics of these nanoparticle films to get properties like color sensitivity and differential stimulation, which are needed for an artificial retina. Creation of an actual implantable artificial retina is, however, a long-range project. But, a variety of less complex applications are enabled by a tiny, versatile light-activated interface with nerve cells, e.g., ways to connect with artificial limbs and new tools for imaging, diagnosis, and therapy. The main advantage of this technology is that remote activation by light is possible without cumbersome wire connections. This type of technology can provide noninvasive connections between the human nervous system and prostheses that are flexible, compact, and reliable. Such tools will provide nanoneurology new capabilities that were not possible with conventional methods.

Nanoneurosurgery

Neurosurgery is an extension of neurology involving surgery, nanodiagnostics, and application of new technologies for treatment of neurological disorders. Advances in nanobiotechnology have already refined many surgical approaches to diseases of the nervous system, and this new field can be called nanoneurosurgery. Examples are applications in brain cancer, neuroregeneration, and CNS implants.

Femtolaser Neurosurgery

Understanding how nerves regenerate is an important step toward developing treatments for human neurological disease, but investigation has so far been limited to complex organisms (mouse and zebrafish) in the absence of precision techniques for severing axons (axotomy). Femtosecond laser surgery has been used for axotomy in the roundworm *Caenorhabditis elegans*, and these axons functionally regenerated after the operation (Yanik et al. 2004). Femtolaser acts like a pair of tiny "nano-scissors," which is able to cut nanosized structures like nerve axons

The pulse has a very short length making the photons in the laser concentrate in one area, delivering a lot of power to a tiny, specific volume without damaging surrounding tissue. Once cut, the axons vaporize and no other tissue is harmed. The researchers cut axons they knew would impair the worms' backward motion. The worms could not move backward after surgery. But within 24 h, most of the severed axons regenerated and the worms recovered backward movement, confirming that laser's cut did not damage surrounding tissue and allowed the neurons to grow a new axon to reach the muscle. Application of this precise surgical technique should enable nerve regeneration to be studied in vivo.

Nanofiber Brain Implants

Several brain probes and implants are used in neurosurgery. Examples are those for the management of epilepsy, movement disorders, and pain. Many of these implants are still investigational. The ideal inert material for such implants has not yet been discovered. Silicon probes are commonly used for recording of electrical impulses and for brain stimulation. The body generally regards these materials as foreign, and the probes get encapsulated with glial scar tissue, which prevents them from making good contact with the brain tissue.

An in vitro study was done to determine cytocompatibility properties of formulations containing carbon nanofibers pertinent to neural implant applications (McKenzie et al. 2004). Substrates were prepared from four different types of carbon fibers, two with nanoscale diameters (nanophase, or less than or equal to 100 nm) and two with conventional diameters (or greater than 100 nm). Within these two categories, both a high and a low surface energy fiber were investigated and tested. Astrocytes (glial scar tissue-forming cells) were seeded onto the substrates for adhesion, proliferation, and long-term function studies (such as total intracellular protein and alkaline phosphatase activity). Results provided the first evidence that astrocytes preferentially adhered and proliferated on carbon fibers that had the largest diameter and the lowest surface energy. Formulations containing carbon fibers in the nanometer regime limited astrocyte functions leading to decreased glial scar tissue formation. Positive interactions with neurons and, at the same time, limited astrocyte functions leading to decreased gliotic scar tissue formation are essential for increased neuronal implant efficacy. Nanotubes, because of the interesting electronic properties and reduction in scar formation, hold great promise for replacing conventional silicone implants.

Nanoparticles as an Aid to Neurosurgery

A research team from Oregon Health and Science University (Portland OR) has shown that an iron oxide nanoparticle can outline not only brain tumors under MRI but also other lesions in the brain that may otherwise have gone unnoticed (Neuwelt et al. 2004). Ferumoxtran-10 (Combidex®, AMAG Pharmaceuticals Inc.), a dextran-coated iron oxide nanoparticle, provides enhancement of intracranial tumors by MRI for more than 24 h and can be imaged histologically by iron staining. Each iron oxide nanoparticle is the size of a small virus and is much smaller than a bacterium but much larger than an atom or standard gadolinium contrast molecule. It is an iron oxide crystal surrounded with a carbohydrate or "sugar" coating called dextran, which gives the particle a longer plasma half-life, allowing it to slowly slip through the BBB. Ferumoxtran-10 can also provide a "stable imaging marker" during surgery to remove brain tumors, and it remains in the brain long enough for postoperative MRI, even after surgical manipulation. These findings have the potential to assist image-guided brain surgery and improve diagnosis of lesions caused by multiple sclerosis, stroke, and other neurological disorders, in addition to residual tumors. Because ferumoxtran-10 can stay in brain lesions for days, it can be administered to patients 24 h before surgery and can image other, noncancerous lesions. It has some advantages over gadolinium, a metal used as an MRI contrast agent for 20 years and which must be administered just before surgery. However, it will complement gadolinium but not replace it. Ferumoxtran-10 gives additional information that cannot be obtained in some patients with gadolinium. Using both the contrast agents, one can get better diagnostic information that has the potential to improve the patient's outcome. In addition, ferumoxtran-10 can be detected with an iron stain in the tissue removed by biopsy or surgery, allowing physicians to see it in brain tissue samples under a microscope. Unlike any other MRI contrast agent, ferumoxtran-10 enables the comparison of images from an MRI scan with the tissue taken out at surgery. Moreover, it is relatively safe when diluted and administered as an infusion.

Nanoscaffold for CNS Repair

There are several barriers that must be overcome to achieve axonal regeneration after injury in the CNS: (1) scar tissue formation, (2) gaps in nervous tissue formed during phagocytosis of dying cells after injury, (3) factors that inhibit axon growth in the mature mammalian CNS, and (4) failure of many adult neurons to initiate axonal extension.

Using the mammalian visual system as a model, a self-assembling peptide nanofiber scaffold was designed, which creates a permissive environment for axons not only to regenerate through the site of an acute injury but also to knit the brain tissue together. In experiments using a severed optic tract in the hamster, it was shown that regenerated axons reconnect to target tissues with sufficient density to promote functional return of vision, as evidenced by visually elicited orienting behavior (Ellis-Behnke et al. 2006). The peptide nanofiber scaffold not only represents a previously undiscovered nanobiomedical technology for tissue repair and restoration but also raises the possibility of effective treatment of CNS and other tissue or organ trauma. This peptide nanofiber scaffold has several advantages over currently available polymer biomaterials: (1) it forms a network of nanofibers that are similar in scale to the native extracellular matrix and therefore provides an "in vivo" environment for cell growth, migration, and differentiation; (2) it can be broken down into natural L-amino acids and metabolized by the surrounding tissue; (3) it is synthetic and free of chemical and biological contaminants that may be present in animal-derived biomaterials such as collagens; and (4) it appears to be immunologically inert, thus avoiding the problem of neural tissue rejection.

Electrospun Nanofiber Tubes for Regeneration of Peripheral Nerves

Several neural prostheses have been used to replace the loss of nervous tissue in peripheral nerve injuries by providing a path for regenerating nerve fibers. Most of these use rigid channel guides that may cause cell loss due to the lack of physiological local stresses exerted over the nervous tissue during the patient's movement. The electrospinning technique makes it possible to spin nanofiber flexible tubular scaffolds, with high porosity and surface/volume ratio. Electrospun tubes made of biodegradable polymers (a blend of PLGA/PCL) have been used to regenerate a 10-mm nerve gap in a rat sciatic nerve (Panseri et al. 2008). In most of the treated animals, the electrospun tubes induced neural regeneration and functional reconnection of the two severed sciatic nerve tracts. Myelination occurred and no significant inflammatory responses were observed. Reestablishment of functional neuronal connections with reinnervation of the affected muscles was demonstrated by neural tracers and evoked potential recordings. These findings show that electrospun tubes, with additional biological coating or incorporated drugs, are promising scaffolds for functional neural regeneration. They can be knitted in meshes, and their mechanical properties can be tuned to provide biomimetic functionalization. Moreover, the conduits can be loaded with neurotrophic factors and seeded with stem cells.

Buckyballs for Brain Cancer

Buckyballs (fullerenes) are under investigation to improve the ability of MRIs to locate brain tumors and deliver a payload of radiation to destroy them. Experiments on rats have shown that buckyballs packed with the MRI contrast metal gadolinium can increase the sensitivity of MRI detection by at least 40-fold. This level of precision is reaching a point at which cancer cells that have spread beyond the margins of the tumor may become visible. Stray cells, left behind after surgery, are thought to be responsible for tumor relapse. Finding and removing these cells could improve a patient's chance of survival. The scientists have created a modified version of the buckyballs with a fluorescent metal atom called terbium, which could guide surgeons to remove tumors with greater precision. Addition of yet another metal, lutetium, would deliver a lethal dose of radiation to the cancer cells, including those missed by the surgeon. The research is a few years away from testing in humans, but the potential is promising.

Application of Nanobiotechnology to Pain Therapeutics

Nanotechnology offers the potential to address multiple, major unmet problems in the diagnosis, treatment, and symptom management of a large variety of diseases and conditions, including cancer. Nanobiotechnology will contribute to improvement of cancer pain therapeutics through facilitation of drug discovery for pain. A more immediate application is in facilitating drug delivery for pain. A transbuccal transmucosal system, Buccal Patch®, has been developed for the administration of remifentanil for the management of breakthrough cancer pain (Sprintz et al. 2005). The nanochannel size of the device permits the diffusion of the drug from its reservoir to the target tissue at a consistent and controlled rate, minimizing the risk of overdosing the patient. Intravenous administration of ibuprofen in lipid nanocapsules formulation has an advantage as an analgesic over oral preparations.

US Army is supporting research to develop nanoparticle-based analgesics that can be injected with a pen-like device by injured soldiers' comrades, or even injured soldiers themselves, on the battlefield. The method will use analgesic drugs coupled to polymers but can be released to provide adequate pain relief as well as antidotes to avoid adverse effects of these drugs. For example, morphine, an analgesic commonly used to treat wounded soldiers, needs to be injected by skilled medical personnel. Patients who receive morphine need to be monitored carefully because the painkiller can cause breathing problems. These requirements restrict the use of morphine on the battlefield. If successful, the nanotechnology approach could markedly improve the treatment of soldiers in the field. Various types of nanoparticles will be designed and tested. The aim is to create nanoparticles that can achieve the following objectives:

- Control the release of morphine over extended periods to ensure pain relief until a soldier can be evacuated to a military acute care facility
- Continuously monitor the soldier's breathing and, if needed, release the drug naloxone, which counters morphine's effects on breathing

Chapter 10
Nanocardiology

Introduction

Nanocardiology is the application of nanobiotechnology to cardiovascular diseases. Recent rapid advances in nanobiotechnology offer a wealth of new opportunities for diagnosis and therapy of cardiovascular diseases (Jain 2011a). As far back as 2003, the National Heart, Lung, and Blood Institute, USA, convened a working group on nanotechnology for translational applications to heart, lung, blood disorders, and cardiovascular complications of sleep apnea to solve clinical problems.

Nanotechnology-Based Cardiovascular Diagnosis

Nanobiotechnology has refined molecular diagnosis, and this applies to detection of cardiovascular diseases also. Availability of genotyping and detection of single nucleotide polymorphisms (SNPs) will provide information on risks of developing genetically linked cardiovascular diseases. Application of nanodiagnostics in pharmacogenetics will be used for selection and guidance of appropriate therapy for an individual patient. This will facilitate the development of personalized medicine.

Biomarkers play an important role in diagnosis of cardiovascular disorders, particularly myocardial infarction (Jain 2010). Detection of biomarkers, particularly using proteomic technologies, has also been refined by nanobiotechnology.

Detection of Biomarkers of Myocardial Infarction in Saliva by a Nanobiochip

The feasibility and utility of saliva as an alternative diagnostic fluid for identifying biomarkers of acute myocardial infarction (AMI) has been investigated. A lab-on-a-chip method was used to assay 21 proteins in serum and unstimulated

K.K. Jain, *The Handbook of Nanomedicine*, DOI 10.1007/978-1-61779-983-9_10,

whole saliva procured from AMI patients within 48 h of chest pain onset and from apparently healthy controls (Floriano et al. 2009). Both established and novel cardiac biomarkers demonstrated significant differences in concentrations between patients with AMI and controls. The saliva-based biomarker panel of C-reactive protein (CRP), myoglobin, and myeloperoxidase showed diagnostic capability, which was better than that of ECG alone. When used in conjunction with ECG, screening capacity for AMI was enhanced and was comparable to that of a panel of brain natriuretic peptide, troponin I, creatine kinase-MB, and myoglobin. To translate these findings into clinical practice, the whole saliva tests were adapted to a nanobiochip platform, which may provide a convenient and rapid screening method for cardiac events at point-of-care.

Nanobiosensors for Detection of Cardiovascular Disorders

Nanobiosensors can be electronically gated to respond to the binding of a single molecule. Prototype sensors have demonstrated detection of nucleic acids, proteins, and ions. These sensors can operate in the liquid or gas phase, opening up an enormous variety of downstream applications. The detection schemes use inexpensive low voltage measurement schemes and detect binding events directly so there is no need for costly, complicated and time-consuming labeling chemistries such as fluorescent dyes or the use of bulky and expensive optical detection systems. As a result, these sensors are inexpensive to manufacture and portable. It may even be possible to develop implantable detection and monitoring devices for cardiovascular disorders based on these detectors.

Use of Magnetic NPs as MRI Contrast Agents for Cardiac Imaging

Magnetic nanoparticles (MNPs) have been used as contrast agent for MRI and have refined molecular imaging. Targeted imaging of vascular inflammation or thrombosis may enable improved risk assessment of atherosclerosis by detecting plaques at high risk of acute complications (Saraste et al. 2009). Cell death in the heart can be imaged in vivo by using annexin-labeled MNPs, particularly AnxCLIO-Cy5.5 (Chen et al. 2011). Experimental studies have shown the feasibility of combination of diagnosis and therapy using MNPs. In a study on mice, MNPs conjugated with plasmid DNA expressing enhanced green fluorescent protein and coated with chitosan were injected into tail vein and directed to the heart by means of an external magnet without the need to functionalize the NPs, and their location was confirmed by fluorescent imaging (Kumar et al. 2010). This approach requires further investigations before clinical applications can be considered.

Perfluorocarbon NPs for Combining Diagnosis with Therapy in Cardiology

Perfluorocarbon (PFC) nanoparticles provide an opportunity for combining molecular imaging and local drug delivery in cardiovascular disorders. Ligands such as MAbs and peptides can be cross-linked to the outer surface of PFCs to enable active targeting to biomarkers expressed within the vasculature. PFC nanoparticles are naturally constrained by size to the circulation, which minimizes unintended binding to extravascular, nontarget tissues expressing similar epitopes. Moreover, their prolonged circulatory half-life of approximately 5 h allows saturation of receptors without addition of PEG or lipid surfactant polymerization. The utility of targeted PFC nanoparticles has been demonstrated for a variety of applications in animal models and phantoms, including the diagnosis of ruptured plaque, the quantification and antiangiogenic treatment of atherosclerotic plaque, and the localization and delivery of antirestenotic therapy following angioplasty (Lanza et al. 2006).

Cardiac Monitoring in Sleep Apnea

Because sleep apnea is a cause of irregular heartbeat, hypertension, heart attack, and stroke, it is important that patients be diagnosed and treated before these highly deleterious sequelae occur. For patients suspected of experiencing sleep apnea, in vivo sensors could constantly monitor blood concentrations of oxygen and cardiac function to detect problems during sleep. In addition, cardio-specific antibodies tagged with nanoparticles may allow physicians to visualize heart movement while a patient experiences sleep apnea to determine both short- and long-term effects of apnea on cardiac function.

Detection and Treatment of Atherosclerotic Plaques in the Arteries

A key feature of the atherosclerotic process is the angiogenic expansion of the vasa vasorum in the adventitia, which extends into the thickening intimal layer of the atheroma in concert with other neovessels originating from the primary arterial lumen. Magnetic resonance molecular imaging of focal angiogenesis with integrin-targeted paramagnetic contrast agents has been reported with PFC nanoparticles and liposomes. Site-targeted PFC nanoparticles also offer the opportunity for local drug delivery in combination with molecular imaging.

The diagnosis and treatment of unstable plaque is an area in which nanotechnology could have an immediate impact. Fibrin-specific PFC nanoparticles may allow the detection and quantification of unstable plaque in susceptible patients, which may be an important feature of future strategies to prevent heart attacks or stroke.

Research is under way using probes targeted to plaque components for noninvasive detection of patients at risk. In an extension of this approach, targeted nanoparticles, multifunctional macromolecules, or nanotechnology-based devices could deliver therapy to a specific site, localized drug release being achieved either passively (by proximity alone) or actively (through supply of energy as ultrasound, near-infrared, or magnetic field). Targeted nanoparticles or devices could also stabilize vulnerable plaque by removing material, e.g., oxidized low-density lipoproteins. Devices able to attach to unstable plaques and warn patients and emergency medical services of plaque rupture would facilitate timely medical intervention.

Monitoring for Disorders of Blood Coagulation

Patients would benefit greatly from nanotechnology devices that could monitor the body for the onset of thrombotic or hemorrhagic events. Multifunctional devices could detect events, transmit real-time biologic data externally, and deliver anticoagulants or clotting factors to buy critical time.

A gold nanoparticle-based simple assay has been described that enables the visual detection of a protease (Guarise et al. 2006). The method takes advantage of the high molar absorptivity of the plasmon band of gold colloids and is based on the color change of their solution when treated with dithiols. Contrary to the native ones, cleaved peptides are unable to induce nanoparticles aggregation; hence, the color of the solution does not change. The assay was used to detect two proteases: thrombin (involved in blood coagulation and thrombosis) and lethal factor (an enzyme component of the toxin produced by *Bacillus anthracis*). The sensitivity of this nanoparticle-based assay is in the low nanomolar range.

Controlled Delivery of Nanoparticles to Injured Vasculature

Optimal size of nanoparticles designed for systemic delivery is approximately 50–150 nm, but this size range confers a high surface area-to-volume ratio, which results in fast diffusive drug release. Spatial control has been achieved by biopanning a phage library to discover materials that target abundant vascular antigens exposed in disease (Chan et al. 2010). Temporal control is achieved by designing 60-nm hybrid nanoparticles with a lipid shell interface surrounding a polymer core, which is loaded with slow-eluting conjugates of paclitaxel for controlled ester hydrolysis and drug release over approximately 12 days. The nanoparticles inhibit human aortic smooth muscle cell proliferation in vitro and showed greater in vivo vascular retention during percutaneous angioplasty as compared to nontargeted controls. This nanoparticle technology may potentially be used toward the treatment of injured vasculature.

IGF-1 Delivery by Nanofibers to Improve Cell Therapy for Myocardial Infarction

Strategies for cardiac repair include injection of cells, but these approaches have been hampered by poor cell engraftment, survival, and differentiation. To address these shortcomings for the purpose of improving cardiac function after injury, a self-assembling peptide nanofiber was designed for prolonged delivery of insulin-like growth factor 1 (IGF-1), a cardiomyocyte growth and differentiation factor, to the myocardium, using a "biotin sandwich" approach (Davis et al. 2006). Biotinylated IGF-1 was complexed with streptavidin and then bound to biotinylated self-assembling peptides. This biotin sandwich strategy enabled binding of IGF-1 but did not prevent self-assembly of the peptides into nanofibers within the myocardium. IGF-1 that was bound to peptide nanofibers activated Akt, decreased activation of caspase-3, and increased expression of cardiac troponin I in cardiomyocytes. In studies on rats, cell therapy with IGF-1 delivery by biotinylated nanofibers improved systolic function after experimental myocardial infarction. This nanobiotechnology approach has the potential to improve the results of cell therapy for myocardial infarction, which is in clinical trials currently.

Injectable Peptide Nanofibers for Myocardial Ischemia

Endothelial cells can protect cardiomyocytes from injury through platelet-derived growth factor (PDGF)-BB signaling. PDGF-BB induces cardiomyocyte Akt phosphorylation in a time- and dose-dependent manner and prevents apoptosis via PI3K/Akt signaling. An experimental study in rats using injectable self-assembling peptide nanofibers, which bound PDGF-BB in vitro, demonstrated sustained delivery of PDGF-BB to the myocardium at the injected sites for 14 days (Hsieh et al. 2006). This blinded and randomized rat study showed that injecting nanofibers with PDGF-BB, but not nanofibers or PDGF-BB alone, decreased cardiomyocyte death and preserved systolic function after myocardial infarction. A separate blinded and randomized study showed that PDGF-BB delivered with nanofibers decreased infarct size after ischemia/reperfusion. PDGF-BB with nanofibers induced PDGFR-β and Akt phosphorylation in cardiomyocytes in vivo. These data demonstrate that PDGF-BB signaling and in vitro finding can be translated into an effective in vivo method of protecting myocardium after infarction. Furthermore, this study shows that injectable nanofibers allow precise and sustained delivery of proteins to the myocardium with potential therapeutic benefits.

Liposomal Nanodevices for Targeted Cardiovascular Drug Delivery

High-affinity ligand-receptor interactions have been exploited in the design and engineering of targeting systems that use a liposomal nanodevice for site-specific cardiovascular drug delivery. An example of application is atherothrombosis, a condition in which platelet activation/adhesion/aggregation is closely associated with vascular thrombotic events. Therefore, the majority of antithrombotic therapies have focused on drugs that impede platelet-activation pathways or block ligand-binding platelet integrins. In spite of reasonable clinical efficacy of these therapies, the magic bullet, a single drug and delivery system that selectively targets pathologically thrombotic environment without affecting hemostatic balance, remains elusive. The use of anti-integrin/anticoagulant/anti-inflammatory drugs in conjunction might be necessary to treat the multifactorial nature of pathological thrombogenesis. For this purpose, a nanoscale device that can carry such a combination selectively to a thrombotic site is being developed at the Department of Biomedical Engineering of Case Western Reserve University (Cleveland, OH). The liposomal nanodevice surface is modified by RGD (arginine-glycine-aspartic acid) motifs that specifically target and bind activated platelets by virtue of the high-affinity interaction between the RGD motif and the integrin GPIIb-IIIa expressed on active platelets, potentially acting as a thrombus-targeted vector. The ability of such liposomes to compete with native ligand fibrinogen in specifically binding activated platelets has been accomplished using both in vitro and in vivo approaches. The results demonstrate feasibility of using liposomes as platelet-targeted devices for delivery of cardiovascular therapeutics. By utilizing a library of synthetic peptide/peptidomimetic ligands having binding affinity toward specific receptors expressed in cardiovascular biology, it is possible to manipulate the liposome surface modification and hence dictate targeting specificity and affinity of the liposomal nanodevices.

Low-Molecular-Weight-Heparin-Loaded Polymeric Nanoparticles

Low-molecular-weight-heparin (LMWH) nanoparticles are available as potential oral heparin carriers. The nanoparticles are formulated using an ultrasound probe by water-in-oil-in-water emulsification and solvent evaporation with polymers. The mean diameter of LMWH-loaded nanoparticles ranges from 240 to 490 nm and is dependent on the reduced viscosity of the polymeric organic solution. The highest encapsulation efficiencies are observed when Eudragit polymers are used in the composition of the polymeric matrix. The in vitro biological activity of released LMWH, determined by the antifactor Xa activity with a chromogenic substrate, is preserved after the encapsulation process, making these nanoparticles good candidates for oral administration.

Nanoparticles for Cardiovascular Imaging and Targeted Drug Delivery

The potential dual use of nanoparticles for both imaging and site-targeted delivery of therapeutic agents to cardiovascular disease offers great promise for individualizing therapeutics. Image-based therapeutics with site-selective agents should enable verification that the drug is reaching the intended target and a molecular effect is occurring. Experimental studies have shown that binding of paclitaxel to smooth muscle cells in culture has no effect in altering the growth characteristics of the cells. If paclitaxel-loaded nanoparticles are applied to the cells, however, specific binding elicits a substantial reduction in smooth muscle cell proliferation, indicating that selective targeting may be a requirement for effective drug delivery for in this situation. Similar behavior has been demonstrated for doxorubicin containing particles. Intravenous delivery of fumagillin (an antiangiogenic agent)-loaded nanoparticles targeted to $\alpha v\beta 3$-integrin epitopes on vasa vasorum in growing plaques results in marked inhibition of plaque angiogenesis in cholesterol-fed rabbits. The unique mechanism of drug delivery for highly lipophilic agents such as paclitaxel contained within emulsions depends on close apposition between the nanoparticle carrier and the targeted cell membrane and has been described as "contact-facilitated drug delivery." In contrast to liposomal drug delivery (generally requiring endocytosis), the mechanism of drug transport in this case involves lipid exchange or lipid mixing between the emulsion vesicle and the targeted cell membrane, which depends on the extent and frequency of contact between two lipidic surfaces. The rate of lipid exchange and drug delivery can be greatly increased by the application of clinically safe levels of ultrasound energy that increase the propensity for fusion or enhanced contact between the nanoparticles and the targeted cell membrane.

The combination of targeted drug delivery and molecular imaging with MRI has the potential to enable serial characterization of the molecular epitope expression based on imaging readouts. Monitoring and confirmation of therapeutic efficacy of the therapeutic agents at the targeted site would facilitate personalized medical regimens.

Nanofiber-Based Scaffolds with Drug-Release Properties

Electrospinning is a versatile technique that enables the development of nanofiber-based scaffolds, from a variety of polymers that may have drug-release properties. Using nanofibers, it is now possible to produce biomimetic scaffolds that can mimic the extracellular matrix for tissue engineering (Ashammakhi et al. 2009). Nanofibers can guide cell growth along their direction. Combining factors like fiber diameter, alignment, and chemicals offers new ways to control tissue engineering. In vivo evaluation of nanomats included their degradation, tissue reactions, and engineering of specific tissues. New advances made in electrospinning, especially in drug delivery, support the massive potential of these nanobiomaterials. Nevertheless, there is already at least one product based on electrospun nanofibers with drug-release properties in a phase III clinical trial, for wound dressing. Hopefully, clinical applications in tissue engineering will follow to enhance the success of regenerative therapies.

NP-Based Systemic Drug Delivery to Prevent Cardiotoxicity

Nanotechnology can have a beneficial effect on cardiovascular health by reducing cardiotoxicity of drugs used to treat noncardiac diseases. Use of halofantrine, an antimalarial drug for treatment of multidrug-resistant malaria, is limited by prolongation of the QT interval (the time between the Q wave and the end of the T wave) as seen on ECG, which can result in bradycardia and hypotension. By encapsulating the drug in polycaprolactone nanocapsules, halofantrine was administered in mice with blunting of the cardiotoxic effects of the drug (Leite et al. 2007). Nanoparticle encapsulation also shows promise for reduction of cardiotoxicity of other drugs. For example, administration of the anticancer drug doxorubicin is limited by cardiotoxicity. However, doxorubicin packaged into 100 nm pegylated liposomes shows comparable efficacy but reduced cardiotoxicity.

Nanotechnology-Based Therapeutics for Cardiovascular Diseases

Nanolipoblockers for Atherosclerotic Arterial Plaques

Nanoscale particles can be synthetically designed to potentially intervene in lipoprotein matrix retention and lipoprotein uptake in cells – processes central to atherosclerosis. These micelles can be engineered to present varying levels of anionic chemistry, which is a key mechanism to induce differential retentivity of low-density lipoproteins (LDLs). Rutgers University scientists have reported on lipoprotein interactions of nanoscale micelles self-assembled from amphiphilic scorpion-like macromolecules based on a lauryl chloride-mucic acid hydrophobic backbone and poly(ethylene glycol) shell. They have used nanoengineered molecules called nanolipoblockers (NLBs) to attack atherosclerotic plaques due to raised levels of LDLs (Chnari et al. 2006). Their approach contrasts with statin drug therapy, which aims to reduce the amount of LDL throughout the body. NLPs compete with oxidized LDLs for a macrophage's attention. The NLBs bind to receptor sites on macrophages, cutting the accumulation of oxidized LDL by as much as 75 %.

Nanotechnology Approach to the Vulnerable Plaque as Cause of Cardiac Arrest

Recent studies have shown that plaque exists in two modes: nonvulnerable and vulnerable. The latter is the probable cause of death in sudden cardiac arrest. Blood passing through an artery exerts a shearing force and can cause vulnerable plaque to rupture, which often leads to occlusion and myocardial infarction. Approximately 60–80 % of sudden cardiac deaths can be attributed to the physical rupture of vulnerable plaque.

There is currently no satisfactory solution to the problem of vulnerable plaque, but it will be tackled by a "Program of Excellence in Nanotechnology" by the National Heart, Lung, and Blood Institute of the NIH. In concert with the NIH's strategy to accelerate progress in medical research through innovative technology and interdisciplinary research, cardiac disease was chosen as the focus of the National Heart, Lung, and Blood Institute's Program of Excellence in Nanotechnology. The program will be a partnership of 25 scientists from the Burnham Institute (La Jolla, CA); University of California, Santa Barbara; and The Scripps Research Institute (San Diego, CA) that will design nanotechnologies to detect, monitor, treat, and eliminate "vulnerable" plaques. By focusing on devising nanodevices, machines at the molecular level, the scientists at these institutions will specifically target vulnerable plaque. It is hoped that this work will lead to useful diagnostic and therapeutic strategies for those suffering from this form of cardiac disease. The project team will work on three innovative solutions to combat vulnerable plaque:

- Building delivery vehicles that can be used to transport drugs and nanodevices to sites of vulnerable plaque
- Designing a series of self-assembling polymers that can be used as molecular nanostents to physically stabilize vulnerable plaque
- Creating nanomachines comprised of human proteins linked to synthetic nanodevices for the purpose of sensing and responding to vulnerable plaque

Nanotechnology for Regeneration of the Cardiovascular System

Nanotechnology may facilitate repair and replacement of blood vessels, myocardium and myocardial valves. It also may be used to stimulate regenerative processes such as therapeutic angiogenesis for ischemic heart disease. Cellular function is integrally related to morphology, so the ability to control cell shape in tissue engineering is essential to ensure proper cellular function in final products. Precisely constructed nanoscaffolds and microscaffolds are needed to guide tissue repair and replacement in blood vessels and organs. Nanofiber meshes may enable vascular grafts with superior mechanical properties to avoid patency problems common in synthetic grafts, particularly small-diameter grafts. Cytokines, growth factors, and angiogenic factors can be encapsulated in biodegradable microparticles or nanoparticles and embedded in tissue scaffolds and substrates to enhance tissue regeneration. Scaffolds capable of mimicking cellular matrices should be able to stimulate the growth of new heart tissue and direct revascularization.

Nanostructures promote formation of blood vessels and bolster cardiovascular function after heart attack. Scientists at the Institute of Bionanotechnology in Medicine at Northwestern University (Evanston, Ill) have shown that injecting nanoparticles into the hearts of mice that suffered heart attacks helped restore cardiovascular function in these animals. The finding is an important research advance that one day could help rapidly restore cardiovascular function in people who have heart disease. The self-assembling nanoparticles – made from naturally occurring polysaccharides and molecules known as peptide amphiphiles – boost chemical signals to nearby cells that induce formation of new blood vessels, and this may be the mechanism through which they restore cardiovascular function. One month following injection, the hearts of the treated mice were capable of contracting and pumping blood almost as well as healthy mice. In contrast, the hearts of untreated mice contracted about 50 % less than normal.

Cellular function is integrally related to morphology, so the ability to control cell shape in tissue engineering is essential to ensure proper cellular function in final products. Precisely constructed nanoscaffolds and microscaffolds are needed to guide tissue repair and replacement in blood vessels and organs. Nanofiber meshes may enable vascular grafts with superior mechanical properties to avoid patency problems common in synthetic grafts, particularly small-diameter grafts. Cytokines, growth factors, and angiogenic factors can be encapsulated in biodegradable microparticles or nanoparticles and embedded in tissue scaffolds and substrates to enhance tissue regeneration. Scaffolds capable of mimicking cellular matrices should be able to stimulate the growth of new heart tissue and direct revascularization.

Nanotechnology-Based Stents

A coronary stent is a tiny expandable mesh tube made of medical grade stainless steel. A stent is delivered on a balloon catheter and implanted in the coronary artery after balloon angioplasty to help keep the artery open. After the plaque is compressed against the arterial wall, the stent is fully expanded into position, thereby acting as miniature "scaffolding" for the artery. The balloon is then deflated and removed and the coronary stent is left behind in the patient's blood vessel. It may be necessary to place more than one stent, depending on the length of the blockage. The inside lining of the artery eventually heals around the stent. Technical advances are providing the development of improved materials for coating of DES. Nanomaterials are the most prominent among these.

Restenosis After Percutaneous Coronary Angioplasty

Restenosis after percutaneous coronary intervention continues to be a serious problem in clinical cardiology. Advances in nanoparticle technology have enabled the delivery of NK911, an antiproliferative drug, selectively to the balloon-injured artery for a longer time (Uwatoku et al. 2003). NK911 is a core-shell nanoparticle of PEG-based block copolymer encapsulating doxorubicin. It accumulates in vascular lesions with increased permeability. In a balloon injury model of the rat carotid artery, intravenous administration of NK911 significantly inhibited the neointimal formation. The effect of NK911 was due to inhibition of vascular smooth muscle proliferation but not to enhancement of apoptosis or inhibition of inflammatory cell recruitment. NK911 was well tolerated without any adverse systemic effects. These results suggest that nanoparticle technology is a promising and safe approach to target vascular lesions with increased permeability for the prevention of restenosis after balloon injury. Coroxane™ (Abraxis), a nanoparticulate microtubule stabilizer, is in phase II clinical trials in conjunction with angioplasty/stents to prevent arterial restenosis.

Biomedical engineers at Purdue University (Lafayette, IN) have shown that vascular stents used to repair arteries might perform better if their surfaces contained "nanobumps" that mimic tiny features found in living tissues. The stents, which are made of titanium and other metals, enable the arteries to grow new tissue after vessel-clogging plaque deposits have been removed. A major problem, however, is that the body often perceives the metal devices as foreign invaders, hindering endothelial cells from attaching to the scaffolding and prompting the creation of scar tissue, which can build up inside blood vessels and interfere with blood flow. If a stent does not attach firmly, it can become loose, and parts of it will actually break off and go down the bloodstream. There is need for new materials that cause the endothelial cells to attach better to these stents without creating as much dangerous scar tissue. The researchers tested disks of titanium containing surface bumps about as wide as 100 nm. The metals used to make conventional stents have features about 10 times larger or none at all. The nanometer-scale bumps mimic surface features of proteins and natural tissues, prompting cells to stick better. Ideally endothelial cells should quickly attach to stents and form a coating only one cell layer thick. The researchers found that nearly three times as many cells stuck to the disks containing the nanobumps, as compared to ordinary titanium. Further research is planned that will replace the titanium disks with tube-shaped pieces of the nano-featured metal, which will resemble the actual shape of real stents.

Currently available stents have problems with imaging within the stent structure, where potential restenosis can occur. Biophan Technologies Inc. has two solutions for stent visibility: a thin-film nanomagnetic particle-coating solution and an anti-antenna solution. These solutions enable the noninvasive, MRI-based, imaging of these devices which today can only be accomplished through more complicated invasive procedures. These approaches will become an important part of the rapidly growing worldwide market for stents and vascular implants.

By using antiproliferative compounds that elute from the surface of a stent, the latest generation of stents has enabled a significant reduction in restenosis rates, i.e., when there is a renarrowing of the vessel after stent implantation. Nanocarrier-based delivery presents a viable alternative to the current stent-based therapies (Brito and Amiji 2007; Feng et al. 2007; Margolis et al. 2007).

Drugs Encapsulated in Biodegradable Nanoparticles

Local delivery of antiproliferative drugs encapsulated in biodegradable nanoparticles has shown promise as an experimental strategy for preventing restenosis development. A novel PDGFR-β-specific tyrphostin, AGL-2043 (Calbiochem), was formulated in polylactide-based nanoparticles and was administered intraluminally to the wall of balloon-injured rat carotid and stented pig coronary arteries (Banai et al. 2005). The antiproliferative effect of nanoencapsulated tyrphostin was found to be considerably higher than that of surface-adsorbed drug. In the pig model, intramural delivery of AGL-2043 resulted in reduced in-stent neointima formation in the coronary arteries as compared to control despite similar degrees of wall injury. The results of this study suggest that locally delivered tyrphostin AGL-2043 formulated in biodegradable nanoparticles may be applicable for antirestenotic therapy independent of stent design or type of injury.

Magnetic Nanoparticle–Coated DES

Biophan Technologies' drug delivery technology (Fig. 10.1), based on tuning magnetic nanoparticles (MNPs) to resonate at a specific frequency, led to their use for selective control of drug release. This technology can be used for reloading drug-eluting coatings for surface elution on demand is active in contrast to the passive drug-eluting polymer coatings. It provides a physician better control over the patient's treatment. Currently, many cardiovascular experts predict the next generation of DES will be comprised of a biocompatible, biodegradable, resorbable material with the strength to acutely open and maintain the confirmation of a vessel. The advantage is that they gradually dissolve while delivering the drug. At the end of a predetermined period, nothing is left at the site where it was introduced.

Magnetic Nanoparticles Encapsulating Paclitaxel Targeted to Stents

Because current DESs lack the capacity for adjustment of the drug dose and release kinetics to the disease status of the treated vessel, attempts have been made to address these limitations by a strategy combining magnetic targeting via a uniform-field-induced magnetization effect and a biocompatible magnetic nanoparticle

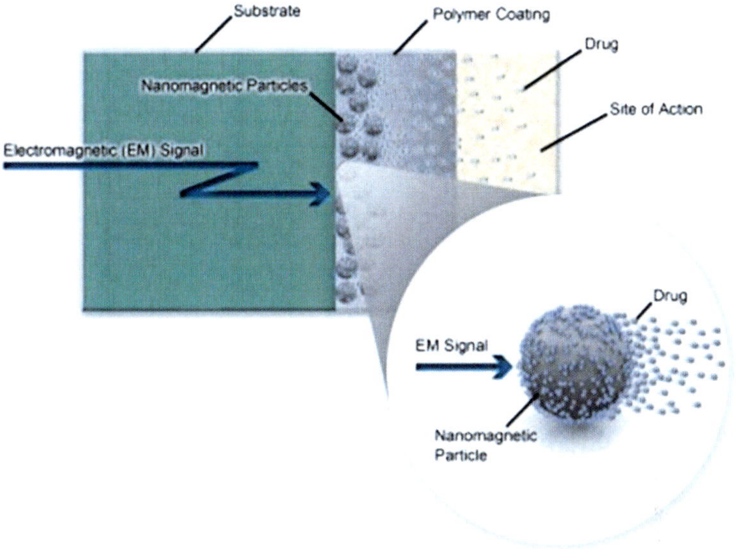

Fig. 10.1 Magnetic nanoparticle–coated stent (Reproduced by permission of Biophan Technologies Inc.)

(MNP) formulation designed for efficient entrapment and delivery of paclitaxel (PTX). Magnetic treatment of cultured arterial smooth muscle cells with PTX-loaded MNPs was shown to inhibit cell growth significantly as compared to non-magnetic conditions (Chorny et al. 2010). Furthermore, significantly higher localization rates of locally delivered MNPs to stented arteries were achieved with uniform-field-controlled targeting compared to nonmagnetic controls in the rat carotid stenting model. The arterial tissue levels of stent-targeted MNPs remained four- to tenfold higher in magnetically treated animals vs. control over 5 days post-delivery. The enhanced retention of MNPs at target sites due to the uniform-field-induced magnetization effect resulted in a significant inhibition of in-stent restenosis with a relatively low dose of MNP-encapsulated PTX. This study demonstrates the feasibility of site-specific drug delivery to implanted magnetizable stents by uniform-field-controlled targeting of MNPs with efficacy for in-stent restenosis.

Nanocoated DES

MIV Therapeutics Inc has developed unique coating technologies that utilize hydroxyapatite (HAp) for application on medical devices and drug delivery systems. The lead product in development is a HAp-coated coronary stent with a nanofilm coating. In 2006, the results of an independently conducted 4-week porcine study, performed by the Department of Cardiology, Thoraxcenter, Erasmus University Medical Center in the Netherlands, indicated that three variations of MIV's polymer-free drug-eluting coatings were at least as effective as or better than Cypher (Johnson & Johnson). The study concluded that MIV's HAp coating, with or without drugs, demonstrated highly promising performance. A pilot clinical trial was launched in 2007, and the first HAp-coated was implanted at the Institute Dante Pazzanese of Cardiology in Sao Paulo, Brazil.

ElectroNanospray™ formulation technology (Nanocopoeia Inc.) produces precise, ultrapure nanoparticles. Particle sizes can be designed from 2 to 200 nm. The device is capable of applying a coating to the particles in a single process step, producing a drug-loaded core. Competitive processes to produce nanoparticles using wet milling and super critical fluid are inherently limited in their ability to produce consistently pure particles within a specified size range and distribution. ElectroNanospray™ technology provides a novel approach for applying challenging materials to the surfaces of medical devices. This process can generate both single- and multiple-phase coatings and apply these with tight control to small, complex surfaces. ElectroNanospray™ process is being developed for applying nanoparticle-based drug-eluting coatings to coronary stents.

Debiotech SA in collaboration with the Laboratory of Powder Technology at Ecole Polytechnique Fédérale de Lausanne (Lausanne, Switzerland) is developing a new type of structured ceramic coatings for drug-eluting stents and other implants. Ceramics offer unique properties compared to polymers. Polymers dissolve over time and residues provoke inflammation, whereas ceramic is stable and inert when in contact with living tissue. With this coating, one can combine an active release of drug during the first weeks after implantation with the long-term stability of the ceramic. Nanostructured ceramics provide novel properties to biomaterials which are not attainable with other materials. The challenge in this project is to process nanosized ceramic powders to reach unique surface structures, which show a controlled porosity over a size range of 2,000 times between the smallest and largest pore. Based on results of fundamental research activities in the field of ordered arrangement of nanosized particles at surfaces, the knowledge of processing particles smaller than 10 nm at large scale has been established as a key competence to achieve that goal.

Nanopores to Enhance Compatibility of DES

Scientists at the Forschungszentrum Dresden-Rossendorf in Germany have developed an innovative method to create a large number of nanopores on the surface of stainless steel. Bombarding the surface of a stent from all sides with a high dose of noble gas ions generates a scaffold of nanopores in the material below the surface. The desired porosity can be precisely engineered by tuning the ion energy, the flux, and the temperature during the process. A larger amount of the highly effective drugs can be deposited on the enlarged noble metal surface, due to this nanoporous structure, which enhances the biocompatibility of the implants in the human body. Thus, this treatment results in the release of drugs over a longer period of time. This method is currently being assessed as a platform technology for the next generation of DES by the Boston Scientific Corporation. The objective of this research collaboration is to further develop this technique for commercialization.

Chapter 11
Nanopulmonology

Introduction

Pulmonology deals with treatment of respiratory diseases, which is a challenging task with rising incidence and limitations of currently available treatments. Application of nanobiotechnology to pulmonology, nanopulmonology, offers NP-based drug and gene delivery for treatment of lung diseases as well as a route for delivery of systemic therapy. Delivery of exogenous genes to the airway epithelium in vivo has been limited by several physiological barriers, resulting in the low success rate of these systems. NP-based drug delivery systems have revolutionized the field of pharmacotherapy by presenting the ability to alter the pharmacokinetics of the conventional drugs to extend the drug retention time, reduce the toxicity, and increase the half-life of the drugs (Swai et al. 2009).

Nanoparticles for Pulmonary Drug Delivery

Pulmonary drug delivery is attractive for both local and systemic drug delivery as a noninvasive route that provides a large surface area, thin epithelial barrier, high blood flow, and the avoidance of first-pass metabolism. Nanoparticles may be used for systemic drug delivery via pulmonary route or for effect on the respiratory system. Nanoparticles can be designed to have several advantages for controlled and targeted drug delivery, including controlled deposition, sustained release, and reduced dosing frequency, as well as an appropriate size for avoiding alveolar macrophage clearance or promoting transepithelial transport (Rytting et al. 2008). The selection of natural or synthetic materials is important in designing particles or nanoparticle clusters with the desired characteristics, such as biocompatibility, size, charge, drug release, and polymer degradation rate.

K.K. Jain, *The Handbook of Nanomedicine*, DOI 10.1007/978-1-61779-983-9_11,

Systemic Drug Delivery via Pulmonary Route

Biodegradable polymers can be used for nanocarrier-based strategies for the systemic delivery of drugs, peptides, proteins, genes, siRNA, and vaccines by the pulmonary route. Chemical modifications to PLGA add several benefits in the selection of a suitable material for nanocarriers in the lung. The introduction of positive or negative charges can enhance the encapsulation efficiency and release profile of oppositely charged drugs, proteins, or genetic material.

An issue that remains surrounded by considerable debate is the question whether the lung should be used as an entry port for systemic drug administration. In this context, the safety of the nanocarriers and a lack of inflammatory and immunogenic potential need to be demonstrated under chronic treatment conditions. Such studies have not been presented with drug-loaded nanocarriers but will be necessary during future clinical trials. Such issues are not limited to pulmonary drug delivery but are also important in oral and intravenous administration.

Nanoparticle Drug Delivery for Effects on the Respiratory System

It is generally accepted in the field of pulmonary delivery that particles must be in the range of 1–3 μm to be delivered effectively to the deep lung. Larger particles have too much energy and hit the walls of the branching upper airways. Smaller particles do not have enough energy and tend to drift and adhere to the walls of the upper airways. However, as particle size is reduced further still, into the nanometer range, an increase in deep lung deposition is seen. Efficient particle deposition in the alveoli can be achieved with nanoparticles in the size range of 50–100 nm as compared to 1–5 μm for the lower airways, but this range of nanoparticles has not been exploited in therapeutic setting (Rogueda and Traini 2007). This is partly due to technological limitations of generating stable aerosols on nanoscale. Nebulizers are the most advanced in using the nanoscale, pressurized metered dose inhalers require further development to realize their potential, and dry powder inhalers are specifically in need of a dry solid nanoparticle generation technique to make it a reality.

Fate and Toxicology of Nanoparticles Delivered to the Lungs

Soluble nanoparticles are absorbed in the blood circulation whereas insoluble nanoparticles have local action in the lung or they can be cleared by lymphatic uptake or by macrophages. Other parameters that influence toxicity of inhaled nanoparticles are chemical composition, bioavailability, surface area, and morphology.

The suitability of nanoparticles, synthesized from porcine gelatin, human serum albumin, and polyalkylcyanoacrylate, as drug and gene carriers for pulmonary application was investigated in vitro on primary airway epithelium cells and the cell line 16HBE14o (Brzoska et al. 2004). Confocal laser scan microscopy and flow cytometry experiments showed that the nanoparticles were incorporated into bronchial epithelial cells provoking little or no cytotoxicity and no inflammation as measured by IL-8 release. Based on their low cytotoxicity and the lack of inflammatory reaction in combination with an efficient uptake in human bronchial epithelial cells, protein-based nanoparticles are suitable drug and gene carriers for pulmonary applications.

Nanoparticle Drug Formulations for Spray Inhalation

Drugs delivered through inhalers are usually either in a suspension (as particles dispersed in liquid) or in a solution (when the drug is dissolved in the liquid). However, there are problems with both methods – a suspension can lead to sediment in the inhaler and less of the drug reaching the target area of the lung, while solutions present problems in dissolving the drug in the inhaler propellant liquid and can make the drug itself less stable. Preparation of the drug in nanoparticle form ensures that the correct dosage reaches the lung and the drug retains its stability, providing the possibility of slowing the release of the drug in the lung for longer therapeutic effect. This could lead to the possibility of more drugs being administered effectively by inhaler, rather than by tablet or injection. Patients suffering from conditions as diverse as asthma and diabetes could benefit from this method of drug delivery.

Nanobiotechnology for Improving Insulin Delivery in Diabetes

One of the important areas of application of nanotechnology-based pulmonary drug delivery is diabetes. Some of the techniques are described in the following paragraphs.

Inhalation of Glucose-Sensitive NP for Regulated Release of Insulin

In an attempt to achieve inhalable self-regulated insulin release, a microparticle agglomerate of nano-sized liposomal particles consisting of a blood sugar–sensing protein named concanavalin A (Con A) has been constructed, which is loaded with insulin, and cross-linkages are capable of cleavage by glucose (Karathanasis et al. 2007). Con A releases the particles to bind independently to the sugars, which then release their insulin. The particles exhibited a small aerodynamic diameter within the human respirable range, but a large geometric diameter that prevents macrophage uptake and clearance. Upon intratracheal instillation of the glucose-sensitive nanoparticle into the lungs of rats, hyperglycemic events triggered an acceleration of the release of insulin achieving normoglycemia shortly after sensing the elevated systemic glucose. This work is a demonstration of an inhalable particle with long residence times in the lungs capable of modulating insulin release based on systemic glucose levels and thus mimic the functions of the pancreas. This approach has the potential of improving management of diabetes by regulated insulin delivery.

Pulmonary Delivery of Insulin by Surface Acoustic Wave Technology

There has been growing interest in the potential for the systematic delivery of drugs and therapeutic agents (e.g., peptides and proteins) via pulmonary (inhalation) means. Surface acoustic wave technology (SAW) enables well-controlled generation of fine nanoparticles and is ideal for this pulmonary drug delivery, particularly for a number of drugs that require frequent dosing. An important application of this technology will be for the development of a device producing insulin nanoparticles delivered across the pulmonary alveoli. There is already evidence for increased efficacy of inhaled insulin compared to injected insulin, due to faster uptake and clearance. An economically viable microdevice for portable pulmonary drug delivery for human use based on SAW has significant commercial potential. Another advantage of using noninvasive techniques as alternatives to frequent injections would be the profound impact on a child's willingness to comply with diabetic treatment.

Nanotechnology-Based Treatment of Pulmonary Disorders

Management of Cystic Fibrosis

Cystic fibrosis (CF) is the most common autosomal recessive disorder in Caucasians. CF affects approximately 1 in 2,000–2,500 live births with a carrier rate in white Americans of 1 in 25. There are about 30,000 CF patients in the USA, and 2,000 babies with CF are born every year. An estimated eight million are carriers. Parents of a child with CF, who are carriers but do not have the condition, are at a one-in-four risk of having a child with CF with each pregnancy. CF is a multisystem disorder of children and adults characterized by an abnormality of exocrine gland function, which manifests as chronic respiratory tract infections and malabsorption. Cystic fibrosis transmembrane conductance regulator (CFTR) gene has been isolated. Determining the sweat chloride concentration is the standard screening test (high level of chloride ion means a positive test). It is not suitable for prenatal diagnosis and cannot detect carriers. The method of choice is identification of DF508 (the most common mutation) by PCR, which detects 85–95 % of CF carriers.

CF is a potentially lethal disease although the current life expectancy has improved to >30 years with advances in the medical treatment. CF also affects the digestive system, resulting in progressive disability and early death. Thick mucous production in the airways leads to distorted mucociliary clearance and weakening of the immune system leading to frequent lung infections. Currently used methods for the treatment of pulmonary complications of CF include physiotherapy, bronchodilator therapy, mucolytic agents, corticosteroids, and lung transplant. These methods are directed at the management of manifestations and none of these addresses the cause of the disease. Because of the devastating clinical sequelae and the lack of definitive therapy, CF is prime candidate for gene therapy.

The goal of gene therapy is correction of the mutant CFTR gene with wild-type (wt) DNA sequences to restore normal CFTR protein and function. Experiments with wtCFTR cDNA expression vectors have shown that Cl ion transport phenotype associated with CF can be corrected to resemble that in normal cells. Several methods of gene transfer are used including those involving nanobiotechnology.

Nanobiotechnology-Based Gene Transfer in CF

Nonviral DNA Nanoparticle–Mediated CFTR Gene Transfer

Nanoparticles have been used for CFTR gene delivery in the nose of CF patients in clinical trials and led to partial correction of the chloride transport defect in nasal epithelium (Griesenbach et al. 2004).

Nanoparticles consisting of single molecules of DNA condensed with PEG-substituted lysine 30-mers have been shown to efficiently transfect lung epithelium following intrapulmonary administration (Fink et al. 2006). Nanoparticles formulated with lysine polymers having different counterions at the time of DNA mixing have distinct geometric shapes: trifluoroacetate or acetate counterions produce ellipsoids or rods, respectively. Based on intracytoplasmic microinjection studies, nanoparticle ellipsoids having a minimum diameter less than the 25-nm nuclear membrane pore efficiently transfect nondividing cells. This 25-nm size restriction corresponds to a 5.8-kbp plasmid when compacted into spheroids, whereas the 8- to 11-nm diameter of rod-like particles is smaller than the nuclear pore diameter. In mice, up to 50 % of lung cells are transfected after dosing with a rod-like compacted expression plasmid, and correction of the CFTR chloride channel was observed in humans following intranasal administration. To further investigate the potential size and shape limitations of DNA nanoparticles for in vivo lung delivery, reporter gene activity of ellipsoidal and rod-like compacted luciferase plasmids ranging in size between 5.3 and 20.2 kbp was investigated. Equivalent molar reporter gene activities were observed for each formulation, indicating that microinjection size limitations do not apply to the in vivo gene transfer setting.

Chitosan-DNA-FAP-B nanoparticles are good candidates for targeted gene delivery to fibronectin molecules (FAP-B receptors) of lung epithelial cell membrane. In a study, aerosol delivery of chitosan-DNA-FAP-B nanoparticles resulted in 16-fold increase of gene expression in the mice lungs compared with chitosan-DNA nanoparticles, suggesting that chitosan-DNA-FAP-B nanoparticle can be a promising carrier for targeted gene delivery to the lung (Mohammadi et al. 2011).

An hCFTR expression plasmid was optimized as a payload for compacted DNA nanoparticles formulated with PEG-substituted 30-mer lysine peptides. Compared to hCFTR cDNA, the codon-optimized version (CO-CFTR) produced a ninefold increased level of hCFTR protein in CF mice, when compacted as DNA nanoparticles (Padegimas et al. 2012).

Liposome-Mediated CFTR Gene Transfer

Lipofection of cells in vitro with CFTR cDNA constructs can, like virally transduced cells, elicit the electrophysiological responses characteristic of the CFTR ion channel. The advantages of liposome-mediated gene transfer are the potential for standardized production of large amount of vector, freedom from risk of viruses, and the possibility of readministration with minimal host reaction. The disadvantages are the lack of sustained expression using current strategies. The safety and efficacy of this technique was demonstrated in rodents, and clinical trials have been conducted in CF patients.

Magnetofection for Enhancing Nonviral Gene Transfer to the Airways

Superparamagnetic nanoparticles with either the therapeutic CF gene or a reported gene attached to them were inhaled and targeted to the airway epithelium via positioning of a strong magnet over the target site, which functions to pull the particles into contact with the cells (Dobson 2006). To improve the in vivo transfection efficiency of DNA delivery of this system, an oscillating magnet array system (TransMAG) is being developed, which will introduce energy and a lateral component to advance the movement and interaction of the particles coupled with Lipofectamine 2000 to form a plasmid DNA (pDNA) liposome complexes, to enhance interaction with the epithelial cells (Xenariou et al. 2006).

NP-Based Delivery of Antibiotics for Treatment of Pulmonary Infections in CF

Pulmonary infections are common in CF and are currently treated with antibiotics that are prescribed on the basis of the infectious agent, but many of these bacteria are resistant to multiple antibiotics and require prolonged treatment with intravenous antibiotics such as tobramycin, ciprofloxacin, and piperacillin. Inhaled therapy with other antibiotics is also followed in some cases to improve lung function by impeding the growth of colonized bacteria. These antibiotics may produce side effects such as hearing loss and kidney failure. To address these shortfalls, a liposomal formulation of ciprofloxacin powder manufactured using a sprayfreeze-drying process with the required mass mean aerodynamic diameter and fine particle fraction has been used (Sweeney et al. 2005). This is administered by inhalation and thus increases the bioavailability of the drug.

Respiratory tract infections are the primary cause of death in persons with CF, and there are no effective therapies for patients infected with bacterial species that are resistant to all known antibiotics. A surfactant-stabilized oil-in-water nanoemulsion, NB-402 (NanoBio Corporation), was found to be bactericidal against all but two of 150 bacterial strains, regardless of their levels of resistance (LiPuma et al. 2009). NB-402 has been shown to be highly efficacious in vitro against *Pseudomonas aeruginosa*, *Burkholderia*, *Acinetobacter*, *Stenotrophomonas*, and other multidrug-resistant bacterial strains from CF patients. In addition, the nanoemulsion retains activity when organisms are growing in biofilms and mucus. Resistance to the nanoemulsion is not anticipated based on its unique mechanism of action of interacting with the bacterial membrane and causing lysis. These results support NB-402's potential role as a novel antimicrobial agent for the treatment of infection due to CF-related opportunistic pathogens. There are plans for this product to enter clinical trials.

Nanotechnology-Based Treatment of Chronic Obstructive Pulmonary Disease

Chronic airway inflammation and mucous hypersecretion are features of chronic obstructive pulmonary disease (COPD), asthma, and CF. One of the major challenges in drug delivery and therapeutic efficacy are airway defense, severe inflammation, and mucous hypersecretion, which are further aggravated by infection. Treatments such as corticosteroids and antibiotics aim to controlling chronic inflammation.

Few of the numerous available nano-based drug delivery systems have been tested for COPD. Targeted nanoparticle-mediated sustained drug delivery is required to control inflammatory cell chemotaxis, fibrosis, protease-mediated chronic emphysema, and/or chronic lung obstruction in COPD. Design and development of nano-based targeted vehicles with integrated therapeutic, imaging and airway-defense penetrating capability are currently being evaluated to treat the underlying cause of CF and COPD lung disease (Vij 2011).

Chapter 12
Nanoorthopedics

Introduction

Nanoorthopedics means the application of nanobiotechnology in orthopedics, the medical specialty dealing with disorders of bones and joints. Two important areas of application are bone implants and joint injuries involving cartilage.

Application of Nanotechnology for Bone Research

There is a significant need and demand for the development of a bone substitute that is bioactive and exhibits material properties (mechanical and surface) comparable with those of natural, healthy bone. Nano-sized ceramics, polymers, metals, and composites are receiving considerable attention for bone tissue engineering.

Nanomechanical heterogeneity is expected to influence elasticity, damage, fracture, and remodeling of bone. The spatial distribution of nanomechanical properties of bone has quantified at the length scale of individual collagen fibrils (Tai et al. 2007). This study sheds new light on how bone absorbs energy by probing its fundamental building block, collagen embedded with tiny nanoparticles of mineral, at nanoscale. The mechanical properties of bone were shown to vary greatly within a single region only two micrometers wide. Because a variety of bone disorders lead to changes in bone structure, the discovery of the nonuniformity of bone's mechanical properties at nanoscale could lead to improved diagnoses of diseases. For example, if specific nanoscale patterns of stiffness within bone structure are tied to disease or aging, these could potentially be identified earlier or provide more conclusive evidence of a disorder. The results of this study could lead to new ways of producing improved structural composites that mimic nature's clever design that enables bones to resist sudden fractures.

K.K. Jain, *The Handbook of Nanomedicine*, DOI 10.1007/978-1-61779-983-9_12,

Reducing Reaction to Orthopedic Implants

Currently used materials for joint replacement are not acceptable for many reasons including the nonbiocompatibility of metallic materials, which produce debris due to wear and, as a result, have a short lifetime requiring several revisions. In orthopedic implants, titanium and/or titanium alloys often become encapsulated with undesirable soft fibrous but not hard bony tissue. There is no ideal material, but use of nanomaterials for this purpose has been explored.

Although possessing intriguing electrical and mechanical properties for neural and orthopedic applications, carbon nanofibers/nanotubes have not been widely considered for these applications previously. A carbon nanofiber-reinforced polycarbonate urethane composite has been developed in an attempt to determine the possibility of using carbon nanofibers (CNs) as orthopedic prosthetic devices. Mechanical characterization studies determined that such composites have properties suitable for orthopedic applications. These materials enhanced osteoblast (bone-forming cell) functions whereas functions of cells that contribute to fibrous-tissue encapsulation events for bone implants (fibroblasts) decreased on PU composites containing increasing amounts of CNs. In this manner, this study provided the first evidence of the future that CN formulations may have toward interacting with bone cells, which is important for the design of successful orthopedic implants.

Control of the nanostructure can provide ceramic materials with better fatigue resistance in static and dynamic conditions, whereas control of the macrostructure can provide flow tolerant materials. Combination of both materials may provide a new generation of longer-lasting implants for joint replacement (Torrecillas et al. 2009).

Enhancing the Activity of Bone Cells on the Surface of Orthopedic Implants

It is very important to increase the activity of bone cells on the surface of materials used in the design of orthopedic implants so that such cells can promote either integration of these materials into surrounding bone or complete replacement with naturally produced bone if biodegradable materials are used. Osteoblasts are bone-producing cells and, for that reason, are the cells of interest in initial studies of new orthopedic implants. If these cells are functioning normally, they lay down bone matrix onto both existing bone and prosthetic materials implanted into the body. It is generally accepted that a successful material should enhance osteoblast function, leading to more bone deposition and, consequently, increased strength of the interface between the material and juxtaposed bone. A study has provided evidence of greater osteoblast function on carbon and alumina formulations that mimic the nanodimensional crystal geometry of hydroxyapatite found in bone (Price et al. 2003).

Synthetic Nanomaterials as Bone Implants

Nanoscale molecular scaffolds have been designed that resemble the basic structure of bone. Design of peptide-amphiphile structures allows nanofibers to be reversibly cross-linked to enhance or decrease their structural integrity. After cross-linking, the fibers are able to direct mineralization of hydroxyapatite to form a composite material. Nanofibers, approximately 8 nm in size, come in the form of a gel that can be injected into a broken bone to help the crystallization process for repair of fracture. This approach recreates the structure of bone at the nanoscale level and has implications beyond bone repair. It could lead to development of a hardening gel that speeds the healing of fractures. It could also help patients avoid conventional surgery or be used to repair bone fractures of soldiers in battlefield.

NanoBone Implants

Another method of repairing bones using nanotechnology is based on bone scaffold material (nanohydroxyapatite (HA)/collagen/PLA composite) produced by biomimetic synthesis. It shows some features of natural bone both in main composition and hierarchical microstructure, which is nanohydroxyapatite and collagen assembled into mineralized fibril (Liao et al. 2004). The 3D porous scaffold materials mimic the microstructure of cancellous bone. Cell culture and animal model tests show that the composite material is bioactive. The osteoblasts that are isolated from the neonatal rat calvaria adhere, spread, and proliferate throughout the pores of the scaffold material within a week. This is implanted into a bone defect model in the radius bone of the rabbit and has been shown to be partially substituted by new bone tissue after 3 months.

The scaffolds or "NanoBones" have successfully implanted in patients in China for repair of bone defects after fractures or tumor removal and also for spinal fusion. The NanoBone material is inserted where the bone needs to heal. The critical material is calcium phosphorus, which is reduced to 30 nm in thickness and 60 nm in width. At this size, the properties of calcium phosphorus change. On a large scale the calcium phosphorus does not degrade, but on a nanoscale it does. The nanoscale material degrades after a minimum of 6 months, and the space is filled by natural bone. This technology is better than current methods that use ceramics or metals because those materials remain in the patient's body and can cause infection, pain, and make the repaired bone more vulnerable to fracture.

The technology has been found to be effective in repairing small bones ranging from 1 to 2 cm in length, making the technology useful after removal of bone tumors. Research is currently being performed on larger bones up to 4 cm in length. The NanoBone technology, which was approved by China's regulatory agency, is available for commercial use in Chinese hospitals. The cost of the NanoBone implant, which is initially high, is expected to be reduced over time to be economically competitive with other technologies available.

Although nanohydroxyapatite (HA) has a wide range of medical applications, particle mobilization and slow resorption limit its use in certain applications, particularly periodontal and alveolar ridge augmentation. However, the rate of resorption of a composite of hydroxyapatite and chitosan is higher than hydroxyapatite and may have a great impact on human health care systems as bioresorbable bone substitute (Murugan and Ramakrishna 2004). A transparent and slight yellow chitosan (CS)/HA nanocomposite with high performed, potential application as internal fixation of bone fracture was prepared by a novel and simple in situ hybridization (Hu et al. 2004). The bending strength and modulus of CS/HA with ratio of 100/5 (wt/wt) was slightly higher than that of pure CS rod.

NanoCeram® Fibers

Other synthetic materials such as NanoBone have been designed that possess the grain size, shape, and porosity similar to HA, the natural mineral present in bone. The HA in bone has a fibrous shape and is less than 100 nm in diameter. A study was performed on NanoCeram® fibers, where its cytocompatibility was tested and where osteoblast (bone) cell adhesion and proliferation were measured. The study showed that after 1, 3, and 5 days of culture, the number of osteoblasts was significantly greater on nanofiber alumina than nano- or micron-sized alumina spheres, metallic titanium, or HA compacts. There were more than two, three, and four times the number of osteoblasts on nanofiber alumina than on titanium after 1, 3, and 5 days, respectively.

NanoBone Versus Bio-Oss

A study has compared the biocompatibility of a synthetic bone substitute, NanoBone®, to the widely used natural bovine bone replacement material Bio-Oss® (Liu et al. 2011). The in vitro response of human osteoblasts to both materials was investigated. Cell performance was assessed using SEM, cell vitality staining, and biocompatibility tests. Both materials showed low cytotoxicity and good biocompatibility as they caused only little damage to human osteoblasts, which can justify their clinical application. However, NanoBone® was able to support and promote proliferation of human osteoblasts slightly better than Bio-Oss®. The results may guide physicians in the choice between a natural and a synthetic biomaterial. Further experiments are necessary to determine the comparison of biocompatibility in vivo.

Carbon Nanotubes as Scaffolds for Bone Growth

Artificial bone scaffolds have been made from a wide variety of materials, such as polymers or peptide fibers. Their drawbacks include low strength and the potential for rejection in the body. Chemically functionalized SWCNTs have been used as scaffolds for the growth of artificial bone material (Zhao et al. 2005). The strength, flexibility, and light weight of SWCNTs enable them to act as scaffolds to hold up regenerating bone. Bone tissue is a natural composite of collagen fibers and crystalline hydroxyapatite, which is a mineral based on calcium phosphate. SWCNTs can mimic the role of collagen as a scaffold for inducing the growth of hydroxyapatite crystals. By chemically treating the nanotubes, it is possible to attract calcium ions and to promote the crystallization process while improving the biocompatibility of the nanotubes by increasing their water solubility. SWCNTs may lead to improved flexibility and strength of artificial bone, to new types of bone grafts, and to inroads in the treatment of osteoporosis and fractures.

A nanowire coating on the surface of biocompatible titanium can be used to create more effective surfaces for hip replacement, dental reconstruction, and vascular stenting. Further, the material can easily be sterilized using ultraviolet light and water or using ethanol, making it useful in hospital settings. The length, the height, the pore openings, and the pore volumes within the nanowire scaffolds can be controlled by varying the time, temperature, and alkali concentration in the reaction. In contrast to the titanium implant, which may fail after some years because of nonadherence to it of muscle tissue and require reoperation, the nanowire-coated joint adheres to the muscle tissue as shown in experimental animals.

Bone cells can grow and proliferate on a scaffold of CNTs because they are not biodegradable but behave like an inert matrix on which cells can proliferate and deposit new living material, which becomes functional, normal bone (Zanello et al. 2006). CNTs carrying neutral electric charge sustained the highest cell growth and production of plate-shaped crystals. There was a dramatic change in cell morphology in osteoblasts cultured on multiwalled CNTs, which correlated with changes in plasma membrane functions. CNTs hold promise in the treatment of bone defects in humans associated with the removal of tumors, trauma, and abnormal bone development and in dental implants. More research is needed to determine how the body will interact with carbon nanotubes, specifically in its immune response.

By using stem cells attached to titanium oxide nanotube implants, precise change in nanotube diameter can be controlled to induce selective differentiation of hMSCs into osteoblasts (Oh et al. 2009). Small (≈30-nm diameter) nanotubes promoted adhesion without noticeable differentiation, whereas larger (≈70- to 100-nm diameter) nanotubes elicited a dramatic stem cell elongation (≈10-fold increased), which induced cytoskeletal stress and selective differentiation into osteoblast-like cells, offering a promising nanotechnology-based route for unique orthopedic-related hMSC treatments. Use of nanostructures is preferable to chemicals for stem cell implants in order to control cell differentiation as chemicals can sometimes have undesirable side effects on the human body. Clinical implication of this research is that if the surgeon uses titanium oxide nanotubes with stem cells, the bone healing could be accelerated following fracture of leg bones, and a patient may be able to walk in 1 month instead of being on crutches for 3 months.

Aligning Nanotubes to Improve Artificial Joints

Artificial joints might be improved by making the implants out of tiny CNTs (diameter 60 nm) and filaments that are all aligned in the same direction, mimicking the alignment of collagen fibers and natural ceramic crystals in real bones. The smaller features stimulate the growth of more new bone tissue, which is critical for the proper attachment of artificial joints once they are implanted. Nanotubes and nanofibers are aligned in the same direction, and this orientation is similar to the way collagen and natural ceramic crystals, called hydroxyapatite, are aligned in bone. One-third more bone-forming cells (osteoblasts) attach to CNTs that possess surface bumps about as wide as 100 nm than to conventional titanium, which has surface features on the scale of microns. The nanometer-scale bumps mimic surface features of proteins and natural tissues, prompting cells to stick better and promoting the growth of new cells. Using such nanometer-scale materials might cause less of a rejection response from the body. Rejection eventually weakens the attachment of implants and causes them to become loose and painful, requiring replacement surgery. Aligning the nanotubes to further mimic natural bone also might provide more strength.

Cartilage Disorders of Knee Joint

The meniscus is the knee's shock absorber. It is a cartilage spacer for preventing friction and absorbing approximately one-third of the impact load that the joint cartilage surface experiences. Cartilage injuries of the knees are one of the common injuries in sports, particularly football and hockey. Unlike other body tissues, the meniscus does not repair itself because only a very small part receives blood. The conventional treatment of a torn cartilage is surgical removal of the loose pieces and repair of the tear where possible to save as much as possible of the cartilage. The procedure has become refined with arthroscopy. Although the results are generally good in terms of relief of pain and recovery of function of the joint, there are long-term effects if the cartilage is lost, and degenerative changes in the joints may occur.

Several methods have been developed to encourage the regeneration of cartilage defects. Procedures such as debridement, lavage, microfracturing, subchondral bone drilling, and abrasion arthroplasty may perhaps alleviate symptoms but cannot restore the hyaline articular cartilage. The regenerated tissue formed in response to these procedures consists of fibrocartilage and does not possess the biomechanical or biochemical properties of hyaline articular cartilage. Nanotechnology and cell therapy are being used as refinements of procedures to replace the torn knee cartilage.

Role of Nanotechnology in Engineering of a Replacement for Cartilage

Use of nanotechnology is being explored to produce viable structural and functional scaffolds capable of promoting the growth of mesenchymal stem cells (MSCs) and differentiate these cells into meniscal tissue. A thorough understanding is needed of how MSCs interact with scaffolds and how to optimize conditions promoting the cell growth around these scaffolds, which should not only encourage cell growth but must also degrade at the correct rate so that all that remains is meniscal tissue.

Apart from the knee, regeneration of intervertebral disc (IVD) may be a useful procedure as an alternative for spinal fusion because there is inherent limitation of hardware-based IVD replacement prostheses, which indicates the importance of biological approaches to disc repair. In one study, multipotent, adult human MSCs were seeded into a novel biomaterial amalgam to develop a biphasic construct that consisted of electrospun, biodegradable nanofibrous scaffold (NFS) enveloping a hyaluronic acid (HA) hydrogel center (Nesti et al. 2008). The cartilaginous HA/NFS construct architecturally resembled a native IVD, with an outer annulus fibrosus-like region and inner nucleus pulposus-like region. Histological and biochemical analyses, immunohistochemistry, and gene expression profiling revealed the time-dependent development of chondrocytic phenotype of the seeded cells. The cells also maintain the microarchitecture of a native IVD. These findings suggest the potential of MSC-seeded HA/NFS constructs for the tissue engineering of biological replacements of degenerated IVD.

Nanotechnology as an Aid to Arthroscopy

Arthroscopy of joints, particularly the knee joint, is an established procedure for diagnosis and treatment. Nanotechnology has been used to refine this procedure. The first step was the study of cartilage by AFM as a basis for the construction of a scanning force arthroscope.

Cartilage stiffness was measured ex vivo at the micrometer and nanometer scales to explore structure-mechanical property relationships at smaller scales than has been done previously. A method has been developed to measure the dynamic elastic modulus, in compression by indentation-type AFM (Stolz et al. 2004). Spherical indenter tips (radius approximately 2.5 μm) and sharp pyramidal tips (radius approximately 20 nm) were employed to probe micrometer-scale and nanometer-scale response, respectively, on subsurface cartilage from porcine femoral condyles. From results of AFM imaging of cartilage, the micrometer-scale spherical tips resolved no fine structure except some chondrocytes, whereas the nanometer-scale pyramidal tips resolved individual collagen fibers and their 67-nm axial repeat distance. The cartilage compressive stiffness was different at the nanometer scale compared to the overall structural stiffness measured at the micrometer and larger scales because of the fine nanometer-scale structure, and enzyme-induced structural changes can affect this scale-dependent stiffness differently. The collagen fibers were seen to coalesce together as evidence of disease state.

Early detection and the ability to monitor the progression of osteoarthritis are important for developing effective therapies. Indentation-type AFM can monitor age-related morphological and biomechanical changes in the hips of normal and osteoarthritic mice (Stolz et al. 2009). Early damage in the cartilage of osteoarthritic patients undergoing hip or knee replacements could similarly be detected using this method. Changes due to aging and osteoarthritis are clearly depicted at the nanometer scale well before morphological changes can be observed using current diagnostic methods. Indentation-type AFM may potentially be developed into a minimally invasive arthroscopic tool to diagnose the early onset of osteoarthritis in situ.

Scanning Force Arthroscope

A prototype of the device constructed at Muller Institute for Structural Biology (Basel, Switzerland) combines both diagnostics and therapeutics in a single tube in contrast to the conventional arthroscopes which have two tubes – one for visualization and the other for manipulation with instruments. There are inflatable balloons to provide an irrigation system. This prototype fulfills the requirements of an ideal arthroscope:

- It is user friendly.
- It provides information not obtainable by conventional methods.
- It is expected to have an affordable price tag.

So far, the device has been tested only in models of knee joint. It is expected to be in the market within a decade.

Chapter 13
Nano-ophthalmology

Introduction

Nanotechnology has many applications in disorders of the eye, which could be included under the heading of nano-ophthalmology. These include drug delivery, study of pathomechanism of eye diseases, regeneration of the optic nerve, and counteracting neovascularization involved in some degenerative disorders. Nanoparticles enable delivery of ocular drugs to specific target sites, and results to date strongly suggest that ophthalmology will benefit enormously from the use of this nanometric scale technology.

Nanocarriers for Ocular Drug Delivery

Approximately 90 % of all ophthalmic drug formulations are applied as eye drops. While eye drops are convenient, about 95 % of the drug contained in the drops is lost through tear drainage, a mechanism for protecting the eye against exposure to noxious substances. Moreover, several barriers impede direct and systemic drug access to the specific site of action. The tight epithelium of the cornea compromises the permeation of drug molecules. An ideal topical drug delivery system should possess certain desirable properties, such as good corneal and conjunctival penetration, prolonged precorneal residence time, easy instillation, non-irritative and comfortable to minimize lachrymation and reflex blinking, and appropriate rheological properties.

Advantages of using nanoparticles include improved topical passage of large, poorly water-soluble molecules such as glucocorticoid drugs or cyclosporine for immune-related diseases that threaten vision. Other large and unstable molecules, such as nucleic acids, delivered using nanoparticles, offer promising results for gene transfer therapy in severe retinal diseases. Nanoparticles enable targeted delivery to specific types of cancer such as melanoma while sparing normal cells (Diebold and Calonge 2010).

K.K. Jain, *The Handbook of Nanomedicine*, DOI 10.1007/978-1-61779-983-9_13,

Table 13.1 Nanoparticles used for drug delivery in ophthalmology

Nanoparticles	Drug delivered	Advantages
Acrylate polymer nanosuspensions	Flurbiprofen	Higher drug levels in the aqueous humor and inhibition of paracentesis-induced miosis
Albumin nanoparticles	Ganciclovir	Enhanced antiviral activity against cytomegalovirus infection
Chitosan-sodium alginate nanoparticles	Antibiotic, gatifloxacin	Enhanced delivery to external ocular tissues without systemic drug exposure and/or affecting the intraocular structures (Motwani et al. 2008)
Dendrimers	Pilocarpine nitrate, tropicamide	Prolonged miotic activity
Discomes	Timolol maleate	Entraps greater amount of drug than niosomes
Liposomes	Acetazolamide	More effective decrease in intraocular pressure (IOP)
Nanoparticles	Amikacin	Improved delivery of drug to cornea and aqueous humor
Niosomes	Cyclopentolate	Enhanced ocular absorption of the drug
Poly(butyl)-cyanoacrylate nanoparticles	Pilocarpine	Enhanced miotic response by 22 % and decreased IOP

Nanocarriers, such as nanoparticles, liposomes, and dendrimers, are used to enhance ocular drug delivery (Vandervoort and Ludwig 2007). Easily administered as eye drops, these systems provide a prolonged residence time at the ocular surface after instillation, thus avoiding the clearance mechanisms of the eye. In combination with a controlled drug delivery, it should be possible to develop ocular formulations that provide therapeutic concentrations for a long period of time at the site of action, thereby reducing the dose administered as well as the instillation frequency. In intraocular drug delivery, the same systems can be used to protect and release the drug in a controlled way, reducing the number of injections required. Another potential advantage is the targeting of the drug to the site of action, leading to a decrease in the dose required and a decrease in side effects. Nanoparticles used for drug delivery in ophthalmology are shown in Table 13.1.

Nanoparticle-Based Topical Drug Application to the Eye

Nanoparticle technology has been used for ophthalmic formulations for a decade, but research is still in progress to improve the delivery and safety of drugs used for treating disorders of the eye. Topical application of nonsteroidal anti-inflammatory drugs on the eye is a common treatment used to treat the inflammatory reaction manifested by narrowing of the pupil (miosis) induced by surgical injury such as cataract extraction. With the aim of improving the availability of sodium ibuprofen (IBU) at the intraocular level, IBU-loaded polymeric nanoparticle suspensions have been made from Eudragit RS100, an inert polymer resin. Particles in nanosuspension have a mean size of ~100 nm and a positive charge making them suitable for ophthalmic applications. In vitro dissolution tests indicated a controlled release profile of IBU from nanoparticles. Drug levels in the aqueous humor are also higher after application of the nanosuspensions; moreover, IBU-loaded nanosuspensions do not show toxicity on ocular tissues.

Chitosan Nanoparticles for Topical Drug Application to the Eye

Use of chitosan (CS) nanoparticles for ocular drug delivery has been investigated by studying their interaction with the ocular mucosa in vivo and also their toxicity in conjunctival cell cultures (de Salamanca et al. 2006). The in vivo interaction of fluorescent CS (CS-fl) nanoparticles with the rabbit cornea and conjunctiva was analyzed by spectrofluorimetry and confocal microscopy. CS-fl nanoparticles were found to be stable upon incubation with lysozyme and did not affect the viscosity of mucin dispersion. In vivo studies showed that the amounts of CS-fl in cornea and conjunctiva were significantly higher for CS-fl nanoparticles than for a control CS-fl solution, these amounts being fairly constant for up to 24 h. Confocal studies suggest that nanoparticles penetrate into the corneal and conjunctival epithelia. Cell survival at 24 h after incubation with CS nanoparticles was high, and the viability of the recovered cells was near 100 %. These findings indicate that CS nanoparticles are promising vehicles for ocular drug delivery.

Chitosan has been modified by covalent coupling to cholesterol (Yuan et al. 2006). These molecules self-aggregate into nanoparticles with a size of approximately 200 nm. Cyclosporin was incorporated with a drug loading of 6.2 %. In vitro tests demonstrated that the drug was gradually released over a period of 48 h. Use of SPECT and scintillation counter demonstrated that 71 % of the drug was still present at the ocular surface after 112 min.

These and other studies indicate that chitosan-based nanostructures are versatile systems that can be tailor-made according to required compositions, surface characteristics, and particle size. Such parameters, which are known to influence their in vivo performance, can be modulated by adjusting the formulation conditions of the nanotechnologies responsible for their formation, by incorporating additional materials in the preparation steps, and/or by using synthetically modified chitosan.

Polylactide Nanoparticles for Topical Drug Application to the Eye

PLA nanoparticles incorporating flurbiprofen have been prepared by the solvent displacement technique using poloxamer 188 as a stabilizer to improve the availability of the drug for the prevention of the inflammation caused by ocular surgery (Vega et al. 2006). Formulations, with particle size of 230 nm, do not show toxicity on ocular tissues. In vivo studies in rabbits have demonstrated that the formulations do not induce toxicity or irritation. Nanoparticle formulations have been compared with commercial eye drops (Ocuflur™) after induction of inflammation by instillation of sodium arachidonate. The commercial eye drop showed a suppression of inflammation, with minimal inflammation reached after 90 min. A comparable nanoparticle formulation demonstrated a higher suppression, which increased throughout the 150-min observation time of the study.

Ophthalmic Drug Delivery Through Nanoparticles in Contact Lenses

Polymeric lens materials that can be loaded with nanoparticulate eye medication for ophthalmic drug delivery applications. The solution to constitute the lens contains a mixture of molecules, which create nanochannels when they set. The channels act as conduits for the drug to be released when the lens comes into contact with eye fluid. The channels also render the lens nanoporous, i.e., tears and gases can cross into and out of lens, making it more compatible with the human eye. The rate of delivery can be controlled by adjusting the channel size. The use of contact lenses loaded with nanoparticles is better than topical application of ophthalmic drugs in the form of eye drops. The drug molecules will have a much longer residence time of several days in the postlens tear film, compared with about 2–5 min in the case of topical application of drugs in the form of eye drops. This method of drug delivery would reduce drug waste and adverse effects of systemic absorption, resulting in increased patient compliance. The duration of drug delivery from contact lenses can be significantly increased if the drug is first entrapped in niosomes, which are nonionic surfactant vesicles, before they are dispersed throughout the contact lens material. This also prevents the interaction of drug with the polymerization mixture and provides additional resistance to drug release, as the drug must first diffuse through the nanoparticle and penetrate the particle surface to reach the hydrogel matrix.

Nanoparticles for Intraocular Drug Delivery

Nanoparticles have also been investigated for intraocular drug delivery to provide controlled drug release, to protect the drug against enzymatic degradation, and to direct the drug to the site of action. The kinetics of PLA nanoparticle localization within the intraocular tissues and their potential to release encapsulated material have been studied in experimental animals (Bourges et al. 2003). Intravitreous injection of PLA nanoparticles appears to result in transretinal movement, with a preferential localization in the retinal pigmented epithelial (RPE) cells. Encapsulated rhodamine dye diffuses from the nanoparticles and stains the neuroretina and the RPE cells. The findings support the idea that specific targeting of these tissues is feasible. Furthermore, the presence of the nanoparticles within the RPE cells 4 months after a single injection shows that a steady and continuous delivery of drugs can be achieved.

Subconjunctivally administered 200-nm and larger PLA nanoparticles can be almost completely retained at the site of injection in male Sprague-Dawley rats for at least 2 months (Amrite and Kompella 2005). The 20-nm particles disappeared more rapidly, with 8 % of the administered dose remaining after 7 days. The neuroprotective effects of PLGA nanospheres to encapsulate pigment epithelium-derived factor (PEDF) have been evaluated in induced retinal ischemic injury (Li et al. 2006). Intravitreal injection of the naked peptides demonstrated a 44 % reduction of cell death of the retinal ganglion cells (RGCs) after 48 h. Injection of the encapsulated peptide showed a very similar protective effect that lasted for at least 7 days. The authors attributed the extended effect to the slow release of PEDF from the PLGA particles and to the protection of the peptide against degradation and rapid clearance.

Besides size, grafting polymers with PEG is another method of controlling particle distribution. A hydrophobic polymer, cyanoacrylate-co-hexadecyl cyanoacrylate, was coupled to hydrophilic PEG chains in order to produce tamoxifen-loaded nanoparticles (De Kozak et al. 2004). Intraocular injection in rats resulted in a significant inhibition of experimentally induced autoimmune uveoretinitis, whereas injection of the free drug did not alter the disease.

DNA Nanoparticles for Nonviral Gene Transfer to the Eye

The eye is an excellent candidate for gene therapy as it is immune privileged and much of the disease-causing genetics are well understood. Compacted DNA nanoparticles have been investigated as a system for nonviral gene transfer to ocular tissues. The compacted DNA nanoparticles have already been shown to be safe and effective in a human clinical trial, have no theoretical limitation on plasmid size, do not provoke immune responses, and can be highly concentrated. An experimental study has shown that DNA nanoparticles can be targeted to different tissues within the eye by varying the site of injection (Farjo et al. 2006). Almost all cell types of the eye can be transfected by nanoparticles and produce robust levels of gene expression that are dose dependent. Subretinal delivery of these nanoparticles transfects nearly all of the photoreceptor population and produces expression levels almost equal to that of rhodopsin, the highest expressed gene in the retina. As no deleterious effects on retinal function have been observed, this treatment strategy appears to be clinically viable and provides a highly efficient nonviral technology to safely deliver and express nucleic acids in the retina and other ocular tissues. These findings have implications for the development of DNA-based therapeutics for various eye disorders, including retinitis pigmentosa, diabetic retinopathy, and macular degeneration.

Nanotechnology for Treatment for Age-Related Macular Degeneration

Although the retina is a fairly accessible portion of the CNS, there are virtually no treatments for early age-related macular degeneration (AMD), a degenerative retinal disease that causes progressive loss of central vision and is the leading cause of irreversible vision loss in persons over the age of 50. Drugs that inhibit vascular endothelial growth factor (VEGF) have proven effective in treating late-stage AMD, but drug delivery is a problem. Nanoparticles show considerable promise for drug delivery to the retina, for gene therapy, and for construction of prosthetic artificial retinas (Birch and Liang 2007).

Nanotechnology-Based Therapeutics for Eye Disorders

Use of Dendrimers in Ophthalmology

One of the main motives of using dendrimers in ophthalmology is to overcome the limitation of agents targeted to a single molecule or receptor. Dendrimers enable polyvalent medicines, larger molecules where several ligands can bind to several receptors in order to get the desired biological response. The use of dendrimers in drug delivery to the eye is also being explored to target multiple pathologies. Dendrimers can be used to prevent scar formation following eye surgery. Another use would be to disrupt inflammation and angiogenesis in the posterior chamber of the eye. Synthetically engineered dendrimers can be tailored to have defined immunomodulatory and antiangiogenic properties; they can be used synergistically to prevent scar tissue formation.

Nanotechnology for Prevention of Neovascularization

Some of the strategies for treatment of eye disorders involve prevention of neovascularization. Examples of how nanotechnology can refine these procedures are as follows.

Photodynamic therapy (PDT) has been used for exudative age-related macular degeneration (AMD). This therapy can be refined by using a supramolecular nanomedical device, i.e., a novel dendritic photosensitizer (DP) encapsulated by a polymeric micelle formulation. The characteristic dendritic structure of the DP prevents aggregation of its core sensitizer, thereby inducing a highly effective photochemical reaction. With its highly selective accumulation on choroidal neovascularization (CNV) lesions, this treatment results in a remarkably efficacious CNV occlusion with minimal unfavorable phototoxicity.

A long-term study was performed into the use of a lipophilic amino-acid dendrimer to deliver an antivascular endothelial growth factor (VEGF) oligonucleotide (ODN-1) into the eyes of rats and inhibit laser-induced CNV (Marano et al. 2005). In addition, the uptake, distribution, and retinal tolerance of the dendrimer plus oligonucleotide conjugates were examined. Analysis of fluorescein angiograms of laser-photocoagulated eyes revealed that dendrimer plus ODN-1 significantly inhibited the development of CNV for 4–6 months by up to 95 % in the initial stages. Eyes similarly injected with ODN-1 alone showed no significant difference. Intravitreally injected ODN-1 was absorbed by a wide area of the retina and penetrated all retinal cell layers to the retinal pigment epithelium. Ophthalmological examinations indicated that the dendrimers plus ODN-1 conjugates were well tolerated in vivo, which was later confirmed using immunohistochemistry, which showed no observable increase in antigens associated with inflammation. The use of such dendrimers may provide a viable mechanism for the delivery of therapeutic oligonucleotides for the treatment of angiogenic eye diseases.

Nanoparticles as Nonviral Vectors for Gene Therapy of Retinal Disorders

DNA nanoparticles have been shown to correct visual defects in a mouse model of retinitis pigmentosa by delivery of normal copies of genes into photoreceptor cells. DNA nanoparticles may also offer the potential to provide effective treatments for more complex eye disorders such as diabetic retinopathy, macular degeneration, and various diseases that injure ganglion cells and the optic nerve. There is a plan to move these studies to a potential human clinical trial.

Nonviral vectors based on solid lipid nanoparticles (SLN) have been investigated for the treatment of X-linked juvenile retinoschisis (XLRS) by gene therapy (Delgado et al. 2012). After ocular administration of the dextran-protamine-DNA-SLN complex to the rat eyes, expression of EGFP was detected in various types of cells depending on the administration route. The vectors were also able to transfect corneal cells after topical application. Results of the study demonstrated the potential usefulness of nonviral vectors loading XLRS1 plasmid and provided evidence for their potential application for the treatment of degenerative retina disorders as well as diseases of ocular surface.

Nanobiotechnology for Treatment of Glaucoma

Glaucoma involves abnormally high pressure of the fluid inside the eye, which, if left untreated, can result in damage to the optic nerve and vision loss. Human carbonic anhydrase (hCAII), a metalloenzyme that catalyzes the reversible hydration of carbon dioxide to bicarbonate, is associated with glaucoma. High pressure occurs, in part, because of a buildup of carbon dioxide inside the eye. Drug therapy is aimed at blocking hCAII. Carbonic anhydrase inhibitors such as acetazolamide, methazolamide, ethoxzolamide, and dichlorophenamide were and still are widely used systemic antiglaucoma drugs. Their mechanism of action consists in inhibition of CA isozymes present in ciliary processes of the eye with the consequent reduction of bicarbonate and aqueous humor secretion and of elevated intraocular pressure.

However, barely 1–3 % of existing glaucoma medicines penetrate into the eye. Earlier experiments with nanoparticles have shown not only high penetration rates but also little patient discomfort. The miniscule size of the nanoparticles makes them less abrasive than some of the complex polymers now used in most eye drops. A specialized cerium oxide nanoparticle has been bound with a compound that has been shown to block the activity of an hCAII (Patil et al. 2007). Carboxybenzenesulfonamide, an inhibitor of the hCAII enzyme, was attached to nanoceria particles using epichlorohydrin as an intermediate linkage. Along with the CA inhibitor, a fluorophore (carboxyfluorescein) was also attached on the nanoparticles to enable the tracking of the nanoparticles in vitro as well as in vivo. X-ray photoelectron spectroscopic studies carried out at each reaction step confirmed the successful derivatization of the nanoceria particles. The attachment of carboxyfluorescein was also confirmed by confocal fluorescence microscopy. Preliminary studies suggest that carboxybenzenesulfonamide-functionalized nanoceria retains its inhibitory potency for hCAII.

Dendrimers and other nanotechnology devices are being used for research on ophthalmic genetics and genomics as applied to glaucoma. Apart from study of the disease, there is a search for improved methods of drug delivery. The investigators are interested in devices that will enable (1) genotyping in real time in clinical or point-of-care situations, (2) assaying levels of gene expression in situ to enable evaluation of effects during attempted interventions, and (3) methods to deliver gene therapy to specific cell types without using viral vectors.

One of the problems in treating glaucoma is to get the drug into the cells rather than in the surrounding space or on cell surface. Trabecular meshwork (TM) cells have phagocytic properties and could be induced to take up a variety of carrier particles. A latrotoxin analog can be used to direct the dendrimer to the latrotoxin receptor on the surface of the TM cells. The presence of a relatively specific protease, PCSK1, in TM cells offers the possibility of genetically engineering the protein to be attached to the particle. This can be done via a cleavable peptide attachment that will result in release from the dendrimer once it enters the cell in which the protease is located.

Chapter 14
Nanomicrobiology

Introduction

Microbiology plays an important role in practice of medicine. Nanodiagnostics have refined the detection of infectious diseases and many new nanotechnology-based therapies, particularly of viral diseases, are in development.

Nanodiagnosis of Infections

Nanobiotechnology-based molecular diagnostic techniques were described in Chap. 4. Examples of specific applications for detection of infectious agents will be given in this chapter.

Detection of Viruses

Several nanotechnology-based methods have already been described in this chapter, including ferrofluid magnetic nanoparticles, ceramic nanospheres, and nanowire sensors for viruses. Role of cantilevers, SWCNTs, QDs, and surface-enhanced Raman scattering (SERS) will be described in this section.

Cantilever Beams for Detection of Single Virus Particles

Microfabrication and application of arrays of silicon cantilever beams as microresonator sensors with nanoscale thickness has been applied to detect the mass of individual virus particles (Gupta et al. 2004). The dimensions of the fabricated cantilever beams were in the range of 4–5 μm in length, 1–2 μm in width, and 20–30 nm in thickness. The virus particles used in the study were vaccinia virus, which is a member of the Poxviridae family and forms the basis of the smallpox vaccine. The frequency spectra of the cantilever beams, due to thermal and ambient noise, were measured using a laser Doppler vibrometer under ambient conditions.

K.K. Jain, *The Handbook of Nanomedicine*, DOI 10.1007/978-1-61779-983-9_14,

The change in resonant frequency as a function of the virus particle mass binding on the cantilever beam surface forms the basis of the detection scheme. This device can detect a single vaccinia virus particle with an average mass of 9.5 fg. Such devices can be very useful as components of biosensors for the detection of airborne virus particles. This technology has been refined as described under nanocantilever biosensors.

Carbon Nanotubes as Biosensors for Viruses

SWCNTs have been functionalized under ambient conditions with either the Knob protein domain from adenovirus serotype 12 (Ad 12 Knob) or its human cellular receptor, the CAR protein, via diimide-activated amidation (Zhang et al. 2007). The biological activity of Knob protein immobilized on the nanotube surfaces was confirmed by using its labeled conjugate antibody. The activity and specificity of bound CAR on SWCNTs was evaluated first, in the presence of fluorescently labeled Knob, which interacts specifically with CAR, and second, with a negative control protein, YieF, which is not recognized by biologically active CAR proteins. In addition, current-gate voltage measurements on a dozen nanotube devices explored the effect of protein binding on the intrinsic electronic properties of the SWCNTs and also demonstrated the devices' high sensitivity in detecting protein activity. All data show that both Knob and CAR immobilized on SWCNT surfaces fully retain their biological activities, suggesting that SWCNT-CAR complexes can serve as biosensors for detecting environmental adenoviruses.

Electric Fields for Accelerating Detection of Viruses

The rapid detection of viruses in biological samples is of increasing interest, particularly with the recent emergence of new viruses. A device that can quickly and easily detect them is difficult to construct because viral particles are present at such low concentrations in biological samples, such as blood. Typical procedures involve using passive diffusion to get the viral particles to bind to an antibody, a slow process that is not feasible for many applications, such as on the battlefield, where quick results are critical. However, liquid crystals are known to amplify signals from low concentrations of viral particles, quickly indicating whether or not a virus is present on a surface. A study to speed up the collection of viral particles, particularly vesicular stomatitis virus, used directed assembly, in which external electrical and fluid flow fields are designed in order to drive nanoparticles to specific locations and in specific concentrations on a substrate (Docoslis et al. 2007). Electrical fields are advantageous because by designing the electrodes in a certain way, engineers can control directionality and intensity of electrical forces acting on nanoparticles. By using electrodes separated by just a few micrometers together with electrothermally induced fluid flow, one can accelerate the transport of viral particles from aqueous suspensions with physiological ionic strength to specific points on a surface, allowing them to reach local concentrations high enough to enable subsequent rapid detection. These observations provide a potentially useful approach for addressing a bottleneck in the development of devices that allow for rapid sampling and "on-the-spot" detection of infectious biological agents such as viruses.

QD Fluorescent Probes for Detection of Respiratory Viral Infections

Respiratory syncytial virus (RSV) causes about one million deaths annually worldwide. RSV mediates serious lower respiratory tract illness in infants and young children and is a significant pathogen of the elderly and immune-compromised. Although it is only life-threatening in one case out of every 100, it infects virtually all children by the time they are 5 years old. Approximately 120,000 children are hospitalized with RSV in the USA each year. Few children in the USA die from RSV, but it causes 17,000–18,000 deaths annually among the elderly.

Rapid and sensitive RSV diagnosis is important for infection control and efforts to develop antiviral drugs. Current RSV detection methods are limited by sensitivity and/or time required for detection, which can take 2–6 days. This can delay effective treatment. Antibody-conjugated nanoparticles rapidly and sensitively detect RSV and estimate relative levels of surface protein expression. A major development is use of dual-color QDs or fluorescence energy transfer nanobeads that can be simultaneously excited with a single light source.

A QD system can detect the presence of particles of the RSV in a matter of hours. It is also more sensitive, allowing it to detect the virus earlier in the course of an infection. When an RSV virus infects lung cells, it leaves part of its coat containing F and G proteins on the cell's surface. QDs have been linked to antibodies keyed to structures unique to RSV's coat. As a result, when QDs come in contact with either viral particles or infected cells, they stick to their surface. In addition, co-localization of these viral proteins was shown using confocal microscopy. The potential benefits for such an early detection system are that it can:

1. Increase the proper use of antiviral medicines. Although such medicines have been developed for some respiratory viruses, they are not used often as therapy because they are only effective if given early in the course of infection. By the time current tests identify the virus, it is generally too late for them to work.
2. Reduce the inappropriate use of antibiotics. Currently, physicians often prescribe antibiotics for respiratory illnesses. However, antibiotics combat respiratory illness caused by bacteria and are ineffective on viral infections. An early virus detection method would reduce the frequency with which doctors prescribe antibiotics for viral infections inappropriately, thereby reducing unnecessary antibiotic side effects and cutting down on the development of antibiotic resistance in bacteria.
3. Allow hospital personnel to isolate RSV patients. RSV is extremely infectious so early detection would allow hospital personnel to keep the RSV patients separate from other patients who are especially susceptible to infection, such as those undergoing bone-marrow transplants.

Currently, there are three diagnostic tests available for identifying respiratory viruses like RSV. The "gold standard" involves incubating an infected sample in a tissue culture for a few days and then using a fluorescent dye to test for the presence of the virus. The main problem with this technique is that the virus is multiplying in the patient at the same time as it is growing in the culture. This has caused many hospitals to switch to real-time PCR, which is extremely sensitive but still takes several hours because of the need for a technician well trained in molecular biologist to conduct the test in a reference laboratory. The third method, the antigen test, takes ~30 min, but it is not sensitive enough to detect the presence of the virus at the early stages of an infection. By comparison, the QD method takes 1–2 h and is even more sensitive than real-time PCR. It can detect the presence of RSV within an hour after the virus is added to a culture. QDs have an advantage over many traditional fluorophores because their fluorescence properties can be finely tuned and they are resistant to photobleaching (Halfpenny and Wright 2010).

Verigene Respiratory Virus Plus Assay

Verigene (Nanosphere Inc.) platform is based on a direct genomic detection technology that uses DNA probes coated with gold nanoparticles to identify a unique oligonucleotide sequence and combines it with biobarcode protein detection technology. Verigene respiratory virus plus assay, which runs on an automated sample-to-result molecular diagnostic instrument, is more sensitive than currently available rapid tests. It combines optimized ease of use and turnaround time not found in either traditional culture methods or the currently available molecular tests for viruses and is cleared by the FDA for detection of influenza and RSV.

Surface-Enhanced Raman Scattering for Detection of Viruses

Although surface-enhanced Raman scattering (SERS) is well known, previous attempts to use spectroscopy to diagnose viruses failed because the signal produced is inherently weak. A spectroscopic assay based on SERS using silver nanorods, which significantly amplify the signal, has been developed for rapid detection of trace levels of viruses with a high degree of sensitivity and specificity (Shanmukh et al. 2006). The technique measures the change in frequency of a near-infrared laser as it scatters viral DNA or RNA. That change in frequency is as distinct as a fingerprint. This novel SERS assay can detect spectral differences between viruses, viral strains, and viruses with gene deletions in biological media. The method provides rapid diagnostics (<1 min) for detection and characterization of viruses generating reproducible spectra without viral manipulation. It is also quite cheap and is very reproducible.

A dual-mode molecular beacon, based on a combined SERS and fluorescent molecular beacon assay that is assembled on nanobarcode particles, has been developed and used to measure unlabeled human viral RNA (Sha et al. 2007). The molecular beacon probe is a single-stranded oligonucleotide that has been designed with a hairpin structure that holds the dye at 3′-end close to the particle surface when the probe is attached through a 5′-thiol group. In this configuration, the SERS spectrum of the label is obtained and its fluorescence quenched because the dye is in very close proximity to a noble metal surface with nanoscale features. The SERS signal decreases and the fluorescence signal increases when target viral RNA is captured by this molecular beacon probe. In addition, a HCV RT-PCR product is detected using this dual-mode beacon. The development of a multiplexed, label-free assay system with the reassurance offered by detection of two distinctly separate signals offers significant benefits for rapid molecular diagnostics.

Detection of Bacteria

The rapid and sensitive detection of pathogenic bacteria is extremely important in diagnosis of infections at POC. Limitations of most of the conventional diagnostic methods are lack of ultrasensitivity or delay in getting results. Nanobiotechnology has made a significant contribution to improvements in detection of bacterial infections.

Nanoparticle-Based Methods for Bacterial Detection

Bioconjugated nanoparticle-based assays for in situ pathogen quantification can detect a single bacterium within minutes. Such nanoparticles provide high fluorescent signals for bioanalysis and can be easily incorporated in a biorecognition molecule such as an antibody. The antibody-conjugated nanoparticles can readily and specifically identify a variety of bacteria such as *Escherichia coli O157:H7* through antibody-antigen interaction and recognition. This method can be applied to multiple bacterial samples with high throughput and has a potential for application in ultrasensitive detection of disease biomarkers and infectious agents.

Verigene gram-positive blood culture (BC-GP) test (Nanosphere Inc.) is a multiplexed, nanoparticle-based automated nucleic acid test for the identification of genus, species, and genetic resistance determinants for a broad panel of the most common gram-positive blood culture isolates. Whereas conventional microbiological methods may require 2–4 days to produce bacterial identification and resistance results, the Verigene BC-GP test provides results within 2.5 h of blood culture positivity. The Verigene system's unique instrumentation with <5 min of user hands-on time per test enables true random access test processing directly from positive blood culture bottles.

Multifunctional magnetic-plasmonic Fe_3O_4-Au core-shell nanoparticles (Au-MNPs) have been prepared for simultaneous fast concentration of bacterial cells by applying an external point magnetic field and sensitive detection and identification of bacteria using SERS (Zhang et al. 2012a). Surrounded by dense uniformly packed Au-MNPs, bacteria can be sensitively and reproducibly detected directly using SERS. This method can be used in molecular diagnostics of bacterial infections.

QDs for Detection of Bacterial Infections

Detection of single-molecule hybridization has been achieved by a hybridization detection method using multicolor oligonucleotide-functionalized QDs as nanoprobes

(Ho et al. 2005). In the presence of various target sequences, combinatorial self-assembly of nanoprobes via independent hybridization reactions leads to the generation of discernible sequence-specific spectral codings. This method can be used for genetic analysis of anthrax pathogenicity by simultaneous detection of multiple relevant sequences.

Fluorescent QDs coated with zinc(ii)-dipicolylamine coordination complexes can selectively stain a rough *E. coli* mutant that lacks an O-antigen element and permit optical detection in a living mouse leg infection model (Leevy et al. 2008). QDs have potential use as labeling agents for bacteriophages associated with bacterial infections. A rapid and simple method has been reported that combines in vivo biotinylation of engineered host-specific bacteriophage and conjugation of the phage to streptavidin-coated QDs (Edgar et al. 2006). The method provides specific detection of as few as 10 bacterial cells per ml in experimental samples, with an approximately 100-fold amplification of the signal over background in 1 h. The method can be applied to any bacteria susceptible to specific phages and would be particularly useful for detection of bacterial strains that are slow growing such as Mycobacterium or are highly infectious such as *B. anthracis*. To monitor the infection of *E. coli* cells by light microscopy, procedures have been developed for the tagging of mature bacteriophages with QDs (Edgar et al. 2008). Fluorescent QDs have been used for detection and sorting of pathogenic bacteria by flow cytometry (Zahavy et al. 2012).

Detection of Fungi

Conventional methods for diagnosis of invasive fungal infections in the clinical microbiology laboratory are time-consuming processes or result in misidentification of the fungus due to low sensitivity or low specificity. There is a need for improvement of methods of detection of fungi, and nanobiotechnology-based techniques have been used.

Nano-amplification Technique for the Detection of Fungal Pathogens

In one method, fungal ribosomal DNA was amplified using PCR and the products were hybridized with the species-specific probes immobilized on the surface of a microarray where hybridizing signals were enhanced with gold nanoparticles and silver deposition (Lu et al. 2010). The probes were designed to detect several different clinical pathogenic fungi using a flatbed scanner or visually. The technique showed higher efficiency, specificity, and sensitivity compared with other methods.

Magnetic Nanoparticle-Based Technique

Magnetic resonance detection (T2 Biosystems) uses magnetic nanoparticles coupled with reagents to quickly detect, within minutes, the presence of specific substances in solution using a miniaturized, portable MRI instrument. Detection of the high-level MR signal from the solution enables the detection of low concentrations of target agents or substances. Unlike most existing diagnostic detection techniques which are based on optical detection methods that require pure samples and multiple processing steps, T2's technology is not optical and therefore does not require purification of biological samples. The significant advantage allows the T2 system to perform single-step processing and rapid turnaround times without the need for trained technicians. Furthermore, the technology can accurately identify almost any specimen, including proteins, nucleic acids, or enzymes; microbes; or small molecule drug compounds within almost any sample, including whole blood, plasma, serum, and urine. This method has been used to analyze whole blood specimens from patients with five different types of *Candida* spp. infections and is currently in clinical trials with an aim for regulatory approval for diagnosis of Candida infection.

Nanobiotechnology and Virology

Study of Interaction of Nanoparticles with Viruses

Scanning surface confocal microscopy, simultaneous recording of high-resolution topography, and cell surface fluorescence in a single scan enables imaging of individual fluorescent particles in the nanometer range on fixed or live cells. This technique has been used to record the interaction of single virus-like particles with the cell surface and demonstrated that single particles sink into the membrane in invaginations reminiscent of pinocytic vesicles. This method enables elucidation of the interaction of individual viruses and other nanoparticles, such as gene therapy vectors, with target cells.

Silver nanoparticles undergo a size-dependent interaction with HIV-1 and particles in the range of 1–10 nm attached to the virus (Elechiguerra et al. 2005). The regular spatial arrangement of the attached nanoparticles, the center-to-center distance between nanoparticles, and the fact that the exposed sulfur-bearing residues of the glycoprotein knobs would be attractive sites for nanoparticle interaction suggest that silver nanoparticles interact with the HIV-1 virus via preferential binding to the gp120 glycoprotein knobs. Due to this interaction, silver nanoparticles inhibit the virus from binding to host cells, as demonstrated in vitro.

Study of Pathomechanism of Viral Diseases

Research in nanobiotechnology may be helpful in understanding the pathomechanism of viral diseases and devising strategies for treatment. An example is the neurotropic herpes simplex virus (HSV), which infects mucosal epithelia and enters nerve terminals, from where it travels in axons to dorsal root ganglia neurons and delivers its genome into the nucleus of the cell body. In the nucleus, the genome may give rise to infectious progeny or become latent with little gene expression. The silenced genome can be reactivated upon stress and establish a productive infection in the peripheral nervous system and, later, also in the mucosal periphery. To achieve this, a virus must elude host restrictions at multiple levels, including entry, cytoplasmic transport, replication, innate and adaptive immune recognition, and egress from the infected cell.

Research on virus nanoparticles has provided cues to the regulation of cytoplasmic transport. Viruses that replicate their genomes in the nucleus make use of the microtubule and the actin cytoskeleton as molecular motors for trafficking toward the nuclear membrane during entry and the periphery during egress after replication. Analyzing the underlying principles of viral cytosolic transport will be helpful in the design of viral vectors to be used in research as well as human gene therapy and in the identification of new antiviral target molecules (Dohner and Sodeik 2005).

Transdermal Nanoparticles for Immune Enhancement in HIV

DermaVir Patch (Genetic Immunity) is a transdermally delivered nanomedicine to enhance de novo HIV-specific memory T cell responses of HIV-infected individuals and improve the ability of their own immune system to control the disease by killing only HIV-infected cells. Mice receiving DermaVir formulated with HIV-1 Gag plasmid in the presence of IL-7- or IL-15-encoding plasmid have significantly enhanced Gag-specific central memory T cells, as measured by a peptide-based cultured IFN-γ ELISPOT (Calarota et al. 2008). In a DermaVir prime/vaccinia vector boost regimen, the inclusion of IL-15 together with DermaVir significantly improved Gag-specific effector-memory T cell responses. This study demonstrates IL-15 is a promising DermaVir adjuvant to enhance antigen-specific central memory type T cells in a prime-boost setting. It is in phase II clinical trials.

Nanofiltration to Remove Viruses from Plasma Transfusion Products

One of the complications of blood transfusion is transmission of viral infections. Nanofiltration, use of nanotechnology in viral removal filtration systems, is an important safety step in the manufacture of plasma-derived coagulation factor concentrates and other biopharmaceutical products from human blood. Nanofiltration of plasma products has already been carried out since the early 1990s to improve margin of viral safety, as a complement to the viral reduction treatments, such as solvent-detergent and heat treatments, which are applied for the inactivation of HIV as well as hepatitis B and C viruses. The main reason for the introduction of nanofiltration was the need to improve product safety against non-enveloped viruses and to provide a possible safeguard against new infectious agents potentially entering the human plasma pool. Nanofiltration has gained quick acceptance as it is a relatively simple manufacturing step that consists in filtering protein solution through membranes with nanopores (pore size typically 15–40 nm) under conditions that retain viruses by a mechanism largely based on size exclusion. Recent large-scale experience throughout the world has now established that nanofiltration is a robust and reliable viral reduction technique that can be applied to essentially all plasma products. Many of the licensed plasma products are currently nanofiltered. The technology has major advantages as it is flexible and it may combine efficient and largely predictable removal of a wide range of viruses. Compared with other viral reduction means, nanofiltration may be the only method to date permitting efficient removal of enveloped and non-enveloped viruses under conditions where 90–95 % of protein activity is recovered. New data indicate that nanofiltration may also remove prions, opening new perspectives in the development of this technique.

Shortcomings of some membranes are that they often form pin-holes and cracks during the fabrication process, resulting in wasted membranes. Specially designed ceramic membranes have been used as nanomesh for nanofiltration as they are less likely to be damaged during manufacture and have the potential to remove viruses from water, air, and blood. Mesh structure, which is the most efficient form of filtration, has been successfully constructed on a nanoscale with ceramic fibers. This modification has increased the rates of flow that pass through the membranes tenfold compared with current ceramic membranes while maintaining the efficiency of capturing over 96 % of the unwanted particles. This technology could be used to filter airborne viruses such as the severe acute respiratory syndrome (SARS) and the avian flu virus. It may be possible to filter HIV from human blood to treat patients with AIDS.

Role of Nanobacteria in Human Diseases

Nanobacteria are mineral-forming, sterile-filterable, slow-growing gram-negative infectious agents. They are detected in bovine/human blood and urine. Nanobacteria-like particles have been detected in synovial fluids of arthritis patients and were shown to gradually increase in number and in size in culture (Tsurumoto et al. 2006). Nanobacteria have been implicated in a variety of human diseases associated with pathological calcification. Their most remarkable characteristic is the formation of carbonate apatite crystals of neutral pH and at physiologic phosphate and calcium concentrations. The extracellular mineralization forms a hard protective shelter for these hardy microorganisms and enables them to survive conditions of physical stress that would be lethal to most other bacterial species. The Olavi Kajander group (Finland) suggests that the apatite produced by nanobacteria may play a key role in the formation of all kidney stones, by providing a central calcium phosphate deposit around which other crystalline components can collect.

Nanobacteria seem to be causative agents of diseases related to biomineralization processes. Nanobacteria are also associated with calcified geological specimens, human kidney stones, and psammoma bodies in ovarian cancer. Much research has focused attention on the potential role these particles may play in the development of urologic pathology, including polycystic kidney disease, renal calculi, and chronic prostatitis. Nanobacteria may be an important etiological factor for type III prostatitis, which was reproduced in rat prostate infection models by infusing nanobacteria suspension transurethrally (Shen et al. 2010). Recent clinical research on agents targeting nanobacteria has proven effective in treating some patients with refractory category III prostatitis.

Nature of Nanobacteria

According to their 16S rDNA structure, nanobacteria belong to the alpha-2 Proteobacteria, subgroup, which includes the Brucella and Bartonella species. Nanobacterium sanguineum (nanobacteria) is the smallest self-replicating organism ever detected – at 50–500 billionths of a meter, 1/1000th the size of the smallest previously known bacteria. Primordial proteins in nanobacteria, only recently identified in the atmosphere, could play a significant role in clouds, accelerating the formation of cloud droplets and interconnecting nanobacteria (and possibly nanobacteria and other microorganisms), thus enhancing their chances to eventually reach the Earth.

Several research studies indicate that nanobacteria are alive, but it is still unclear whether they represent novel life forms, overlooked nanometer-size bacteria, or some other primitive self-replicating microorganisms. A study has shown that $CaCO_3$ precipitates prepared in vitro are remarkably similar to purported nanobacteria in terms of their uniformly sized, membrane-delineated vesicular shapes, with cellular division-like formations and aggregations in the form of colonies (Martel and Young 2008). The gradual appearance of nanobacteria-like particles in incubated human serum as well as the changes seen with their size and shape can be influenced and explained by introducing varying levels of CO_2 and $NaHCO_3$ as well as other conditions known to influence the precipitation of $CaCO_3$. Western blotting reveals that the monoclonal antibodies, claimed to be specific for nanobacteria, react in fact with serum albumin. Furthermore, nanobacteria-like particles obtained from human blood are able to withstand high doses of irradiation up to 30 kGy, and no bacterial DNA is found by performing broad-range PCR amplifications. These findings provide a more plausible abiotic explanation for the unusual properties of purported nanobacteria.

Nanobacteria and Kidney Stone Formation

Approximately 12 % of men and 5 % of women develop kidney stones by the time they reach the age of 70 years, but exactly how kidney stones form is not known. Kidney stones can be debilitating and recur in 50 % of patients within 5 years. Kidney stone formation is considered to be a multifactorial disease in which the defense mechanisms and risk factors are imbalanced in favor of stone formation. One theory is that if nanoparticles accumulate in the kidney, they can form the focus of subsequent growth into larger stones over months to years. Other factors, such as physical chemistry and protein inhibitors of crystal growth, also play a role.

Mineral-forming nanobacteria are active nidi that attach to, invade, and damage the urinary epithelium of collecting ducts and papilla forming the calcium phosphate center(s) found in most kidney stones. Scientists at NASA have used multiple techniques to determine that nanobacteria infection multiplies faster in spaceflight-simulated conditions than on earth. Nanobacteria are considered to initiate kidney stone formation as they grow faster in a microgravity environment and may explain why astronauts get kidney stones on space missions. This discovery may prove to be critical for future exploratory missions to the moon and Mars. For further proof to this hypothesis, screening of the nanobacterial antigen and antibody level in flight crew before and after flight would be necessary. This concept also opens the door for new diagnostic and therapeutic techniques addressing nanobacterial infection in kidney stones.

Nanoparticles have been isolated and cultured from the majority of renal stones obtained at the time of surgical resection (Kumar et al. 2006). Isolates were susceptible to selected metabolic inhibitors and antibiotics and contained conserved bacterial proteins and DNA. These results suggest that renal stone formation is unlikely to be driven solely by physical chemistry; rather, it is critically influenced by specific proteins and cellular responses, and understanding these events will provide clues toward novel therapeutic targets. Using high-spatial and energy resolution near-edge x-ray absorption fine structure at the 25 nm spatial scale, it is possible to define a biochemical signature for cultured calcified bacteria, including proteins, polysaccharides, nucleic acids, and hydroxyapatite (Benzerara et al. 2006). These preliminary studies suggest that nanoparticles isolated from human samples share spectroscopic characteristics with calcified proteins.

Nanobacteria in Cardiovascular Disease

Nanometer-scale objects, spherical in shape and ranging in size from 30 to 100 nm with a spectral pattern of calcium and phosphorus (high-energy dispersive spectroscopy), have been identified with positive immunostaining in surgical specimens from patients with cardiovascular pathology. Nano-sized particles cultured from calcified but not from non-calcified aneurysms were recognized by a DNA-specific dye and incorporated radiolabeled uridine and, after decalcification, appeared via electron microscopy to contain cell walls. Nanometer-scale particles similar to those described as nanobacteria isolated from geological specimens and human kidney stones can be visualized in and cultured from human-calcified cardiovascular tissue. In further studies, nanoparticles were found near plaque-filled arteries in animal models. These observations suggest that nanoparticles potentially represent a previously unrecognized factor in the development of arteriosclerosis and calcific arterial disease.

Nanotechnology-Based Microbicidal Agents

Nanoscale Bactericidal Powders

Certain formulations of nanoscale powders possess antimicrobial properties. These formulations are made of simple, nontoxic metal oxides such as magnesium oxide (MgO) and calcium oxide (CaO, lime) in nanocrystalline form, carrying active forms of halogens, e.g., MgO.Cl2 and MgO.Br2. When these ultrafine powders contact vegetative cells of *E. coli*, *Bacillus cereus*, or *Bacillus globigii*, over 90 % are killed within a few minutes. Likewise, spore forms of the Bacillus species are decontaminated within several hours. Dry contact with aflatoxins and contact with MS2 bacteriophage (surrogate of human enterovirus) in water also cause decontamination in minutes.

A nanopowder of MgO can scour contaminated rooms of anthrax spores. Unlike antibacterial gases and foams, which are messy, and corrosive and ruin electrical equipment, the powder can be sprayed into rooms and swept or vacuumed up. The chemical specks attract oppositely charged spores. The particles then chemically break down the spores' tough outer shell. Based on this technology, NanoScale Corporation markets a dry powder dubbed FAST-ACT® (First Applied Sorbent Treatment-Against Chemical Threats) that decomposes toxic chemicals. The powder contains reactive nanoparticles that attract and then break down at least 24 commonly transported toxic chemicals, including some acids. Unlike foams, the powder need not be wet to be effective and works on liquids and vapors.

Nanotubes for Detection and Destruction of Bacteria

A simple molecule has been synthesized from a hydrocarbon and an ammonium compound to produce a unique nanotube structure with antimicrobial capability (Lee et al. 2004). The quaternary ammonium compound is known for its ability to disrupt cell membranes and cause cell death, whereas the hydrocarbon diacetylene can change colors when appropriately formulated; the resulting molecule would have the desired properties of both a biosensor and a biocide.

The self-assembled nanotubes are perfectly uniform and organize themselves into an expanse of upright clusters that when magnified a million times resemble the fibers of a shag rug leading to the name "nanocarpet." The self-assembling nanotubes had all the same diameter (89 nm) and wall thickness (27 nm). The nanocarpet measures about 1 μm in height, approximately the same height as the free-form nanotubes. This alignment of nanotubes in the absence of a template is unprecedented and represents an important step toward rational design of bioactive nanostructures. In addition, because they form within hours under room-temperature conditions, the significant costs of synthesizing carbon nanotubes can be reduced. Normally a neutral color, when exposed to ultraviolet light, the nanotubes changed to a permanent deep blue. The process also chemically altered the nanotubes so that they became polymerized, giving them a more firm structure. Polymerized, these nanotubes could change from blue to other colors, depending on its exposure to different materials. For instance, in tests with acids and detergents, they turned red or yellow.

Because they display sensitivity to different agents by changing color, these nanotubes can be trained to kill bacteria. In the presence of *E. coli*, some strains of which are food-borne pathogens, the nanotubes turned shades of red and pink. Moreover, with the aid of an electron microscope, the researchers observed the tubes piercing the membranes of the bacteria like a needle being inserted into the cell. Both the polymerized (those that can change color) and the unpolymerized nanotube structures were effective antimicrobials, completely killing all the *E. coli* within an hour's time. The findings have implications for developing products that can simultaneously detect and kill biological weapons. The research, funded by the Department of Defense's Army Research Office, has as its goal the development of a paint that in the event of biological or chemical agents being deployed would change color and simultaneously destroy the deadly substances.

Carbon Nanotubes as Antimicrobial Agents

CNTs have the potential to address the challenges of combating infectious agents by both minimizing toxicity by dose reduction of standard therapeutics and allowing a multiple payload capacity to achieve both targeted activity and combating infectious strains, resistant strains in particular (Rosen and Elman 2009). One of their unique characteristics is the network of carbon atoms in the nanometer scale, allowing the creation of nanochannels via cellular membranes.

Attempts have been made to destroy anthrax spores by antimicrobial agents targeted to bind to carbohydrates on the spore surface but with limited success. SWCNTs have been successfully used as a truly unique scaffold for displaying multivalent monosaccharide ligands that bind effectively to anthrax spores with divalent cation mediation to cause significant spore aggregation (Wang et al. 2006). The work should have far-reaching implications in development of technologies to counteract bioterrorism such as by use of anthrax. For anthrax to be effective, it has to be made into a fine powder that can easily enter the lungs when inhaled. That nanotechnology-based agent clings to the anthrax spores to make their inhalation into the lungs difficult. Similar approach using sugar-coated carbon nanotubes to stop the spread of *E. coli* bacteria was tested successfully.

Nanoemulsions as Microbicidal Agents

The antimicrobial nanoemulsions (NanoBio) are emulsions that contain water and soya bean oil with uniformly sized droplets in the 200–400 nm range. These droplets are stabilized by surfactant and are responsible for the microbicidal activity. In concentrated form, the nanoemulsions appear as a white milky substance with a taste and consistency of cream. They can be formulated in a variety of carriers allowing for gels, creams, liquid products, etc. In most applications, the nanoemulsions become largely water-based and in some cases such as a beverage preservative, comprise 0.01 % or less of the resultant mixture. Laboratory results indicate a shelf life of at least 2 years and virtually no toxicity. NanoBio Corporation's nanoemulsions destroy microbes effectively without toxicity or harmful residual effects. The nanoparticles fuse with the membrane of the microbe, and the surfactant disrupts the membrane, killing the microbe. The classes of microbes eradicated are virus (e.g., HIV, herpes), bacteria (e.g., *E. coli*, *Salmonella*), spores (e.g., anthrax), and fungi (e.g., *Candida albicans*, *Byssochlamys fulva*). NB-402 (NanoBio), a nanoemulsion antimicrobial agent for the treatment of infection due to CF-related opportunistic pathogens, is in development (see Chap. 11). Clinical trials have shown efficacy in healing cold sores due to herpes simplex virus 1 and toenail fungus. The nanoemulsions also can be formulated to kill only one or two classes of microbes. Due in large part to the low toxicity profile, the nanoemulsions are a platform technology for any number of topical, oral, vaginal, cutaneous, preservative, decontamination, veterinary, and agricultural antimicrobial applications.

Since it is nontoxic and non-corrosive, nanoemulsion can be used to decontaminate personnel, equipment, terrain, structures, and water. Further, tests by DTRA (Defense Threat Reduction Agency), an agency of the US Department of Defense, have demonstrated that the nanoemulsion is a chemical decontaminating agent. The US Army tested the nanoemulsion and nine other biodecontamination technologies at against an anthrax surrogate. The nanoemulsion was one of four technologies that proved effective.

Silver Nanoparticle Coating as Prophylaxis Against Infection

The Institute for New Materials (Saarbrucken, Germany), a research institute specializing in applied nanotechnology applications, has developed a silver nanoparticles surface coating that is deadly to fungi and bacteria. The researchers added the germicidal ability by sprinkling copious amounts of silver nanoparticles through the coating material (every square centimeter contains more than one billion of the invisible particles) and aligning them so that they release a tiny number of silver ions. These ions are the death knell for fungus and bacteria that might have succeeded in gathering on the surface despite its already dirt-repellent qualities.

Applications include any surface where germs can gather and possibly endanger people's health. That includes surfaces in hospitals, public buildings, factories, or in the home. The coating could be applied to almost any surface that people touch often such as metal, glass or plastic and would remove the need for constant cleaning with liquid disinfectants, especially in areas where hygienic conditions are crucial. People who normally cannot use hearing aids that lie inside the ear because of the risk of infection of the auditory canal can safely wear nanocoated appliances.

Bio-Gate (Nürnberg, Germany) produces NanoSilver BG, a nanoporous silver powder with particle size ranging from 50 to 100 nm. It has a homogeneous distribution of nanoparticles in the material and anti-infective properties.

Silver nanoparticles have been incorporated in preparations for wound care to prevent infection. Acticoat bandages (Smith & Nephew) contain nanocrystal silver, which is highly toxic to pathogens in wounds.

AcryMed's silver nanoparticle technology, SilvaGard, involves coating with silver nanoparticles with size range of 2–20 nm in a stable solution and antimicrobial treatment levels last for more than a year. With other technologies, nano-based silver coatings must be applied through vapor deposition, which coats only on one side, whereas AcryMed technology is a solution that provides a complete surface treatment rather than a coating.

Nanotechnology-Based Antiviral Agents

Silver Nanoparticles as Antiviral Agents

Silver nanoparticles possess many unique properties that make them attractive for use in biological applications. Silver nanoparticles are used as surface coatings for prophylaxis of infections. It has been shown that 10 nm silver nanoparticles are bactericidal, and possible use of silver nanoparticles as an antiviral agent is being explored.

Silver nanoparticles undergo a size-dependent interaction with HIV-1 and particles in the range of 1–10 nm attached to the virus (Elechiguerra et al. 2005). The regular spatial arrangement of the attached nanoparticles, the center-to-center distance between nanoparticles, and the fact that the exposed sulfur-bearing residues of the glycoprotein knobs would be attractive sites for nanoparticle interaction suggest that silver nanoparticles interact with the HIV-1 virus via preferential binding to the gp120 glycoprotein knobs. Due to this interaction, silver nanoparticles inhibit the virus from binding to host cells, as demonstrated in vitro.

Silver nanoparticles are capable of inhibiting a prototype arenavirus, Tacaribe virus, at nontoxic concentrations and effectively inhibit arenavirus replication when administered prior to viral infection or early after initial virus exposure (Speshock et al. 2010). This suggests that the mode of action of viral neutralization by silver nanoparticles occurs during the early phases of viral replication.

Fullerenes as Antiviral Agents

A series of bis-fulleropyrrolidines bearing two ammonium groups have been synthesized and their activities against HIV-1 and HIV-2 have been evaluated (Marchesan et al. 2005). Two *trans* isomers were found to have interesting antiviral properties, confirming the importance of the relative positions of the substituent on the C60 cage. None of the compounds showed any inhibitory activity against a variety of DNA and RNA viruses other than HIV.

Cationic, anionic, and amino acid–type fullerene derivatives have shown inhibitory effect against HIV-reverse transcriptase and HCV (Mashino et al. 2005). Out of all derivatives of fullerenes, anionic fullerenes were found to be the most active. All the tried fullerene derivatives were more active than the non-nucleoside analog of HIV-RT inhibitor. The effect of long alkyl chains on fullerenes was not significant; rather, it depressed the inhibition strength. The two important targets for anti-HIV characteristics are the HIV-protease and HIV-reverse transcriptase. The molecular modeling experimental designs exhibit that C60 core could penetrate into hydrophobic binding site of HIV protease. However, the mechanism of this anti-HIV activity is through HIV-protease inhibition, which has not been experimentally demonstrated.

Gold Nanorod-Based Delivery of RNA Antiviral Therapeutics

The emergence of the pandemic 2009 H1N1 influenza virus has become a worldwide health concern. As drug resistance appears, a new generation of therapeutic strategies will be required. Use of RNA immune activator molecule is limited by their instability when delivered into cells, but this can be overcome by using a nanobiotechnology-based delivery system. Gold nanorods protect the RNA from degrading once inside cells while allowing for more selected targeting of cells. Usefulness of delivery of a biocompatible gold nanorod, GNR-5′PPP-ssRNA nanoplex, has been demonstrated for innate immune activation against type A influenza virus (Chakravarthy et al. 2010). In human respiratory bronchial epithelial cells, this nanoplex containing the single-stranded RNA molecule activated the retinoic acid-inducible gene I pathogen recognition pathway, resulting in increased expression of IFN (interferon)-β and other IFN-stimulated genes (e.g., PKR, MDA5, IRF1, IRF7, and MX1), which resulted in a decrease in the replication of H1N1 influenza viruses. The novelty of this approach is that most of RNA viruses share a common host-response immune pathway, and enhancement of the host immune response reduces the ongoing viral resistance generated through mutations. Diseases that could be effectively targeted with this new approach include any viruses that are susceptible to the innate immune response triggered by IFN. Animal studies have started based on these in vitro results, and further evaluation of biocompatible nanoplexes as unique antivirals for treatment of seasonal and pandemic influenza viruses is warranted.

Nanocoating for Antiviral Effect

Laboratory testing of the permanent nanocoating SERQET™ (LaamScience Inc.) showed that coating kills 99.9 % of influenza viruses and 99.99 % of vaccinia viruses that cause rash, fever, and head and body aches. This technology may enable one to protect oneself from virtually all viruses and bacteria by simply exposing a surface to light.

In 2006, Mass Transit Railway (MTR), the corporation that runs Hong Kong subway, announced that nano silver-titanium dioxide coating (NSTDC, a nontoxic disinfectant) will be applied to surfaces that passengers commonly touch in order to enhance hygiene levels in MTR stations and trains. The coating is manufactured using nanotechnology, which maximizes coverage and effectiveness of NSTDC. Developed in Japan, NSTDC is certified to be effective in killing a wide range of bacteria, viruses, and mold, including the H1N1 Influenza Virus A. It is used in hospitals, offices, and homes in Japan. NSTDC's main component, titanium dioxide (TiO2), has been approved for use in foods by the FDA and under the Public Health and Municipal Services Ordinance in Hong Kong.

Nanoviricides

Nanoviricides (NanoViricides Inc.) are polymeric micelles, which act as nanomedicines to destroy viruses. As defined by NanoViricides Inc., "a nanoviricide is a polymeric single chemical chain with covalently attached ligands that specify the virus target. The antiviral spectrum of the drug is determined by the specificity of the set of ligands attached to the chain, in addition to other functionally important aspects inherent in the chemistries." Nanoviricide is designed to seek a specific virus type; attach to the virus particle; engulf or coat the virus particle, thereby neutralizing the virus's infectivity; and destabilize and possibly dismantle the virus particle; and optionally it may also be made capable of attacking the viral genome, thereby destroying the virus completely. Active pharmaceutical ingredients are optional and can be hidden in the core of the nanoviricide missile.

In contrast to other approaches, a NanoViricide™ micelle can recognize and bind to more than one type of binding site on the virus. The NanoViricide™ system enables design of a drug that binds to more than one type of site – currently as many as three different sites, on the virus – for a highly effective attack. NanoViricides Inc. terms this as "multi-specific targeting." A NanoViricide™ drug goes much further than just blocking all of the binding sites of the virus. The base material of a NanoViricide™ is a specially designed polymeric micelle material. It has the ability to disassemble an HIV particle by itself. Thus, after coating the virus particle, the NanoViricide™ loosens the virus particle and weakens it. Some virus particles will even fall apart (uncoat). This provides a further therapeutic benefit. NanoViricides plans to enhance the viral disassembly capabilities of the NanoViricides™ by attaching specially designed "molecular chisels" to the NanoViricide™. Once the

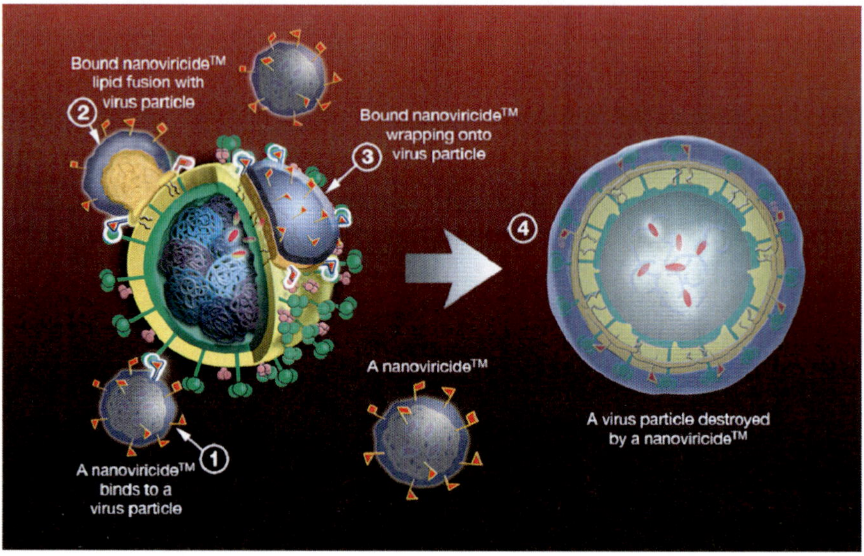

Fig. 14.1 Schematic representation of NanoViricide attacking a virus particle (Reproduced by permission of NanoViricide Inc.)

NanoViricide™ micelles coat the virus particle, the attached "molecular chisels" will go to work. They literally insert themselves into the virus coat at specific vulnerable points and pry apart the coat proteins so that the virus particle falls apart readily. The mechanism of action of NanoViricide is depicted schematically in Fig. 14.1.

NanoViricides have been compared to current approaches to viral diseases, which are seldom curative, and some of the advantages include the following:

- Specific targeting of the virus with no metabolic adverse effects on the host.
- The biological efficacy of NanoViricides drugs may be several orders of magnitude better than that of usual chemical drugs. This in itself may limit the potential for mutant generation.
- There are also other key aspects of the design of NanoViricides that are expected to lead to minimizing mutant generation.
- Nanoviricides are safe because of their unique design and the fact that they are designed to be biodegradable within the body.
- The new technology enables rapid drug development against an emerging virus, which would be important for global biosecurity against natural as well as manmade (bioterrorism) situations. It is possible to develop a research drug against a novel life-threatening viral disease within 3–6 weeks after the infection is found, i.e., as soon as an antibody from any animal source is available.
- It is possible to make a single NanoViricide drug that responds to a large number of viral threats by using targeting ligands against the desired set of viruses

in the construction of the drug. It is possible to "tune" the specificity and range (spectrum) of a NanoViricide drug within a virus type, subtype, or strain, by appropriate choices of the targeting ligand(s).
- The safety of NanoViricide drugs is proven now as they specifically attack the virus and not the host.
- A variety of formulations, release profiles, and routes of administration are possible.
- Low cost of drug development, manufacture, distribution.

Targets for this approach include influenzas, HIV, HCV, rabies, and other viruses. NanoViricide drug candidates are currently in preclinical studies. Clinical trials are planned. Initially injectable products are considered to be most effective but alternative routes of administrations such as nasal sprays and bronchial aerosols can also be developed.

NanoViricides Inc. developed and evaluated NanoViricides against influenza and avian flu H5N1 for efficacy and safety. FluCide™ nanoviricide is designed as a polymeric surfactant micelle which has covalently attached to its ligands that bind to the influenza A virus on conserved sites. The current drug candidate was effective against both H1 and H5 and different strains of H5. This is a direct result of using a conserved binding strategy, very similar to that used by zanamivir. However, FluCide is directed to bind to HA rather than to inhibit neuraminidase. Binding to HA followed by putative engulfment of the virus particle should lead to viral load reduction and therapeutic benefit. Preclinical studies have shown that FluCide does more than an antibody does, in that it completes the task of encapsulating and possibly dismantling the virus, rather than merely tagging it for the immune system as a foreign particle. In mouse model of common murine-adapted influenza A/H1N1, it would require about eight times greater dosage of oseltamivir as compared to FluCide to achieve the same survival advantage results. In cell cultures (MDCK as well as PMKC) against influenza A/H5N1 bird flu strains (Clade 1 as well as Clade 2), these drug candidates have shown as high as 70 % CPE inhibition. An intravenous preparation for systemic administration is in development with a ready-to-use preloaded syringe. Other routes of administration are being explored. A pre-IND briefing document about FluCide has been submitted to the FDA.

Chapter 15
Miscellaneous Healthcare Applications of Nanobiotechnology

Introduction

Nanobiotechnology impacts nearly all aspects of healthcare. Separate chapters were devoted to major therapeutic areas. Other areas which are not well defined or specialties where the use of nanobiotechnology is still limited are all included in this chapter.

Nanoimmunology

Allergic and immune disorders are leading cause of illness. Although various treatments have been developed to control allergy, no cure has yet been found. Nanobiotechnology is now being applied to tackle allergic and immune disorders to advance the emerging field of medicine known as nanoimmunology. The immune system can protect as well as cause harm, so there is need to help manage the harmful effects.

Mast cells are responsible for causing allergic response and are stuffed with granules containing histamine. They are present in nearly all tissues except blood. When mast cells are triggered, inflammatory substances such as histamine, heparin, and a number of cytokines are quickly released into the tissues and blood, promoting an allergic response. Fullerenes (buckyballs) are able to interrupt the allergy/immune response by suppressing a fundamental process in the mast cells that leads to the release of histamine. Human mast cells and peripheral blood basophils exhibit a significant inhibition of IgE-dependent mediator release when preincubated with C60 fullerenes (Ryan et al. 2007). Protein microarray demonstrated that inhibition of mediator release involves profound reductions in the activation of signaling molecules involved in mediator release and oxidative stress. Follow-up studies demonstrated that the tyrosine phosphorylation of Syk was dramatically inhibited in Ag-challenged cells first incubated with fullerenes. In addition, fullerene preincubation significantly inhibited IgE-induced elevation in cytoplasmic reactive oxygen species levels. Furthermore, fullerenes prevented the in vivo release of histamine and drop in core body temperature in vivo using a mast cell-dependent model of

K.K. Jain, *The Handbook of Nanomedicine*, DOI 10.1007/978-1-61779-983-9_15,

anaphylaxis. These findings identify a new biological function for fullerenes and may represent a novel way to control mast cell-dependent diseases including asthma, inflammatory arthritis, heart disease, and multiple sclerosis.

Nanohematology

Artificial Red Cells

The artificial mechanical red blood cell, called respirocyte, measures about 1 μm in diameter and just flows along the bloodstream (Freitas 1998). It is a spherical nanorobot made of 18 billion atoms. The respirocyte is equipped with a variety of chemical, thermal, and pressure sensors and an onboard nanocomputer. This device is intended to function as an artificial erythrocyte, duplicating the oxygen and carbon dioxide transport functions of red cells, mimicking the action of natural hemoglobin-filled red blood cells. It is expected to be capable of delivering 236 times more oxygen per unit volume than a natural red cell. Specially installed equipment enables this device to display many complex responses and behaviors. Additionally, it has been designed to draw power from abundant natural serum glucose supplies and thus is capable of operating intelligently and virtually indefinitely, while red blood cells have a natural life span of 4 months.

Feraheme

Feraheme (ferumoxytol) is a superparamagnetic iron oxide nanoparticle coated with a low molecular weight semi-synthetic carbohydrate. It helps to isolate the bioactive iron from plasma components until the iron-carbohydrate complex enters the reticuloendothelial system macrophages of the liver, spleen, and bone marrow. The iron is released from the iron-carbohydrate complex within vesicles in the macrophages. Iron then either enters the intracellular storage iron pool (e.g., ferritin) or is transferred to plasma transferrin for transport to erythroid precursor cells for incorporation into hemoglobin. Feraheme is approved by the FDA and specifically indicated for the treatment of iron-deficiency anemia in adult patients with chronic kidney disease.

Nanoparticles for Targeted Therapeutic Delivery to the Liver

The liver is an essential organ because it metabolizes various waste products. It is affected by hepatitis viruses and cancer. Liver dysfunction can cause hepatitis, cirrhosis, hyperlipemia, hyperuricemia, type II diabetes, and infarction. Liver is protected by a major immune defense system, the reticuloendothelial system, which captures micro- and nanoscaled materials from the bloodstream in the liver and provides obstacles to the development of liver-specific nanoparticle-based medicines. Several nanoparticle-based therapeutic materials have been developed for delivery of proteins, genes, and siRNAs to the human liver to treat various diseases. Liposomes (including lipoplex), polymer micelles, and polymers (including polyplex) are modified to enhance liver specificity and delivery efficiency. Drug and gene delivery systems specific to the human liver have used bionanocapsules comprising hepatitis B virus (HBV) envelope L protein, which has a pivotal role in vaccination against HBV infection (Kasuya and Kuroda 2009).

Nanonephrology

Nephrology is the branch of medicine that deals with diseases of the kidney. Nanonephrology is the application of nanobiotechnology for the study of renal structure and function as well as for treatment of renal disorders. Nanobiotechnology applications for renal cancer management follow the same patterns as cancer management of other organs. For example, imaging of renal cancer and chemotherapy may be nanobiotechnology-based. One serious problem is renal failure, which is usually managed with renal dialysis.

Nanobiotechnology-Based Renal Dialysis

Renal dialysis is used to provide an artificial replacement for lost kidney function due to renal failure. It is a life-support treatment and not treatment of the kidney disease, which is the cause of renal failure. Dialysis provides filtration of toxins in the blood that would have normally been the function of the kidneys. Approximately one million patients worldwide suffer from end-stage renal disease and require treatment through dialysis or transplantation. There are approximately 480,000 patients in the USA alone who suffer from kidney failure. The number is expected to more than double by 2010, placing considerable stress on healthcare systems throughout the world. Despite the availability of various forms of renal replacement therapy for nearly four decades, mortality and morbidity is high, and patients often have a poor quality of life. The average survival rate of a dialysis patient is only 5 years. Renal transplantation is not available for all the patients as there are not enough kidney donors. Classical dialysis techniques remained in practice for more than 40 years. Within the last decade, there have been some innovations in techniques for renal dialysis. Those involving nanotechnology will be mentioned here.

Nanotechnology-Based Human Nephron Filter for Renal Failure

A human nephron filter (HNF) development could eventually enable a continuously functioning, portable, or implantable artificial kidney (Nissenson et al. 2005). The HNF is the first application in developing a renal replacement therapy to potentially eliminate the need for dialysis or kidney transplantation in end-stage renal disease patients. The HNF utilizes a unique membrane system created through applied nanotechnology. The ideal renal replacement device should mimic the function of natural kidneys, continuously operating, and should be adjustable to individual patient needs. No dialysis solution would be used in this device. Operating 12 h a day, 7 days a week, the filtration rate of the HNF is double that of conventional hemodialysis administered three times a week. The HNF system, by eliminating dialysate and utilizing a novel membrane system, represents a breakthrough in renal replacement therapy based on the functioning of native kidneys. The enhanced solute removal and wearable design should substantially improve patient outcomes and quality of life. Animal studies using this technology are scheduled.

Blood-Compatible Membranes for Renal Dialysis

A novel heparin- and cellulose-based biocomposite membrane has been fabricated with nanopores by exploiting the enhanced dissolution of polysaccharides in room temperature ionic liquids (Murugesan et al. 2006a). Using this approach, it is possible to fabricate the biomaterials in any form, such as films or membranes, nanofibers, nanospheres, or any shape using templates. Surface morphological studies on this biocomposite film showed the uniformly distributed presence of heparin throughout the cellulose matrix. Activated partial thromboplastin time and thromboelastography demonstrate that this composite is superior to other existing heparinized biomaterials in preventing clot formation in human blood plasma and in human whole blood. Membranes made of these composites allow the passage of urea while retaining albumin, representing a promising blood-compatible biomaterial for renal dialysis, with a possibility of eliminating the systemic administration of heparin to the patients undergoing renal dialysis.

Ceramic Filter for Renal Dialysis

A new ceramic filter has the potential to make kidney dialysis much more efficient and to reduce by 30 min to 1 hour the time required for a dialysis treatment. Specifically, the new filter promises to double the amount of toxins removed during dialysis and to double the glomerular filtration rate (GFR), or rate of toxin removal. GFR is 100% in a normal person but only 15% at best for a dialysis patient, a rate that has changed little in the past 30 years. The ceramic filter's secret lies in its nanopores, which are organized in orderly rows and columns and correspond more closely to the nanosized toxins in the blood than do the larger pores of the standard dialysis filter.

Nanotechnology for Wound Healing

Several nanotechnology-based products have been used for wound care. Polyurethane membrane, produced via electrospinning (a process by which nanofibers can be produced by an electrostatically driven jet of polymer solution), is particularly useful as a wound dressing because of the following properties: it soaks fluid from the wound so that it does not build up under the covering and does not cause wound desiccation. Water loss by evaporation is controlled, there is excellent oxygen permeability, and exogenous microorganism invasion is inhibited because of the ultrafine pore size. Histological examination of the wound shows that the rate of epithelialization is increased and the dermis becomes well organized if wounds are covered with electrospun nanofibrous membrane. This membrane has potential applications for wound dressing.

Chronic wounds are associated with poor epidermal and dermal remodeling. Efficacy of keratinocyte growth factor (KGF) in reepithelialization and elastin in dermal wound healing is well known. A fusion protein comprising of elastin-like peptides and KGF has been fabricated that retains the performance characteristics of KGF and elastin as evidenced by its enhancement of keratinocyte and fibroblast proliferation (Koria et al. 2011). It was also shown to self-assemble into nanoparticles at physiological temperatures. When applied to full-thickness wounds in diabetic mice, these particles enhanced reepithelialization and granulation by two- and threefold, respectively, as compared to the controls. These findings suggest that self-assembled nanoparticles may be beneficial in the treatment of chronic wounds resulting from diabetes or other underlying circulatory conditions.

Nanotechnology-Based Products for Skin Disorders

Cubosomes for Treating Skin Disorders of Premature Infants

When surfactants are added to water at high concentrations, they self-assemble to form thick fluids called liquid crystals. The most viscous liquid crystal is bicontinuous cubic phase, a unique material that is clear and resembles stiff gelatin. When cubic phase is dispersed into small particles, these nanoparticles are termed cubosomes. Cubosomes are potential drug delivery vehicles. Among many other applications, cubosomes have been investigated to create new treatments for skin disorders of premature infants. The cubosomes permit a “breathing layer” for skin at the nano-level, which is due to their bicontinuous structure of oil and water interweaved together but never crossing each other. Unlike Vaseline, which forms a protective barrier layer over the skin, cubosomes can both protect skin from outside elements and at the same time let the skin “breathe” and exchange moisture with its environment. Skin Science Institute of the University Children’s Hospital (Cincinnati, OH) once worked on “artificial vernix” – cubosome-based outer protective layer that will help premature babies born without a fully developed outer skin layer. Efforts are being made to find a way to make large-scale manufacture of cubosomes more efficient. The only way known to manufacture cubosomes initially was to use very high energy processes like ultrasound to fragment bulk cubic phase into cubosomes for use in cosmetic products for the skin. There is no commercial product available in the market that is based on cubosomes.

Nanoparticles for Improving Targeted Topical Therapy of Skin

Long-term topical glucocorticoid treatment can induce skin atrophy by the inhibition of fibroblasts. Therefore, investigators have looked for the newly developed drug carriers that may contribute to a reduction of this risk by an epidermal targeting. Prednicarbate (PC, 0.25%) was incorporated into solid lipid nanoparticles of various compositions, and studies were conducted where conventional PC cream of 0.25% and ointment served as reference. Local tolerability as well as drug penetration and metabolism was studied in excised human skin and reconstructed epidermis. With the latter, drug recovery from the acceptor medium was about 2% of the applied amount following PC cream and ointment but 6.65% following nanoparticle dispersion. Moreover, PC incorporation into nanoparticles appeared to induce a localizing effect in the epidermal layer which was pronounced at 6 h and declined later. Dilution of the PC-loaded nanoparticle preparation with cream did not reduce the targeting effect, while adding drug-free nanoparticles to PC cream did not induce PC targeting. Therefore, the targeting effect is closely related to the PC nanoparticles and not a result of either the specific lipid or PC adsorbance to the surface of the formerly drug-free nanoparticles. Lipid nanoparticle-induced epidermal targeting may increase the benefit/risk ratio of topical therapy.

Since nanoparticles can penetrate the skin barrier along the transfollicular route, chitosan nanoparticles loaded with minoxidil sulfate have been studied for their ability to sustain the release of the drug, in a targeted delivery system for the topical treatment of alopecia (Gelfuso et al. 2011). Results of these studies revealed that the chitosan nanoparticles were able to sustain about three times the release rate of minoxidil indicating the potential to target and improve topical therapy of alopecia with minoxidil.

Nanoparticle-Based Sunscreens

Zinc oxide offers the best broad-spectrum protection from the sun. Unlike titanium dioxide, another commonly used inorganic sunscreen, zinc oxide offers protection from both UVB and the more harmful UVA rays. In spite of the advantage of zinc oxide as a natural UV filter, its marketability was lacking because of its whiteness. This has been overcome with the use of nanoparticle technology to create an invisible screen. ZinScreen is such a product that has been marketed successfully in Australia since 2003.

NanoArc® zinc oxide (Nanophase Technologies Corporation), which is produced under cGMP conditions and is FDA approved for use as an active ingredient in personal care products, can also provide effective UV attenuation. The application of solid lipid nanoparticles as physical sunscreens and as active carriers for molecular sunscreens has been investigated. The amount of molecular sunscreen could be decreased by 50% while maintaining the protection level compared to a conventional emulsion.

Solaveil XT-40 W (Croda) based on Oxonica Ltd.'s OPTISOL™ technology is a photostable UV absorber that has applications in skin-care products and other materials. The OPTISOL technology works by absorbing UVA radiation without the concurrent formation of free radicals. Furthermore, while providing balanced UVA and UVB protection, OPTISOL also provides additional benefits in absorbing free radicals that may be generated by other components of the sunscreen formulation and also enhances formulation stability. OPTISOL is based on ultrafine titanium dioxide with the inclusion in the crystal of a small amount (<1%) of manganese. This causes a reconfiguration of the crystal's internal electronic structure that allows absorbed UV energy to be dissipated, virtually eliminating the generation of free radicals. Secondly, manganese near the crystal surface can catalyze free radicals that have been generated by other sunscreen components into harmless chemical species. A reduction in free radical load has benefits for both the skin and for the sunscreen formulation. Free radicals are implicated in photoaging of the skin, photocarcinogenesis, and organic component degradation.

Nanoengineered Bionic Skin

NanoApplications Center at Oak Ridge National Laboratory (ORNL), in collaboration with NASA, is developing flexible integrated lightweight multifunctional skin (FILMskin), a revolutionary concept for the skin used in human prosthetic devices. Nanotechnology is used to create a water-resistant skin composite, which is shaped by lasers to be lifelike. The nano-enabled FILMskin will contain pressure- and temperature-sensing capabilities, like human skin, yet will be tough and flexible. ORNL scientists and engineers are using the novel properties of carbon nanotubes and aligned nanotube arrays to enhance electroactive polymers to provide multifunctional capabilities to next-generation prosthetic devices. Scientists at ORNL are also providing robotic and cognitive capabilities for the prosthetics. It may be used for skin replacement in burns patients.

Topical Nanocreams for Inflammatory Disorders of the Skin

Inflammatory skin diseases, including atopic dermatitis and psoriasis, are common. The current treatment is unsatisfactory, although several topical and systemic therapies, including steroids and immunomodulators, are available. The efficacy is not durable, and they are associated with adverse effects. Efforts continue to develop safer alternative treatment for these disorders.

Nanocrystalline silver has been demonstrated to have exceptional antimicrobial properties and has been successfully used in wound healing. Studies conducted by Nucryst Pharmaceuticals have revealed that topical application of nanocrystalline silver cream (0.5% and 1%) ointment produced significant suppressive effects on allergic contact dermatitis in a guinea pig model. There was a clear concentration-response relationship to the decrease of inflammation as lower concentrations were not effective. The effects were equivalent to the immunosuppressant tacrolimus ointment. This study suggests that nanocrystalline silver cream has therapeutic potential for treating inflammatory skin diseases.

Nanobiotechnology for Disorders of Aging

The human life expectancy has nearly doubled during the last century. Aging is not a disease, but certain diseases are associated with aging. The incidence of these diseases, many of which are incurable at the present state of knowledge, has spurred research activity. In the academic sector, there is an increase in the research activity to unravel the biology of aging, and several companies are developing products for managing disorders associated with aging. Several factors play a role in aging process. These include mitochondria and telomeres.

Telomeres are proteins that function like caps on the ends of chromosomes and ensure successful DNA replication when a cell divides. However, every time a cell divides, the telomeres shorten and eventually become exhausted. In general, aging cells become progressively less able to form and maintain tissue. This dysfunction plays a key role in a variety of presently incurable age-associated diseases such as macular degeneration, arteriosclerosis, atherosclerosis, osteoporosis, skin atrophy, progeria, and others.

Stanford University has developed and patented DNA nanocircles that can be used to simply and efficiently synthesize long telomere repeats on human chromosomes without the need for telomerase. Because telomere length acts as a cellular clock to define the life span of a cell population, this technology could potentially extend the life of cells used for ex vivo cell therapy without cancerous transformation. The applications include enhancing the growth and delaying or preventing senescence in cultured primary tissues such as islet cells, bone marrow, skin, hepatic tissues, and stem cells. Currently, it is not clear if this technology has any potential applications in disorders associated with aging.

Personal Care Products Based on Nanotechnology

Effects on skin hydration and viscoelasticity are important criteria during the development of novel cosmetic formulations. Solid lipid nanoparticles represent a promising compound for hydrating new cosmetic formulations. Addition of solid lipid nanoparticles to conventional creams leads to an increase of skin hydration after repeated applications for a few weeks.

Nanotechnology is used in several personal care products such as sun creams and deodorants. Nanomaterials can act as sun blockers to protect human skin in formulations that eliminate unnecessary exposure to the harmful UV rays of the sun. Nanomaterials can impart antibacterial and anti-odor functionality on human skin in powder, gel, stick, or spray underarm products that can be applied smoothly without plugging the spray nozzle, caking, or staining while maintaining clarity. The market trends for deodorants and antiperspirants are toward clear and highly effective formulations that are mild and non-irritating.

Nanomaterials can be introduced into a variety of oral care products to impart antimicrobial and anti-irritant properties without sacrificing flow, texture, or color. Dental pastes, creams, and cleaners need to thoroughly clean teeth and gums and leave a clean and crisp feeling in the mouth afterward.

Nanotechnology for Hair Care

Human hair is a nanocomposite biological fiber with well-characterized microstructures. Nanomechanical characterization of human hair can help to evaluate the effect of cosmetic products on hair surface, can provide a better understanding of the physicochemical properties of a wide variety of composite biological systems, and can provide the dermatologists with some useful markers for the diagnosis of hair disorders. A systematic study of nanomechanical properties of human hair, including hardness, elastic modulus, and creep, was carried out using the nanoindentation technique (Wei et al. 2005). The samples include Caucasian, Asian, and African hair at virgin, chemo-mechanically damaged, and treated conditions. Hair morphology was studied using scanning electron microscopy (SEM). Indentation experiments were performed on both the surface and cross-section of the hair, and the indents were studied using SEM. The techniques were used to test a new high-tech hair conditioner. Ultimately, the same techniques could be used to improve lipstick, nail polish, and other beauty products. Application of nanotribology – the measurement of very small things such as the friction between moving parts in microelectronics – is important for study of hair as friction is a major issue. Everyday activities like washing, drying, combing, and brushing all cause hairs to rub against objects and against each other. Over time, the friction causes wear and tear. If damaged hair is exposed to humidity, the hairs plump up, and the cuticles stick out even further, leading to more friction – a fact confirmed by the AFM when a tiny needle is moved across the surface. This research is being utilized by manufacturers of hair care products for developing a new formula with additives to make the conditioner coat the hair evenly. In the future, the AFM techniques could be used to develop wear-resistant nail polishes and lipsticks.

Nanodentistry

Nanodentistry will make possible the maintenance of comprehensive oral health by involving the use of nanomaterials, biotechnology (including tissue engineering), and, ultimately, dental nanorobotics. The first dental nanorobots could be constructed by the year 2015 and will enable precisely controlled oral analgesia, dental replacement therapy using autologous cell teeth, and rapid nanometer-scale precision restorative dentistry.

Bonding Materials

Nano-Bond Universal Bonding System (Pentron Clinical Technologies) is based on Hybrid Plastics' POSS® technology (polyhedral oligomeric silsesquioxanes). It results in strengthened resin while it infiltrates the etched surface and provides strong interface between the tooth and the restorative material. The system consists of a uniquely formulated self-etch primer and adhesive system that are said to work together for great bonding to dentin and cut enamel. The kit also contains a dual-cure activator that promotes reliable bonding to self- and dual-cured materials. The Nano-Bond System greatly alleviates the problem of post-bonding sensitivity by keeping tubules occluded during the self-etching step.

Adper™ Single Bond Plus Adhesive (3M ESPE) is a high bond strength dental adhesive. The improved adhesive incorporates a nanofiller technology that contributes to higher dentin bond strength performance. Adper is ideal for bonding all classes of direct composite restorations, as well as root surface desensitization and porcelain veneers. The nanofiller particles in Adper are added in a manner that does not allow them to cluster together. The particles are stable and will not settle out of dispersion. Therefore, no shaking is needed prior to use.

Dental Caries

The conventional treatment of dental caries involves mechanical removal of the affected part and filling of the hole with a resin or metal alloy. However, this method is not suitable for small early cavities because a disproportionate amount of healthy tooth must be removed to make the alloy or resin hold in place. A dental paste has been produced from synthetic enamel that rapidly and seamlessly repairs early caries lesions by nanocrystalline growth, with minimal wastage of the natural enamel (Yamagishi et al. 2005).

The application of surfactants as reverse micelles or microemulsions for the synthesis and self-assembly of nanoscale structures is one of the most widely adopted methods in nanotechnology. The resulting synthetic nanostructure assemblies sometimes have an ordered arrangement. These developments in nanotechnology have been used to mimic the natural biomineralization process to create dental enamel – the hardest tissue in the human body. This is the outermost layer of the teeth and consists of enamel prisms, highly organized micro-architectural units of nanorod-like calcium hydroxyapatite (HA) crystals arranged roughly parallel to each other. The hydroxyapatite nanorod surface was synthesized and modified with monolayers of surfactants to create specific surface characteristics, which enable the nanorods to self-assemble into an enamel prism-like structure at a water/air interface (Chen et al. 2005a). The size of the synthetic hydroxyapatite nanorods can be controlled, and nanorods similar in size to human enamel were synthesized. The prepared nanorod assemblies were examined using TEM and AFM and were shown to be comprised of enamel prism-like nanorod assemblies with a Ca/P ratio between 1.6 and 1.7. It is possible that an enamel-like composite would be available within a year and crowns suitable for repairing decayed teeth within about 4 years. Application of synthetic enamel will not be limited to filling cavities. It has potential for use in bone repair and bone augmentation.

New treatment opportunities based on nanobiotechnology may include detection of dental decay spots prior to formation of cavities and repair of these, improved nanomaterials for covering dental enamel, and continuous oral health maintenance through the use of mechanical dentifrobots.

Nanospheres for Dental Hypersensitivity

Dental hypersensitivity, a painful condition due to exposure of the dentine of the tooth, affects millions of people worldwide. The dentine contains tiny fluid-filled channels which radiate outward from the nerve terminals at the center of the tooth. Heat or cold and some chemicals can cause the fluid in these channels to move in or out irritating the nerve endings and causing sharp pain. Nanospheres of hydroxyapatite, a ceramic material, could be a long-term solution or cure for sensitive teeth. Commercially available silica nanospheres are approximately 40 nm in diameter. Nanospheres could help dentists fill the tiny holes in the teeth that make them incredibly sensitive. If these channels are fully or partially blocked, the flow can be reduced and the pain stopped or significantly reduced. The next stage of the research will be to synthesize nanospheres combining hydroxyapatite and fluorine, which would fill the holes and encourage remineralization at the same time and also provide a powerful repair tool for dentists.

Nanomaterials for Dental Filling

The standard composite resin filling, a natural-looking restoration, is the method of choice when appearance is an issue. A dentist creates the filling by mixing the pure liquid resin with a powder that contains coloring, reinforcement, and other materials, packing the resulting paste into the cavity, and illuminating the tooth with a light that causes the paste to polymerize and harden. For decay-fighting composite fillings, the problem arises from an additive that is included in the powder to provide a steady release of calcium and phosphate ions. These ions are essential to the long-term success of the filling because they not only strengthen the crystal structure of the tooth itself but buffer it against the decay-causing acid produced by bacteria in the mouth. Yet the available ion-releasing compounds are structurally quite weak, to the point where they weaken the filling as a whole.

Nanotechnology has the potential to produce tooth restorations that are both stronger and more effective at preventing secondary decay than any decay-fighting fillings available currently. The new spray-drying technique yields particles of several compounds, one of which being dicalcium phosphate anhydrous (DCPA), which are about 50 nm across, 20 times smaller than the 1 μm particles in a conventional DCPA powder. Because these nanoscale particles have a much higher surface-to-volume ratio, they are much more effective at releasing ions, which means that much less of the material is required to produce the same effect. That, in turn, leaves more room in the resin for reinforcing fibers that strengthen the final filling. To exploit that opportunity, the Paffenbarger researchers also have developed nanoscale silica-fused fibers that produce a composite resin nearly twice as strong as the currently available commercial variety.

Nanomaterials for Dental Implants

Dental manufacturers are incorporating nanotechnology into their dental implant surface designs because the technology is purported to cut healing time in half and improve osseointegration. These include 3i and Bicon, both of which have branded their nanotechnology-based dental implant surfaces as "NanoTite," Astra Tech with its OsseoSpeed, and Straumann with its SLActive.

Nanomedical Aspects of Oxidative Stress

Free radical reactions involving reactive oxygen species (ROS) and reactive nitrogen species (RNS) contribute to the pathogenesis and progression of several human diseases. Antioxidants, such as vitamins C and E, 21-aminosteroids, and other free radical scavengers, have met with only limited success in clinical applications. This is partly due to our inability to design efficient antioxidants with site-directed, controlled activity. Nanotechnology has provided dramatic improvement in controlling or eliminating oxidation reactions in materials applications, which may provide a new basis for pharmacological treatment of diseases related to oxidative stress.

Nanoparticle Antioxidants

Nanotechnology has made significant advances in the reduction of free radical damage in the field of materials science. Cross-disciplinary interactions and the application of this technology to biological systems have led to the elucidation of novel nanoparticle antioxidants. Three of the most-studied nanoparticle redox reagents at the cellular level are rare earth oxide nanoparticles (particularly cerium), fullerenes, and carbon nanotubes.

Fullerene-Based Antioxidants

Water-soluble derivatives of buckminsterfullerene C60 derivatives are a unique class of nanoparticle compounds with potent antioxidant properties. Studies on one class of these compounds, the malonic acid C60 derivatives (carboxyfullerenes), indicated that they are capable of eliminating both superoxide anion and H_2O_2 and were effective inhibitors of lipid peroxidation, as well. Carboxyfullerenes demonstrated robust neuroprotection against excitotoxic, apoptotic, and metabolic insults in cortical cell cultures. They were also capable of rescuing mesencephalic dopaminergic neurons from both MPP(+)- and 6-hydroxydopamine-induced degeneration. Although there is limited in vivo data on these compounds, systemic administration of the C3 carboxyfullerene isomer has been shown to delay motor deterioration and death in a mouse model familial amyotrophic lateral sclerosis. Ongoing studies in other animal models of CNS disease states suggest that these novel antioxidants are potential neuroprotective agents for other neurodegenerative disorders including Parkinson's disease.

Ceria Nanoparticles as Neuroprotective Antioxidants

Ceria nanoparticles from anthanide series have several unique properties that make them highly efficient redox reagents. Several studies have reported the ability of ceria nanoparticles to mitigate oxidative stress at the biological level. Ceria nanoparticles also protect neurons from free radical-mediated damage initiated by ultraviolet (UV) light, H_2O_2, and excitotoxicity, leading to the hypothesis that the mechanism of action is one of free radical scavenging (Rzigalinski et al. 2006). When compared with single doses of other free radical scavengers, such as vitamin E, melatonin, and N-acetylcysteine, ceria nanoparticles demonstrated significantly greater neuroprotection after a 5- and 15-min UV insult. A single dose of nanoparticles delivered up to 3 h post-injury also afforded neuroprotection. Ceria nanoparticles were also effective in reducing cell death associated with γ-irradiation. In another study, nanoparticles were shown to directly decrease free radical production (Schubert et al. 2006). No toxicity was observed with ceria nanoparticle sizes of 6 and 12 nm, and yttrium oxide nanoparticles were even more effective than ceria. Ceria nanoparticles larger than 30 nm or nitrates and sulfates of cerium did not have any significant effects.

Several studies also suggest that ceria nanoparticles are potent anti-inflammatory agents. Microglia, the immune cells of the brain, are "activated" in response to neuronal damage and show an inflammatory response with release NO as well as IL-1β. Treatment of injured organotypic cultures with ceria nanoparticles reduced their ability to activate microglia. Further, treatment of activated microglia with ceria nanoparticles reduces production of soluble factors that promoted death in uninjured neurons, including NO and IL-1β. Delivery of nanoparticles to the uninjured neurons also directly affords neuroprotection from the damaging effects of activated microglia. Thus, it appears that nanoparticles may blunt the inflammatory response in immune cells, as well as reduce inflammatory injury to nonimmune cells.

Antioxidant Nanoparticles for Treating Diseases Due to Oxidative Stress

The prospects for the use of nanoparticles for free radical scavenging in diseases due to oxidative stress are promising. However, further studies in animals and clinical trials will be needed to ascertain this beneficial effect. Other nanoparticles such as fullerenes also show biological antioxidant activity and potent neuroprotective effects, which need to be investigated. Some studies indicate that there may be an optimal level of free radical scavenging above which antioxidant nanoparticles may interfere with the beneficial roles of free radicals within the cell and have harmful effects. This is important for establishing the safety and proper doses of antioxidant nanomedicines.

Nanotechnology and Homeopathic Medicines

Homeopathy was founded in Germany in 1789. The basic principle is that "like cures like." The materia medica of this system is based on the description of symptoms induced by a large number of substances including metals and plant derivatives. Extreme dilutions of the same substance are considered to be effective for the treatment of the symptoms. Currently, there is a resurgence of interest in homeopathy both in Europe and the USA. Clinical trials have produced mixed results.

Homeopathy is controversial because medicines in high potencies designated as 30c and 200c involve huge dilution factors (10^{60} and 10^{400}, respectively) which are many orders of magnitude greater than Avogadro's number, so that theoretically there should be no measurable remnants of the starting materials. No hypothesis which predicts the retention of properties of starting materials has been proposed nor has any physical entity been shown to exist in these high-potency medicines. A study has used TEM, electron diffraction, and chemical analysis by inductively coupled plasma-atomic emission spectroscopy (ICP-AES) to examine market samples of metal-derived medicines from reputable homeopathic manufacturers and demonstrated for the first time the presence of physical entities in these extreme dilutions, in the form of nanoparticles of the starting metals and their aggregates (Chikramane et al. 2010). Further investigations are in progress to determine if presence of nanoparticles is related to therapeutic effect.

Nanoparticles as Antidotes for Poisons

Currently, antidotes are not available to treat many harmful, even life-threatening reactions. Drug removal differs from drug delivery in that some drugs are given in encapsulated form to prolong action whereas drug removal has to be accomplished rapidly. Removal agents must reduce the available drug concentration to below the toxicity threshold, and they must be biocompatible.

In one study, emulsion-based nanoparticles with diameter of 118.4 nm extracted bupivacaine from the aqueous phase in a physiological salt solution and attenuated the drug's cardiotoxicity in guinea pig heart to a greater extent than did a macroemulsion with particle diameter of 432 nm (Morey et al. 2004). Additionally, nanoparticles sequestered bupivacaine from the aqueous phase of human blood and merit further investigation in animal models of intoxication.

Synthetic polymer NPs that bind poisonous molecules and neutralize their effect in vivo have been investigated as plastic antidotes. Although techniques are now available for synthesizing polymer NPs with affinity for target peptides, their performance in vivo is a far greater challenge. Particle size, surface charge, and hydrophobicity affect not only the binding affinity and capacity to the target toxin but also the toxicity of NPs and the creation of a layer of proteins around an NP that can alter and/or suppress the intended performance. Design rationale of a plastic antidote for in vivo applications has been reported, which optimizes the choice and ratio of functional monomers incorporated in the NP and maximizes the binding affinity to a target peptide (Hoshino et al. 2012). Biocompatibility tests of the NPs in vitro and in vivo revealed the importance of tuning surface charge and hydrophobicity to minimize NP toxicity and prevent aggregation induced by nonspecific interactions with plasma proteins. The toxin neutralization capacity of NPs in vivo showed a strong correlation with binding affinity and capacity in vitro. Furthermore, in vivo imaging experiments established the NPs accelerate clearance of the toxic peptide and eventually accumulate in macrophages in the liver. These results provide a platform to design plastic antidotes and reveal the potential and possible limitations of using synthetic polymer nanoparticles as plastic antidotes.

Magnetic nanoparticles conjugated with β-cyclodextrin adsorb diazepam, which can then be removed from blood by an external magnetic field; this has potential applications in treatment of diazepam overdose and other poisonings (Cai et al. 2011).

Nanoparticles for Chemo-Radioprotection

Chemotherapy and radiotherapy are the standard treatments for cancer, but they have severe adverse effects on the body. Radiation can damage epithelial cells and lead to permanent hair loss, among other effects, and certain types of systemic chemotherapy can produce hearing loss and damage to a number of organs, including the heart and kidneys. Only one drug, amifostine, has been approved to date by the FDA to help protect normal tissue from the side effects of chemotherapy and radiation, and there is a need for new and improved agents.

Animal experimental studies have shown that the nanoparticle, fullerene CD60_DF1 (C Sixty), can help fend off damage to normal tissue from radiation. It acts like an "oxygen sink," binding to dangerous oxygen radicals produced by radiation. Fullerenes can be considered as a potentially "new class of radioprotective agents." CD60_DF1 given before and even immediately after exposure to X-rays reduces organ damage by one-half to two-thirds, which is equal to the level of protection given by amifostine. Moreover, the fullerene provides organ-specific protection, e.g., the kidney as well as certain parts of the nervous system.

Role of Nanobiotechnology in Biodefense

Nanobiotechnology provides several devices for the diagnosis of agents used in biological warfare and bioterrorism. Because of its ability to create structures of nanoscale dimension with large aggregate particle surface area-to-volume ratios, nanotechnology offers new opportunities to treat drug poisonings. Some examples from experimental studies support this potential.

Nanoparticles to Combat Microbial Warfare Agents

Nanomaterials could play a role as an anthrax antibiotic. Antibodies that latch on to dormant anthrax spores and drugs that destroy anthrax could be linked to the spherical carbon molecule fullerene to make an antibiotic. The drug would kick in when an inhaled spore germinates, killing the anthrax before it releases deadly amounts of toxin. This could be effective if one has just been exposed, or the exposure is expected within 24 h following administration of the drug.

Preventing the interaction of toxins with their cellular receptors CMG2 and ATR/TEM8 is an important goal for anthrax therapy. Nanotechnology approaches have been used for the multivalent display of engineered receptor decoys and their efficacy against anthrax lethal toxin in vitro and in vivo.

Several concepts of targeting of nanosuspension dosage forms for treatment of bioweapon-mediated diseases have been developed at the Baxter Healthcare Corporation. Alterations of pharmacokinetic profiles of existing antibiotics can lead to enhanced efficacy with reduced side effects. This has been shown for a nanosuspension formulation of the antifungal agent itraconazole. Secondly, viral sanctuaries breed resistance and often include the brain and lymphatics. These may be targeted by loading nanoparticulate drug into macrophages which target these organs, increasing antiviral drug concentration in these typically inaccessible regions. Finally, a strategy for dendritic cell vaccines has been developed for use against bioweapons.

Removal of Toxins from Blood

An important part of treatment of poisoning is removal of toxins circulating in the blood. An even more critical need is for removal of biological and radiological toxic material from the blood after exposure to bioterrorist attacks or biochemical warfare. The Argonne National Laboratory and the University of Chicago are collaborating to develop a magnetic nanoparticle-based technology that removes biological, radiological, and, in some cases, chemical toxins from blood. Their goal is to devise a portable system that is fast and thorough using magnetic nanoparticles coated with antibodies or chemicals that complement toxins. Once injected into the patient, the particles latch onto toxins and are later removed when the blood is pumped through tubing into a magnetic separator. Cleansed blood is then returned into the body. The approach is particularly safe because the closed loop system never exposes blood to the outside environment. Regardless of the type of exposure, if the toxins are removed before they accumulate in tissues, then organs will not fail, and the patients will survive. This method may not be effective in case of rapidly acting agents such as nerve gas where the time between exposure and death is a matter of minutes. However, some chemical and many biological and radiological agents need hours or even days to cause fatal damage, allowing a wide enough window for the particle treatment to be effective. Some requirements for the development of this technology are as follows:

- The particles, which start with magnetic cores in the 8- to 12-nm range, must be the right size to navigate within the body. If they are too small, they may pass out of the kidneys, and if they are too large, they may get trapped.
- The particles also need to be biocompatible, so the body accepts them.
- They should be biodegradable in case some remain after treatment.

Several companies are already developing magnetic nanoparticles for medical applications. Their use of FDA-approved antibodies, reagents, and off-the-shelf medical components could remove some regulatory hurdles. That could pave the way for not only military and civilian defense applications but clinical treatments such as overdoses. However, it will take a couple of years for this development.

Nanobiotechnology for Public Health

Emphasis on preventive medicine and public health is increasing. High-technology medicines will benefit a limited number of population and mostly in the developed countries. Measures to improve public health will have a much larger impact on the future healthcare for most of the people on this earth. One of the major problems in developing countries is sanitary water supply.

Nanotechnology for Water Purification

Nanotechnology has the potential to provide novel nanomaterials for treatment of surface water, groundwater, and wastewater contaminated by toxic metal ions, organic and inorganic solutes, and microorganisms. These consist of nanomaterials for water filtration, nanotechnologies for water remediation, and NPs for disinfection of water.

Nanofiltration to Remove Viruses from Water

Nanofiltration is a relatively simple and reliable procedure that consists in filtering water through membranes with nanopores (size 15–40 nm) that retain viruses by size exclusion.

Shortcomings of some membranes are that they often form pinholes and cracks during the fabrication process, resulting in wasted membranes. Scientists at the Queensland University of Technology (QUT) in Australia have developed specially designed ceramic membranes used as nano-mesh for nanofiltration, which are less likely to be damaged during manufacture and have the potential to remove viruses from water. This modification has increased the rates of flow that pass through the membranes tenfold compared with current ceramic membranes while maintaining the efficiency of capturing over 96% of the unwanted particles.

Nanostructured Membranes for Water Purification

Current methods for the purification of contaminated water sources are chemical-intensive and energy-intensive and/or require posttreatment due to unwanted by-product formation. Integration of nanostructured materials and Fe-catalyzed free radical reactions enables detoxification of water. Harmful organic contaminants can be degraded through the addition of a substrate, glucose, which is enzymatically converted to H_2O_2 without adding harmful chemicals (Lewis et al. 2011). Application of these technologies can be extended to disinfection and/or virus inactivation.

Nanotechnologies for Water Remediation

Advantages of use of nanomaterials for water remediation are their enhanced reactivity, surface area, and sequestration characteristics. Several nanomaterials are in development for this purpose including the following:

- Biopolymers
- Carbon nanotubes
- Iron nanoparticles
- Zeolites

Cyanobacterial metabolites – microcystin, cylindrospermopsin (CYN), 2-methylisoborneol (MIB) and geosmin (GSM) – are a major problem for the water industry. Low-molecular-weight cutoff (MWCO), or "tight" NF, membranes afford average removals above 90% for CYN, while removal by higher MWCO, or "loose" NF, membranes is lower. MIB and GSM are removed effectively (>75%) by tight NF but less effectively by loose NF. Microcystin variants are removed to above 90% by tight NF membranes; however, removal using loose NF membranes depends on the hydrophobicity and charge of the variant. Natural organic matter concentration in the waters treated with this method had no effect on the removal of cyanobacterial metabolites (Dixon et al. 2011).

Several NPs have been shown to have antibacterial effects and are used as disinfectants, e.g., silver NP coated on surfaces. A study has shown that lanthanum calcium manganate (LCMO) NPs have greater antibacterial efficacy against *Pseudomonas aeruginosa*-ATCC 27853, a soil- and waterborne pathogenic bacterium, as compared to Eu3+-doped lanthanum calcium manganate (LECMO) NPs (De et al. 2010). Size of synthesized NPs was 50–200 nm, and X-ray diffraction pattern showed the formation of a single-phase LCMO or LECMO of an orthorhombic crystal structure after annealing the precursor at 10,000 °C for 2 h in air. LCMO NPs can offer future applications as antimicrobial drugs and for water purification.

Iron oxide (α-Fe_2O_3) nanoparticles, 5 nm in size, have been used to remove arsenic ions from natural water samples (Tang et al. 2011a). Iron nanoparticles maintained their arsenic adsorption capacity even at very high competing anion concentrations. This method was used to purify contaminated natural lake water sample to meet the US Environmental Protection Agency's drinking water standard for arsenic.

Nanotechnology-Based Photochemical Water Purification

Nanotechnology-based photochemical water purification is also feasible. Bioactive nanoparticles can be used for disinfection of water, e.g., metal-oxide NPs, particularly silver, and titanium dioxide for photocatalytic disinfection can provide alternative to chlorination of water. Multiple wavelengths of a light-emitting device or natural light illuminate a high-surface-area nanotechnology coating to cause photochemical reactions. In development by Puralytic Inc., this process effectively removes the broadest range of contaminants including the following:

- Organic compounds: pharmaceuticals and petrochemicals
- Heavy metals: lead, mercury, arsenic, and selenium
- Microorganisms: viruses, bacteria, protozoa, and cysts

The process is environment friendly and cost-effective with no chemicals or additives, wastewater, or pressure loss.

Nanobiotechnology and Nutrition

Nanotechnologies will have an impact on nutrition research in many ways. Nanodevices can be used for real-time optical intracellular sensing (Ross et al. 2004). These technologies may be particularly useful in obtaining accurate spatial information and low-level detection of essential and nonessential bioactive food components (nutrients) and their metabolites and in enhancing the understanding of the impact of nutrient/metabolite and biomolecular interactions. Nanobiotechnology will have an impact on food production as well as improved nutrition. Nanotechnologies can provide many benefits as shown in Table 15.1.

Table 15.1 Applications of nanotechnologies in food and nutrition sciences

Food manufacture
Nanoparticles and nanocrystals of essential nutrients to improve bioavailability
Use of self-assembly in nature and materials on a nanoscale with bottom-up approach
Use of nanoparticles to increase material strength barriers
Product research and innovation
Development of new products based on nanoscience research of natural foods
Control of bioavailability
Products based on simulation of customer preferences based on taste and smell nanosensing
Study of molecular physiology and genomics of taste cells
Testing of food effects via biomarkers
Product marketing
Unique nanobarcodes on proprietary products
Shelf life indicators
Quality control and testing
Nanodiagnostics for food contaminants and microorganisms
Nanosensors for quality control of food
Nutrition
Development of personalized nutrition based on metabolic needs of individual
Development of foods based on personalized sensory needs of the individual
Development of nutriceuticals
Development of nutricosmetics

Nanobiotechnology and Food Industry

A major challenge in food production lies in the translation of established technologies into food production for delivering the optimal health and sensory benefits such as taste and smell. The major company involved in this area is Nestle SA. The company believes that nutrition, consumers, and use of new technologies in food science will be key drivers for future product innovation.

One example is that of lycopene, the carotenoid that gives tomatoes and other fruits and vegetables their red color. Health benefits of lycopene are well recognized. Lycopene has been of particular interest recently as regards its role in prostate cancer. Lycopene from fresh and unprocessed tomatoes is poorly absorbed by humans. Absorption of lycopene is higher from processed foods such as tomato paste and tomato juice heated in oil. Nestle has developed a food-grade lycopene formulation that is bioavailable in humans. Called "lactolycopene," it is made by trapping lycopene with whey proteins. However, bioavailability of lycopene in lactolycopene is no higher than that of tomato paste. Lycopene crystallizes in aqueous solution and forms nanocrystals. Micelles provide a convenient, inexpensive, and nontoxic vehicle for dissolving and stabilizing as lycopene in tissue culture media and then delivering it to cells growing in culture. Nestle is researching this area for improving bioavailability of lycopene.

BioDelivery Sciences International has applied nanotechnology to food processing to encochleate sensitive and easily degraded nutrients (beta carotene, antioxidants, and others) for addition to processed food and beverages. Nanoencochleation's all-natural process encochleates and preserves essential nutrients like antioxidants into a protected "shell" for high-temperature/pressure canning and bottling applications.

Another consideration is absorption of minerals and other essential nutrient supplements. Reducing the size of low-solubility iron (Fe)-containing compounds to nanoscale has the potential to improve their bioavailability. Because Fe and zinc (Zn) deficiencies often coexist in populations, combined Fe/Zn-containing nanostructured compounds may be useful for nutritional applications (Hilty et al. 2009). Phosphates and oxides of Fe and atomically mixed Fe/Zn-containing (primarily $ZnFe_2O_4$) nanostructured powders were produced by flame spray pyrolysis. Solubility of the nanostructured compounds is dependent on their particle size. Nanostructured powders produce minimal color changes when added to dairy products containing chocolate or fruit compared to the changes produced when ferrous sulfate or ferrous fumarate are added to these foods. Flame-made Fe- and Fe/Zn-containing nanostructured powders have solubilities comparable to ferrous and Zn sulfate but may produce fewer color changes when added to difficult-to-fortify foods. Therefore, these powders are promising for fortification of food and other nutritional applications.

Role of Nanobiotechnology in Personalized Nutrition

Nutrition plays a crucial role in health as well as disease. With advances in molecular biology, there is a shift in focus from epidemiology and biochemistry to an understanding of how nutrients act at molecular level. Advances in genomics have led to recognition of the importance of genes in human nutrition. Genetic predisposition is an important factor in mortality linked to diet such as cardiovascular disease.

Technologies such as high-density microarrays enable the simultaneous study of the whole transcriptome relevant to nutrition. Advances in proteomic and metabolomic technologies will also enable the analysis of the whole system at proteomic and metabolomic levels as well. Introduction of nanotechnologies will further improve and enable practical personalization of nutrition.

Chapter 16
Nanobiotechnology and Personalized Medicine

Introduction

Personalized medicine simply means the prescription of specific therapeutics best suited for an individual. It is usually based on pharmacogenetic, pharmacogenomic, transcriptomic, pharmacoproteomic, and pharmacometabolomic information. Other individual variations in patients and environmental factors are also taken into consideration (Jain 2009). Personalized medicine means improving healthcare by incorporating early detection of disease, preventive medicine, rational drug discovery and development, and monitoring of therapy. Concept of personalized medicine as systems medicine is the best way of integrating new technologies and translating them into clinical application for improving healthcare. Application of nanobiotechnology is described for personalized management of cancer and cardiovascular disorders. Advances in nanobiotechnology will facilitate the development of personalized medicine by the following:

- Nanodiagnostics will improve the sensitivity and extend the present limits of molecular diagnostics/molecular imaging of CNS disorders.
- Nanotechnology can be integrated in detection of biomarkers, POC devices, biochips, and biosensors.
- Biomarkers discovered by use of nanodiagnostics will facilitate the development of new personalized drugs for various disorders.
- Nanobiotechnology will facilitate integration of diagnosis and therapy, which is an important part of personalized medicine.

Figure 16.1 shows the broad scope and interrelationships of personalized medicine.

K.K. Jain, *The Handbook of Nanomedicine*, DOI 10.1007/978-1-61779-983-9_16,

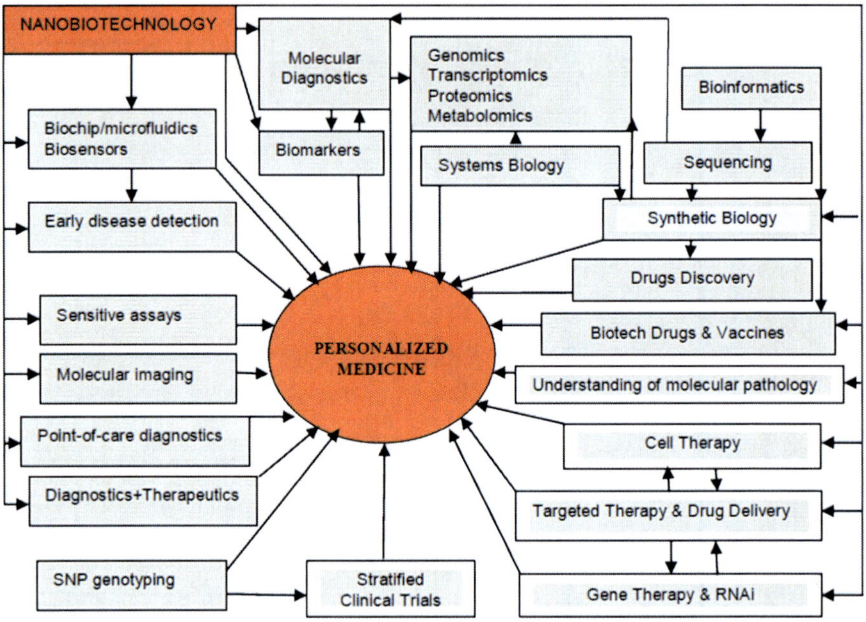

Fig. 16.1 Relationship of nanobiotechnology to personalized medicine (© Jain PharmaBiotech.)

Role of Nanobiotechnology in Personalized Management of Cancer

In case of cancer, the variation in behavior of cancer of the same histological type from one patient to another is also taken into consideration in addition to variations among patients. Personalization of cancer therapies is based on a better understanding of the disease at the molecular level, and nanotechnology will play an important role in this area (Jain 2010a). Various components of personalized therapy of cancer that are relevant to nanobiotechnology are shown in Fig. 16.2.

Nanobiotechnology, by enabling early detection of cancer, refinement of cancer diagnosis, and monitoring of cancer therapy, will contribute to the development of personalized therapy of cancer (Jain 2012h). Nanobiotechnology will facilitate combination of diagnostics with therapeutics, which is an important feature of personalized medicine approach to cancer. QDs, which play an important role in cancer diagnosis and monitoring of therapy in combination with cancer biomarkers, will provide useful tools for personalizing cancer therapy.

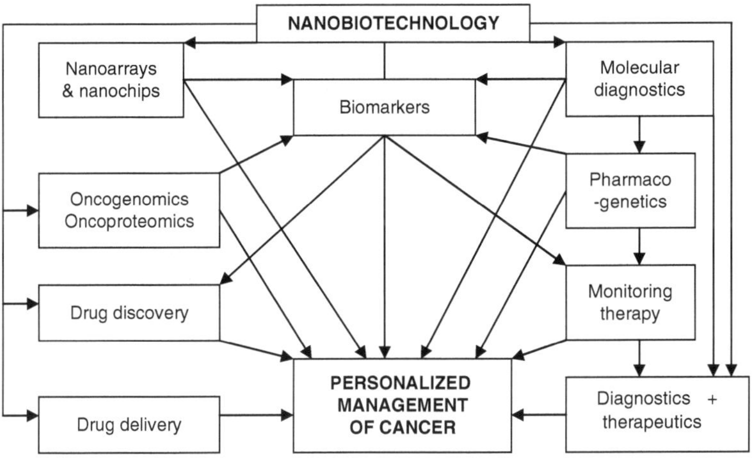

Fig. 16.2 Role of nanobiotechnology in personalized management of cancer (© Jain PharmaBiotech.)

Nanotechnology-Based Personalized Medicine for Cardiology

The future of cardiovascular diagnosis already is being impacted by nanosystems that can both diagnose pathology and treat it with targeted delivery systems (Wickline et al. 2006). The potential dual use of nanoparticles for both imaging and site-targeted delivery of therapeutic agents to cardiovascular disease offers great promise for individualizing therapeutics. Image-based therapeutics with site-selective agents should enable verification that the drug is reaching the intended target and a molecular effect is occurring. Experimental studies have shown that binding of paclitaxel to smooth muscle cells in culture has no effect in altering the growth characteristics of the cells. If paclitaxel-loaded nanoparticles are applied to the cells, however, specific binding elicits a substantial reduction in smooth muscle cell proliferation, indicating that selective targeting may be a requirement for effective drug delivery in this situation. Similar behavior has been demonstrated for doxorubicin-containing particles. Intravenous delivery of fumagillin (an antiangiogenic agent)-loaded nanoparticles targeted to $\alpha v\beta 3$-integrin epitopes on vasa vasorum in growing plaques results in marked inhibition of plaque angiogenesis in cholesterol-fed rabbits. The unique mechanism of drug delivery for highly lipophilic agents such as paclitaxel contained within emulsions depends on close apposition between the nanoparticle carrier and the targeted cell membrane and has been described as "contact-facilitated drug delivery." In contrast to liposomal drug delivery (generally requiring endocytosis), the mechanism of drug transport in this case involves lipid exchange or lipid mixing between the emulsion vesicle and the targeted cell membrane, which depends on the extent and frequency of contact between two lipidic surfaces. The rate of lipid exchange and drug delivery can be greatly increased by the application of clinically safe levels of ultrasound energy that increase the propensity for fusion or enhanced contact between the nanoparticles and the targeted cell membrane.

The combination of targeted drug delivery and molecular imaging with MRI has the potential to enable serial characterization of the molecular epitope expression based on imaging readouts. Monitoring and confirmation of therapeutic efficacy of the therapeutic agents at the targeted site would permit personalized medical regimens.

Nanobiotechnology for Therapeutic Design and Monitoring

Current therapeutic design involves combinatorial chemistry and system biology-based molecular synthesis and bulk pharmacological assays. Therapeutic delivery is usually nonspecific to disease targets and requires excessive dosage. Efficient therapeutic discovery and delivery would require molecular-level understanding of the therapeutics-effectors (e.g., channels and receptors) interactions and their cell and tissue responses. A multidimensional nanobiotechnology-based approach to personalized medicine starts with scanning probe techniques, especially AFM to identify potential targets for drug discovery (Lal and Arnsdorf 2010). AFM can be integrated with nanocarriers and implantable vehicles for controlled delivery. Characterization of nanocarrier-based drug delivery can enable high efficiency of in vivo or topical administration of a small dosage of therapeutics. High-throughput parallel nanosensors, comprising integrated cantilevered microarrays, TIRF, microfluidics, and nanoelectronics, can be used for rapid diagnosis of diseases, detection of biomarkers, as well as for therapeutics design. Therapeutic efficacy can be assessed by monitoring biomechanics.

Chapter 17
Nanotoxicology

Introduction

Toxicology is the branch of medicine that deals with the study of the adverse effects of chemicals and biological agents on the human body. It is the study of symptoms, mechanisms, treatments, and detection of poisoning. The broad scope of toxicology covers not only the adverse effects of therapeutics but also environmental agents and poisons. Nanotoxicology covers safety issues relevant to nanomaterials.

The success of nanomaterials is due to their small size, which enables us to get them into parts of the body where usual inorganic materials cannot enter because of their large particle size. There is an enormous advantage in drug delivery systems or cancer therapeutics. Current research is trying to find simple ways to control the degree of a particle's toxicity. This control means that the particle will be toxic only under certain desirable circumstances, such as for curing cancer. This also raises questions about unintentional effects of such powerful agents on the human body. This, however, would not be an issue for the use of nanoparticles for in vitro diagnostics.

Effects of particles on human health have been studied by toxicologists previously. Effects of larger particles generated by wearing down of implants in the body and aerosolized particles of all sizes on have been studied. However, there is little information on health impacts of very small, nano-engineered particles under 20 nm. The main concern will be about particles less than 50 nm, which can enter the cells. There are still many unanswered questions about their fate in the living body. Because of the huge diversity of materials used and the wide range in size of nanoparticles, these effects will vary a lot. It is conceivable that particular sizes of some materials may turn out to have toxic effects. At this stage, no categorical statement can be made about the safety of nanoparticles, i.e., one cannot say that nanoparticles are entirely safe or that they are dangerous. Further investigations will be needed.

K.K. Jain, *The Handbook of Nanomedicine*, DOI 10.1007/978-1-61779-983-9_17,

Toxicity of Nanoparticles

The biological impacts of nanoparticles are dependent on size, chemical composition, surface structure, solubility, shape, and aggregation. These parameters can modify cellular uptake, protein binding, translocation from portal of entry to the target site, and the possibility of causing tissue injury. Effects of nanoparticles depend on the routes of exposure that include gastrointestinal tract, skin, lung, and systemic administration for diagnostic and therapeutic purposes. Nanoparticles' interactions with cells, body fluids, and proteins play a role in their biological effects and ability to distribute throughout the body. Nanoparticle binding to proteins may generate complexes that are more mobile and can enter tissue sites that are normally inaccessible. Accelerated protein denaturation or degradation on the nanoparticle surface may lead to functional and structural changes, including interference in enzyme function. Nanoparticles also encounter a number of defenses that can eliminate, sequester, or dissolve them.

Testing for Toxicity of Nanoparticles

A mouse spermatogonial stem cell line has been used as a model to assess nanotoxicity in the male germ line in vitro (Braydich-Stolle et al. 2005). The effects of different types of nanoparticles on these cells were evaluated using light microscopy, cell proliferation, and standard cytotoxicity assays. The results demonstrated a concentration-dependent toxicity for all types of particles tested, while the corresponding soluble salts had no significant effect. Silver nanoparticles were the most toxic, while MoO3 nanoparticles were the least toxic. These results suggest that this cell line provides a valuable model to assess the cytotoxicity of nanoparticles in the germ line in vitro.

Experiments have been conducted to test silica, silica/iron oxide, and gold nanoparticles for their effects on the growth and activity of *Escherichia coli* (Williams et al. 2006). TEM and dynamic light scattering were used to characterize the morphology and quantify size distribution of the nanoparticles, respectively. TEM was also used to verify the interactions between composite iron oxide nanoparticles and *E. coli*. The results from DLS indicated that the inorganic nanoparticles formed small aggregates in the growth media. Growth studies measured the influence of the nanoparticles on cell proliferation at various concentrations, showing that the growth of *E. coli* in media containing the nanoparticles indicated no overt signs of toxicity. Although the in vitro study has its limitations, it does indicate the relative safety of the nanoparticles tested under certain conditions.

In Vitro Testing of Nanoparticle Toxicity

Screening nanomaterials by means of in vitro studies has been suggested as a fast and economical approach to distinguish between low and high toxicity nanomaterials. However, to maximize the use of in vitro assays for this purpose, their values and limitations need to be revealed. Even in risk assessment frameworks for regular chemicals, in vitro studies play a minor role. A comparative analysis of published in vitro data with nanomaterials demonstrates that there are a number of issues that need resolving before in vitro studies can play a role in the risk assessment of nanomaterials (Park et al. 2009). A major limitation of in vitro assays is that the exposure and dispersion methods most often used do not adequately reflect the exposure as it occurs in vivo. As more in vivo studies with nanomaterials become available, the values and limitations of in vitro studies as predictive tools in risk assessment of nanomaterials will come to light. Over the next 10 years, a balance will be sought between decreasing the use of in vivo studies by replacing them with in vitro studies and reliably predicting the risks of nanomaterials.

Variations in Safety Issues of Different Nanoparticles

Carbon Nanotube Safety

In contrast to the use of nanoparticles, the use of carbon nanotubes (CNTs) in life sciences is more recent. Toxicity of multiwalled CNTs, carbon nanofibers, and carbon nanoparticles was tested in vitro on lung tumor cells and clearly showed that these materials are toxic while the hazardous effect was size-dependent (Magrez et al. 2006). Moreover, cytotoxicity is enhanced when the surface of the particles is functionalized after an acid treatment.

Water-soluble, single-walled CNTs (SWCNTs) have been functionalized with the chelating molecule diethylentriaminepentaacetic (DTPA) and labeled with indium (^{111}In) for imaging purposes (Singh et al. 2006). Intravenous administration of these functionalized SWCNTs (f-SWCNTs) followed by radioactivity tracing using gamma scintigraphy indicated that f-SWCNTs are not retained in any of the reticuloendothelial system organs (liver or spleen) and are rapidly cleared from systemic blood circulation through the renal excretion route. The observed rapid blood clearance and half-life (3 h) of f-SWCNTs have major implications for all potential clinical uses. Moreover, urine excretion studies using both f-SWCNT and functionalized multiwalled CNT followed by electron microscopy analysis of urine samples revealed that both types of nanotubes were excreted as intact nanotubes. The next step for this research is to prolong the blood circulation of CNTs in order to give them enough time before excretion to get to a target tissue. The researchers will also be considering pharmaceutical development of functionalized CNTs for drug delivery.

In one study, carbon nanoparticles including carbon nanotubes, except C60 fullerenes, stimulated platelet aggregation and accelerated the rate of vascular thrombosis in rat carotid arteries (Radomski et al. 2005). All particles resulted in upregulation of GPIIb/IIIa in platelets. In contrast, particles differentially affected the release of platelet granules, as well as the activity of thromboxane-, ADP, matrix metalloproteinase- and protein kinase C-dependent pathways of aggregation. Furthermore, particle-induced aggregation was inhibited by prostacyclin and S-nitroso-glutathione, but not by aspirin. Thus, some carbon nanoparticles have the ability to activate platelets and enhance vascular thrombosis. These observations are of importance for the pharmacological use of carbon nanoparticles and support the safety of C60 fullerenes.

Manufactured SWCNT usually contain significant amounts of iron as impurities that may act as a catalyst of oxidative stress. Because macrophages are the primary responders to different particles that initiate and propagate inflammatory reactions and oxidative stress, interaction of SWCNT (0.23 wt.% of iron) with macrophages has been studied (Kagan et al. 2006). Nonpurified SWCNT more effectively converted superoxide radicals generated by xanthine oxidase/xanthine into hydroxyl radicals as compared to purified SWCNT. Iron-rich SWCNT caused significant loss of intracellular low molecular weight thiols (GSH) and accumulation of lipid hydroperoxides in macrophages. Catalase was able to partially protect macrophages against SWCNT-induced elevation of biomarkers of oxidative stress (enhancement of lipid peroxidation and GSH depletion). Thus, the presence of iron in SWCNT may be important in determining redox-dependent responses of macrophages.

Raman spectroscopic signatures of SWCNTs were measured following intravenous injection in mice (Liu et al. 2008). SWCNTs were detected in various organs and tissues over a period of 3 months. Functionalization of SWCNTs by branched PEG chains prolonged their blood circulation up to 1 day, reduced uptake in the reticuloendothelial system, and near-complete clearance from the main organs in ~2 months. Raman spectroscopy detected SWCNT in the intestine, feces, kidney, and the bladder, suggesting excretion and clearance via the biliary and renal pathways. No toxic side effect of SWCNTs to mice was observed at necropsy, histology, and blood chemistry. These findings clear the way to future biomedical applications of SWCNTs.

Fullerene Toxicity

Recent toxicology studies suggest that nanosized aggregates of fullerene molecules can enter cells and alter their functions and also cross the BBB. Computer simulations have been used to explore the translocation of fullerene clusters through a model lipid membrane and the effect of high fullerene concentrations on membrane properties (Wong-Ekkabut et al. 2008). The fullerene molecules rapidly aggregate in water but disaggregate after entering the membrane interior. The permeation of a solid-like fullerene aggregate into the lipid bilayer is thermodynamically favored and occurs on the microsecond timescale. High concentrations of fullerene induce changes in the structural and elastic properties of the lipid bilayer, but these are not large enough to mechanically damage the membrane. These results suggest that mechanical damage is an unlikely mechanism for membrane disruption and fullerene toxicity.

Gold Nanoparticle Toxicity

Toxicity has been observed at high concentrations using gold nanoparticles. Studies using 2-nm core gold nanoparticles have shown that cationic particles are moderately toxic, whereas anionic particles are quite nontoxic (Goodman et al. 2004). Concentration-dependent lysis mediated by initial electrostatic binding was observed in dye release studies using lipid vesicles, providing the probable mechanism for observed toxicity with the cationic particles.

Quantum Dot Safety Issues

To increase the stability, QDs are made from cadmium selenide (CdSe) and zinc sulfide for use as fluorescent labels. These QDs may release potentially toxic cadmium and zinc ions into cells. While cytotoxicity of bulk CdSe is well documented, CdSe QDs are generally cytocompatible, at least with some immortalized cell lines. Using primary hepatocytes as a liver model, CdSe-core QDs were found to be acutely toxic under certain conditions (Derfus et al. 2004). Although previous in vitro studies had not shown significant toxicity as the cell line used in these studies were not sensitive to heavy metals or the exposed to short-time QD labeling. The authors found that the cytotoxicity of QDs was modulated by processing parameters during synthesis, exposure to ultraviolet light, and surface coatings. These data further suggest that cytotoxicity correlates with the liberation of free Cd^{2+} ions due to deterioration of the CdSe lattice. When appropriately coated, CdSe-core QDs can be rendered nontoxic and used to track cell migration and reorganization in vitro. These results provide information for design criteria for the use of QDs in vitro and especially in vivo, where deterioration over time may occur. Capping QDs with ZnO effectively prevented Cd^{2+} formation upon exposure to air but not to ultraviolet radiation, and attempts have continued to find better coating materials.

To solve this problem, scientists at the US Department of Energy's Lawrence Berkeley National Laboratory have coated QDs with a protective layer of PEG, which is a very nonreactive and stable compound that is used extensively by the pharmaceutical industry in drug formulation. This layer is designed to prevent the dots from leaking heavy metal ions into cells once they are inside. The tool used test the safety of QDs is a gene chip packed with 18,400 probes of known human genes, and it is a comprehensive method to measure the toxicity of nanoscale particles. This chip is designed to enable the researchers to expose the human genome QDs and determine the extent to which the compound forces the genes to express themselves abnormally.

A high-throughput gene expression test determined that specially coated QD fluorescent nanoprobes affect only 0.2 % of the human genome, dispelling the concern that the mere presence of these potentially toxic sentinels disrupts a cell's function (Zhang et al. 2006a). The number of genes affected is very small given the large dose of QDs used in the study, which is up to 1,000 times greater than the dose that would typically be used in human applications. Moreover, the affected genes are not related to heavy metal exposure, which would be the case if the cells had been exposed to cadmium or zinc ions. Because of their protective coating, QDs have minimal impact on cells; the only gene changes are in transporter proteins, which are expected because the dots have to be transported into and within the cell.

Skin penetration is one of the major routes of exposure for nanoparticles to gain access to a biological system. Biological interactions of QD nanoparticles with skin have been studied. QD621 are nail-shaped nanoparticles that contain a cadmium/

selenide core with a cadmium sulfide shell coated with PEG and are soluble in water. QD were topically applied to porcine skin flow-through diffusion cells to assess penetration, which was found to be minimal and limited primarily to the outer stratum corneum layers (Zhang et al. 2008). QD655 and QD565, coated with carboxylic acid, were studied for 8 and 24 h in flow-through diffusion cells with flexed, tape-stripped, and abraded rat skin to determine if these mechanical actions could perturb the barrier and affect penetration (Zhang and Monteiro-Riviere 2008). Barrier perturbation by tape stripping did not cause penetration, but abrasion allowed QD to penetrate deeper into the dermal layers. While the study shows that QDs of different sizes, shapes, and surface coatings do not penetrate rat skin unless there is an abrasion, it shows that even minor cuts or scratches could potentially allow these nanoparticles to penetrate deep into the viable dermal layer – or living part of the skin – and potentially reach the bloodstream. These findings indicate safety concerns for workers handling QDs.

Fate of Nanoparticles in the Human Body

Following inhalation, ultrafine and fine particles can penetrate through the different tissue compartments of the lungs and eventually reach the capillaries and circulating cells or constituents, e.g., erythrocytes. These particles are then translocated by the circulation to other organs including the liver, the spleen, the kidneys, the heart, and the brain, where they may be deposited. In one study, a series of NIR fluorescent nanoparticles were systematically varied in chemical composition, shape, size, and surface charge, and their biodistribution and elimination were quantified in rat models after lung instillation (Choi et al. 2010b). Nanoparticles with hydrodynamic diameter <34 nm and a noncationic surface charge translocate rapidly from the lung to mediastinal lymph nodes. Nanoparticles of <6 nm can traffic rapidly from the lungs to lymph nodes and the bloodstream and are subsequently cleared by the kidneys.

A kinetic study was performed to determine the influence of particle size on the in vivo tissue distribution of spherical-shaped gold nanoparticles in the rat (De Jong et al. 2008). For all gold nanoparticle sizes, the majority of the gold was demonstrated to be present in liver and spleen. A clear difference was observed between the distribution of the 10-nm particles and the larger particles. The 10-nm particles were present in various organ systems including blood, liver, spleen, kidney, testis, thymus, heart, lung, and brain, whereas the larger particles were only detected in blood, liver, and spleen. The results demonstrate that tissue distribution of gold nanoparticles is size dependent with the smallest 10-nm nanoparticles showing the most widespread organ distribution.

Smaller particles apparently circulate for much longer and in some cases can cross the BBB to lodge in the brain. They can leak out of capillaries and get into the fluids between cells. So they can go to places in the body that an average inorganic mineral cannot. Such effects may not be a concern in case of targeted delivery of nanoparticle-based therapy in cancer. The eventual decision to use nanoparticle-based therapy may depend on a risk-versus-benefit assessment.

Pulmonary Effects of Nanoparticles

Modern humans breathe in considerable numbers of nanoparticles on a daily basis in traffic fumes and even from cooking. Nanoparticles are used increasingly in industrial processes and have been hypothesized to be an important contributing factor in the toxicity and adverse health effects of particulate air pollution. Small size, a large surface area, and an ability to generate reactive oxygen species play a role in the ability of nanoparticles to induce lung injury. In some individuals, they can trigger asthma by setting off an inflammatory response from the body's immune system. In one study, rats were instilled with fine and ultrafine carbon black and titanium dioxide (Renwick et al. 2004). Ultrafine particles induced more polymorphonuclear recruitment, epithelial damage, and cytotoxicity than their fine counterparts, exposed at equal mass. Both ultrafine and fine particles significantly impaired the phagocytic ability of alveolar macrophages. Only ultrafine particle treatment significantly enhanced the sensitivity of alveolar macrophages to chemotact toward C5a. It was concluded that ultrafine particles of two very different materials induced inflammation and epithelial damage to a greater extent than their fine counterparts. In general, the effect of ultrafine carbon black was greater than ultrafine titanium dioxide, suggesting that there are differences in the likely harmfulness of different types of ultrafine particles. Epithelial injury and toxicity were associated with the development of inflammation after exposure to ultrafines. Increased sensitivity to a C5a chemotactic gradient could make the ultrafine exposed macrophages more likely to be retained in the lungs, so allowing dose to accumulate.

The experience of researchers at DuPont who tested CNTs was different (Warheit et al. 2004). When the researchers injected CNTs into the lungs of rats, the animals unexpectedly began gasping for breath and 15 % of them quickly died. Yet surprisingly, all the surviving rats seemed completely normal within 24 h. What initially looked like disaster pointed to a possible safety feature: the CNTs' tendency to clump rapidly led to suffocation for some rats exposed to huge doses, but it also kept most tubes from reaching deep regions of the lung where they could not be expelled by coughing and could cause long-term damage. Now researchers see the clumping of CNTs and other nanomaterials as a new field for inquiry. Other findings of Dupont scientists were as follows:

- Exposures to quartz particles produced significant increases versus controls in pulmonary inflammation, cytotoxicity, and lung cell parenchymal cell proliferation indices.
- Exposures to carbon nanotubes produced transient inflammatory and cell injury effects. They produced a non-dose-dependent series of multifocal granulomas, which were evidence of a foreign tissue body reaction and were nonuniform in distribution and not progressive beyond 1 month following exposure.

In further studies, DuPont scientists observed that exposures to the various alpha-quartz particles produced differential degrees of pulmonary inflammation and cytotoxicity, which were not always consistent with particle size. The results of their studies demonstrate that the pulmonary toxicities of alpha-quartz particles appear to correlate better with surface activity than particle size and surface area (Warheit et al. 2007).

Polyamidoamine (PAMAM) dendrimer can induce acute lung injury in vivo as it triggers autophagic cell death by deregulating the Akt-TSC2-mTOR signaling pathway. The autophagy inhibitor 3-methyladenine was shown to rescue PAMAM-induced cell death and ameliorate acute lung injury caused by PAMAM in mice (Li et al. 2009).

It has been suggested that the nanoparticulate component of 10 μM (PM10) is capable of translocating into the circulation with the potential for direct effects on the vasculature and is a potential risk factor for cardiovascular disease. A study was conducted in healthy volunteers to determine the extent to which inhaled technetium-99m (99mTc)-labeled CNTs (Technegas) were able to access the systemic circulation (Mills et al. 2006). Technegas particles were 4–20 nm in diameter and aggregated to a median particle diameter of ~100 nm. Radioactivity was immediately detected in blood, with levels increasing over 60 min. Thin layer chromatography of whole blood identified a species that moved with the solvent front, corresponding to unbound 99mTc-pertechnetate, which was excreted in urine. There was no evidence of particle-bound 99mTc at the origin. Gamma camera images demonstrated high levels of Technegas retention in the lungs, with no accumulation of radioactivity detected over the liver or spleen. Thus, the majority of 99mTc-labeled carbon nanoparticles remain within the lung up to 6 h after inhalation. In contrast to previous published studies, thin layer chromatography did not support the hypothesis that inhaled Technegas carbon nanoparticles pass directly from the lungs into the systemic circulation.

The physiological relevance of various findings should ultimately be determined by conducting an inhalation toxicity study. As yet, no one has created a realistic test for the effects of inhaled nanoparticles; such a test could easily cost more than $1 million to design and carry out. Some studies have been planned to study any possible adverse effects of nanoparticles on the lungs. One of those studies is a multi-year project led by the US National Institute of Environmental Health Sciences that will examine the potential toxic and carcinogenic effects of inhalation exposure to nanomaterials.

Neuronanotoxicology

Neuronanotoxicology is the study of potential toxic effects of nanoparticles on the nervous system. The concern has arisen because nanoparticles can cross the BBB to enter the brain following introduction into the systemic circulation and are not cleared out. The effect depends on the type of nanoparticle introduced as some are neuroprotective whereas others are neurotoxic.

Nanoparticle Deposits in the Brain

Passage of nanoparticles across the BBB to enter the brain has already been documented. There is a possible risk in inhaling nanoparticles that are so small that they can slip through membranes inside the lungs, enter systemic circulation and lodge in the brain. Research on rats has shown nanoparticles deposited in the nose can migrate to the brain and move from the lungs into the bloodstream. They can also change shape as they move from liquid solutions to the air, making it harder to draw general conclusions about their potential impact on living things. More experiments are needed to establish the impact of nanoparticles on the brain if they remain there.

Fullerenes (buckyballs) are lipophilic and localize into lipid-rich regions such as cell membranes in vitro, and they are redox active. Other nanosize particles and soluble metals have been shown to selectively translocate into the brain via the olfactory bulb in mammals and fish. A preliminary study found rates of brain damage 17 times higher in largemouth bass exposed to a form of water-soluble buckyballs than unexposed fish (Oberdorster 2004). Significant lipid peroxidation was found in brains of largemouth bass after 48 h of exposure to 0.5 ppm uncoated nano-C60. Buckyballs are also toxic in vitro, causing 50 % of the cultured human cells to die at a concentration of 20 parts per billion (Sayes et al. 2004). With the addition of an antioxidant, L-ascorbic acid, the oxidative damage and resultant toxicity of nano-C60 was completely prevented (Sayes et al. 2005).

Nanoparticles and Neurodegeneration

There is the potential for neurodegenerative consequence of nanoparticle entry to the brain. Histological evidence of neurodegeneration has been reported in both canine and human brains exposed to high ambient particulate matter levels, which may be caused by the oxidative stress pathway. Thus, oxidative stress due to nutrition, age, and genetics, among others, may increase the susceptibility for neurodegenerative diseases (Peters et al. 2006).

A worldwide project called NeuroNano is investigating if engineered nanoparticles could constitute a significant risk to humans for neurodegenerative diseases. NeuroNano partners include universities of Ulster, Dublin, Cork, Edinburgh, and Munich in Europe; universities of California, Rochester, and Rice in the USA; and the National Institute of Materials Science in Japan. The University of Ulster experts, funded by a 2009 grant from the European Commission, is specifically looking at nanoparticles present in chemicals found in sunscreens and an additive in some diesel fuels – titanium dioxide and cerium oxide – and their connection to Alzheimer's and Parkinson's diseases. Nanoparticles can have highly significant impact on the rate of misfolding of key proteins associated with neurodegenerative diseases.

Effect of Nanoparticles on the Heart

Some nanoparticles may influence the heart function, whereas others do not. The effect of nanoparticles is being studied on the Langendorff Heart or "isolated perfused heart," which is an in vitro technique used in pharmacological and physiological research (Stampfl et al. 2011). The modified Langendorff heart is a particularly good test object as it has its own impulse generator, the sinus node, enabling it to function outside the body for several hours. This model enables observation and analysis of electrophysiological parameters over a minimal time period of 4 h without influence by systemic effects complications of an intact animal while enabling the determination of stimulated release of substances under influence of NPs. A significant dose- and material-dependent increase in heart rate up to 15 % was found accompanied by arrhythmia evoked by NPs made of flame soot (Printex 90), spark discharge generated soot, anatas (TiO2), and silicon dioxide (SiO2). However, flame-derived SiO2 (Aerosil) and monodisperse polystyrene lattices exhibited no effects. The increase in heart rate is attributed to catecholamine release from adrenergic nerve endings within the heart. This new heart model may prove to be particularly useful in medical research and could serve as a test organ to help select nanoparticles that do not affect the heart adversely. The manufacturing process as well as the shape of a NP may play an important role. Therefore, further studies will examine the surfaces of different types of NPs and their interactions with the cells of the cardiac wall.

Blood Compatibility of Nanoparticles

Given that the majority of nanoparticles are intended to travel to tumors through the bloodstream, the effects of nanoparticles on blood cells are of particular concern to those developing nanoparticle-based therapeutic and imaging agents. The blood compatibility of nanoparticles depends on the material used.

Carbon Nanoparticle-Induced Platelet Aggregation

To determine the potential for blood platelet-nanoparticle interactions, the effects of engineered and combustion-derived carbon nanoparticles were studied on human platelet aggregation in vitro and rat vascular thrombosis in vivo (Radomski et al. 2005). Multiple-wall and single-wall nanotubes, C60 fullerenes and mixed carbon nanoparticles were compared with standard urban particulate matter (average size 1.4 μm). Carbon particles, except C60 fullerenes, stimulated platelet aggregation and accelerated the rate of vascular thrombosis in rat carotid arteries. All particles resulted in upregulation of GPIIb/IIIa in platelets. The particle-induced aggregation was inhibited by prostacyclin and S-nitroso-glutathione, but not by aspirin. It is concluded that some carbon nanoparticles and microparticles have the ability to activate platelets and enhance vascular thrombosis. These observations are of importance for the pharmacological use of carbon nanoparticles and pathology of urban particulate matter.

Compatibility of Lipid-Based Nanoparticles with Blood and Blood Cells

Pegylated and nonpegylated cetyl alcohol/polysorbate nanoparticles (E78 NPs) are being tested as drug carriers for specific tumor and brain targeting. Because these nanoparticle formulations are designed for systemic administration, the compatibility of these lipid-based NPs with blood and blood cells was tested with a particular focus on hemolytic activity, platelet function, and blood coagulation (Koziara et al. 2005). E78 NPs did not cause in vitro red blood cell lysis at concentrations up to 1 mg/mL. In addition, under conditions tested, E78 and polyethylene glycol (PEG)-coated E78 NPs (PEG-E78 NPs) did not activate platelets. In fact, both NP formulations very rapidly inhibited agonist-induced platelet activation and aggregation in a dose-dependent manner. It was concluded that PEG-coated and nonpegylated E78 NPs have potential blood compatibility at clinically relevant doses. Based on the calculated nanoparticle-to-platelet ratio, the concentration at which E78 NPs could potentially affect platelet function in vivo was approximately 1 mg/mL.

Transfer of Nanoparticles from Mother to Fetus

The toxicopathology research group at the University of Liverpool (Liverpool, UK) has investigated the fate of injected gold nanoparticles into pregnant rats to determine whether they can be transferred across the placenta to the fetus. The findings, reported in 2004, show unidentified particles in the fetus but follow-up results have not been reported. This research on possible transfer of nanoparticles to the fetus could indicate a new, particular, hazard of nanoparticles that would be a cause for concern. Scientists at the Center for Biological and Environmental Nanotechnology of Rice University are of the opinion that it is still too early to establish the transfer of nanoparticles to the fetus. They believe that first they have to establish whether nanoparticles can be accumulated in the body and then they can investigate chronic effect.

A study has shown that QDs may be transferred from female mice to their fetuses across the placental barrier (Chu et al. 2010). Smaller QDs are more easily transferred than larger QDs and the number of QDs transferred increases with increasing dosage. Capping with an inorganic silica shell or organic PEG reduces QD transfer but does not eliminate it. These results suggest that the clinical utility of QDs could be limited in pregnant women.

Cytotoxicity of Nanoparticles

Cytotoxicity refers to toxic effects on individual cells. In cytotoxicological studies, identical cell cultures are exposed to various forms and concentrations of toxins. In order to compare the toxicity of different compounds, scientists determine the concentration, measured in parts per million or parts per billion, of materials that lead to the death of 50 % of the cells in a culture within 48 h.

A particle's surface chemistry may determine how it interacts with the tissues of the body. AFM was used to show that aqueous solutions of poly(amidoamine) dendrimers cause the formation of holes 15–40 nm in diameter in previously intact lipid bilayers (Mecke et al. 2004). In contrast, carboxyl-terminated core-shell tectodendrimer clusters do not create holes in the lipid membrane but instead show a strong affinity to adsorb to the edges of existing bilayer defects. Multiwalled CNTs, not derivatized nor optimized for biological applications, are capable of both localizing within and initiating an irritation response in human epidermal keratinocytes, which may occur in the skin of workers as occupational exposure during manufacture of nanotubes (Monteiro-Riviere et al. 2005).

Indirect DNA Damage Caused by Nanoparticles Across Cellular Barriers

Cobalt-chromium nanoparticles (29.5 +/− 6.3 nm in diameter) have been shown to damage human fibroblast cells across an intact cellular barrier without having to cross the barrier (Bhabra et al. 2009). The damage is mediated by a novel mechanism involving transmission of purine nucleotides (such as ATP) and intercellular signaling within the barrier through connexin gap junctions or hemichannels and pannexin channels. The outcome, which includes DNA damage without significant cell death, is different from that observed in cells subjected to direct exposure to nanoparticles. These results suggest the importance of indirect effects when evaluating the safety of nanoparticles. The potential damage to tissues located behind cellular barriers needs to be considered when using nanoparticles for targeting diseased states

Measures to Reduce Toxicity of Nanoparticles

Scientists at the Rice University's Center for Biological and Environmental Nanotechnology are able to significantly lower the toxicity level of buckyballs when exposed to liver and skin cells in a petri dish. They have accomplished this by attaching other molecules to the surface of buckyballs. This simple chemical modification could lower potential exposure risks during disposal of a product like a fuel cell or within a manufacturing plant. Removing attached molecules and enhancing toxicity could also be useful in chemotherapy treatments, for instance.

There is a huge potential for a new generation of gold nanoparticle (AuNP)-based nanomedicinal products, nontoxic AuNP constructs and formulations that can be readily administered site-specifically through the intravenous mode for diagnostic imaging by CT scan or for therapy via various modalities. The use of gum arabic (GA) was explored as it has been used for a long time to stabilize foods such as yogurt and hamburgers. It has unique structural features, including a highly branched polysaccharide structure consisting of a complex mixture of potassium, calcium, and magnesium salts derived from arabic acid. GA can be used to absorb and assimilate metals and create a "coating" that makes gold nanoparticles stable and nontoxic. A study has described the synthesis and stabilization of AuNPs within the nontoxic phytochemical gum arabic matrix (GA-AuNPs) and has presented detailed in vitro analysis and in vivo pharmacokinetics studies of GA-AuNPs in pigs to gain insight into the organ-specific localization of this new generation of AuNP vector (Kattumuri et al. 2007). X-ray CT contrast measurements of GA-AuNP vectors were carried out for potential application in molecular imaging. The results demonstrate that naturally occurring GA can be used as a nontoxic phytochemical excipient in the production of readily administrable biocompatible AuNPs for diagnostic and therapeutic applications in nanomedicine.

Reducing Toxicity of Carbon Nanotubes

Water-soluble SWCNTs are significantly less toxic to begin with and can be rendered nontoxic with minor chemical modifications. SWCNTs can be rendered soluble via the attachment of the chemical subgroups hydrogen sulfite, sodium sulfite, and carboxylic acid. The research is a continuation of pioneering efforts to both identify and mitigate potential nanotechnology risks. The cytotoxicity of undecorated SWCNTs is 200 parts per billion, which compares to the level of 20 parts per billion for undecorated buckyballs (Sayes et al. 2006). Competitive bindings of blood proteins on the SWCNT surface can greatly alter their cellular interaction pathways and result in much reduced cytotoxicity for these protein-coated SWCNTs (Ge et al. 2011). For medical applications, it is encouraging to see that the cytotoxicity of nanotubes is low and can be further reduced with simple chemical changes.

SWCNTs can be catalytically biodegraded over several weeks by the plant-derived enzyme, horseradish peroxidase. A study has shown that hypochlorite and reactive radical intermediates of the human neutrophil enzyme myeloperoxidase catalyze the biodegradation of SWCNTs in vitro, in neutrophils and, to a lesser degree, in macrophages (Kagan et al. 2010). Molecular modeling suggests that interactions of basic amino acids of the enzyme with the carboxyls on the carbon nanotubes position the nanotubes near the catalytic site. Importantly, the biodegraded nanotubes do not generate an inflammatory response when aspirated into the lungs of mice. These findings suggest that the extent to which carbon nanotubes are biodegraded may be a major determinant of the scale and severity of the associated inflammatory responses in exposed individuals. These findings will open the door to using SWCNTs as a safe drug delivery tool and lead to the development of a natural treatment for people exposed to nanotubes.

A Screening Strategy for the Hazard Identification of Nanomaterials

The International Life Sciences Institute Research Foundation/Risk Science Institute convened an expert working group to develop a screening strategy for the hazard identification of engineered nanomaterials. The working group report presented the elements of a screening strategy rather than a detailed testing protocol (Oberdorster et al. 2005). Based on an evaluation of the limited data currently available, the report presents a broad data gathering strategy applicable to this early stage in the development of a risk assessment process for nanomaterials. Oral, dermal, inhalation, and injection routes of exposure are included recognizing that, depending on use patterns, exposure to nanomaterials may occur by any of these routes. The three key elements of the toxicity screening strategy are: (1) physicochemical characteristics, (2) in vitro assays (cellular and noncellular), and (3) in vivo assays. There is a strong likelihood that biological activity of nanoparticles will depend on physicochemical parameters not routinely considered in toxicity screening studies. Physicochemical properties that may be important in understanding the toxic effects of test materials include particle size and size distribution, agglomeration state, shape, crystal structure, chemical composition, surface area, surface chemistry, surface charge, and porosity. In vitro techniques allow specific biological and mechanistic pathways to be isolated and tested under controlled conditions, in ways that are not feasible in in vivo tests. Tests are suggested for portal-of-entry toxicity for lungs, skin, and the mucosal membranes, and target organ toxicity for endothelium, blood, spleen, liver, nervous system, heart, and kidney. Noncellular assessment of nanoparticle durability, protein interactions, complement activation, and prooxidant activity is also considered. The report focuses on the likely toxic impact of nanoparticles in the body but does not comment on the actual risk of exposure because currently there are few situations where people are directly exposed to the nanoparticles.

Concluding Remarks on Safety Issues of Nanoparticles

There is no consensus on the real risks of nanomaterials. Risk evaluation presents challenges due to a lack of data, the complexity of nanomaterials, measurement difficulties, and undeveloped hazard assessment frameworks. There is a paucity of published material on this topic, which could provide scientific guidance; less than 800 journal articles on health risks of engineered nanomaterials have been published. Until the risk assessment is evaluated further, some precautionary measures should be considered to reduce risks, such as exposure control. It is recommended that manufacturers of nanomaterials should inventory all products and applications to potential exposures across the product life cycle. The risk of each application should be characterized based on exposure and available knowledge about hazard. The risk of exposure should be mitigated through additional testing and product redesign.

Research into Environmental Effects of Nanoparticles

Research strategies for safety evaluation of nanomaterials have been planned in the USA, Europe, and Japan. An important component of these programs is the development of reliable risk and safety evaluations for these materials to ensure their safety for human health and the environment. The scope of each of these programs includes efforts to assess the hazards posed by nanomaterials in realistic exposure conditions (Thomas et al. 2006a). University of Wisconsin-Madison's Nanotechnology in Society Project has published a summary of some of the key data gaps, uncertainties, and unknowns that need to be addressed to develop adequate risk assessments for nanomaterials and to take timely and appropriate public health precautions (Powell and Kanarek 2006).

Environmental Safety of Aerosols Released from Nanoparticle Manufacture

Det Norske Veritas (Oslo, Norway), a classification society, conducted the EU project NANOTRANSPORT (2006–2008), which addressed the behavior of aerosols released to ambient air from nanoparticle manufacturing. The project brought together leading expert organizations in risk management, aerosol monitoring, filtration, nanoparticle technology, and online particle characterization fields. Key conclusions of the study are as follows:

- There is considerable evolution of nanoaerosols over time: their average size increases, while their concentration decreases.
- Natural background aerosols are scavengers for nanoparticles.
- The time scale for size evolution depends on concentration and primary size of the nanoparticles and that of the background aerosol – it may range from a matter of a few minutes up to half an hour.
- Nanoparticles will be physically/chemically present in size classes other than those in which they were originally emitted.
- Filtration efficiency of primary nanoparticles <80 nm is usually sufficiently high, but their agglomerates may be in the most penetrating particle size range of 80–200 nm.

Role of US Government Agencies in Research on Safety of Nanoparticles

Nanotechnology advocates say they support faster and broader environmental research, but paying for it has not been a priority for businesses or the government. The Environmental Protection Agency, which had previously focused on supporting research into how nanotechnology could help clean or protect the environment, is seeking grant proposals from researchers looking at potential risks. But the amounts awarded are only a fraction of those allocated for nanotechnology research and development. The difficulty and cost of researching risk are influencing business decisions. L'Oréal, the cosmetics company, for instance, dropped its research on the characteristics of nanoparticles after outside researchers raised questions about toxicity. Some smaller nanotechnology start-ups say they simply do not have the resources to push into promising areas that pose health questions.

Work at Nanosafety Laboratories Inc UCLA

UCLA (University of California, Los Angeles, CA) has developed a new testing method that would help manufacturers monitor and test the safety and health risks of engineered nanomaterials (Nel et al. 2006). The testing model developed at UCLA is based on toxicity testing for occupational and air pollution particles, which include nanoparticles. The strong scientific foundation of air pollution particle testing is used to help understand the health impact of engineered nanoparticles and ensure safe manufacturing of nanoproducts. The impact of nanoparticle interactions with the body is dependent on their size, chemical composition, surface structure, solubility, shape, and how the individual nanoparticles amass together. Nanoparticles may modify the way cells behave and potential routes of exposure, including the gastrointestinal tract, skin, and lungs. The three key elements of the toxicity screening strategy include (1) the physical and chemical characterization of nanomaterials, (2) tissue cellular assays, and (3) animal studies.

A mature toxicological science has emerged from the study of these particles, providing a framework for a predictive testing strategy applicable to engineered nanomaterials. A predictive strategy is one in which a series of simple but high-quality tests can be employed to predict which materials could be hazardous and therefore speed up the process of classifying materials into those that are safe and those that could pose toxicity problems. This type of approach is similar to that used by the National Toxicology Program for evaluation of chemical agents. The UCLA model predicts toxicity according to the ability of some nanoparticles to generate toxic oxygen radicals that can cause tissue injury, including inflammation and other toxic effects. For air pollution particles, this injury can translate into asthma and atherosclerotic heart disease. Using this model, the UCLA laboratory has developed a series of tests to assess nanoparticle toxicity in nonbiological environments as well as in tissue cultures and animal models. Funding for the research on air pollution particles that contributed to this paper came from the National Institute of Environmental Health Sciences and the US Environmental Protection Agency.

Center for Biological and Environmental Nanotechnology

Rice University's Center for Biological and Environmental Nanotechnology (CBEN) has played an active role in informing the public, lawmakers, and industry about potential unintended environmental consequences of nanotechnology. CBEN's research aims to understand how nanomaterials function in water-based environments such as living organisms and ecosystems. In its first 5 years since its founding in 2000, CBEN helped produce groundbreaking research in nanomedicine, nanobiotechnology, nanotoxicology, and nanoscale methods for environmental remediation. Further details of CBEN can be viewed at its web site (http://cben.rice.edu/).

European Nest Project for Risk Assessment of Exposure to Nanoparticles

A study, led by the Institute of Occupational Medicine (UK), is investigating the safety of new and emerging science and technologies (NEST) that have the potential to generate particulates, which can enter the body via inhalation, ingestion, or dermal absorption. Information is needed regarding the possible risks from exposure to these particles, including the routes of exposure and subsequent disposition, their potential toxicity, appropriate toxicological testing procedures, and susceptible subpopulations. The institute will acquire a bank of five particles potentially generated by NEST (NESTP) and will assess the health risk from exposure to these materials through air or the food supply with a work program, integrating in vitro experiments, animal models of healthy/susceptible individuals, and exposure/risk assessment.

Public Perceptions of the Safety of Nanotechnology

Past and current experience in biotechnology has shown how an environmentally concerned public reacts to new technologies. One can predict antinanotechnology groups in the future similar to the antibiotechnology groups. Apart from the unknown long-term effects of nanomaterials in the human body, a much greater concern is expressed about the environmental effects of the release of nanoparticles from the industry. A question that is being already asked is about the possibility of accumulation of nanomaterials in water or the earth and the risk if this takes place. Some of these issues are already under investigation. So far, the public's outlook on nanotechnology remains positive despite a lack of knowledge, but press coverage and agitation from various groups indicates that nanotechnology industry will not be able to dodge these questions much longer. Instead of remaining silent on this issue, companies need a communications strategy to share their safety studies, collaborate with trusted partners, and explain the benefits nanotechnology can bring. The public relations departments of research institutes and companies involved in nanotechnology will have a major task of educating the public about the safety of nanotechnology.

As an example, it was anticipated that exploration of the human genome could result in public concerns – ethical, legal, and cultural. So 3–5 % of federal research money was set aside to fund the study of these issues and to communicate with the public and encourage lots of openness and transparency. This is now the model for a proactive approach to new technology development.

The largest and most comprehensive survey of public perceptions of nanotechnology products finds that US consumers are willing to use specific nano-containing products – even if there are health and safety risks – when the potential benefits are high (Currall et al. 2006). The study, which was conducted by researchers at Rice

University's Center for Biological and Environmental Nanotechnology (Houston, TX), University College London, and the London Business School, also finds that US consumers rate nanotechnology as less risky than everyday technologies like herbicides, chemical disinfectants, handguns, and food preservatives. The research was based on more than 5,500 survey responses. One survey polled consumers about how likely they would be to use four specific, nano-containing products: a drug, skin lotion, automobile tires, and refrigerator gas coolant. This is the first large-scale study to experimentally gauge the public's reaction to specific, nano-containing products, and the use of scenarios about plausible, specific products yielded results that challenge the assumption that the public focuses narrowly on risk. The greater the potential benefits, the more risks people are willing to tolerate.

Evaluation of Consumer Exposure to Nanoscale Materials

Although there are numerous likely consumer advantages from products containing nanoscale materials, there is very little information available regarding consumer exposure to the nanoscale materials in these products or any associated risks from these exposures. The products include cosmetics, sunscreen, textiles, and sporting goods. An important component in addressing potential health risks is the potential exposure to the consumer. The presence of a toxic substance in a consumer product does not constitute a health hazard if the product design or use prevents the consumer from being exposed to the substance. For consumer product applications, if the nanomaterial is attached to the product in a manner that minimizes its release, the exposure potential will be minimal. If the nanomaterials are released in significant quantities during reasonably foreseeable product use or foreseeable misuse, exposure may result via dermal contact, ingestion, or inhalation.

Toxic Substances Control Act gives the US Environmental Protection Agency (EPA) authority to regulate the manufacture, use, distribution in commerce, and disposal of chemical substances. This act authorizes the agency to regulate both new and existing compounds and is currently undergoing scrutiny by the agency to determine to what extent it can incorporate engineered nanomaterials (Thomas et al. 2006). FDA regulates very few materials but many types of products. Cosmetics do not require premarket approval from the FDA, but if the FDA considers that there is a safety concern resulting from the use of any cosmetic ingredient, including nanoparticles, then it has several options to prohibit the marketing and sale of those products. Sunscreens are considered to be cosmetics in Europe; in the USA, they are considered to be drugs. Currently, the FDA is involved in studies of marketed sunscreens in collaboration with the National Toxicology Program, Rice University, and the National Institute for Standards and Technology. These studies will help identify those sunscreens that contain nanoscale particles of titanium dioxide and zinc oxide and characterize the size ranges for these nanoscale particles.

US Consumer Product Safety Commission (CPSC) is charged with protecting the public from unreasonable risks of serious injury or death from over 15,000 types of consumer products under the agency's jurisdiction. Nanotechnology-derived products entering commerce, containing materials with novel chemical, physical, biological, optical, and electronic properties, will require assessment to determine if there may be exposure to a potential health risk that might negatively impact consumer safety. The potential health risk of nanomaterials can be assessed with existing CPSC statutes, their administering regulations, and interpretative guidelines.

Safety of Nanoparticle-Based Cosmetics

Regulations in the European Union

Under cosmetics regulations in the European Union (EU), ingredients (including those in the form of nanoparticles) can be used for most purposes without prior approval, provided they are not on the list of banned or restricted use chemicals and that manufacturers declare the final product to be safe. Given the concerns about toxicity of any nanoparticles penetrating the skin, the Royal Society of UK recommends that their use in products be dependent on a favorable opinion by the relevant European Commission scientific safety advisory committee. A favorable opinion has been given for the nanoparticulate form of titanium dioxide (because chemicals used as UV filters must undergo an assessment by the advisory committee before they can be used), but insufficient information has been provided to allow an assessment of zinc oxide. In the meantime, it is recommended that manufacturers publish details of the methods they have used in assessing the safety of their products containing nanoparticles that demonstrate how they have taken into account that properties of nanoparticles may be different from larger forms. Because nanoparticles of zinc oxide are not used extensively in cosmetics in Europe, this is not a major problem. If chemicals produced in the form of nanoparticles are to be treated as new chemicals, the product label would identify that nanoparticles have been used in manufacture.

Nanotechnology-Based Sunscreens

In 2006, a petition was filed by consumer, health and environmental groups in the USA that asked the FDA to recall sunscreens that contain nanoparticles unless they are proven safe. The petition also called for premarket safety testing of nano sunscreens and for nano-specific toxicity testing and mandatory labeling of nano products.

According to the sunscreen industry, sun creams are safe, and academic experts do not have enough evidence for harmful effects to justify a recall of sunscreens, although some recommend labeling of the products and more public access to information about safety studies done by industry. The FDA is participating in studies of skin absorption of nano-sized titanium dioxide and zinc oxide preparations used in sunscreens.

Friends of the Earth, one of the petitioners, published a list of 116 personal care products, cosmetics, and sunscreens that contain nanomaterials. The sunscreen ingredients the petitioners warned about are nanoparticles of titanium dioxide and zinc oxide smaller than 100 nm, the upper size limit of what is usually called nanoparticles. It is unclear whether such particles can enter intact skin. The FDA has previously classified 16 sunscreen ingredients as safe and effective, and particle size did not affect the classification of these ingredients. But the petitioners argue that size does matter, in that nanoparticles are likely more harmful than larger particles of the same material, and products that contain them should therefore be recalled and tested for safety.

According to the International Council on Nanotechnology at Rice University, a recall of sun creams at this stage would be premature because no health risks have manifested themselves yet. Toxicity studies on nanoparticles are in progress at Rice University, but their tests cannot be designed unless the exact contents are stated on the labels of sun creams. The Cosmetic, Toiletry, and Fragrance Association, an industry group, issued a statement that sunscreens use small microparticles, which are larger than 100 nm (not nanoparticles by definition), have been deemed safe by the FDA. Some manufacturers call their products "nano" only for promotional reasons. However, some face creams and moisturizers contain fullerenes that have a potential for toxic effects due to elevation of free radicals in the brain as a reaction to the particles.

Cosmetic Industry's White Paper on Nanoparticles in Personal Care

The Cosmetic, Toiletry, and Fragrance Association (www.ctfa.org) has released a white paper on the application of nanotechnology in personal care products, including cosmetics and certain over-the-counter drug products, specifically sunscreens. The report discusses the advantages of the use of nanomaterials, the regulatory evaluation of personal care products using nanotechnology, particular properties of nanoparticles, the potential for dermal absorption of nanoparticles used in topical lotions or creams, and what it characterizes as the general scientific consensus and toxicology conclusions about the use of nanotech in personal care products. The report specifically addresses the issue of titanium dioxide and zinc oxide used in nanoparticle form in sunscreens. The industry-supported report argues that nanoparticles applied topically to the skin in lotions or creams are safe.

Skin Penetration of Nanoparticles Used in Sunscreens

Nanoparticles are commonly used in sunscreens and other cosmetics, and since consumer use of sunscreen is often applied to sun damaged skin, the effect of ultraviolet radiation (UVR) on nanoparticle skin penetration is a concern due to potential toxicity. A study has investigated nanoparticle skin penetration by employing an in vivo semiconductor QD nanoparticle model system, which improves imaging capabilities (Mortensen et al. 2008). In these experiments, carboxylate QD were applied to the skin of SKH-1 mice in a glycerol vehicle with and without UVR exposure. The skin collection and penetration patterns were evaluated 8 and 24 h after QD application using tissue histology, confocal microscopy, and transmission electron microscopy. Low levels of penetration were seen in both the non-UVR-exposed mice and the UVR-exposed mice. Qualitatively higher levels of penetration were observable in the UVR-exposed mice. The particles accumulate around the hair follicles and in tiny skin folds. These results provide important insight into the ability of QD to penetrate intact and UVR-compromised skin barrier. Part of the explanation likely lies with the complex reaction of skin when it is assaulted by the UVR. The cells proliferate, and molecules in the skin known as tight-junction proteins loosen so that new cells can migrate to where they are needed. Those proteins normally act as gatekeepers that determine which molecules to allow through the skin and into the body and which molecules to block. When the proteins loosen up, they become less selective than usual, possibly giving nanoparticles an opportunity to pass through the barrier. In the future, the investigators plan to study titanium dioxide and zinc oxide, two materials that are widely used in sunscreens, and other cosmetic products to help block the damaging effects of UVR.

Chapter 18
Ethical and Regulatory Aspects of Nanomedicine

Introduction

Ethical and regulatory aspects are important in practice of medicine, and the same applies to nanomedicine. As has happened with introduction of all new technologies, in healthcare, these issues need to be considered. The FDA is formulating specific regulations relevant to nanobiotechnology products. The development of pharmaceuticals containing nanoparticles and methods of drug delivery, however, will be regulated by the FDA like any other biopharmaceutical product.

Ethical and Social Implications of Nanobiotechnology

Nanotechnology's impact has been mostly in engineering, communications, electronics, and consumer products. Now that nanobiotechnology is being applied to pharmaceuticals, food sciences, and human medicine, there is awareness of the consequences, good or bad, that are still not well understood. Most technologies have potential for good or evil. New technologies may have disruptive influence on the society as evidenced by the introduction of biotechnology in agriculture. Genetically modified foods continue to disturb trade between the USA and Europe, and the products find a mixed reception in supermarkets. In 2005, the Swiss public voted to place a moratorium on application of biotechnology in agriculture for a period of 5 years. There is a concern that the same thing could happen to nanotechnology.

Communication with the public is important at this stage. This will involve education of the public and consideration of their opinion. Greenpeace argues that current research priorities need to shift in favor of environmental and health protection to engender public support and/or an ongoing need to remain sensitive to emerging societal preferences (Parr 2005). US federal grants have been awarded to universities for research on ethical, legal, and social issues arising from introduction of nanotechnology.

K.K. Jain, *The Handbook of Nanomedicine*, DOI 10.1007/978-1-61779-983-9_18,

Nanoethics

Ethical aspects are important for all new technologies, and nanobiotechnology is no exception. Although nanotechnology has not raised any new ethical issues, it is worthwhile to keep these considerations in mind while developing and applying nanobiotechnologies to medicine. Ethical, social, and legal issues arising from the application of nanotechnology to medicine have been reviewed recently (Resnik and Tinkle 2007).

In the ethical debate on nanotechnology (nanoethics), there has been a strong tendency to strongly focus on extremes either the upside or the downside with the result that ethical assessments tend to diverge radically. Many of the extreme views are based on simplified and outdated visions of a nanotechnology dominated by self-replicating assemblers and nanomachines. There is a need for development of more balanced and better-informed assessments (Gordijn 2005). Various pitfalls of nanoethics are (1) the restriction of ethics to prudence understood as rational risk management, (2) the reduction of ethics to cost/benefit analysis, and (3) the confusion of technique with technology and of human nature with the human condition (Dupuy 2007). Once these points have been clarified, it is possible to take up some philosophical and metaphysical questions about nanobiotechnologies.

In 2006, the Nanoethics Group (http://www.nanoethics.org/), located in Cal Poly State University (San Luis Obispo, California), was awarded grants by the National Science Foundation to study ethical issues related to human enhancement and nanotechnology. The role of nanotechnology in enhancement procedures is not clear. Enhancement may mean use of hormones or cosmetic surgeries or other procedures to enhance human performance (mental or physical) or appearance. The group is concerned that the accelerating pace of new technology may lead to some fantastic scenarios such as advanced cybernetic body parts and computers imbedded in the brains, which will raise ethical issues. In 2007, the Nanoethics Group released a collection of papers to address both urgent and distant issues related to nanotechnology's impact on society (Allhoff et al. 2007). It tackles a full range of issues facing nanotechnology, such as those related to benefits, risk, environment, health, human enhancement, privacy, military, democracy, education, humanitarianism, molecular manufacturing, space exploration, artificial intelligence, life extension, and more (Allhoff et al. 2010).

In 2007, EGE, the European Group on Ethics in Science and New Technologies (http://ec.europa.eu/european_group_ethics/activities/index_en.htm) issued a draft report that recognizes the potential of nanomedicine in terms of developing new diagnostics and therapies. The group proposes that measures be established to verify the safety of nanomedical products and devices and calls on the relevant authorities to carry out a proper assessment of the risks and safety of nanomedicine. It also recommended that there should be an EU website on ethics and nanomedicine, where citizens can find information and pose questions to researchers. Academic and public debates should be held on the issues raised by forthcoming developments in nanomedicine. The report also placed a strong emphasis on the importance of carrying out more research into the ethical, legal, and social implications (ELSI) of nanomedicine. They recommend that up to 3 % of the nanotech research budget be set aside for ELSI research. They also call on the EC to set up a dedicated European network on nanotechnology ethics. This network would bring together experts from a range of fields, promote deeper understanding of the ethical issues arising from nanotechnology and nanomedicine, promote education in these fields, and work to ensure that ethics become embedded in research practices in nanomedicine and nanotechnology. The group also suggests that the EC fund a study on the social effects on nanomedicine in developing countries. On the legal front, EGE does not believe that structures set up specifically to deal with nanomedicine are needed right now. However, they suggest monitoring existing regulatory systems to ensure they do address all nanomedical products.

Nanotechnology Patents

Because of the tremendous potential of application of nanobiotechnology in healthcare, i.e., nanomedicine, and opportunities for commercialization, securing valid and defensible patent protection will be important in the future. Medical device companies seeking to implement nanotechnology in their products need to be aware of the emerging intellectual property (IP) trends in nanotechnology. USA holds approximately 70 % of the world patents in nanotechnology. The number of issued patents in nanotechnology has grown from less than 200 in 2000 to over 6,000 by the end of 2010, and there is a plethora of patent claims with backlog at the US Patent and Trademark Office (USPTO). One of the problems is an inadequate patent classification system and lack of differentiation between the terms "nanotechnology," which is used to search patent databases, and "nanobiotechnology." Various types of patents covered include nanomaterial, nanostructure, nanofiber, nanowire, nanoparticle, fullerene, quantum dot, nanotube, dendrimer, or nanocrystal. Patents are not classified according to areas of application relevant to healthcare.

The USPTO has created a preliminary classification for nanotechnology, designated as Class 977 nanotechnology cross-reference art collection, and its purpose is described on its website (http://www.uspto.gov/web/patents/biochempharm/crossref.htm). It is the first step in a multiphase nanotechnology classification project and will serve to facilitate the searching of prior art related to nanotechnology, function as a collection of issued US patents and published pre-grant patent applications relating to nanotechnology across the technology centers, and assist in the development of an expanded, more comprehensive, nanotechnology cross-reference art collection classification schedule. It is important to note that this digest should not be construed as an exhaustive collection of all patent documents that pertain to nanotechnology. Nanoparticles have been patented for diagnostic use as well as combined diagnostic and therapeutic use.

Quantum Dot Patents Relevant to Healthcare Applications

Most of the patents about QDs relevant to healthcare are owned by Life Technologies, which acquired Quantum Dot Corporation, 2005. Life Technologies owns or has licensed over 160 QD patents or international patent applications currently under examination. The most important of these is Qdot™, which were originally developed at Lawrence Berkeley National Laboratory and the University of Melbourne in Australia and licensed to Quantum Dot Corporation. Qdot nanocrystals enable powerful new approaches to genetic analysis, drug discovery, and clinical diagnostics.

In 2008, Life Technologies sued Evident Technologies for allegedly infringing the three QD patent and in 2009, the latter filed for chapter 11 bankruptcy. The three patents at issue in the case were US patent nos. 6423551, 6699723, and 6927069. All three patents cover nanocrystal probe technology for biological applications and were issued to the University of California, which exclusively licensed the patents to Invitrogen (now part of Life Technologies) and its two subsidiaries, Quantum Dot and Molecular Probes. Evident Technologies currently develops quantum dot semiconductor nanocrystals for thermoelectric applications.

Challenges and Future Prospects of Nanobiotechnology Patents

Over the past decade, universities and companies have been engaged in an intense race to patent their nanotechnology inventions, seeking a source of future licensing revenue and control of an emerging technology, which has led to overlapping patent rights in nanotechnology. According to one estimate, approximately 20 % of nanotechnology patents are owned by universities, a disproportionately large number considering that universities typically hold about 1–2 % of the patents issued in the USA each year (Mouttet 2006). Patent overlap can partially be attributed to the complex nature of nanotechnology itself and to the fact that much of the field is the result of cumulative innovation, where innovations are built on many previous innovations. Because multiple patents from competing groups may cover each incremental innovation to some degree, a large number of overlapping patents are inevitable as complex technologies become commercialized. Nanotechnology is fundamentally a multidisciplinary field that overlaps a wide range of scientific and technical disciplines: materials science, biotechnology, synthetic chemistry, electrical engineering, and physical chemistry. There is some confusion about the validity and enforceability of numerous issued patents, and reforms are urgently needed at the USPTO to address problems about poor patent quality and questionable examination practices. A robust patent system is needed for the development of competitive and commercially viable nanomedicine products.

One of the conclusions of a review of this topic is that robust patent system will aid nanomedicine companies that are striving to develop commercially viable products (Bawa 2007). Valid patents stimulate market growth and innovation, generate revenue, prevent unnecessary licensing, and greatly reduce infringement lawsuits.

Legal Aspects of Nanobiotechnology

Like any other new technology, nanobiotechnology is likely to raise some legal issues. Legal aspects of nanobiotechnology in healthcare, particularly cell therapy and tissue engineering, are discussed in a separate article (Jain and Jain 2006). An early publication had already anticipated some of these issues (Fiedler and Reynolds 1994). The authors of this publication suggested that appropriate controls, in the form of regulations and legislation, must be tailored to fit the risk/benefit ratio of nanotechnology. Active measures in anticipation of development of technology should be considered as passive waiting for regulations to develop may allow unnecessary harm to society from unregulated technology. This will involve discussion of likely directions nanobiotechnology will take and preparation of flexible legislation to provide appropriate regulatory schemes even before the products arrive in the market place. Considering that these suggestions were made more than a decade ago and considerable advances and applications have taken place in nanobiotechnology, no legislation has been implemented yet to control nanobiotechnology. Regulatory authorities such as the FDA are just discussing the possible regulation of nanobiotechnology.

There is no law until now concerning nanotechnology. Because of safety concerns of exposure to nanoparticles in the environments and at workplace, as well as applications in human healthcare, there is a risk of lawsuits in the future. Therefore, it would be necessary to use existing regulations about the use of drugs and therapies as a model. But this field of technology requires a specified ruling because many problems will arise in the future, which are not ruled until now, and the safety of human being should be guaranteed as a primary aim, which could be achieved through an extensive legal regulation. As example serves the application of a nanorobot, there is no similar machine existing, neither with the same function nor the same aim. A situation could arise in which a surgeon uses a nanorobot, which is either afflicted with a technical problem or applied wrongly. The question followed by such an act would as usual contain the search after a person to hold liable and result in a product's liability, a negligent employment of nanodevices or a personal responsibility. But to handle problems in such a new technical field with its specified terms and characteristics, it is critical to achieve an optimal dialog between the technologists and legal experts using common terms.

Standards (see following section) will play an important role in nanotechnology law. They are needed for consistent measurement and characterization of various nanomaterials. Guidelines about nanotechnology do not exist yet, but would be very helpful for potential legal problems such as tort liability. To avoid or simplify possible lawsuits of or against consumers, patients, or companies, it would be desirable to require a declaration for nanoparticles, which does not exist yet. Anybody who gets in contact with nanoparticles should get fully informed about the possible effects and critical nature. Companies marketing nanoparticle-based products for healthcare should inform the consumers about potential safety issues.

Legal aspects of nanobiotechnology are complicated by interaction and combination with other new technologies. Cell/gene therapies have their own ethical and legal issues. For example, genetic modification of cells with products incorporating nanobiotechnology may raise issues combining those of nanoparticles and genetic modification.

Nanotechnology Standards

The first meeting of the American National Standards Institute (ANSI) Nanotechnology Standards Panel (NSP) was held in 2004 with participation from academia, government, industry, and non-governmental with the aim of defining the needs and priorities of nanotechnology standards. In considering nanotechnology nomenclature and terminology, the panel participants reached consensus on several important issues, but there was debate about the more general use of the terms "nanotechnology," "nanomaterial," and "nano" generally. Some felt that keeping the definitions broad allowed the most relevant topics within this area and is in accordance with the spirit of the National Nanotechnology Initiative definition. Others preferred to narrow such terms more substantially, and where possible, and draw distinctions in the names between artificial and naturally occurring nanomaterials and science and technology, among other issues. The group also addressed the need for future standards activity beyond terminology and nomenclature. The three broad classes identified were:

1. Measurement and metrology.
2. Environmental, health, and safety guidelines. This is most relevant to nanobiotechnology and requires development of reference standards and testing methods for toxicity.
3. Processes and manufacturing.

It appears nanobiotechnology would require specific set of standards. None have been set so far. Apart from ANSI, there should be participation by the biopharmaceutical industry and academic researchers in nanobiotechnology. Organizations interested in participants are invited to contact ANSI-NSP at www.ansi.org/surveybank.

Preclinical Testing of Nanomaterials for Biological Applications

Nanotechnology Characterization Laboratory (NCL) at Science Applications International Corporation (Frederick, Maryland) provides services for preclinical testing of nanomaterials for biomedical applications as a free national resource available to investigators from academia, industry (domestic as well as foreign), and government. The aim of this initiative is to characterize nanoparticles using standardized methods, conduct structure activity relationships studies, and facilitate regulatory review of nanoconstructs. NCL is a formal collaboration between the NCI, FDA, and NIST of the USA. NCL assay cascade includes physicochemical characterization of nanoparticles as well as in vitro and in vivo studies. Adequate immunological characterization is considered difficult without comprehensive physicochemical characterization. Immunotoxicity of nanoparticles is determined by the size, charge, hydrophobicity, and targeting. Even slight differences in a nanoparticle's properties can greatly influence its interaction with the immune system. Biocompatibility depends on surface charge, which is important for protein binding and uptake by the reticuloendothelial system (RES). PEGylation of nanoparticle surface, i.e., coating with polyethylene glycol (PEG), prevents uptake by the RES. Because there are no formal regulatory guidelines for immunotoxicity assessment, proper characterization of nanoparticles is difficult. NCL works with investigators to find solutions for some of these problems.

FDA Regulation of Nanobiotechnology Products

The FDA regulates a wide range of products, including foods, cosmetics, drugs, devices, and veterinary products, some of which may utilize nanotechnology or contain nanomaterials. The FDA has not established its own formal definition, though the agency participated in the development of the National Nanotechnology Initiative (NNI) definition of nanotechnology (see Chap. 1). Using that definition, nanotechnology relevant to the FDA might include research and technology development that both satisfies the NNI definition and relates to a product regulated by FDA. A review has discussed the drugs and medical devices approved by the FDA to date with observations about the emerging trends (Strickland 2007). Several medical products reportedly comprised of nanomaterials have already been approved by the FDA for prescription use in humans. To date, only one new nanopharmaceutical has gone through the entire rigorous FDA premarket-approval process. Six other drugs and three medical devices using nanomaterials have also been approved through abbreviated review procedures. Data on FDA-approved nanotechnology-based drugs is shown in Table 18.1.

The first generation of nanomedicines (liposomal preparations) were approved more than a decade ago before a real awareness existed about a number of issues related to safety concerns of nanomaterials, and with a demonstrable relative success, in terms of their clinical safety assessment and safe use in cancer. However, nanomaterials such as phospholipids or biodegradable/bioerodible polymers are of a completely different nature from other anticipated materials that will be produced in the near future from the research pipeline. Carbon nanotubes, quantum dots, and other nonbiodegradable and potentially harmful materials should be given different and closer attention, looking at their toxicological potential impact in a number of different applications. By the same standards and in the new context, already existing nanopharmaceuticals, when administered for the same or new therapeutic indications making use of different administration routes (e.g., pulmonary), should not be waived of a full assessment of their differential potential toxicology impact, particularly in the proinflammatory area (Gaspar 2007).

The FDA approval is essential for clinical applications of new technologies, and substantial regulatory problems may be encountered in the approval of nanotechnology-based products. Some of the previously approved products with particles in the nanosize range were not considered to be nanotechnology products and were subject to the same testing requirements as all other products reviewed by the agency.

Table 18.1 FDA-approved nanotechnology-based drugs

Name	Company	Particle size	Full PMA	Animal studies	Human studies	Claimed benefit relevant to nanotechnology
Emend (aprepitant)	Merck & Co	<1,000 nm	Yes	Yes	Phase II/III	Improved bioavailability and reduced food effect
Rapamune (sirolimus)	Wyeth	<1,000 nm	No	No	Phase III	Improved solubility and reduction of effective dose
Estrasorb	Novavax	?	No	?	Phase III	Transdermal delivery
TriCor®145 fenofibrate	Abbott	<1,000 nm	No	No	Phase III	Improved bioavailability and reduced food effect
Doxil® doxorubicin	Alza	100 nm	No	?	Phase III	Improved bioavailability, cellular penetration, and accumulation at target site
Abraxane (paclitaxel)	Abraxis/AstraZeneca	130 nm	No	Yes	Phase III	Improved solubility and elimination of toxicity of solvents
Megace® (megestrol)	Par Pharma	<1,000 nm	No	No	Phase III (stopped)	Improved bioavailability

But some of the novel platforms being developed, such as the multifunctional dendrimers, may require a multifaceted approach toward their review and evaluation. There is every expectation that some novel products utilizing nanotechnology will be combination products (i.e., drug-device, drug-biologic, or device-biologic) and will likely undergo a relevant review process. In order to insure that nanotechnology products are regulated in a coordinated fashion across all product types, the FDA has established a NanoTechnology Interest Group on which all FDA centers and offices that report to the Office of the Commissioner participate. Centers have established multidisciplinary working groups in order to share information and help coordinate the review for the various product types. The groups are charged with identifying and defining the scientific and regulatory challenges in the various review disciplines and to propose a path forward. However, the appropriate review divisions will conduct the review of nanotechnology applications submitted to FDA. The Working Group on Nanotechnology has discussed the need for a central resource to acquire basic safety data such as biodistribution, pharmacokinetics, efficacy, and toxicity for nanoparticles and other macromolecules. Information on FDA and its regulation of nanotechnology products can be viewed at the FDA web site (http://www.fda.gov/nanotechnology/regulation.html). Some of the questions that FDA considers internally are:

- What are the standard tools used to characterize nanoparticle properties?
- How to determine the short- and long-term stability of nanomaterials in various environments?
- What are the critical physical and chemical properties of nanomaterials including residual solvents, impurities, and excipients and how do these affect product quality and performance?
- What are the critical steps in the scale-up and manufacturing process of nanotechnology products?
- What is the residence time of nanoparticles in body tissues, their clearance from the body, and their effects on cell and tissue function?
- Are current methods used for measuring drug levels in blood and tissues adequate for assessing nanoparticle levels?
- What methods would identify the nature, quantity, and extent of nanoparticle release in the environment and what would be the impact of this on other species?

While the significant impact of nanotechnology and its applications is expected to be in the future, FDA has already approved many products such as imaging agents and nanoparticle ingredients in sunscreens. There are also cosmetics currently on the market that claim to contain nanoparticles. However, cosmetics do not undergo premarket approval, as do drugs and devices. Finally, there are products on the market that are reformulated to contain nanoparticles of previously approved products, in order to improve product performance. The position of the FDA not to require labels indicating that products contain nanomaterials has been controversial for some advocacy groups. Some of the existing cosmetic-labeling requirements have been examined in the context of recent calls by advocacy groups for special labels for cosmetics containing nanoscale materials (Monica 2008). Although the FDA has made a serious attempt to address cosmetic nano-labeling issues, a more rigorous analysis of some nano-labeling arguments is required.

While sponsors of nanotechnology products will be subject to the same testing requirements as for non-nanotechnology products, there likely will be certain challenges prior to commercialization. Specifically, there will need to be an understanding of the physical and chemical parameters that are crucial to product performance. Additionally, appropriate test methods and specifications to control the product or the manufacturing processes may need to be developed.

Finally, because much of the data currently available on nanotechnology products results from pilot batches produced in universities or small laboratories, there is very little known about what might be the challenges of scale-up to mass production. While testing during the investigational phase of the product may be conducted with pilot batches, bridging the investigational data to cover the scaled-up batches that will be commercialized may pose challenges for some of the novel nanotechnology formulations. However, these challenges are not considered to be insurmountable.

FDA and Nanotechnology-Based Medical Devices

Medical devices are handled by the Center for Devices and Radiological Health. There is no available evidence that any of the three currently approved medical devices containing nanoparticles had any additional premarket safety or efficacy review, whether nano-specific or not. Two of these devices, nanOss and Vitoss, are to be used as filler in damaged bone, providing a framework to support new bone growth, and then be absorbed into the body. TiMesh is classified as a bone fixation device. It appears that all were permitted to go straight to market on the basis of the sponsors' 510(k) claims that the products performed essentially the same function as devices using traditionally engineered materials have done for years. The regulatory documents for nanOss bone void filler are representative of this group. Angstrom Medica filed its rule 510(k) premarket notification in 2005, describing a device made of calcium-phosphate-preformed pellets intended for gentle packing into bony voids or gaps not intrinsic to the stability of the structure. The sponsor claimed that nanOss is substantially equivalent in indications and design principles to several devices on the market prior to 1976. The device utilizes nanocrystalline processing, which are translucent and uniform in density and strength. FDA's approval letter for this device makes no reference whatever to nanotechnology. The documents supporting approval of the 510(k) application for the other bone void filler, Vitoss, mention a canine study showing that 80 % of the vVitoss scaffold of nanoparticles was reabsorbed within 12 weeks. This study seems to have identified a novel property – faster reabsorption – that the sponsor attributed to the use of nanoparticles. Still, FDA accepted the claim that the product was substantially equivalent to bone void fillers that had been approved before, without considering the possibility that it might present new risks.

NanOss and TiMesh were considered class II medical devices, and therefore could have been subject to special conditions. None were actually imposed by FDA. Unlike drugs, therefore, it appears that medical devices utilizing nanotechnology have not been subject to additional testing to establish that the new nano-products are safe and effective. The assumption that nanotechnology devices are substantially equivalent to products made of traditional materials and marketed prior to 1976 is scientifically unproven.

FDA's Nanotechnology Task Force

The FDA has an internal Nanotechnology Task Force, which is charged with the task of determining regulatory approaches that encourage the continued development of innovative, safe, and effective FDA-regulated products that use nanotechnology materials (http://www.fda.gov/ScienceResearch/SpecialTopics/Nanotechnology/ucm257926.htm). The FDA internal task force on nanotechnology's report was published in 2007. The summary of the report is as follows:

The report addresses scientific issues as distinct from regulatory policy issues in recognition of the important role of the science in developing regulatory policies in this area, rapid growth of the field of nanotechnology, and evolving state of scientific knowledge relating to this field. Rapid developments in the field mean that attention to the emerging science is needed to enable the agency to predict and prepare for the types of products FDA may see in the near future.

A general finding of the report is that nanoscale materials present regulatory challenges similar to those posed by products using other emerging technologies. However, these challenges may be magnified both because nanotechnology can be used in, or to make, any FDA-regulated product and because, at this scale, properties of a material relevant to the safety and (as applicable) effectiveness of FDA-regulated products might change repeatedly as size enters into or varies within the nanoscale range. In addition, the emerging and uncertain nature of the science and potential for rapid development of applications for FDA-regulated products highlight the need for timely development of a transparent, consistent, and predictable regulatory pathway.

The task force's initial recommendations relating to scientific issues focus on improving scientific knowledge of nanotechnology to help ensure the agency's regulatory effectiveness, particularly with regard to products not subject to premarket authorization requirements. The report also addresses the need to evaluate whether the tools available to describe and evaluate nanoscale materials are sufficient and the development of additional tools where necessary.

The task force also assessed the agency's regulatory authorities to meet any unique challenges that may be presented by FDA-regulated products containing nanoscale materials. This assessment focused on such broad questions as whether FDA can identify products containing nanoscale materials, the scope of FDA's authorities to evaluate the safety and effectiveness of such products, whether FDA should require or permit products to be labeled as containing nanoscale materials, and whether the use of nanoscale materials in FDA-regulated products raises any issues under the National Environmental Policy Act.

The task force concluded that the agency's authorities are generally comprehensive for products subject to premarket authorization requirements, such as drugs, biological products, devices, and food and color additives, and that these authorities give FDA the ability to obtain detailed scientific information needed to review the safety and, as appropriate, effectiveness of products. For products not subject to premarket authorization requirements, such as dietary supplements, cosmetics, and food ingredients that are generally recognized as safe (GRAS), manufacturers are generally not required to submit data to FDA prior to marketing, and the agency's oversight capacity is less comprehensive.

The task force has made various recommendations to address regulatory challenges that may be presented by products that use nanotechnology, especially regarding products not subject to premarket authorization requirements, taking into account the evolving state of the science in this area. A number of recommendations deal with requesting data and other information about effects of nanoscale materials on safety and, as appropriate, effectiveness of products. Other recommendations suggest that FDA provides guidance to manufacturers about when the use of nanoscale ingredients may require submission of additional data, change the product's regulatory status or pathway, or merit taking additional or special steps to address potential safety or product quality issues. The task force also recommends seeking public input on the adequacy of FDA's policies and procedures for products that combine drugs, biological products, and/or devices containing nanoscale materials to serve multiple uses, such as both a diagnostic- and a therapeutic-intended use. The task force also recommends encouraging manufacturers to communicate with the agency early in the development process for products using nanoscale materials, particularly with regard to such highly integrated combination products.

The guidance that the task force is recommending would give affected manufacturers and other interested parties timely information about FDA's expectations, so as to foster predictability in the agency's regulatory processes, thereby enabling innovation and enhancing transparency while protecting the public health.

In 2011, FDA announced that it is issuing a draft guidance on considering whether an FDA-regulated product contains nanomaterials or otherwise involves the use of nanotechnology. FDA's issuance of this guidance is a first step toward providing regulatory clarity on FDA's approach to nanotechnology. Over time, the agency plans to issue more specific recommendations tailored to particular products or classes of products. These actions are consistent with the 2007 FDA Nanotechnology Task Force's science and policy recommendations to the commissioner.

FDA Collaboration with Agencies/Organizations Relevant to Nanotechnology

With the advent of nanotechnology, the regulation of many products will involve more than one center, e.g., a "drug" delivery "device." In these cases, the assignment of regulatory lead is the responsibility of the Office of Combination Products. To facilitate the regulation of nanotechnology products, the agency has formed a

NanoTechnology Interest Group (NTIG), which is made up of representatives from all the centers. The NTIG meets quarterly to ensure there is effective communication between the centers. Most of the centers also have working groups that establish the network between their different components. There are also a wide range of products involving nanotechnologies, which are regulated by other federal agencies. The breadth of products regulated by FDA and the other agencies is shown below.

The only viable approach to providing the public with innovative and beneficial novel therapies is to maintain an open dialog with the developers of such products. As such, the FDA has partnered with NIST and NCI. However, this partnership does not create a "fast track" through the back door to product approval. It is intended to create a straight track or an efficient and direct track through the front door.

By working together (FDA, academia, and industry) during the early stages of product development and evaluation, the appropriate test methodologies can be identified to insure that the correct tests are done at the outset. These early discussions also are critical to insure that the most efficient and predictive testing is done on the final commercial form of the product. If the right questions can be asked early, then the process can move forward. Additionally, if some of the test methods used can be standardized, then many of the regulatory hurdles may be overcome.

Within FDA, the Office of Science and Health Coordination (OC/OSHC) coordinates regular discussions on nanotechnology among the major experts from every organizational entity within the agency. In addition, the centers within FDA, e.g., Drugs and Medical Devices, have organized similar regular discussion groups. The purpose of these meetings is to share experiences with the review of the products, insure that each center is aware of product guidance that may be developing elsewhere within the agency, and generally educate staff and policy makers about nanotechnology. Safety issues are identified and studied.

In a similar manner, FDA coordinates knowledge and policy with the other US government agencies as a member of the Nanoscale Science, Engineering, and Technology (NSET) Subcommittee of the National Science and Technology Council (NSTC) Committee on Technology. Also, FDA and NIOSH co-chair the NSET working group on Nanomaterials Environmental and Health Implications (NEHI) to define new test methods/protocols and to define safety of these products. Finally, FDA is a direct contributor to the evaluations of the toxicity of materials supported by the NIEHS and the National Toxicology Program (NTP).

The National Institute of Standards and Technology, the FDA, and the National Cancer Institute have established the Nanotechnology Characterization Laboratory to perform preclinical efficacy and toxicity testing of nanoscale materials.

Regulation of Nanotechnology in the European Union

The current impression is that the European Union (EU) will adopt a cautious approach to regulation of nanotechnology. Like in the USA and elsewhere, there is no existing regulatory framework for nanotechnology in the EU. Some fear that registration, evaluation, authorization, and restriction of chemicals (REACH), the new EU's chemical policy, may be used as a source of reference for the regulation of nanotechnology, which might imply a qualified shift of the burden of proof with regard to safety, from the authorities to the manufacturer. Product liability law is less likely to play a preponderant role, at least at the EU level (as opposed to within the individual member states) because the EU's harmonization in this field of practice is limited. An overview of current and future EU regulation of nanotechnology, with some comparisons between the EU and US regulatory frameworks, has been published (Geert van Calster 2006).

The scope of nanotechnology-based medicinal products for human use reflects current thinking and initiatives taken by the European Medicines Agency (EMEA) following recent development of nanotechnology-based medicinal products. Nanotechnology is an emerging scientific research field with wide applicability and, in the context of medical science, is expected to contribute in developing a more proactive paradigm for the diagnosis and therapy of disease. Medicinal products containing nanoparticles have already been authorized both in EU and the USA under existing regulatory frameworks.

Although nanosizing does not necessarily imply novelty, it is expected that nanotechnology will yield innovative products. Such products could span the regulatory boundaries between medicinal products and medical devices, challenging current criteria for classification and evaluation. Appropriate expertise will need to be mobilized for the evaluation of the quality, safety, efficacy, and risk management of nanomedicinal products, and the need for new or updated guidelines will be reviewed in the light of accumulated experience. EMEA has created the Innovation Task Force (ITF) to ensure EMEA-wide coordination of scientific and regulatory competence in the field of emerging therapies and technologies, including nanotechnologies, and to provide a forum for early dialog with applicants on regulatory, scientific, or other issues that may arise from the development.

Safety Recommendations of the Royal Society of UK

The Royal Society of UK has issued a report "Nanoscience and nanotechnologies: opportunities and uncertainties" (http://www.nanotec.org.uk/report/summary.pdf), contains a section on the safety issues of nanotechnology. This study is the first of its kind, and responses are expected from organizations within the UK as well as from other countries. Some comments and recommendations in the report are:

- Most nanotechnologies pose no new risks to health, and almost all the concerns relate to the potential impacts of manufactured nanoparticles and nanotubes that are free rather than fixed to or within a material.
- It is very unlikely that new manufactured nanoparticles could be introduced into humans in doses sufficient to cause the health effects that have been associated with nanoparticles in polluted air.
- Until more is known about the environmental impacts of nanoparticles and nanotubes, the release of manufactured nanoparticles into the environment should be avoided as far as possible.

- "The chemicals in the form of nanoparticles or nanotubes should be treated as new substances under the existing notification of new substances (NONS) regulations and in the registration, evaluation, authorization, and restriction of chemicals".
- Overall, given the appropriate regulation and research along the lines just indicated, there is no need for the moratorium, which some have advocated on the laboratory or the commercial production of manufactured nanomaterials.

European Commission and Safety of Nanocosmetics

The European Commission has requested the Scientific Committee on Consumer Products (SCCP) to prepare an opinion on "safety of nanomaterials in cosmetic products." The preliminary version of the opinion can be found online (http://ec.europa.eu/health/ph_risk/committees/04_sccp/docs/sccp_o_099.pdf). The results obtained with nanosized delivery systems were not consistent. The following list of potential properties was considered:

- Nanomaterial constituents (such as lipids or surfactants) may act as penetration enhancers by penetrating individually into the stratum corneum (after particle disruption on skin surface) and subsequently altering the intercellular lipid lamellae within this skin layer.
- Nanomaterials may serve as a depot for sustained release of dermally active compounds.
- Nanomaterials may serve as a rate-limiting membrane barrier for the modulation of systemic absorption, hence providing a controlled transdermal delivery system.

TiO2 used as a mineral UV filter in sunscreen cosmetic product does not penetrate through the stratum corneum of healthy skin. It poses no local or systemic risk to human health from cutaneous exposure (Borm et al. 2006; Gamer et al. 2006). Little information is available concerning other nanoparticles. Current investigations of nanoparticle penetration into the skin using static imaging technology are unable to detect small fractions of nanoparticles reaching the dermis, vascular bed of the dermis, and hence, the blood stream. However, that if the administered dose of nanoparticles is very large, as for instance could be the case for TiO2 in sunscreens, a possible minute uptake of nanoparticles may be of relevance. A specific feature of nanoparticles is that not only the dose to the intake organ needs to be considered but also the dose in secondary target organs as a result of nanoparticle biokinetic distribution. In addition, nanoparticles may affect more cell types than larger particles because of use of endocytotic and non-endocytotic pathways.

Although cosmetic products are meant to be used on normal skin, it is known that they also are applied on non-healthy skin where the barrier properties may be impaired. There is no published information yet available on the potential penetration of nanomaterials through atopic or sunburnt human skin. The possible uptake of nanosized materials from cosmetics via inhalation has also been considered.

The SCCP adopted a preliminary report on the risk assessment of nanomaterials in 2007. The report provides a review of the applicability of currently available risk assessment methods to nanomaterials in cosmetic products, recommends a general approach in order to assess the health risks of nanomaterials in cosmetic products, and identifies data and methodological gaps where further research and development is needed.

Chapter 19
Research and Future of Nanomedicine

Introduction

Research is an important activity in nanobiotechnology both in the academic and commercial sectors. Major portion of the research activity is in the commercial sector and is focused on translation into clinical applications as the number of products in the market is still limited. Two important segments of research in the commercial sector are nanodiagnostics and nanoparticle-based drug delivery. Whereas most of the academic research in the US is funded by government agencies, research in the commercial sector is funded by venture capital and other private sources. Research activities at various companies involved in nanobiotechnology are described in a commercial report on nanobiotechnology (Jain 2012). There are numerous collaborations between the academia and the industry, and many discoveries made in universities are commercialized by the companies.

Nanobiotechnology Research in the Academic Centers

Almost every university and academic research organization has involvement in research on nanobiotechnology. Since it is a relatively new area, most of the established scientists come from backgrounds such as physics, chemistry, engineering, etc. The younger generation of scientists is receiving training and gets involved in research at the start of their careers in nanotechnology. Some of the noncommercial institutes, where research is conducted in nanobiotechnology, are shown in Table 19.1.

Future Potential of Nanomedicine

Disease and other disturbances of function are caused largely by damage at the molecular and cellular level, but current surgical tools are large and crude. Even a fine scalpel is a weapon more suited to tear and injure than heal and cure. It would

K.K. Jain, *The Handbook of Nanomedicine*, DOI 10.1007/978-1-61779-983-9_19,

Table 19.1 Academic institutes/laboratories involved in nanobiotechnology

Center/program	Parent institutes	Areas of interest
Applied NanoBioscience Center at Biodesign Institute	State University of Arizona (Tempe, AZ)	Nanoscale processing technologies for improving molecular diagnostics
Australian Institute for Bioengineering and Nanotechnology	University of Queensland (Brisbane, Australia)	Cell and tissue engineering Systems biotechnology Biomolecular nanotechnology
Biomedical Engineering Center	Industrial Technology Research Institute (Taiwan)	In vivo nanodevices, biomimetic sensing, nanobiolabeling/diagnosis
Biomolecular Engineering Group	University of Missouri (Columbia, MO)	Engineered membrane protein channels used to make single-molecule biosensors
BioSecurity and NanoSciences Laboratory	Lawrence Livermore National Laboratory (Livermore, CA)	Nanoscience to detect even the single smallest molecule of harmful substances
Birck Nanotechnology Center	Purdue University (West Lafayette, IN).	Nanocantilever biosensors for detection of microorganisms and use of bacteria for delivery of nanoparticles into the cell
Center for Nanomedicine, Sanford-Burnham Medical Research Institute	University of California (Santa Barbara, CA)	Nanoparticles that target tumors and bind to their blood vessels to destroy them
California Nanosystems Institute	UCLA (Los Angeles, CA)	Developing nanomedicine
Carolina Center of Cancer Nanotechnology Excellence	University of North Carolina (Chapel Hill, NC)	Self-assembling nanoparticles for imaging and therapy of cancer
Center for Bio/Molecular Science and Engineering	US Naval Research Laboratory (Washington, DC)	FRET-derived structure of a quantum dot-protein bioconjugate nanoassembly
Center for Functional Nanomaterials	Brookhaven National Laboratory (Upton, NY)	Study of interaction of nanomaterials with biosystems at level of single molecules
Center for Molecular Imaging Research	Massachusetts General Hospital (Boston, MA)	Nanoparticles for in vivo sensing and imaging of molecular events
Center for Nanotechnology	Wake Forest University (Winston-Salem, NC)	Controlling cellular function through nanoscale engineering, e.g., insertion of complex nanostructures into human monocytes
Center for Nanotechnology	University of Washington (Seattle, WA)	Bionanotechnology for cancer diagnostics and therapeutics
Center for Nanotechnology	NASA Ames Research Center (Moffett Field, CA)	Carbon nanotubes and nanowires for biological sensing
Center for Photonics and Optoelectronic Materials	Princeton University (Princeton, NJ)	Interphase of nanotechnology and biological systems

(continued)

Table 19.1 (continued)

Center/program	Parent institutes	Areas of interest
Cornell NanoScale Science and Technology Facility	Cornell University (Ithaca, NY)	Biosensors, drug-delivery systems, microarrays
Centre for Nanoscale Science and Technology	Queen's University (Belfast, Ireland)	Nanostructured materials as templates for tissue engineering
Center for Nanoscience and Nanotechnology	Georgia Institute of Technology (Atlanta, GA)	Nanodevicies and nanosensors for biotechnology
Clinatec (a clinic specializing in nanotechnology-based treatment)	University of Grenoble (France)/Minatec	Nanoneurosurgey of degenerative neurological disorders
FIRST (Frontiers in Research, Space and Time)	Swiss Federal Inst of Technol (Zurich, Switzerland)	AFM as a nanolithography tool
Heath Group	California Institute of Technology (Pasadena, CA)	Nanobiology: nanolab combines several assays on a cm^2 silicon chip resembling a miniature cell farm with rows of cells
IMTEK – Institute of Microsystem Technology	University of Freiburg, (Freiburg, Germany)	Nanoparticles for biosensors
INSERM	Paris, France	Nanodetection, drug delivery
Institute of NanoScience and Engineering	University of Pittsburgh (Pittsburgh, PA)	Nanotubes for molecular diagnostics and nanocarpet to detect and destroy bacteria
Institute for Nanotechnology	Northwestern University (Evanston, IL)	Nanoparticles and biosensors, nano bar code for detection of proteins, nanofibers for neuroregeneration
Institute of Microtechnology	University of Neuchatel, Switzerland	Biological applications of nanotechnology
Institute of Micro- and Nanotechnology	Technical University of Denmark, Denmark	Study of nanoscale structures with in situ scanning tunneling microscopy
Institute of Physical Chemistry	National Centre for Scientific Research (Athens, Greece)	Polymer-based nanosponges, nanotubes, drug delivery
Interdisciplinary Nanoscience Center (iNANO)	University of Aarhus, Denmark	Polyplexes for delivery of bioactive agents: plasmid DNA and siRNA
Kavli Nanoscience Institute	California Institute of Technology (Pasadena, CA)	Nanoproteomics: single-molecule nanomechanical mass spectrometry
Laboratory for Micro- and Nanotechnology	Paul Scherrer Institute (Villigen, Switzerland)	Nanopore membranes, biosensors, and artificial noses
Laboratory for Photonics and Nanostructures	CNRS (Marcoussis, France)	Separation methods for DNA sequencing, protein analysis and on-chip detection, microfluidic systems for cell sorting
Lerner Research Institute	The Cleveland Clinic (Cleveland, OH)	Nanometer-scale tissue engineering, diagnostics, nanosensors for surgery

(continued)

Table 19.1 (continued)

Center/program	Parent institutes	Areas of interest
London Center for Nanotechnology	University College (London, UK)	Use of nanotechnology to develop low-cost diagnostics and drug-delivery systems, and personalized medicine
MacDiarmid Institute, BioNanotechnology Network	University of Canterbury (Christchurch, New Zealand)	Development of biochip for AFM imaging, biosensors by splicing polymers with QDs
Michigan Nanotechnology Institute	University of Michigan (Ann Arbor, MI)	Nanoemulsions as antimicrobial agents, dendrimers for drug delivery in cancer, magnetic nanoparticle MRI agents, dendritic polymer-based nanosensors
Micro and Nano Biosystems Laboratory	Boston University (Boston, MA)	Application of nanotechnology to tissue engineering and cell/drug encapsulation
NanoApplications Center	Oak Ridge National Laboratory (Oak Ridge, TN)	Nanostructured devices for controlled gene expression and nano-enabled FILMskin (bionic): Flexible Integrated Lightweight Multifunctional skin
Nanotechnology Research Institute	University of Ulster (Jordanstown, UK)	In vivo nanobiosensors
National Center for Competence in Research Nanoscale Science	Biozentrum, University of Basel (Basel, Switzerland)	To bring nanotechnology from the bench to the patient by developing new tools
National Center for Nanoscience and Technology	Institute of High Energy Physics, Chinese Academy of Sciences, Beijing, China	Biomedical effects of nanomaterials and nanosafety
Nanobioengineering Laboratory	National University of Singapore (Singapore)	Nanohydroxyapatite/chitosan as resorbable bone paste
Nano-Biomolecular Engineering Group	University of California (Berkeley, CA)	BioCOM cantilever chip for cancer diagnosis, DNA-based self-assembly/replication of inorganic nanostructures, and electrophoretic separation microchip
Nano-Mechanical Technology Lab	Massachusetts Institute of Technology (Cambridge, MA)	Study of changes in human cells for research projects on infectious diseases like malaria and sickle-cell anemia, and cancers of the liver and pancreas
Nanomedicine Development Center	Emory University/ Medical College of Georgia, Atlanta	Focus on DNA damage repair by protein complexes
NanoRobotics Laboratory	Carnegie Mellon University (Pittsburgh, PA)	Nano-enabled imaging capsule to look inside the small intestine
NanoRobotics Laboratory	École Polytechnique de Montréal, Canada	Magnetic resonance targeting for guiding nanobots to targets in vivo

(continued)

Table 19.1 (continued)

Center/program	Parent institutes	Areas of interest
NanoBioTechnology Initiative	Ohio University (Athens, OH)	Diagnosis/treatment: cancer and diabetes
Nanoscale Research Team	University of California (Davis, CA)	Artificial cell membrane to study single protein interaction with cell membrane
Nanoscale Science Research Group: Biomedical Research	University of North Carolina (Chapel Hill, NC)	Cystic fibrosis Fibrin and blood clotting Gene therapy and viruses Bacterial motility Molecular motors
Nanoscience Center	University of Copenhagen (Copenhagen, Denmark)	Boron carbide nanoparticles for boron neutron capture therapy of cancer
Nano-Bio Research Center	Korean Institute of Science and Technology	Collaboration with Purdue University, USA for use of nanobiotechnology to integrate diagnostics and therapeutics
Nanosystems Biology	California Institute of Technology (Pasadena, CA)	Nanowire biosensors for early detection of cancer biomarkers
Nanosystems Laboratory	University of Washington (Seattle, WA)	Nanoprobes based on thin film technology for rapid and cheap sequencing DNA
National Institute of Nanotechnology of Canada	University of Alberta (Edmonton, Canada)	X-ray scattering beamline to determine the size of biomolecules on nanoscale
Pharmaceutical Bioengineering and Nanotechnology Group	University of London (London, UK)	Bridging the gap between nanomaterials engineering and pharmaceutical science to develop nanomedicines
Purdue Nanomedicine Development Center	Purdue University (West Lafayette, IN)	NIH-supported center for research on phi29 nanomotor for potential use in the diagnosis and treatment of diseases
Richard E. Smalley Institute for Nanoscale Science and Technology	Rice University (Houston, Texas)	Carbon nanotechnology: improved delivery of bioactive molecules, nanoscale sensory systems, biochips
Roukes Group	California Institute of Technology (Pasadena, CA)	Nanotechnology for neurophysiology, nanodevices for molecular biosensing
Sandia National Laboratories (Albuquerque, NM)	Dept of Energy, US Government	Nanodevices: biosensors to detect biological agents
Siteman Center of Cancer Nanotechnology Excellence	Washington University School of Medicine (St. Louis, MO)	Molecular imaging using nanoparticle tags and MRI, combined with therapy

(continued)

Table 19.1 (continued)

Center/program	Parent institutes	Areas of interest
Swiss Nanoscience Institute	University of Basel, Switzerland	Rapid and sensitive detection of disease- and treatment-relevant genes Toxicity of nanoparticles
USC Nanocenter	University of South Carolina (Columbia, SC)	Nanomedicine as well as social and ethical implications of nanotechnology
Winship Cancer Institute	Emory University (Atlanta, GA)	Cancer nanotechnology: nanoparticles for molecular and cellular imaging
Yale Institute for Nanoscience and Quantum Engineering	Yale University (New Haven, CT)	Smart nanoparticles: a new class of nanomaterials with properties that mimic biological vectors like bacteria and viruses, for vaccine delivery

make more sense to operate at the cell level to correct the cause of disease, rather than remove large lesions as a result of the disturbances at cell level.

Nanotechnology-based approaches can be used to remove obstructions in the circulatory system, kill cancer cells, or take over the function of subcellular organelles. Instead of transplanting artificial hearts, a surgeon of the future would be transplanting artificial mitochondrion.

Nanotechnology will also provide devices to examine tissue in minute detail. Biosensors that are smaller than a cell would give us an inside look at cellular function. Tissues could be analyzed down to the molecular level, giving a completely detailed "snapshot" of cellular, subcellular, and molecular activities. Such a detailed diagnosis would guide the appropriate treatment.

It is expected that within the next few years, we will have a better understanding of how to coat or chemically alter nanoparticles to reduce their toxicity to the body, which will allow us to broaden their use for disease diagnosis and for drug delivery. Biomedical applications are likely to be some of the earliest. The first clinical applications are in cancer therapy.

US Federal Funding for Nanobiotechnology

The US National Nanotechnology Initiative was signed into a law in 2010 and authorized \$3.7 billion over the following 4 years for the program. The bill also requires the creation of research centers, education and training efforts, research into the societal and ethical consequences of nanotechnology, and efforts to transfer technology into the marketplace.

Nanomedicine Initiative of NIH

The US National Institutes of Health (NIH) started a nanomedicine initiative in 2004 by soliciting comments from the scientific community to help shape the research project aimed at developing new tools to improve human health. Further information about this initiative is available at the following web site: http://nihroadmap.nih.gov/nanomedicinelaunch/.

The initiative, which could last a decade, is a broad program that seeks to catalog molecules and understand molecular pathways and networks. Nanomedicine is one of nine initiatives that make up NIH's roadmap, a long-term plan for improving and accelerating biomedical research. This is a program oriented toward addressing biological issues of health and clinical applications in a context of the overall mission of the NIH.

Unlike many research projects, NIH did not predetermine specific areas of study. Instead, it called for proposals aimed at helping to fulfill the project's goals. To start with, there were debates to find the best way to proceed with the nanomedicine initiative. Much of the initial research took place at a few Nanomedicine Development Centers established by the initiative. The number of nanomedicine centers has increased over the years. Some examples of areas of study that have been funded by the NIH are:

- Probing of molecular events inside cells on biologically relevant time scales that may be on the order of milliseconds or microseconds or even nanoseconds
- To design systems to engineer within living cells
- To ensure the biocompatibility of some nanodevices in humans and develop devices that may eventually reduce the cost of health care

NIH Nanomedicine Center for Nucleoprotein Machines

In 2006, the NIH awarded Georgia Institute of Technology, Emory University, and Medical College of Georgia a grant to partner on the Nanomedicine Center for Nucleoprotein Machines. The new center will initially focus on understanding how the body repairs damage to DNA, a problem that lies at the heart of many diseases and illnesses. As cells replicate, mistakes are created in the DNA that, if not repaired, cause defects that lead to illness. DNA breakage can also occur from ionizing radiation, which is found in the environment, cosmic rays, radon gas, and even the soil, as well as in our bodies, primarily from potassium and carbon. Learning how protein complexes repair DNA damage could be the key to understanding structure-function relationships in the cell nucleus' protein machines, called nucleoprotein machines that synthesize, modify, and repair DNA and RNA. This could someday be used to reverse genetic defects, cure disease, or delay aging. By studying the way natural machines are engineered by the body, researchers will develop the general principles that will enable engineering of artificial machines that could carry out these processes for therapeutic purposes, e.g., to fix genetic defects. The center received over $6 million from the NIH over the following

5 years and approximately $3 million from the Georgia Research Alliance, a public-private partnership of Georgia universities, businesses, and government created to build the state's technology industry.

NCI Alliance for Nanotechnology in Cancer

One of the most important applications of nanotechnology will be in cancer. In 2004, the National Cancer Institute (NCI) launched a $144-million, 5-year plan to apply nanoscale technology for research and treatment of cancer. This brought together researchers, clinicians, and public as well as private organizations to translate cancer-related nanotechnology research for the benefit of the patient.

Formation of the NCI Alliance for Nanotechnology in Cancer has brought together researchers, clinicians, and organizations to develop and translate cancer-related nano research into clinical practice. The alliance has created nano-research centers within existing public facilities and a laboratory for preclinical testing that will help boost regulatory review and translation of nanomaterials and devices into the clinical realm. More detailed information on the NCI Alliance for Nanotechnology in Cancer is available on the web site (http://nano.cancer.gov).

The alliance is designed as one of the first steps in crafting a Cancer Nanotechnology Plan, which will include milestones to measure success over two time periods. Within the first 3 years, the plan calls for accelerating projects promising for near-term clinical application. After 3 years, the plan will focus on solutions to more difficult technological and biological problems that could affect detection and treatment.

Research in Cancer Nanotechnology Sponsored by the NCI

NCI (National Cancer Institute, a part of the NIH) made awards totaling $26.3 million to help establish the following eight Centers of Cancer Nanotechnology Excellence in 2006:

1. Carolina Center of Cancer Nanotechnology Excellence at the University of North Carolina (Chapel Hill, NC). This center will focus on the fabrication of "smart" or targeted nanoparticles and other nanodevices for cancer therapy and imaging.
2. Center of Nanotechnology for Treatment, Understanding, and Monitoring of Cancer at University of California (San Diego, CA). This center will focus on a smart, multifunctional, all-in-one platform capable of targeting tumors and delivering payloads of therapeutics.
3. Emory-Georgia Tech Nanotechnology Center for Personalized and Predictive Oncology (Atlanta, GA). This center will aim to innovate and accelerate the development of nanoparticles attached to biological molecules for cancer molecular imaging, molecular profiling, and personalized therapy.
4. MIT-Harvard Center of Cancer Nanotechnology Excellence (Cambridge, MA). This center will focus on diversified nanoplatforms for targeted therapy, diagnostics, noninvasive imaging, and molecular sensing.

5. Nanomaterials for Cancer Diagnostics and Therapeutics at Northwestern University (Evanston, IL). This center plans to design and test nanomaterials and nanodevices to improve cancer prevention, detection, diagnosis, and treatment.
6. Nanosystems Biology Cancer Center at California Institute of Technology (Pasadena, CA). This center will focus on the development and validation of tools for early detection and stratification of cancer through rapid and quantitative measurement of panels of serum and tissue-based biomarkers.
7. The Siteman Center of Cancer Nanotechnology Excellence at Washington University (St. Louis, MO). This center has a comprehensive set of projects for the development of nanoparticles for in vivo imaging and drug delivery, with special emphasis on translational medicine.
8. Stanford University School of Medicine (Palo Alto, CA). It will aim its efforts at imaging diseases in vivo and determining what is going on within patients' bodies through blood or tissue sample analysis.

Global Enterprise for Micro-Mechanics and Molecular Medicine

It is an international collaboration to use nanotechnology tools for global health and medical research. The members include the National University of Singapore and Institut Pasteur of France. The collaboration, called GEM4 or Global Enterprise for Micro-Mechanics and Molecular Medicine, represents an ambitious effort to apply global sourcing principles to research at the intersection of engineering and life sciences. It will use tools like AFM, laser tweezers, and nanoscale plate stretchers to study changes in human cells such as in sickle-cell anemia and for research projects on infectious diseases like malaria. Further information can be obtained from the web site: http://www.gem4.org/.

Nano2Life

Nano2Life is the first European Network of Excellence supported by the European Commission under the Sixth Framework Program. Its objective is to merge existing European expertise and knowledge in the field of nanobiotechnology in order to keep Europe as a competitive partner and to make it a leader in nanobiotechnology transfer. Nano2Life is tackling the fragmentation of European nanobiotechnology by joining 23 previously unconnected dynamic, highly specialized and competent regions and centers with experience in initiating and running nanobiotechnology programs. A pool of 21 high-tech companies is associated to the network.

Nano2Life aims at setting the basis for a virtual European Nanobiotech Institute, focused on the understanding of the nanoscale interface between biological and nonbiological entities and its possible application in the area of complex and integrated novel sensor technologies for health care, pharmaceuticals, environment, security, food safety, etc.

The partners have agreed on a joint program of activity with actions in:

- Joint research projects
- Education and training
- Sharing of resources
- Communication and dissemination

Nano2Life will contribute to ensure the development of nanobiotech devices, material, and services in agreement with international social and ethical standards and according to the needs of European industry. The network started operation in 2004 and has integrated over 170 researchers from 12 countries since then.

European Technology Platform on Nanomedicine

The European Technology Platform on Nanomedicine is an industry-led consortium, which is bringing together the key European stakeholders in the sector and is supported by the European Commission (www.etp-nanomedicine.eu). In 2005, it delivered a common vision of this technologically and structurally multifaceted area and defined the most important objectives in this Strategic Research Agenda (SRA) that addresses the member states of the European Union, its candidate countries, and associated states to the EU Framework Programs for research and technological development, as well as the European Commission itself. Its main aim is to put forward a sound basis for decision-making processes for policy makers and funding agencies, providing an overview of needs and challenges, existing technologies, and future opportunities in nanomedicine. The SRA also takes into consideration education and training, ethical requirements, benefit/risk assessment, public acceptance, regulatory framework, and intellectual property issues, thus representing a possible reference document for regulatory bodies. The proposed disease-oriented priority setting of this SRA is based on several parameters such as mortality rate, the level of suffering that an illness imposes on a patient, the burden put on society, the prevalence of the disease, and the impact that nanotechnology might have to diagnose and overcome certain illnesses. The scientific and technical approach is horizontal and exploits the benefits of interdisciplinarity and convergence of relevant technologies via breakthrough developments in the areas of diagnosis, targeted delivery systems, and regenerative medicine. The effective implementation of the SRA is expected to provide a major step forward in patient-oriented affordable health care.

Unmet Needs in Nanomedicine

Since nanomedicine is at an early stage of development, one does not expect that many needs would be met at this stage. However, a rough idea of the amount of development needed in major areas of application can be formulated. Figure 19.1 shows the current achievements in various areas of application as a percentage of the total desired development. This identified the areas of greatest need as more desirable targets for further development.

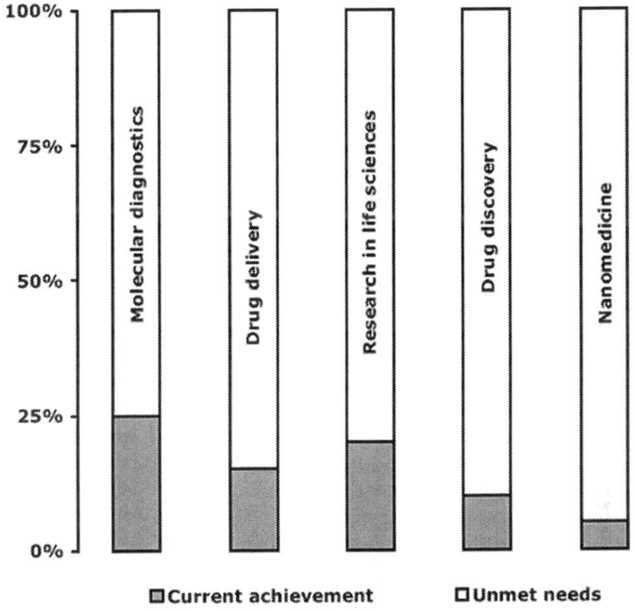

Fig. 19.1 Unmet needs in nanobiotechnology applications (© Jain PharmaBiotech)

Table 19.2 Drivers for the development of nanomedicine

Driver	Example
Molecular diagnostics is a growing field	Early detection of disease Point-of-care diagnostics Non-PCR methods
Growing importance of drug delivery	For improving drug therapy, e.g., delivery across blood-brain barrier facilitated by nanobiotechnology
Nanobiotechnology facilitates understanding of disease mechanism	Important component of personalized medicine along with pharmacogenomics
Future trends in medicine for minimally invasive procedures and correction of disease pathology	Nanobiotechnology will facilitate the development of nanoscale devices for performing such procedures
Regenerative medicine	Improved materials for tissue reconstruction and implants

Drivers for the Development of Nanomedicine

Drivers for the development of nanomedicine are shown in Table 19.2.

References

Abdel-Mottaleb MM, Neumann D, Lamprecht A. Lipid nanocapsules for dermal application: a comparative study of lipid-based versus polymer-based nanocarriers. Eur J Pharm Biopharm 2011;79:36–42.

Ackerson CJ, Sykes MT, Kornberg RD. Defined DNA/nanoparticle conjugates. Proc Natl Acad Sci U S A 2005;102:13383–85.

Agüeros M, Espuelas S, Esparza I, et al. Cyclodextrin-poly(anhydride) nanoparticles as new vehicles for oral drug delivery. Expert Opin Drug Del 2011;8:721–34.

Ahmed F, Pakunlu RI, Srinivas G, et al. Shrinkage of a Rapidly Growing Tumor by Drug-Loaded Polymersomes: pH-Triggered Release through Copolymer Degradation. Mol Pharm 2006;3:340–50.

Ainbinder D, Touitou E. Testosterone ethosomes for enhanced transdermal delivery. Drug Deliv 2005;12:297–303.

Akin D, Sturgis J, Ragheb K, et al. Bacteria-mediated delivery of nanoparticles and cargo into cells. Nature Nanotechnology 2007;2:441–44.

Algar WR, Tavares AJ, Krull UJ. Beyond labels: a review of the application of quantum dots as integrated components of assays, bioprobes, and biosensors utilizing optical transduction. Anal Chim Acta 2010;673:1–25.

Al-Jamal KT, Al-Jamal WT, Akerman S, et al. Systemic antiangiogenic activity of cationic poly-L-lysine dendrimer delays tumor growth. Proc Natl Acad Sci U S A 2010;107:3966–71.

Al-Jamal WT, Kostarelos K. Liposome-nanoparticle hybrids for multimodal diagnostic and therapeutic applications. Nanomedicine 2007:2:85–98.

Allhoff F, et al (eds). Nanoethics: The Ethical and Social Implications of Nanotechnology. John Wiley & Sons, New York, 2007.

Allhoff F, Lin P, Moore D. What Is Nanotechnology and Why Does It Matter?: From Science to Ethics. Wiley-Blackwell, Chichester UK, 2010.

Almutairi A, Rossin R, Shokeen M, et al. Biodegradable dendritic positron-emitting nanoprobes for the noninvasive imaging of angiogenesis. Proc Natl Acad Sci U S A. 2009;106:685–90.

Alsberg E, Feinstein E, Joy MP, et al. Magnetically-Guided Self-Assembly of Fibrin Matrices with Ordered Nano-Scale Structure for Tissue Engineering. Tissue Eng 2006;12:3247–3256.

Amrite AC, Kompella UB. Size-dependent disposition of nanoparticles and microparticles following subconjunctival administration. J Pharm. Pharmacol 2005;57:1555–63.

Anderson EA, Isaacman S, Peabody DS, et al. Viral Nanoparticles Donning a Paramagnetic Coat: Conjugation of MRI Contrast Agents to the MS2 Capsid. Nano Lett 2006;6:1160–1164.

Anderson SA, Glod J, Arbab AS, et al. Noninvasive MR imaging of magnetically labeled stem cells to directly identify neovasculature in a glioma model. Blood 2005;105:420–5.

Andreev OA, Dupuy AD, Segala M, et al. Mechanism and uses of a membrane peptide that targets tumors and other acidic tissues in vivo. Proc Natl Acad Sci U S A 2007;104:7893–8.

Andresen TL, Jensen SS, Jorgensen K. Advanced strategies in liposomal cancer therapy: problems and prospects of active and tumor specific drug release. Prog Lipid Res 2005;44:68–97.

K.K. Jain, *The Handbook of Nanomedicine*, DOI 10.1007/978-1-61779-983-9,

Arias JL, Reddy LH, Othman M, et al. Squalene based nanocomposites: a new platform for the design of multifunctional pharmaceutical theragnostics. ACS Nano 2011;5:1513–21.

Arora HC, Jensen MP, Yuan Y, et al. Nanocarriers Enhance Doxorubicin Uptake in Drug-Resistant Ovarian Cancer Cells. Cancer Res 2012;72:769–78.

Arvizo RR, Rana S, Miranda OR, et al. Mechanism of anti-angiogenic property of gold nanoparticles: role of nanoparticle size and surface charge. Nanomedicine 2011;7:580–7.

Ashammakhi N, Wimpenny I, Nikkola L, Yang Y. Electrospinning: methods and development of biodegradable nanofibres for drug release. J Biomed Nanotechnol 2009;5:1–19.

Ashcroft JM, Tsyboulski DA, Hartman KB, et al. Fullerene (C60) immunoconjugates: interaction of water-soluble C60 derivatives with the murine anti-gp240 melanoma antibody. Chem Commun 2006;28:3004–6.

Atanasijevic T, Shusteff M, Fam P, Jasanoff A. Calcium-sensitive MRI contrast agents based on superparamagnetic iron oxide nanoparticles and calmodulin. Proc Natl Acad Sci U S A 2006;103:14707–12.

Bagalkot V, Zhang L, Levy-Nissenbaum E, et al. Quantum Dot-Aptamer Conjugates for Synchronous Cancer Imaging, Therapy, and Sensing of Drug Delivery Based on Bi-Fluorescence Resonance Energy Transfer. Nano Lett 2007;7:3065–70.

Bai J, Wang J, Zeng XC. Multiwalled ice helixes and ice nanotubes. Proc Natl Acad Sci U S A 2006;103:19664–7.

Bailey VJ, Easwaran H, Zhang Y, et al. MS-qFRET: A quantum dot-based method for analysis of DNA methylation. Genome Res 2009;19: 1455–61.

Bajaj A, Miranda OR, Kim IB, et al. Detection and differentiation of normal, cancerous, and metastatic cells using nanoparticle-polymer sensor arrays. Proc Natl Acad Sci U S A 2009;106:10912–6.

Baldessari F, Santiago JG. Electrophoresis in nanochannels: brief review and speculation. J Nanobiotechnology 2006;4:12.

Ballou B, Ernst LA, Andreko S et al. Sentinel lymph node imaging using quantum dots in mouse tumor models. Bioconjug Chem 2007;18:389–96.

Banai S, Chorny M, Gertz SD, et al. Locally delivered nanoencapsulated tyrphostin (AGL-2043) reduces neointima formation in balloon-injured rat carotid and stented porcine coronary arteries. Biomaterials 2005;26:451–61.

Banatao DR, Cascio D, Crowley CS, et al. An approach to crystallizing proteins by synthetic symmetrization. Proc Natl Acad Sci U S A 2006;103:16230–5.

Bar Sadan M, Houben L, Enyashin AN, et al. Atom by atom: HRTEM insights into inorganic nanotubes and fullerene-like structures. Proc Natl Acad Sci U S A 2008;105:15643–8.

Baral TN, Magez S, Stijlemans B, et al. Experimental therapy of African trypanosomiasis with a nanobody-conjugated human trypanolytic factor. Nat Med 2006;12:580–4.

Barbu E, Molnàr E, Tsibouklis J, et al. The potential for nanoparticle-based drug delivery to the brain: overcoming the blood–brain barrier. Expert Opin Drug Deliv 2009;6:553–65.

Bardhan R, Chen W, Perez-Torres C, et al. Nanoshells with Targeted Simultaneous Enhancement of Magnetic and Optical Imaging and Photothermal Therapeutic Response. Advanced Functional Materials 2009;19:3901–9.

Bartlett DW, Su H, Hildebrandt IJ, et al. Impact of tumor-specific targeting on the biodistribution and efficacy of siRNA nanoparticles measured by multimodality in vivo imaging. Proc Natl Acad Sci U S A 2007;104:15549–554.

Basu S, Harfouche R, Soni S, et al. Nanoparticle-mediated targeting of MAPK signaling predisposes tumor to chemotherapy. Proc Natl Acad Sci U S A 2009;106:7957–61.

Battah S, Balaratnam S, Casas A, et al. Macromolecular delivery of 5-aminolaevulinic acid for photodynamic therapy using dendrimer conjugates. Mol Cancer Ther 2007;6:876–85.

Batten TF, Hopkins CR. Use of protein A-coated colloidal gold particles for immunoelectronmicroscopic localization of ACTH on ultrathin sections. Histochemistry 1979;60:317–20.

Bawa R, Fung SY, Shiozaki A, et al. Self-assembling peptide-based nanoparticles enhance cellular delivery of the hydrophobic anticancer drug ellipticine through caveolae-dependent endocytosis. Nanomedicine 2011 Sep 1. [Epub ahead of print]

Bawa R. Patents and nanomedicine. Nanomed 2007;2:351–74.

Beier JP, Klumpp D, Rudisile M, et al. Collagen matrices from sponge to nano: new perspectives for tissue engineering of skeletal muscle? BMC Biotechnology 2009;9:34.

Benezra M, Penate-Medina O, Zanzonico PB, et al. Multimodal silica nanoparticles are effective cancer-targeted probes in a model of human melanoma. J Clin Invest 2011;121:2768–80.

Benzerara K, Miller VM, Barell G, et al. Search for Microbial Signatures within Human and Microbial Calcifications Using Soft X-Ray Spectromicroscopy. J Investig Med 2006;54:367–79.

Bertin PA, Gibbs JM, Shen CK, et al. Multifunctional polymeric nanoparticles from diverse bioactive agents. J Am Chem Soc 2006;128:4168–9.

Betzig E, Patterson GH, Sougrat R, et al. Imaging Intracellular Fluorescent Proteins at Nanometer Resolution. Science 2006;313:1642–5.

Bewersdorf J, Bennett BT, Knight KL. H2AX chromatin structures and their response to DNA damage revealed by 4Pi microscopy. Proc Natl Acad Sci U S A 2006;103:18137–42.

Bhabra G, Sood A, Fisher B, et al. Nanoparticles can cause DNA damage across a cellular barrier. Nat Nanotech 2009;4:876–83.

Bharali DJ, Klejbor I, Stachowiak EK, et al. Organically modified silica nanoparticles: A nonviral vector for in vivo gene delivery and expression in the brain. 2005;102:11539–44.

Bhaskar K, Anbu J, Ravichandiran V, et al. Lipid nanoparticles for transdermal delivery of flurbiprofen: formulation, in vitro, ex vivo and in vivo studies. Lipids Health Dis 2009;8:6.

Birch DG, Liang FQ. Age-related macular degeneration: a target for nanotechnology derived medicines. Int J Nanomedicine 2007;2:65–77.

Bishnoi SW, Rozell CJ, Levin CS, et al. All-Optical Nanoscale pH Meter. Nano Lett 2006;6:1687–92.

Bisht S, Feldmann G, Soni S, et al. Polymeric nanoparticle-encapsulated curcumin (nanocurcumin): a novel strategy for human cancer therapy. J Nanobiotechnol 2007;5:3.

Blanchard SC, Gonzalez RL, Kim HD, et al. tRNA selection and kinetic proofreading in translation. Nat Struct Mol Biol 2004a;11:1008–14.

Blanchard SC, Kim HD, Gonzalez RL Jr, et al. tRNA dynamics on the ribosome during translation. PNAS 2004b;101:12893–8.

Blazej RG, Kumaresan P, Mathies RA. Microfabricated bioprocessor for integrated nanoliter-scale Sanger DNA sequencing. Proc Natl Acad Sci U S A 2006;103:7240–5.

Bonoiu AC, Mahajan SD, Ding H, et al. Nanotechnology approach for drug addiction therapy: Gene silencing using delivery of gold nanorod-siRNA nanoplex in dopaminergic neurons. Proc Natl Acad Sci U S A 2009;106:5546–50.

Borm PJ, Robbins D, Haubold S, et al. The potential risks of nanomaterials: a review carried out for ECETOC. Part Fibre Toxicol 2006;3:11.

Boukany PE, Morss A, Liao WC, et al. Nanochannel electroporation delivers precise amounts of biomolecules into living cells. Nat Nanotechnol 2011;6:747–54.

Bourges JL, Gautier SE, Delie F, et al. Ocular drug delivery targeting the retina and retinal pigment epithelium using polylactide nanoparticles. Invest Ophthalmol Vis Sci 2003;44:3562–9.

Braydich-Stolle L, Hussain S, Schlager J, Hofmann MC. In vitro cytotoxicity of nanoparticles in mammalian germ-line stem cells. Toxicol Sci 2005;88:412–9.

Breitenstein M, Holzel R, Bier FF. Immobilization of different biomolecules by atomic force microscopy. Journal of Nanobiotechnology 2010;8:10.

Brito L, Amiji M. Nanoparticulate carriers for the treatment of coronary restenosis. Int J Nanomedicine 2007;2:143–61

Bruchez M Jr, Moronne M, Gin P, Weiss S, Alivisatos AP. Semiconductor nanocrystals as fluorescent biological labels. Science 1998;281:2013–6.

Brunker SE, Cederquist KB, Keating CD. Metallic barcodes for multiplexed bioassays. Nanomedicine (Lond) 2007;2:695–710.

Bryson JM, Fichter KM, Chu WJ, et al. Polymer beacons for luminescence and magnetic resonance imaging of DNA delivery. Proc Natl Acad Sci U S A 2009;106;16913–8.

Brzoska M, Langer K, Coester C, et al. Incorporation of biodegradable nanoparticles into human airway epithelium cells-in vitro study of the suitability as a vehicle for drug or gene delivery in pulmonary diseases. Biochem Biophys Res Commun 2004;318:562–70.

Butler TZ, Pavlenok M, Derrington IM, et al. Single-molecule DNA detection with an engineered MspA protein nanopore. Proc Natl Acad Sci U S A 2008;105:20647–52.

Byers RJ, Hitchman ER. Quantum dots brighten biological imaging. Prog Histochem Cytochem 2011;45:201–37.

Cai L, Friedman N, Xie XS. Stochastic protein expression in individual cells at the single molecule level. Nature 2006;440:358–62.

Cai K, Li J, Luo et al. β-Cyclodextrin conjugated magnetic nanoparticles for diazepam removal from blood. Chem Commun (Camb) 2011;47:7719–21.

Calarota SA, Dai A, Trocio JN, et al. IL-15 as memory T-cell adjuvant for topical HIV-1 DermaVir vaccine. Vaccine 2008;26:5188–95.

Callera F, de Melo C. Magnetic resonance tracking of magnetically labeled autologous bone marrow CD34+ cells transplanted into the spinal cord via lumbar puncture technique in patients with chronic spinal cord injury: CD34+ cells' migration into the injured site. Stem Cells and Development 2007;16:461–466.

Cao Z, Ma Y, Yue X, et al. Stabilized liposomal nanohybrid cerasomes for drug delivery applications. Chem Commun (Camb) 2010;46:5265–7.

Carbonaro A, Mohanty SK, Huang H, et al. Cell characterization using a protein-functionalized pore. Lab Chip 2008;8:1478–85.

Castaneda RT, Khurana A, Khan R, Daldrup-Link HE. Labeling stem cells with ferumoxytol, an FDA-approved iron oxide nanoparticle. J Vis Exp 2011;57:e3482.

Cedervall T, Lynch I, Lindman S, et al. Understanding the nanoparticle-protein corona using methods to quantify exchange rates and affinities of proteins for nanoparticles. Proc Natl Acad Sci U S A 2007;104:2050–2055.

Chai J, Wong LS, Giam L, Mirkin CA. Single-molecule protein arrays enabled by scanning probe block copolymer lithography. Proc Natl Acad Sci U S A 2011;108:19521–5.

Chakraborty B, Sha R, Seeman NC. Colloquium Paper: A DNA-based nanomechanical device with three robust states. Proc Natl Acad Sci USA 2008;105:17245–9.

Chakravarthy KV, Bonoiud AC, Davis WG, et al. Gold nanorod delivery of an ssRNA immune activator inhibits pandemic H1N1 influenza viral replication. Proc Natl Acad Sci U S A 2010;107:10172–7.

Chakravarty P, Marches R, Zimmerman NS, et al. Thermal ablation of tumor cells with antibody-functionalized single-walled carbon nanotubes. Proc Natl Acad Sci U S A 2008;105:8697–702.

Chambers E, Mitragotri S. Long Circulating Nanoparticles via Adhesion on Red Blood Cells: Mechanism and Extended Circulation. Exp Biol Med 2007;232:958–966.

Chan JM, Zhang L, Tongc R, et al. Spatiotemporal controlled delivery of nanoparticles to injured vasculature. Proc Natl Acad Sci U S A 2010;107:2213–8.

Chanda N, Kattumuri V, Shukla R, et al. Bombesin functionalized gold nanoparticles show in vitro and in vivo cancer receptor specificity. Proc Natl Acad Sci U S A 2010;107:8760–5.

Chansin GA, Mulero R, Hong J, et al. Single molecule spectroscopy using nanoporous membranes, Nano Lett 2007;7:2901–6.

Charalambous A, Andreou M, Skourides PA. Intein-mediated site-specific conjugation of Quantum Dots to proteins in vivo. J Nanobiotechnology 2009;7(1):9.

Chatterjee DK, Rufaihah AJ, Zhang Y. Upconversion fluorescence imaging of cells and small animals using lanthanide doped nanocrystals. Biomaterials 2008;29:937–43.

Checot F, Rodriguez-Hernandez J, Gnanou Y, Lecommandoux S. pH-responsive micelles and vesicles nanocapsules based on polypeptide diblock copolymers. Biomol Eng 2007;24:81–5.

Chen AA, Derfus AM, Khetani SR, Bhatia SN. Quantum dots to monitor RNAi delivery and improve gene silencing. Nucleic Acids Res 2005;33:e190.

Chen H, Clarkson BH, Sun K, Mansfield JF. Self-assembly of synthetic hydroxyapatite nanorods into an enamel prism-like structure. Journal of Colloid and Interface Science 2005a;288:97–103.

Chen HH, Josephson L, Sosnovik DE. Imaging of apoptosis in the heart with nanoparticle technology. Wiley Interdiscip Rev Nanomed Nanobiotechnol 2011;3:86–99.

Cheng Y, Zhao L, Li Y, Xu T. Design of biocompatible dendrimers for cancer diagnosis and therapy: current status and future perspectives. Chem Soc Rev 2011;40:2673–703.

Chikramane PS, Suresh AK, Bellare JR, Kane SG. Extreme homeopathic dilutions retain starting materials: A nanoparticulate perspective. Homeopathy 2010;99:231–42.

Chilkoti A, Christensen T, MacKay JA. Stimulus responsive elastin biopolymers: Applications in medicine and biotechnology. Curr Opin Chem Biol 2006;10:652–7.

Chiou PY, Ohta AT, Wu MC. Massively parallel manipulation of single cells and microparticles using optical images. Nature 2005;436:370–372.

Chiu DT. Interfacing droplet microfluidics with chemical separation for cellular analysis. Anal Bioanal Chem 2010;397:3179–83.

Chiu SW, Leake MC. Functioning nanomachines seen in real-time in living bacteria using single-molecule and super-resolution fluorescence imaging. Int J Mol Sci 2011;12:2518–42.

Chnari E, Nikitczuk JS, Uhrich KE, et al. Nanoscale Anionic Macromolecules Can Inhibit Cellular Uptake of Differentially Oxidized LDL. Biomacromolecules 2006;7:597–603.

Choi B, Zocchi G, Canale S, et al. Artificial allosteric control of maltose binding protein. Phys Rev Lett 2005;94:038103.

Choi CH, Alabi CA, Webster P, Davis ME. Mechanism of active targeting in solid tumors with transferrin-containing gold nanoparticles. Proc Natl Acad Sci U S A 2010;107:1235–40.

Choi HS, Ashitate Y, Lee JH, et al. Rapid translocation of nanoparticles from the lung airspaces to the body. Nat Biotechnol 2010b;28:1300–3.

Choi J, Jun Y, Yeon S, et al. Biocompatible Heterostructured Nanoparticles for Multimodal Biological Detection. JACS 2006;128:15982–15983.

Choi KY, Chung H, Min KH, et al. Self-assembled hyaluronic acid nanoparticles for active tumor targeting. Biomaterials 2010a;31:106–14.

Choi SJ, Oh JM, Choy JH. Safety aspect of inorganic layered nanoparticles: size-dependency in vitro and in vivo. J Nanosci Nanotechnol 2008;8:5297–301.

Choi YS, Wood TD. Polyaniline-coated nanoelectrospray emitters treated with hydrophobic polymers at the tip. Rapid Communications in Mass Spectrometry 2007;21:2101–2108.

Chorny M, Polyak B, Alferiev IS, et al. Magnetically driven plasmid DNA delivery with biodegradable polymeric nanoparticles. FASEB J 2007;21:2510–9.

Chorny M, Fishbein I, Yellen BB, et al. Targeting stents with local delivery of paclitaxel-loaded magnetic nanoparticles using uniform fields. PNAS 2010;107:8346–51.

Chouhan R, Bajpai AK. Real time in vitro studies of doxorubicin release from PHEMA nanoparticles. Journal of Nanobiotechnology 2009;7:5.

Chowdhury EH. pH-sensitive nano-crystals of carbonate apatite for smart and cell-specific transgene delivery. Expert Opinion on Drug Delivery 2007;4:193–196.

Chu M, Wu Q, Yang H, et al. Transfer of quantum dots from pregnant mice to pups across the placental barrier. Small 2010;6:670–8.

Cinteza LO, Ohulchanskyy TY, Sahoo Y, et al. Diacyllipid Micelle-Based Nanocarrier for Magnetically Guided Delivery of Drugs in Photodynamic Therapy. Mol Pharm 2006;3:415–23.

Coggan JS, Bartol TM, Esquenazi E, et al. Evidence for Ectopic Neurotransmission at a Neuronal Synapse. Science 2005;309:446–51.

Cognet L, Tardin C, Boyer D, et al. Single metallic nanoparticle imaging for protein detection in cells. PNAS 2003;100:11350–5.

Cohen SM, Mukerji R, Cai S, et al. Subcutaneous delivery of nanoconjugated doxorubicin and cisplatin for locally advanced breast cancer demonstrates improved efficacy and decreased toxicity at lower doses than standard systemic combination therapy in vivo. Am J Surg 2011;202:646–53.

Corstjens PL, de Dood CJ, van der Ploeg-van Schip JJ, et al. Lateral flow assay for simultaneous detection of cellular- and humoral immune responses. Clin Biochem 2011;44:1241–6.

Cortez-Retamozo V, Backmann N, Senter PD, et al. Efficient cancer therapy with a nanobody-based conjugate. Cancer Res 2004;64:2853–7.

Cotí KK, Belowich ME, Liong M, et al. Mechanised nanoparticles for drug delivery. Nanoscale 2009;1:16–39.

Crane JM, Van Hoek AN, Skach WR, Verkman AS. Aquaporin-4 dynamics in orthogonal arrays in live cells visualized by quantum dot single particle tracking. Mol Biol Cell 2008;19:3369–78.

Csaki A, Garwe F, Steinbrück A, et al. A parallel approach for subwavelength molecular surgery using gene-specific positioned metal nanoparticles as laser light antennas. Nano Lett 2007;7:247–53.

Cunningham BT. Photonic Crystal Surfaces as a General Purpose Platform for Label-Free and Fluorescent Assays. JALA Charlottesv Va 2010;15:120–135.

Curl RF, Kroto H, Smalley RE. Nobel lectures in chemistry. Reviews of Modern Physics 1997;69:691–730.

Currall SC, King EB, Lane N, et al. What drives public acceptance of nanotechnology? Nat Nanotech 2006;1:153–5.

Davidson M, Karlsson M, Sinclair J, et al. Nanotube-Vesicle Networks with Functionalized Membranes and Interiors. J Am Chem Soc 2003;125:374–78.

Davis ME, Hsieh PC, Takahashi T, et al. Local myocardial insulin-like growth factor 1 (IGF-1) delivery with biotinylated peptide nanofibers improves cell therapy for myocardial infarction. Proc Natl Acad Sci U S A 2006;103:8155–60.

De D, Mandal SM, Gauri SS, et al. Antibacterial effect of lanthanum calcium manganate ($La_{0.67}Ca_{0.33}MnO_3$) nanoparticles against Pseudomonas aeruginosa ATCC 27853. J Biomed Nanotechnol 2010;6:138–44.

De Jong WH, Hagens WI, Krystek P, et al. Particle size-dependent organ distribution of gold nanoparticles after intravenous administration. Biomaterials 2008;29:1912–19.

de Kozak Y, Andrieux K, Villarroya H et al. Intraocular injection of tamoxifen-loaded nanoparticles: a new treatment of experimental autoimmune uveoretinitis. Eur J Immunol 2004;34:3702–12.

de Salamanca AE, Diebold Y, Calonge M et al. Chitosan nanoparticles as a potential drug delivery system for the ocular surface: toxicity, uptake mechanism and in vivo tolerance. Invest Ophthalmol Vis Sci 2006; 47:1416–25.

de Silva AP. Molecular logic gate arrays. Chem Asian J 2011;6:750–66.

Debbage P, Jaschke W. Molecular imaging with nanoparticles: giant roles for dwarf actors. J Histochem Cell Biol 2008;130:845–875.

Degen CL, Poggio M, Mamin HJ, et al. Nanoscale magnetic resonance imaging. Proc Natl Acad Sci U S A 2009;106:1313–7.

Delgado D, Del Pozo-Rodríguez A, Solinís MA, et al. Dextran and protamine-based solid lipid nanoparticles as potential vectors for the treatment of X linked juvenile retinoschisis. Hum Gene Ther 2012 Feb 1. [Epub ahead of print]

Denardo SJ, Denardo GL, Natarajan A, et al. Thermal Dosimetry Predictive of Efficacy of 111In-ChL6 Nanoparticle AMF-Induced Thermoablative Therapy for Human Breast Cancer in Mice. J Nucl Med 2007;48:437–44.

Deng H, Xu Y, Liu Y, et al. Gold nanoparticles with asymmetric polymerase chain reaction for colorimetric detection of DNA sequence. Anal Chem 2012;84:1253–8.

Derfus AM, Chan CW, Bhatia SN, et al. Probing the Cytotoxicity of Semiconductor Quantum Dots Nano Letters 2004;4:11–18.

Derfus AM, von Maltzahn G, Harris TJ, et al. Remotely Triggered Release from Magnetic Nanoparticles. Advanced Materials 2007;9:3932–36.

Dhas NA, Suslick KS. Sonochemical Preparation of Hollow Nanospheres and Hollow Nanocrystals. J Am Chem Soc (Communication) 2005;127;2368–69.

Di Bucchianico S, Poma AM, Giardi MF, et al. Atomic Force Microscope nanolithography on chromosomes to generate single-cell genetic probes. J Nanobiotechnol 2011;9:27

Di Giusto DA, Wlassoff WA, Gooding JJ, et al. Proximity extension of circular DNA aptamers with real-time protein detection. Nucleic Acids Res 2005;33:e64.

Diebold Y, Calonge M. Applications of nanoparticles in ophthalmology. Prog Retin Eye Res 2010;29:596–609.

Ding H, Inoue S, Ljubimov AV, et al. Inhibition of brain tumor growth by intravenous poly(β-Lmalic acid) nanobioconjugate with pH-dependent drug release. Proc Natl Acad Sci U S A 2010;107: 18143–8.

Ditto AJ, Shah PN, Yun YH. Non-viral gene delivery using nanoparticles. Expert Opinion on Drug Delivery 2009;6:1149–60.

Dixon MB, Falconet C, Ho L, et al. Removal of cyanobacterial metabolites by nanofiltration from two treated waters. J Hazard Mater 2011;188:288–95.

Dobrovolskaia MA, Patri AK, Zheng J, et al. Interaction of colloidal gold nanoparticles with human blood: effects on particle size and analysis of plasma protein binding profiles. Nanomedicine 2009;5:106–17.

Dobson J. Magnetic nanoparticles-based gene delivery. Gene Ther 2006;13:283–7.

Docoslis A, Espinoza LA, Zhang B, et al. Using nonuniform electric fields to accelerate the transport of viruses to surfaces from media of physiological ionic strength. Langmuir 2007;23:3840–8.

Dohner K, Sodeik B. The role of the cytoskeleton during viral infection. Curr Top Microbiol Immunol 2005;285:67–108.

Dosev D, Nichkova M, Kennedy IM. Inorganic lanthanide nanophosphors in biotechnology. J Nanosci Nanotechnol 2008;8:1052–67.

Douglas SM, Bachelet I, Church GM. A logic-gated nanorobot for targeted transport of molecular payloads. Science 2012;335:831–4.

Drexler KE. Engines of creation, the coming era of nanotechnology. Anchor, New York, 1987.

Drexler KE. Molecular engineering: an approach to the development of general capabilities for molecular manipulation. Proc Natl Acad Sci U S A 1981;78:5275–8.

D'Souza GG, Weissig V. Subcellular targeting: a new frontier for drug-loaded pharmaceutical nanocarriers and the concept of the magic bullet. Expert Opin Drug Deliv 2009;6:1135–48.

Dubach JM, Das S, Rosenzweig A, Clark HA. Visualizing sodium dynamics in isolated cardiomyocytes using fluorescent nanosensors. Proc Natl Acad Sci U S A 2009;106:16145–50.

Duceppe N, Tabrizian M. Advances in using chitosan-based nanoparticles for in vitro and in vivo drug and gene delivery. Expert Opin Drug Deliv 2010;7:1191–207.

Dufes C, Keith WN, Bilsland A, et al. Synthetic Anticancer Gene Medicine Exploits Intrinsic Antitumor Activity of Cationic Vector to Cure Established Tumours. Cancer Res 2005;65:8079–8084.

Dupuy JP. Some pitfalls in the philosophical foundations of nanoethics. J Med Philos 2007;32:237–61.

Eastman PS, Ruan W, Doctolero M, et al. Qdot Nanobarcodes for Multiplexed Gene Expression Analysis. Nano Lett 2006;6:1059–1064.

Ebner A, Kienberger F, Kada G, et al. Localization of single avidin-biotin interactions using simultaneous topography and molecular recognition imaging. Chemphyschem 2005;6:897–900.

Edgar R, McKinstry M, Hwang J, et al. High-sensitivity bacterial detection using biotin-tagged phage and quantum-dot nanocomplexes. Proc Natl Acad Sci U S A 2006;103:4841–5.

Edgar R, Rokney A, Feeney M, et al. Bacteriophage infection is targeted to cellular poles. Mol Microbiol 2008;68:1107–16.

Edwards KA, Wang Y, Baeumner AJ. Aptamer sandwich assays: human α-thrombin detection using liposome enhancement. Anal Bioanal Chem 2010;398:2645–54.

Eigler DM, Schweizer EK. Positioning single atoms with a scanning tunneling microscope. Nature 1990;344:524–6.

Elbayoumi TA, Torchilin V. Liposomes for targeted delivery of antithrombotic drugs. Expert Opinion on Drug Delivery 2008;5:1185–1198.

Elechiguerra JL, Burt JL, Morones JR, et al. Interaction of silver nanoparticles with HIV-1. Journal of Nanobiotechnology 2005;3:6.

Ellis-Behnke RG, Liang YX, You SW, et al. Nano neuro knitting: Peptide nanofiber scaffold for brain repair and axon regeneration with functional return of vision. Proc Natl Acad Sci U S A 2006;103:5054–9.

El-Sayed IH, Huang X, El-Sayed M. Selective laser photo-thermal therapy of epithelial carcinoma using anti-EGFR antibody conjugated gold nanoparticles. Cancer Letters 2006;239:129–35.

El-Sayed IH, Huang X, El-Sayed MA. Surface Plasmon Resonance Scattering and Absorption of anti-EGFR Antibody Conjugated Gold Nanoparticles in Cancer Diagnostics: Applications in Oral Cancer. Nano Lett 2005;5:829–834.

Everts M, Saini V, Leddon JL, et al. Covalently Linked Au Nanoparticles to a Viral Vector: Potential for Combined Photothermal and Gene Cancer Therapy. Nano Lett 2006;6:587–591.

Fan R, Karnik R, Yue M, et al. DNA Translocation in Inorganic Nanotubes. Nano Lett 2005;5:1633–1637.

Fang RH, Hu CM, Zhang L. Nanoparticles disguised as red blood cells to evade the immune system. Expert Opin Biol Ther 2012 Feb 15; doi: 10.1517/14712598.2012.661710.

Fantner GE, Schumann W, Barbero RJ, et al. Use of self-actuating and self-sensing cantilevers for imaging biological samples in fluid. Nanotechnology 2009;20:434003 (E publication).

Farjo R, Skaggs J, Quiambao AB, et al. Efficient non-viral ocular gene transfer with compacted DNA nanoparticles. PLoS ONE 2006;1:e38.

Farokhzad OC, Jon S, N Khademhosseini A, et al. Nanoparticle-Aptamer Bioconjugates: A New Approach for Targeting Prostate Cancer Cells. Cancer Research 2004;64:7668–7672.

Farrer RA, Butterfield FL, Chen VW, Fourkas JT. Highly Efficient Multiphoton-Absorption-Induced Luminescence from Gold Nanoparticles. Nano Lett 2005: 1139–42.

Fatouros DG, Ioannou PV, Antimisiaris SG. Arsonoliposomes: novel nanosized arsenic-containing vesicles for drug delivery. J Nanosci Nanotechnol 2006;6:2618–37.

Feng SS, Zeng W, Teng Lim Y, et al. Vitamin E TPGS-emulsified poly(lactic-co-glycolic acid) nanoparticles for cardiovascular restenosis treatment. Nanomed 2007;2:333–44.

Festag G, Schüler T, Steinbrück A, et al. Chip-based molecular diagnostics using metal nanoparticles. Expert Opinion on Medical Diagnostics 2008;2:813–28.

Feynman R. There's plenty of room at the bottom: an invitation to enter a new filed of physics. Reprinted in: Crandall BC, Lewis J (eds) Nanotechnology: research and perspectives. The MIT Press, Cambridge, MA, 1992:347–363.

Fichter KM, Flajolet M, Greengard P, Vu TQ. Kinetics of G-protein–coupled receptor endosomal trafficking pathways revealed by single quantum dots. Proc Natl Acad Sci U S A 2010;107:18658–63.

Fiedler FA, Reynolds GH. Legal Problems of Nanotechnology: An Overview. Southern California Interdisciplinary Law Journal, Winter 1994.

Fink TL, Klepcyk PJ, Oette SM, et al. Plasmid size up to 20 kbp does not limit effective in vivo lung gene transfer using compacted DNA nanoparticles. Gene Ther 2006;13:1048–51.

Fleischli FD, Dietiker M, Borgia C, Spolenak R. The influence of internal length scales on mechanical properties in natural nanocomposites: a comparative study on inner layers of seashells. Acta Biomater 2008;4:1694–706.

Floriano PN, Christodoulides N, Miller CS, et al. Use of saliva-based nano-biochip tests for acute myocardial infarction at the point of care: a feasibility study. Clin Chem 2009;55:1530–8.

Fortina P, Kricka LJ, Graves DJ, et al. Applications of nanoparticles to diagnostics and therapeutics in colorectal cancer. Trends Biotechnol 2007;25:145–52.

Fortina P, Kricka LJ, Graves DJ, et al. Applications of nanoparticles to diagnostics and therapeutics in colorectal cancer. Trends Biotechnol 2007;25:145–52.

Fortner JD, Lyon DY, Sayes CM, et al. C60 in Water: Nanocrystal Formation and Microbial Response. Environ Sci Technol 2005;39:4307–4316.

Franzen S. A comparison of peptide and folate receptor targeting of cancer cells: from single agent to nanoparticle. Expert Opin Drug Deliv 2011;8:281–28.

Freitas RA Jr. Exploratory design in medical nanotechnology: a mechanical artificial red cell. Artif Cells Blood Substit Immobil Biotechnol 1998;26:411–30.

Fu J, Mao P, Han J. Nanofilter array chip for fast gel-free biomolecule separation. Appl Phys Lett 2006;87:263902 (online).

Galanzha EI, Shashkov EV, Kelly T, et al. In vivo magnetic enrichment and multiplex photoacoustic detection of circulating tumour cells. Nat Nanotechnol 2009;4:855–60.

Gamer AO, Leibold E, van Ravenzwaay B. The in vitro absorption of microfine zinc oxide and titanium dioxide through porcine skin. Toxicol In Vitro 2006;20:301–7.

Gao D, Xu H, Philbert MA, et al. Ultrafine Hydrogel Nanoparticles: Synthetic Approach and Therapeutic Application in Living Cells. Angew Chem Int Ed Engl 2007;46:2224–7.

Gao H, Shi W, Freund LB. Mechanics of receptor-mediated endocytosis. PNAS 2005;102:9469–74.

Gao X, Cui Y, Levenson RM, et al. In vivo cancer targeting and imaging with semiconductor quantum dots. Nat Biotech 2004;22:969–76.

Garg G, Saraf S, Saraf S. Cubosomes: an overview. Biol Pharm Bull 2007;30:350–3.

Gaspar R. Regulatory issues surrounding nanomedicines: setting the scene for the next generation of nanopharmaceuticals. Nanomedicine 2007;2:143–147.

Gassman NR, Nelli JP, Dutta S, et al. Selection of bead-displayed, PNA-encoded chemicals. J Mol Recognit 2010;23:414–22.

Gaster RS, Hall DA, Nielsen CH, et al. Matrix-insensitive protein assays push the limits of biosensors in medicine. Nat Med 2009;15:1327–32.

Ge C, Du J, Zhao L, et al. Binding of blood proteins to carbon nanotubes reduces cytotoxicity. Proc Natl Acad Sci U S A 2011;108:16968–73.

Geert van Calster. Regulating Nanotechnology in the European Union. Nanotechnology Law & Business 2006;3:359–72.

Geho DH, Jones CD, Petricoin EF, Liotta LA. Nanoparticles: potential biomarker harvesters. Curr Opin Chem Biol 2006;10:56–61.

Gelfuso GM, Gratieri T, Simão PS, et al. Chitosan microparticles for sustaining the topical delivery of minoxidil sulphate. J Microencapsul 2011;28:650–8.

Geng Y, Dalhaimer P, Cai S, et al. Shape effects of filaments versus spherical particles in flow and drug delivery. Nat Nanotech 2007;2:249–255.

Georganopoulou DG, Chang L, Nam JM, et al. Nanoparticle-based detection in cerebral spinal fluid of a soluble pathogenic biomarker for Alzheimer's disease. Proc Natl Acad Sci U S A 2005;102:2273–6.

Ghoroghchian PP, Therien MJ, Hammer DA. In vivo fluorescence imaging: a personal perspective. Wiley Interdiscip Rev Nanomed Nanobiotechnol 2009;1:156–67.

Gibson JD, Khanal B, Zubarev E. Paclitaxel-Functionalized Gold Nanoparticles. J Am Chem Soc 2007;129:11653–11661.

Glover DJ, Ng SM, Mechler A, et al. Multifunctional protein nanocarriers for targeted nuclear gene delivery in nondividing cells. The FASEB Journal 2009;23:2996–3006.

Godin B, Touitou E. Ethosomes: new prospects in transdermal delivery. Crit Rev Ther Drug Carrier Syst 2003;20:63–102.

Godin B, Touitou E. Mechanism of bacitracin permeation enhancement through the skin and cellular membranes from an ethosomal carrier. J Control Release 2004;94:365–79.

Golub E, Pelossof G, Freeman R, et al. Electrochemical, photoelectrochemical, and surface plasmon resonance detection of cocaine using supramolecular aptamer complexes and metallic or semiconductor nanoparticles. Anal Chem 2009;81:9291–8.

Goluch ED, Stoeva SI, Lee JS, et al. A microfluidic detection system based upon a surface immobilized biobarcode assay. Biosens Bioelectron 2009;24:2397–403.

Gong X, Li J, Xu K, et al. A controllable molecular sieve for Na+ and K+ ions. J Am Chem Soc 2010;132:1873–7.

Goodman CM, McCusker CD, Yilmaz T, Rotello VM. Toxicity of gold nanoparticles functionalized with cationic and anionic side chains. Bioconjug Chem 2004;15:897–900.

Gordijn B. Nanoethics: from utopian dreams and apocalyptic nightmares towards a more balanced view. Sci Eng Ethics 2005;11:521–33.

Gordon EM, Levy JP, Reed RA, et al. Targeting metastatic cancer from the inside: A new generation of targeted gene delivery vectors enables personalized cancer vaccination in situ. Int J Oncol 2008 ;33:665–75.

Gourlay PL, Hendricks JK, McDonald AE, et al. Mitochondrial correlation microscopy and nanolaser spectroscopy – new tools for biphotonic detection of cancer in single cells. TCRT 2005;4:585–592.

Gourley PL, McDonald AE. Ultrafast NanoLaser Device for Detecting Cancer in a Single Live Cell. Sandia National Laboratories, Albuquerque, New Mexico, Report # SAND2007-7456, 2007.

Gradishar WJ, Tjulandin S, Davidson N, et al. Superior Efficacy of Albumin-Bound Paclitaxel, ABI-007, Compared With Polyethylated Castor Oil-Based Paclitaxel in Women With Metastatic Breast Cancer: Results of a Phase III Trial. J Clin Oncol 2005;23:7794–803.

Graf A, Jack KS, Whittaker AK, et al. Protein delivery using nanoparticles based on microemulsions with different structure-types. Eur J Pharm Sci 2008;33:434–44.

Gratton SE, Pohlhaus PD, Lee J, et al. Nanofabricated particles for engineered drug therapies: a preliminary biodistribution study of PRINT nanoparticles. J Control Release 2007;121:10–8.

Griesenbach U, Geddes DM, Alton EW. Advances in cystic fibrosis gene therapy. Curr Opin Pulm Med 2004;10:542–6.

Guarise C, Pasquato L, De Filippis V, Scrimin P. Gold nanoparticles-based protease assay. Proc Natl Acad Sci U S A 2006;103:3978–82.

Guarneri V, Dieci MV, Conte PF. Enhancing intracellular taxane delivery: current role and perspectives of nanoparticle albumin-bound paclitaxel in the treatment of advanced breast cancer. Expert Opin Pharmacother 2012;13:395–406.

Gupta A, Akin D, Bashir R. Single virus particle mass detection using microresonators with nanoscale thickness. Appl Phys Lett 2004;84: 1976–1978.

Gupta AK, Nair PR, Akin D, et al. Anomalous resonance in a nanomechanical biosensor. Proc Natl Acad Sci U S A 2006;103:13362–7.

Guzman R, Uchida N, Bliss TM, et al. Long-term monitoring of transplanted human neural stem cells in developmental and pathological contexts with MRI. Proc Natl Acad Sci USA 2007;104:10211–6.

Hahm J, Lieber CM. Direct Ultrasensitive Electrical Detection of DNA and DNA Sequence Variations Using Nanowire Nanosensors. Nano Letters 2004;4:51–54.

Halder J, Kamat AA, Landen CN Jr, et al. Focal Adhesion Kinase Targeting Using In vivo Short Interfering RNA Delivery in Neutral Liposomes for Ovarian Carcinoma Therapy. Clin Cancer Res 2006;12:4916–24.

Halfpenny KC, Wright DW. Nanoparticle detection of respiratory infection. Wiley Interdiscip Rev Nanomed Nanobiotechnol 2010;2:277–90.

Hamada K, Hirose M, Yamashita T, Ohgushi H. Spatial distribution of mineralized bone matrix produced by marrow mesenchymal stem cells in self-assembling peptide hydrogel scaffold. J Biomed Mater Res A 2008;84:128–36.

Han MS, Lytton-Jean AK, Mirkin CA. A gold nanoparticle based approach for screening triplex DNA binders. J Am Chem Soc 2006;128:4954–5.

Happel P, Dietzel ID. Backstep scanning ion conductance microscopy as a tool for long term investigation of single living cells. J Nanobiotechnol 2009;7:7.

Harde H, Das M, Jain S. Solid lipid nanoparticles: an oral bioavailability enhancer vehicle. Expert Opin Drug Deliv 2011;8:1407–24.

Hashi CK, Zhu Y, Yang GY, et al. Antithrombogenic property of bone marrow mesenchymal stem cells in nanofibrous vascular grafts. Proc Natl Acad Sci U S A 2007;104:11915–20.

Haun JB, Yoon TJ, Lee H, Weissleder R. Magnetic nanoparticle biosensors. WIREs Nanomed Nanobiotechnol 2010;2:291–304.

He Y, Wu J, Zhao Y. Designing Catalytic Nanomotors by Dynamic Shadowing Growth. Nano Lett 2007;7:1369–75.

Heidel JD, Liu JY, Yen Y, et al. Potent siRNA Inhibitors of Ribonucleotide Reductase Subunit RRM2 Reduce Cell Proliferation In vitro and In vivo. Clin Cancer Res 2007b;13:2207–15.

Heidel JD, Yu Z, Liu J, et al. Administration in non-human primates of escalating intravenous doses of targeted nanoparticles containing ribonucleotide reductase subunit M2 siRNA. Proc Natl Acad Sci U S A 2007a;104:5715–5721.

Heller DA, Jeng ES, Yeung TK, et al. Optical Detection of DNA Conformational Polymorphism on Single-Walled Carbon Nanotubes. Science 2006;311:508–511.

Heller DA, Jin H, Martinez BM, et al. Multimodal optical sensing and analyte specificity using single-walled carbon nanotubes. Nat Nanotechnol 2009;4:114–20.

Higaki M, Kameyama M, Udagawa M, et al. Transdermal delivery of CaCO3-nanoparticles containing insulin. Diabetes Technol Ther 2006;8:369–74.

Hild W, Pollinger K, Caporale A, et al. G protein-coupled receptors function as logic gates for nanoparticle binding and cell uptake. Proc Natl Acad Sci U S A 2010;107:10667–72.

Hilty FM, Teleki A, Krumeich F, et al. Development and optimization of iron- and zinc-containing nanostructured powders for nutritional applications. Nanotechnology 2009;20:475101.

Hiratsuka Y, Miyata M, Tada T, Uyeda T. A microrotary motor powered by bacteria. Proc Natl Acad Sci U S A 2006;103:13618–23.

Ho YP, Kung MC, Yang S, et al. Multiplexed Hybridization Detection with Multicolor Colocalization of Quantum Dot Nanoprobes. Nano Lett 2005;5:1693–1697.

Hoare T, Santamaria J, Goya GF, et al. A Magnetically Triggered Composite Membrane for On-Demand Drug Delivery. Nano Lett 2009;9:3651–7.

Hohng S, Zhou R, Nahas MK, et al. Fluorescence-force spectroscopy maps two-dimensional reaction landscape of the holliday junction. Science 2007;318:279–83.

Hong J, Edel JB, deMello AJ. Micro- and nanofluidic systems for high-throughput biological screening. Drug Discovery Today 2009;14:134–146.

Hoshino Y, Koide H, Furuya K, et al. The rational design of a synthetic polymer nanoparticle that neutralizes a toxic peptide in vivo. Proc Natl Acad Sci U S A 2012;109:33–8.

Howard KA. Delivery of RNA interference therapeutics using polycation-based nanoparticles. Adv Drug Deliv Rev 2009;61:710–20.

Howarth M, Takao K, Hayashi Y, Ting AY. Targeting quantum dots to surface proteins in living cells with biotin ligase. PNAS 2005;102:7583–8.

Hsieh PCH, Davis ME, Gannon J, et al. Controlled delivery of PDGF-BB for myocardial protection using injectable self-assembling peptide nanofibers. J Clin Invest 2006;116:237–48.

Hu M, Qian L, Brinas RP, et al. Assembly of Nanoparticle-Protein Binding Complexes: From Monomers to Ordered Arrays. Angew Chem Int Ed Engl 2007;46:5111–5114.

Hu Q, Li B, Wang M, Shen J. Preparation and characterization of biodegradable chitosan/hydroxyapatite nanocomposite rods via in situ hybridization: a potential material as internal fixation of bone fracture. Biomaterials 2004;25:779–85.

Huang X, Ren J. Gold nanoparticles based chemiluminescent resonance energy transfer for immunoassay of alpha fetoprotein cancer marker. Anal Chim Acta 2011;686:115–20.

Huang H, Delikanli S, Zeng H, et al. Remote control of ion channels and neurons through magnetic-field heating of nanoparticles. Nat Nanotechnol 2010;5:602–6.

Huang X, El-Sayed IH, Qian W, El-Sayed MA. Cancer cell imaging and photothermal therapy in the near-infrared region by using gold nanorods. J Am Chem Soc 2006;128:2115–20.

Huff TB, Tong L, Zhao Y, et al. Hyperthermic effects of gold nanorods on tumor cells. Nanomed 2007;2:125–32.

Huschka R, Neumann O, Barhoumi A, Halas NJ: Visualizing light-triggered release of molecules inside living cells. Photodynamic therapy of cancer using nanoparticles Nano Lett 2010;10:4117–22.

Hush NS. An overview of the first half-century of molecular electronics. Ann N Y Acad Sci 2003;1006:1–20.

Iijima S, Ajayan PM, Ichihashi T. Growth model for carbon nanotubes. Phys Rev Lett 1992;69:3100–3103.

Injac R, Perse M, Boskovic M, et al. Cardioprotective Effects of Fullerenol C60(Oh)24 on a Single Dose Doxorubicin-induced Cardiotoxicity in Rats With Malignant Neoplasm. Technol Cancer Res Treat 2008;7:15–26.

Ivanovska IL, de Pablo PJ, Ibarra B, et al. Bacteriophage capsids: Tough nanoshells with complex elastic properties. Proc Natl Acad Sci U S A 2004;101:7600–5.

Iwata F, Mizuguchi Y, Ko H, Ushiki T. Nanomanipulation of biological samples using a compact atomic force microscope under scanning electron microscope observation. J Electron Microsc (Tokyo) 2011;60:359–66.

Jain KK. Handbook of Laser Neurosurgery. Charles C. Thomas, Springfield, Illinois, 1983.

Jain KK. Nanodiagnostics: application of nanotechnology in molecular diagnostics. Expert Rev Mol Diagn 2003;4:153–161.

Jain KK. Current status of molecular biosensors. Medical Device Technology 2003a;14:10–5.

Jain KK. The role of nanobiotechnology in drug discovery. Drug Discovery Today 2005;10:1435–42.

Jain KK, Jain V. Impact of Nanotechnology on Healthcare. Nanotechnology Law & Business 2006;3:411–8.

Jain NK, Asthana A. Dendritic systems in drug delivery applications. Expert Opinion on Drug Delivery 2007;4:495–512.

Jain KK. Applications of Nanobiotechnology in Clinical Diagnostics. Clin Chem 2007;53:2002–9.

Jain KK. Use of nanoparticles for drug delivery in glioblastoma multiforme. Expert Rev Neurother 2007a;7:363–72.

Jain KK. Recent advances in nanooncology. TCRT 2008;7:1–13.

Jain KK. Textbook of Personalized Medicine. Springer, New York, 2009.

Jain KK. Current Status and Future Prospects of Nanoneurology. J Nanoneuroscience 2009a;1:56–64.

Jain KK. Handbook of Biomarkers. Springer, New York, 2010.

Jain KK. Recent advances in nanooncology. BMC Medicine, 2010a;8:83

Jain KK. Handbook of Neuroprotection. Springer, New York, 2011.

Jain KK. Applications of Biotechnology in Cardiovascular Therapeutics. Springer, New York, 2011a.

Jain KK. Molecular Diagnostics: technologies, markets and companies. Jain PharmaBiotech Publications, Basel, 2012a.

Jain KK. Nanobiotechnology: applications, markets and companies. Jain PharmaBiotech Publications, Basel, 2012.

Jain KK. Biochips & Microarrays. technologies, markets and companies. Jain PharmaBiotech Publications, Basel, 2012b

Jain KK. Gene Therapy: technologies, markets and companies. Jain PharmaBiotech Publications, Basel, 2012c.

Jain KK. Transdermal Drug Delivery: technologies, markets and companies. Jain PharmaBiotech Publications, Basel, 2012d.

Jain KK. Drug Delivery in Central Nervous System Disorders: technologies, markets and companies. Jain PharmaBiotech Publications, Basel, 2012e.

Jain KK. Nanobiotechnology-based strategies for crossing the blood–brain barrier. Nanomedicine (Lond) 2012f July (advance online).

Jain KK. Regenerative Therapy for Central Nervous System Trauma, Chapter 25, In, Steinhoff G (ed) Regenerative Medicine, 2nd ed, Springer, London, 2012g.

Jain KK. Role of nanodiagnostics in personalized cancer therapy. Clin Lab Med 2012h;32: 15–31.

Jain KK. Applications of Biotechnology in Neurology. Springer, New York, 2012i.

Jain TK, Morales MA, Sahoo SK, et al. Iron oxide nanoparticles for sustained delivery of anticancer agents. Mol Pharm 2005;2:194–205.

Janovjak H, Kedrov A, Cisneros DA, et al. Imaging and detecting molecular interactions of single transmembrane proteins. Neurobiology of Aging 2006;27:546–561.

Jasmin, Torres AL, Nunes HM, et al. Optimized labeling of bone marrow mesenchymal cells with superparamagnetic iron oxide nanoparticles and in vivo visualization by magnetic resonance imaging. J Nanobiotechnology 2011;9:4.
Jia N, Lian Q, Shen H, et al. Intracellular Delivery of Quantum Dots Tagged Antisense Oligodeoxynucleotides by Functionalized Multiwalled Carbon Nanotubes. Nano Lett 2007;7:2976–2980.
Jia-Ming L, Hui ZG, Aihong W, et al. Determination of human IgG by solid substrate room temperature phosphorescence immunoassay based on an antibody labeled with nanoparticles containing Rhodamine 6G luminescent molecules. Spectrochim Acta A Mol Biomol Spectrosc 2005;61:923–7.
Jo K, Dhingra DM, Odijk T, et al. A single-molecule barcoding system using nanoslits for DNA analysis. Proc Natl Acad Sci U S A 2007;104:2673–8.
John G, Zhu G, Li J, Dordick JS. Enzymatically derived sugar-containing self-assembled organogels with nanostructured morphologies. Angew Chem Int Ed Engl 2006;45:4772–5.
John R, Rezaeipoor R, Adie SG, et al. In vivo magnetomotive optical molecular imaging using targeted magnetic nanoprobes. Proc Natl Acad Sci U S A 2010;107:8085–90.
Johnston A, Caruso F. A Molecular Beacon Approach to Measuring the DNA Permeability of Thin Films. Am Chem Soc 2005;127:10014–15.
Jose GP, Santra S, Mandal SK, Sengupta TK. Singlet oxygen mediated DNA degradation by copper nanoparticles: potential towards cytotoxic effect on cancer cells. J Nanobiotechnol 2011;9:9
Kagan VE, Konduru NV, Feng W, et al. Carbon nanotubes degraded by neutrophil myeloperoxidase induce less pulmonary inflammation. Nat Nanotechnol 2010;5:354–9.
Kagan VE, Tyurina YY, Tyurin VA, et al. Direct and indirect effects of single walled carbon nanotubes on RAW 264.7 macrophages: Role of iron. Toxicol Lett 2006;165:88–100.
Kaittanis C, Naser SA, Perez JM. One-Step, Nanoparticle-Mediated Bacterial Detection with Magnetic Relaxation. Nano Lett 2007;7:380–383.
Kam N, O'Connell M, Wisdom JA, Dai H. Carbon nanotubes as multifunctional biological transporters and near-infrared agents for selective cancer cell destruction. Proc Natl Acad Sci U S A 2005;102:11600–5.
Kamau SW, Hassa PO, Steitz B, et al. Enhancement of the efficiency of non-viral gene delivery by application of pulsed magnetic field. Nucleic Acids Res 2006;34:e40.
Kang B, Mackey MA, El-Sayed MA. Nuclear targeting of gold nanoparticles in cancer cells induces DNA damage, causing cytokinesis arrest and apoptosis. J Am Chem Soc 2010;132:1517–9.
Kang C, Yuan X, Zhong Y, et al. Growth Inhibition Against Intracranial C6 Glioma Cells by Stereotactic Delivery of BCNU by Controlled Release from poly(D,L-lactic acid) Nanoparticles. Technol Cancer Res Treat 2009;8:61–70.
Kang X, Xie Y, Kniss DA. Adipose tissue model using three-dimensional cultivation of preadipocytes seeded onto fibrous polymer scaffolds. Tissue Eng 2005;11:458–68.
Kano MR, Bae Y, Iwata C, et al. Improvement of cancer-targeting therapy, using nanocarriers for intractable solid tumors by inhibition of TGF-{beta} signaling. Proc Natl Acad Sci U S A 2007;104:3460–5.
Karathanasis E, Bhavane E, Annapragada AV. Glucose-sensing pulmonary delivery of human insulin to the systemic circulation of rats. Int J Nanomedicine 2007;2:501–13.
Karchemski F, Zucker D, Barenholz Y, Regev O. Carbon nanotubes-liposomes conjugate as a platform for drug delivery into cells. J Control Release 2012 Jan 5. [Epub ahead of print].
Kasuya T, Kuroda S. Nanoparticles for human liver-specific drug and gene delivery systems: in vitro and in vivo advances. Expert Opinion on Drug Delivery 2009;6:39–52.
Kato S, Aoshima H, Saitoh Y, Miwa N. Fullerene-C60 incorporated in liposome exerts persistent hydroxyl radical-scavenging activity and cytoprotection in UVA/B-irradiated keratinocytes. J Nanosci Nanotechnol 2011;11:3814–23.

Kattumuri V, Katti K, Bhaskaran S, et al. Gum Arabic as a Phytochemical Construct for the Stabilization of Gold Nanoparticles: In Vivo Pharmacokinetics and X-ray-Contrast-Imaging Studies. Small 2007;3:333–41.

Kazakov S, Levon K. Liposome-nanogel structures for future pharmaceutical applications. Curr Pharm Des 2006;12:4713–28.

Kennedy LC, Bear AS, Young JK, et al. T cells enhance gold nanoparticle delivery to tumors in vivo. Nanoscale Research Letters 2011;6:283

Khademhosseini A, Langer R, Borenstein J, Vacanti JP. Microscale technologies for tissue engineering and biology. Proc Natl Acad Sci U S A 2006;103:2480–7.

Kim SH, Jeong JH, Lee SH, et al. Local and systemic delivery of VEGF siRNA using polyelectrolyte complex micelles for effective treatment of cancer. J Control Release 2008;129:107–16.

Kim W, Ng JK, Kunitake ME, et al. Interfacing silicon nanowires with mammalian cells. J Am Chem Soc 2007;129:7228–9.

Koch M, Schmid F, Zoete V, Meuwly M. Insulin: a model system for nanomedicine? Nanomedicine 2006;1:373–378.

Kohli AK, Alpar HO. Potential use of nanoparticles for transcutaneous vaccine delivery: effect of particle size and charge. Int J Pharm 2004;275:13–7.

Kojima C. Design of stimuli-responsive dendrimers. Expert Opin Drug Deliv 2010;7:307–19.

Kommareddy S, Amiji M. Antiangiogenic gene therapy with systemically administered sFlt-1 plasmid DNA in engineered gelatin-based nanovectors. Cancer Gene Ther 2007;14:488–98.

Kong X, Zhang W, Lockey RF, et al. Respiratory syncytial virus infection in Fischer 344 rats is attenuated by short interfering RNA against the RSV-NS1 gene. Genet Vaccines Ther 2007;5:4.

Koo OM, Rubinstein I, Onyuksel H. Camptothecin in sterically stabilized phospholipid nanomicelles: a novel solvent pH change solubilization method. J Nanosci Nanotechnol 2006;6:2996–3000.

Kopke RD, Wassel RA, Mondalek F, et al. Magnetic nanoparticles: inner ear targeted molecule delivery and middle ear implant. Audiol Neurootol 2006;11:123–33.

Kosaka N, Mitsunaga M, Bhattacharyya S, et al. Self-illuminating in vivo lymphatic imaging using a bioluminescence resonance energy transfer quantum dot nano-particle. Contrast Media Mol Imaging 2011;6:55–9.

Kose AR, Koser H. Ferrofluid mediated nanocytometry. Lab Chip 2012;12:190–6.

Koster DA, Palle K, Bot ES, et al. Antitumour drugs impede DNA uncoiling by topoisomerase I. Nature 2007;448:213–7.

Kottke PA, Degertekin FL, Fedorov AG. Scanning mass spectrometry probe: a scanning probe electrospray ion source for imaging mass spectrometry of submerged interfaces and transient events in solution. Anal Chem 2010;82:19–22.

Koria P, Yagi H, Kitagawa Y, et al. Self-assembling elastin-like peptides growth factor chimeric nanoparticles for the treatment of chronic wounds. PNAS 2011;108:1034–9.

Koziara JM, Oh JJ, Akers WS, Ferraris SP, Mumper RJ. Blood compatibility of cetyl alcohol/polysorbate-based nanoparticles. Pharm Res 2005;22:1821–8.

Kransnoslobodtsev AV, Shlyakhtenko LS, Ukraintsev E, et al. Nanomedicine and Protein Misfolding Diseases. Nanomedicine 2005;1:300–305.

Kreuter J, Gelperina S. Use of nanoparticles for cerebral cancer. Tumori 2008;94:271–7.

Kreuter J. Drug targeting with nanoparticles. Eur J Drug Metab Pharmacokinet 1994;19:253–6.

Krishnamurthy V, Monfared SM, Cornell B. Ion-Channel Biosensors—Part I: Construction, Operation, and Clinical Studies. IEEE Transactions Nanotechnology 2010;9:303–12.

Krishnan M, Monch I, Schwille P. Spontaneous stretching of DNA in a two-dimensional nanoslit. Nano Lett 2007;7:1270–5.

Kukowska-Latallo JF, Candido KA, Cao Z, et al. Nanoparticle Targeting of Anticancer Drug Improves Therapeutic Response in Animal Model of Human Epithelial Cancer. Cancer Research 2005;65:5317–5324.

Kumar A, Jena PK, Behera S, et al. Multifunctional magnetic nanoparticles for targeted delivery. Nanomedicine 2010;6:64–9.

Kumar V, Farell G, Yu S, Harrington S, et al. Cell biology of pathologic renal calcification: contribution of crystal transcytosis, cell-mediated calcification, and nanoparticles. J Investig Med 2006;54:412–24.

Kural C, Kim H, Syed S, et al. Kinesin and dynein move a peroxisome in vivo: a tug-of-war or coordinated movement? Science 2005;308:1469–72.

Ladewig K, Xu ZP, Lu GQ. Layered double hydroxide nanoparticles in gene and drug delivery. Expert Opin Drug Deliv 2009;6:907–22.

Laforge FO, Carpino J, Rotenberg SA, Mirkin MV. Electrochemical attosyringe. Proc Natl Acad Sci U S A 2007;104:11895–900.

Lai SK, O'hanlon DE, Harrold S, et al. Rapid transport of large polymeric nanoparticles in fresh undiluted human mucus. Proc Natl Acad Sci U S A 2007;104:1482–1487.

Lal R, Arnsdorf MF. Multidimensional atomic force microscopy for drug discovery: a versatile tool for defining targets, designing therapeutics and monitoring their efficacy. Life Sci 2010;86:545–62.

Lang MJ, Fordyce PM, Block SM. Combined optical trapping and single-molecule fluorescence. Journal of Biology 2003;2(1):6 (on line).

Lam R, Chen M, Pierstorff E, et al. Nanodiamond-embedded microfilm devices for localized chemotherapeutic elution. ACS Nano 2008;2:2095–102.

Lam R, Ho D. Nanodiamonds as vehicles for systemic and localized drug delivery. Expert Opin Drug Deliv 2009;6:883–95.

Lanza GM, Winter PM, Caruthers SD, et al. Nanomedicine opportunities for cardiovascular disease with perfluorocarbon nanoparticles. Nanomedicine 2006;1:321–9.

Larina IV, Evers BM, Ashitkov TV, et al. Enhancement of drug delivery in tumors by using interaction of nanoparticles with ultrasound radiation. Technol Cancer Res Treat 2005;4:217–26.

Larsen A, Kolind K, Pedersen DS, et al. Gold ions bio-released from metallic gold particles reduce inflammation and apoptosis and increase the regenerative responses in focal brain injury. Histochem Cell Biol 2008;130:681–92.

le Masne de Chermont Q, Chaneac C, Seguin J, et al. Nanoprobes with near-infrared persistent luminescence for in vivo imaging. Proc Natl Acad Sci U S A 2007;104:9266–9271.

Lee CC, Gillies ER, Fox ME. A single dose of doxorubicin-functionalized bow-tie dendrimer cures mice bearing C-26 colon carcinomas. Proc Natl Acad Sci U S A 2006;103:16649–54.

Lee SB, Koepsel R, Stolz DB, et al. Self-Assembly of Biocidal Nanotubes from a Single-Chain Diacetylene Amine Salt. J Am Chem Soc 2004;126:13400–5.

Lee Y, Park SY, Kim C, Park TG. Thermally triggered intracellular explosion of volume transition nanogels for necrotic cell death. J Control Release 2009;135:89–95.

Leevy WM, Lambert TN, Johnson JR, et al. Quantum dot probes for bacteria distinguish Escherichia coli mutants and permit in vivo imaging. Chem Commun (Camb) 2008;20:2331–3.

Lechene C, Hillion F, McMahon G, et al. High-resolution quantitative imaging of mammalian and bacterial cells using stable isotope mass spectrometry. J Biol 2006;5:20. doi:10.1186/jbiol42.

Lehn JM. Supramolecular chemistry – scope and perspectives: molecules, supermolecules, and molecular devices. Ang Chem Int Ed Engl 1988;27:89–112.

Leite EA, Grabe-Guimarães A, Guimarães HN, et al. Cardiotoxicity reduction induced by halofantrine entrapped in nanocapsule devices. Life Sci 2007;80:1327–34.

Levi N, Hantgan RR, Lively MO. C60-Fullerenes: Detection of intracellular photoluminescence and lack of cytotoxic effects. Journal of Nanobiotechnology 2006;4:14 doi:10.1186/1477-3155-4-14.

Lewis SR, Datta S, Gui M, et al. Reactive nanostructured membranes for water purification. Proc Natl Acad Sci U S A 2011;108:8577–82.

Li C, Liu H, Sun Y, et al. PAMAM nanoparticles promote acute lung injury by inducing autophagic cell death through the Akt-TSC2-mTOR signaling pathway. J Mol Cell Biol 2009;1:37–45.

Li H, Tran V, Hu Y, et al. A PEDF N-terminal peptide protects the retina from ischemic injury when delivered in PLGA nanospheres. Exp Eye Res 2006;83:824–833.

Li K, Stockman MI, Bergman DJ. Self-similar chain of metal nanospheres as an efficient nanolens. Phys Rev Lett 2003;91:227402.

Li L, Wartchow CA, Danthi SN, et al. A novel antiangiogenesis therapy using an integrin antagonist or anti-Flk-1 antibody coated 90Y-labeled nanoparticles. Int J Radiat Oncol Biol Phys 2004;58:1215–27.

Li YJ, Perkin AL, Su Y, et al. Gold nanoparticles as a platform for creating a multivalent poly-SUMO chain inhibitor that also augments ionizing radiation. Proc Natl Acad Sci U S A 2012 Mar 2;doi: 10.1073/pnas.1109131.

Liao JC, Mastali M, Gau V, et al. Use of Electrochemical DNA Biosensors for Rapid Molecular Identification of Uropathogens in Clinical Urine Specimens. Journal of Clinical Microbiology 2006;44:561–570.

Liao SS, Cui FZ, Zhang W, Feng QL. Hierarchically biomimetic bone scaffold materials: nano-HA/collagen/PLA composite. J Biomed Mater Res 2004;69B(2):158–65.

Lin C, Rinker S, Wang X, et al. In vivo cloning of artificial DNA nanostructures. Proc Natl Acad Sci U S A 2008;105:17626–31.

Lin S, Xie X, Patel MR, et al. Quantum dot imaging for embryonic stem cells. BMC Biotechnology 2007;7:67 doi:10.1186/1472-6750-7-67

Liopo AV, Stewart MP, Hudson J, et al. Biocompatibility of Native and Functionalized Single-Walled Carbon Nanotubes for Neuronal Interface. J Nanosci Nanotechnol 2006;6:1365–1374.

LiPuma JJ, Rathinavelu S, Foster BK, et al. In vitro activities of a novel nanoemulsion against Burkholderia and other multidrug-resistant cystic fibrosis-associated bacterial species. Antimicrob Agents Chemother 2009;53:249–55.

Litos IK, Ioannou PC, Christopoulos TK, et al. Genotyping of single-nucleotide polymorphisms by primer extension reaction in a dry-reagent dipstick format. Anal Chem 2007;79:395–402.

Liu J, Levine AL, Mattoon JS; et al Nanoparticles as image enhancing agents for ultrasonography. 2006 Phys Med Biol 2006;51:2179–89.

Liu Q, Douglas T, Zamponi C, et al. Comparison of in vitro biocompatibility of NanoBone® and BioOss® for human osteoblasts. Clinical Oral Implants Research 2011;22:1259–64.

Liu Q, Han M, Bao J, et al. CdSe quantum dots as labels for sensitive immunoassay of cancer biomarker proteins by electrogenerated chemiluminescence. Analyst 2011a;136:5197–203.

Liu Z, Chen K, Davis C, et al. Drug Delivery with Carbon Nanotubes for In vivo Cancer Treatment. Cancer Res 2008a;68:6652–60.

Liu Z, Davis C, Cai W, et al. Circulation and long-term fate of functionalized, biocompatible single-walled carbon nanotubes in mice probed by Raman spectroscopy. Proc Natl Acad Sci U S A 2008;105:1410–1415.

Llinas RR, Walton KD, Nakao M, et al. Neuro-vascular central nervous recording/stimulating system: Using nanotechnology probes. Journal of Nanoparticle Research 2005;7:111–127.

Loh OY, Ho AM, Rim JE, et al. Electric field-induced direct delivery of proteins by a nanofountain probe. Proc Natl Acad Sci U S A 2008;105:16438–43.

Lopez PJ, Gautier C, Livage J, Coradin T. Mimicking Biogenic Silica Nanostructures Formation. Current Nanoscience 2005;1:73–83.

Losic D, Simovic S. Self-ordered nanopore and nanotube platforms for drug delivery applications. Expert Opin Drug Deliv 2009;6:1363–81.

Lu J, Liong M, Zink JI, Tamanoi F. Mesoporous Silica Nanoparticles as a Delivery System for Hydrophobic Anticancer Drugs. Small 2007;3:1341–6.

Lu W, Gu D, Chen X, et al. Application of an oligonucleotide microarray-based nano-amplification technique for the detection of fungal pathogens. Clin Chem Lab Med 2010;48:1507–14.

Lu W, Sun Q, Wan J, She Z, Jiang XG. Cationic Albumin-Conjugated Pegylated Nanoparticles Allow Gene Delivery into Brain Tumors via Intravenous Administration. Cancer Res 2006;66: 11878–87.

M AK, Jung S, Ji T. Protein biosensors based on polymer nanowires, carbon nanotubes and zinc oxide nanorods. Sensors (Basel) 2011;11:5087–111.

Ma Y, Manolache S, Denes FS, et al. Plasma synthesis of carbon magnetic nanoparticles and immobilization of doxorubicin for targeted drug delivery. J Biomater Sci Polym Ed 2004;15:1033–49.

Ma Z, Kotaki M, Inai R, et al. Potential of Nanofiber Matrix as Tissue-Engineering Scaffolds. Tissue Engineering 2005;11:101–109.

MacDiarmid JA, Amaro-Mugridge NB, Madrid-Weiss J, et al. Sequential treatment of drug-resistant tumors with targeted minicells containing siRNA or a cytotoxic drug. Nat Biotechnol 2009;27:643–51.

MacDiarmid JA, Mugridge NB, Weiss JC, et al. Bacterially Derived 400 nm Particles for Encapsulation and Cancer Cell Targeting of Chemotherapeutics. Cancer Cell 2007;11: 431–45.

MacKay JA, Chen M, McDaniel JR, et al. Self-assembling chimeric polypeptide-doxorubicin conjugate nanoparticles that abolish tumours after a single injection. Nat Mat 2009;8:993–9.

Maeda M, Kuroda CS, Shimura T, et al. Magnetic carriers of iron nanoparticles coated with a functional polymer for high throughput bioscreening. Journal of Applied Physics 2006;99: 08H103.

Magrez A, Kasas S, Salicio V, et al. Cellular Toxicity of Carbon-Based Nanomaterials. Nano Lett 2006;6:1121–5.

Malashikhina N, Pavlov V. DNA-decorated nanoparticles as nanosensors for rapid detection of ascorbic acid. Biosens Bioelectron 2012 Jan 21. [Epub ahead of print]

Marano RJ, Toth I, Wimmer N, et al. Dendrimer delivery of an anti-VEGF oligonucleotide into the eye: a long-term study into inhibition of laser-induced CNV, distribution, uptake and toxicity. Gene Ther 2005;12:1544–50.

Marchesan S, Da Ros T, Spalluto G, et al. Anti-HIV properties of cationic fullerene derivatives. Bioorg Med Chem Lett 2005;15:3615–8.

Margolis J, McDonald J, Heuser R, et al. Systemic nanoparticle paclitaxel (nab-paclitaxel) for in-stent restenosis I (SNAPIST-I): a first-in-human safety and dose-finding study. Clin Cardiol 2007;30:165–70.

Martel J, Young JD. Purported nanobacteria in human blood as calcium carbonate nanoparticles. Proc Natl Acad Sci U S A 2008;105:5549–54.

Martel S, Mohammadi M, Felfoul O, et al. Controlled MRI-trackable Propulsion and Steering Systems for Medical Nanorobots Operating in the Human Microvasculature. Int'l J Robotics Res 2009;28:571–82.

Martin-Banderas L, Holgado MA, Venero JL, et al. Nanostructures for drug delivery to the brain. Curr Med Chem 2011;18:5303–21.

Mashino T, Shimotohno K, Ikegami N, et al. Human immunodeficiency virus-reverse transcriptase inhibition and hepatitis C virus RNA-dependent RNA polymerase inhibition activities of fullerene derivatives. Bioorg Med Chem Lett 2005;15:1107–9.

May F, Peter M, Hütten A, et al. Synthesis and Characterization of Photoswitchable Fluorescent SiO(2) Nanoparticles. Chemistry 2012;18:814–21.

McBain SC, Yiu HH, Dobson J. Magnetic nanoparticles for gene and drug delivery. Int J Nanomedicine 2008;3:169–80.

McCarthy JR, Perez M, Brückner C, Weissleder R. Polymeric Nanoparticle Preparation that Eradicates Tumors. Nano Lett 2005;5:2552–2556.

McDevitt MR, Chattopadhyay D, Kappel BJ, et al. Tumor targeting with antibody-functionalized, radiolabeled carbon nanotubes. J Nucl Med 2007;48:1180–9.

McDonagh C, Stranik O, Nooney R, Maccraith BD. Nanoparticle strategies for enhancing the sensitivity of fluorescence-based biochips. Nanomed 2009;4:645–56.

McKenzie JL, Waid MC, Shi R, Webster TJ. Decreased functions of astrocytes on carbon nanofiber materials. Biomaterials 2004;25:1309–17.

McMurray RJ, Gadegaard N, Tsimbouri PM, et al. Nanoscale surfaces for the long-term maintenance of mesenchymal stem cell phenotype and multipotency. Nat Mater 2011;10:637–44.

Mecke A, Uppuluri S, Sassanella TM, et al. Direct observation of lipid bilayer disruption by poly(amidoamine) dendrimers. Chem Phys Lipids 2004;132:3–14.
Medda R, Jakobs S, Hell SW, Bewersdorf J. 4Pi microscopy of quantum dot-labeled cellular structures. J Struct Biol 2006;156:517–23.
Mehrmohammadi M, Yoon KY, Qu M, et al. Enhanced pulsed magneto-motive ultrasound imaging using superparamagnetic nanoclusters. Nanotechnology 2011;22:045502.
Merrifield CJ, Perrais D, Zenisek D. Coupling between Clathrin-Coated-Pit Invagination, Cortactin Recruitment, and Membrane Scission Observed in Live Cells. Cell 2005:121:593–606.
Michalet X, Pinaud FF, Bentolila LA, et al. Quantum Dots for Live Cells, in Vivo Imaging, and Diagnostics. Science 2005;307:538–544.
Miller J, Lam M, Lebovitz R. Derivatized Fullerenes: A New Class of Therapeutics and Imaging Agents. Nanotechnology Law & Business 2007;4:423–431.
Mills NL, Amin N, Robinson SD, et al. Do Inhaled Carbon Nanoparticles Translocate Directly Into the Circulation in Man? Am J Respir Crit Care Med 2006;173:426–31.
Misra N, Martinez JA, Huang SC, et al. Bioelectronic silicon nanowire devices using functional membrane proteins. Proc Natl Acad Sci U S A 2009;106:13780–4.
Missirlis D, Kawamura R, Tirelli N, Hubbell JA. Doxorubicin encapsulation and diffusional release from stable, polymeric, hydrogel nanoparticles. Eur J Pharm Sci 2006;29:120–129.
Mohammadi Z, Dorkoosh FA, Hosseinkhani S, et al. In vivo transfection study of chitosan-DNA-FAP-B nanoparticles as a new non viral vector for gene delivery to the lung. Int J Pharm 2011;421:183–8.
Monica JC. FDA Labeling of Cosmetics Containing Nanoscale Materials, Nanotechnology Law & Business 2008;5:63–72.
Monteiro-Riviere NA, Nemanich RJ, Inman AO, et al. Multi-walled carbon nanotube interactions with human epidermal keratinocytes. Toxicol Lett 2005;155:377–84.
Moon JJ, Suh H, Li AV, et al. Enhancing humoral responses to a malaria antigen with nanoparticle vaccines that expand Tfh cells and promote germinal center induction. Proc Natl Acad Sci U S A 2012;109:1080–5.
Morey TE, Varshney M, Flint JA, et al. Treatment of Local Anesthetic-Induced Cardiotoxicity Using Drug Scavenging Nanoparticles. Nano Letters 2004;4:757–9.
Mortensen LJ, Oberdörster G, Pentland AP, Delouise LA. In Vivo Skin Penetration of Quantum Dot Nanoparticles in the Murine Model: The Effect of UVR. Nano Lett 2008;8:2779–87.
Mortensen MW, Sorensen PG, et al. Preparation and characterization of Boron carbide nanoparticles for use as a novel agent in T cell-guided boron neutron capture therapy. Appl Radiat Isot 2006;64:315–24.
Motwani SK, Chopra S, Talegaonkar S, et al. Chitosan-sodium alginate nanoparticles as submicroscopic reservoirs for ocular delivery: formulation, optimisation and in vitro characterisation. Eur J Pharm Biopharm 2008;68:513–25.
Mouttet B. Nanotechnology and U.S. Patents: A Statistical Analysis. Nanotechnology Law & Business 2006;3:309–16.
Mulder WJ, Koole R, Brandwijk RJ. Quantum Dots with a Paramagnetic Coating as a Bimodal Molecular Imaging Probe. Nano Lett 2006;6:1–6.
Muller-Borer BJ, Collins MC, Gunst PR, et al. Quantum dot labeling of mesenchymal stem cells. J Nanobiotechnology 2007;5:9.
Murakami T, Sawada H, Tamura G, et al. Water-dispersed single-wall carbon nanohorns as drug carriers for local cancer chemotherapy. Nanomed 2008;3:453–63.
Murugan R, Ramakrishna S. Bioresorbable composite bone paste using polysaccharide based nano hydroxyapatite. Biomaterials 2004;25:3829–35.
Murugesan S, Mousa S, Vijayaraghavan A, et al. Ionic liquid-derived blood-compatible composite membranes for kidney dialysis. J Biomed Mater Res B Appl Biomater 2006a;79:298–304.
Murugesan S, Park TJ, Yang H, et al. Blood Compatible Carbon Nanotubes - Nano-based Neoproteoglycans. Langmuir 2006;22:3461–3463.
Muthana M, Scott SD, Farrow N, et al. A novel magnetic approach to enhance the efficacy of cell-based gene therapies. Gene Therapy 2008;15:902–10.

Myc A, Majoros IJ, Thomas TP, Baker JR Jr. Dendrimer-based targeted delivery of an apoptotic sensor in cancer cells. Biomacromolecules 2007;8:13–8.

Na HB, Lee JH, An K, et al. Development of a T1 Contrast Agent for Magnetic Resonance Imaging Using MnO Nanoparticles. Angew Chem Int Ed Engl 2007;46:5397–5401.

Naik AK, Hanay MS, Hiebert WK, et al. Towards single-molecule nanomechanical mass spectrometry. Nat Nanotechnol 2009;4:445–50.

Nair BG, Nagaoka Y, Morimoto H, et al. Aptamer conjugated magnetic nanoparticles as nanosurgeons. Nanotechnology 2010;21:455102.

Nam JM, Stoeva SI, Mirkin CA. Bio-Bar-Code-Based DNA Detection with PCR-like Sensitivity. J Am Chem Soc 2004;126:5932–5933.

Nel A, Xia T, Madler L, Li N. Toxic Potential of Materials at the Nanolevel. Science 2006;311:622–27.

Nesti LJ, Li WJ, Shanti RM, et al. Intervertebral disc tissue engineering using a novel hyaluronic acid-nanofibrous scaffold (HANFS) amalgam. Tissue Eng Part A 2008;14:1527–37.

Neuwelt EA, Varallyay P, Bago AG, et al. Imaging of iron oxide nanoparticles by MR and light microscopy in patients with malignant brain tumours. Neuropathology and Applied Neurobiology 2004;30:456–71.

Nguyen TD, Tseng HR, Celestre PC, et al. A reversible molecular valve. Proc Natl Acad Sci U S A 2005;102:10029–34.

Nissenson AR, Ronco C, Pergamit G, et al. The Human Nephron Filter: Toward a Continuously Functioning, Implantable Artificial Nephron System. Blood Purification 2005;23:269–74.

Noble CO, Krauze MT, Drummond DC, et al. Novel Nanoliposomal CPT-11 Infused by Convection-Enhanced Delivery in Intracranial Tumors: Pharmacology and Efficacy. Cancer Res 2006;66:2801–6.

Nune SK, Gunda P, Thallapally PK, et al. Nanoparticles for biomedical imaging. Expert Opin Drug Deliv 2009;6:1175–94.

Nutiu R, Li Y. A DNA-Protein Nanoengine for On-Demand Release and Precise Delivery of Molecules. Angewandte Chemie 2005;44:5464–5467.

Nyk M, Kumar R, Ohulchanskyy TY, et al. High Contrast in Vitro and in Vivo Photoluminescence Bioimaging Using Near Infrared to Near Infrared Up-Conversion in Tm(3+) and Yb(3+) Doped Fluoride Nanophosphors. Nano Lett 2009;8:3834–8.

Obataya I, Nakamura C, Han SW, et al. Nanoscale Operation of a Living Cell Using an Atomic Force Microscope with a Nanoneedle. Nano Lett 2005;5:27–30.

Oberdorster E. Manufactured nanomaterials (fullerenes, C60) induce oxidative stress in the brain of juvenile largemouth bass. Environ Health Perspect 2004;112:1058–62.

Oberdorster G, Maynard A, Castranova V, et al. Principles for characterizing the potential human health effects from exposure to nanomaterials: elements of a screening strategy. Particle and Fibre Toxicology 2005;2:8

O'Brien JA, Lummis SC. Nano-biolistics: a method of biolistic transfection of cells and tissues using a gene gun with novel nanometer-sized projectiles. BMC Biotechnology 2011;11:66.

Oh S, Brammer KS, Lib YS, et al. Stem cell fate dictated solely by altered nanotube dimension. Proc Natl Acad Sci U S A 2009;106:2130–5.

Ohulchanskyy TY, Roy I, Goswami LN, Organically modified silica nanoparticles with covalently incorporated photosensitizer for photodynamic therapy of cancer. Nano Lett 2007;7: 2835–42.

Ossipov DA. Nanostructured hyaluronic acid-based materials for active delivery to cancer. Expert Opin Drug Deliv 2010;7:681–703.

Osterfeld SJ, Yu H, Gaster RS, et al. Multiplex protein assays based on real-time magnetic nanotag sensing. Proc Natl Acad Sci U S A 2008;105:20637–40.

Ozin GA, Arsenault AC, Cademartiri L. Nanochemistry: A Chemical Approach to Nanomaterials, 2nd ed, Royal Society of Chemistry, UK 2009 .

Ozin GA. Nanochemistry – synthesis in diminishing dimensions. Adv Mat 1992;4: 612–649.

Padegimas L, Kowalczyk TH, Adams S, et al. Optimization of hCFTR lung expression in mice using DNA nanoparticles. Mol Ther 2012;20:63–72.

Palakurthi S, Yellepeddi VK, Vangara KK. Recent trends in cancer drug resistance reversal strategies using nanoparticles. Expert Opin Drug Deliv 2012;9:287–301.

Paleos CM, Tziveleka LA, Sideratou Z, Tsiourvas D. Gene delivery using functional dendritic polymers. Expert Opinion on Drug Delivery 2009;6:27–38.

Paleos CM, Tsiourvas D, Sideratou Z, Tziveleka LA. Drug delivery using multifunctional dendrimers and hyperbranched polymers. Expert Opin Drug Deliv 2010;7:1387–98.

Pan D, Caruthers SD, Hu G, et al. Ligand-directed nanobialys as theranostic agent for drug delivery and manganese-based magnetic resonance imaging of vascular targets. J Am Chem Soc 2008;130:9186–7.

Panseri S, Cunha C, Lowery J, et al. Electrospun micro- and nanofiber tubes for functional nervous regeneration in sciatic nerve transections. BMC Biotechnology 2008;8:39.

Papasani MR, Wang G, Hill RA. Gold nanoparticles: the importance of physiological principles to devise strategies for targeted drug delivery. Nanomedicine 2012 Feb 1 [Epub ahead of print]

Pappas TC, Wickramanyake WM, Jan E, et al. Nanoscale Engineering of a Cellular Interface with Semiconductor Nanoparticle Films for Photoelectric Stimulation of Neurons. Nano Lett 2007;7:513–9.

Paraskar AS, Soni S, Chin KT, et al. Harnessing structure-activity relationship to engineer a cisplatin nanoparticle for enhanced antitumor efficacy. Proc Natl Acad Sci U S A 2010;107:12435–40.

Park K, Jang J, Irimia D, et al. 'Living cantilever arrays' for characterization of mass of single live cells in fluids. Lab Chip 2008;8:1034–41.

Park MV, Lankveld DP, van Loveren H, de Jong WH. The status of in vitro toxicity studies in the risk assessment of nanomaterials. Nanomed 2009;4:669–85.

Parr D. Will nanotechnology make the world a better place? Trends Biotechnol 2005;23:395–8.

Partha R, Lackey M, Hirsch A, et al. Assembly of amphiphilic C60 fullerene derivatives into nanoscale supramolecular structures. J Nanobiotech 2007;5:6.

Partlow KC, Chen J, Brant JA, et al. 19F magnetic resonance imaging for stem/progenitor cell tracking with multiple unique perfluorocarbon nanobeacons. FASEB J 2007;21:1647–54.

Patel VR, Agrawal YK. Nanosuspension: An approach to enhance solubility of drugs. J Adv Pharm Technol Res 2011;2:81–7.

Patil S, Reshetnikov S, Haldar MK, et al. Surface-Derivatized Nanoceria with Human Carbonic Anhydrase II Inhibitors and Fluorophores: A Potential Drug Delivery Device. J Phys Chem C 111 2007;24:8437–42.

Patolsky F, Zheng G, Hayden O, et al. Electrical detection of single viruses. Proc Natl Acad Sci U S A 2004;101:14017–14022.

Patolsky F, Zheng G, Lieber CM. Nanowire sensors for medicine and the life sciences. Nanomedicine 2006;1:51–65.

Patra CR, Bhattacharya R, Patra S, et al. Inorganic phosphate nanorods are a novel fluorescent label in cell biology. J Nanobiotechnology 2006;4:11.

Pauzauskie PJ, Radenovic A, Trepagnier E, et al. Optical trapping and integration of semiconductor nanowire assemblies in water. Nat Mater 2006;5:97–101.

Pavkov-Keller T, Howorka S, Keller W. The structure of bacterial S-layer proteins. Prog Mol Biol Transl Sci 2011;103:73–130.

Peng C, Palazzo RE, Wilke I. Laser intensity dependence of femtosecond near-infrared optoinjection. Phys Rev E 2007a;75:041903.

Peng W, Anderson DG, Bao Y, et al. Nanoparticulate delivery of suicide DNA to murine prostate and prostate tumors. Prostate 2007b;67:855–62.

Peters A, Veronesi B, Calderon-Garciduenas L, et al. Translocation and potential neurological effects of fine and ultrafine particles: A critical update. Particle and Fibre Toxicology 2006;3:13.

Petersen S, Soller JT, Wagner S, et al. Co-transfection of plasmid DNA and laser-generated gold nanoparticles does not disturb the bioactivity of GFP-HMGB1 fusion protein. Journal of Nanobiotechnology 2009;7:6 doi:10.1186/1477-3155-7-6.

Petry KG, Boiziau C, Dousset V, Brochet B. Magnetic resonance imaging of human brain macrophage infiltration. Neurotherapeutics 2007;4:434–42.

Pfeifer MA, Williams GJ, Vartanyants IA, et al. Three-dimensional mapping of a deformation field inside a nanocrystal. Nature 2006;442:63–6.

Pille JY, Li H, Blot E, Bertrand JR, et al. Intravenous Delivery of Anti-RhoA Small Interfering RNA Loaded in Nanoparticles of Chitosan in Mice: Safety and Efficacy in Xenografted Aggressive Breast Cancer. Hum Gene Ther 2006;17:1019–1026.

Piner RD, Zhu J, Xu F, et al. Dip-Pen" nanolithography. Science 1999;283:661–3.

Pirollo KF, Rait A, Zhou Q, et al. Materializing the Potential of Small Interfering RNA via a Tumor-Targeting Nanodelivery System. Cancer Res 2007;67:2938–43.

Posadas I, Guerra FJ, Ceña V. Nonviral vectors for the delivery of small interfering RNAs to the CNS. Nanomedicine 2010;5:1219–36.

Powell MC, Kanarek MS. Nanomaterial health effects--Part 2: Uncertainties and recommendations for the future. WMJ 2006;105:18–23.

Price RL, Haberstroh KM, Webster TJ. Enhanced functions of osteoblasts on nanostructured surfaces of carbon and alumina. Med Biol Eng Comput 2003;41:372–5.

Prudhomme RK, Saad WS. Mayer L. Paclitaxel Conjugate Block Copolymer Nanoparticle Formation by Flash NanoPrecipitation. In, Technical Proceedings of the 2006 NSTI Nanotechnology Conference and Trade Show. Nanotech 2006;2:824–826.

Puri A, Kramer-Marek G, et al. HER2-specific affibody-conjugated thermosensitive liposomes (Affisomes) for improved delivery of anticancer agents. J Liposome Res 2008;18:293–307.

Qi L, Gao X. Emerging application of quantum dots for drug delivery and therapy. Expert Opin Drug Deliv 2008;5:263–7.

Qian H, Zhu Y, Jin R. Atomically precise gold nanocrystal molecules with surface plasmon resonance. Proc Natl Acad Sci U S A 2012 Jan 3; doi: 10.1073/pnas.1115307.

Qian L, Winfree E, Bruck J. Neural network computation with DNA strand displacement cascades. Nature 2011;475:368–72.

Qian X, Peng XH, Ansari DO, et al. In vivo tumor targeting and spectroscopic detection with surface-enhanced Raman nanoparticle tags. Nat Biotech 2008;26:83–90.

Qiao Y, Chen J, Guo X, et al. Fabrication of nanoelectrodes for neurophysiology: cathodic electrophoretic paint insulation and focused ion beam milling. Nanotechnology 2005;16:1598–1602.

Qing Q, Pal SK, Tian B, et al. Nanowire transistor arrays for mapping neural circuits in acute brain slices. Proc Natl Acad Sci U S A 2010;107:1882–7.

Qu Q, Zhu Z, Wang Y, et al. Rapid and quantitative detection of Brucella by up-converting phosphor technology-based lateral-flow assay. J Microbiol Methods 2009;79:121–3.

Rabin O, Manuel Perez J, Grimm J, et al. An X-ray computed tomography imaging agent based on long-circulating bismuth sulphide nanoparticles. Nat Mater 2006; 5:118–22.

Rabinow B, Kipp J, Papadopoulos P, et al. Itraconazole IV nanosuspension enhances efficacy through altered pharmacokinetics in the rat. Int J Pharm 2007;339:251–60.

Radomski A, Jurasz P, Alonso-Escolano D, et al. Nanoparticle-induced platelet aggregation and vascular thrombosis. Br J Pharmacol 2005;146:882–93.

Rapoport N, Gao Z, Kennedy A. Multifunctional Nanoparticles for Combining Ultrasonic Tumor Imaging and Targeted Chemotherapy. JNCI 2007;99:1095–106.

Rasmussen JW, Martinez E, Louka P, Wingett DG. Zinc oxide nanoparticles for selective destruction of tumor cells and potential for drug delivery applications. 2010;7:1063–77.

Raviv U, Nguyen T, Ghafouri R, et al. Microtubule protofilament number is modulated in a stepwise fashion by the charge density of an enveloping layer. Biophys J 2007;92:278–87.

Ravizzini G, Turkbey B, Barrett T, et al. Nanoparticles in sentinel lymph node mapping. WIREs Nanomed Nanobiotechnol 2009;1:610–23.

Reddy GR, Bhojani MS, McConville P, et al. Vascular targeted nanoparticles for imaging and treatment of brain tumors. Clin Cancer Res 2006;12:6677–86.

Reddy ST, Rehor A, Schmoekel HG, et al. In vivo targeting of dendritic cells in lymph nodes with poly(propylene sulfide) nanoparticles. J Control Release 2006a;112:26–34.

Reddy MK, Wu L, Kou W, et al. Superoxide Dismutase-Loaded PLGA Nanoparticles Protect Cultured Human Neurons Under Oxidative Stress. Appl Biochem Biotechnol 2008;151:565–77.

Renwick LC, Brown D, Clouter A, Donaldson K. Increased inflammation and altered macrophage chemotactic responses caused by two ultrafine particle types. Occup Environ Med 2004;61:442–7.

Resnik DB, Tinkle SS. Ethics in nanomedicine. Nanomedicine 2007;2:345–50.

Riviere CN, Patronik NA, Zenati MA. Prototype epicardial crawling device for intrapericardial intervention on the beating heart. Heart Surg Forum 2004;7:E639–43.

Rivkin I, Cohen K, Koffler J, et al. Paclitaxel-clusters coated with hyaluronan as selective tumor-targeted nanovectors. Biomaterials 2010;31:7106–14.

Rocha R, Zhou Y, Kundu S, et al. In vivo observation of gold nanoparticles in the central nervous system of Blaberus discoidalis. J Nanobiotechnol 2011;9:5.

Roco MC. Nanotechnology: convergence with modern biology and medicine. Curr Opin Biotechnol 2003;14:337–46.

Robertson JW, Rodrigues CG, Stanford VM, et al. Single-molecule mass spectrometry in solution using a solitary nanopore. PNAS 2007;104:8207–11.

Rogueda P, Traini D. The nanoscale in pulmonary delivery: parts I and II. Expert Opin Drug Deliv 2007;4:595–620.

Rosen Y, Elman. NM Carbon nanotubes in drug delivery: focus on infectious diseases. Expert Opinion on Drug Delivery 2009;6:517–30.

Rosenholm JM, Mamaeva V, Sahlgren C, Lindén M. Nanoparticles in targeted cancer therapy: mesoporous silica nanoparticles entering preclinical development stage. Nanomedicine (Lond) 2012;7:111–20.

Rosi NL, Giljohann DA, Thaxton CS, et al. Oligonucleotide-modified gold nanoparticles for intracellular gene regulation. Science 2006;312:1027–30.

Ross JL, Wallace K, Shuman H, et al. Processive bidirectional motion of dynein-dynactin complexes in vitro. Nat Cell Biol 2006;8:562–70.

Ross SA, Srinivas PR, Clifford AJ, et al. New technologies for nutrition research. J Nutr 2004;134:681–5.

Rouse JG, Yang J, Ryman-Rasmussen JP, et al. Effects of mechanical flexion on the penetration of fullerene amino acid-derivatized peptide nanoparticles through skin. Nano Lett 2007;7:155–60.

Roy S, Glueckert R, Johnston AH, et al. Strategies for drug delivery to the human inner ear by multifunctional nanoparticles. Nanomedicine (Lond) 2012;7:55–63.

Ryan JJ, Bateman HR, Stover A, et al. Fullerene nanomaterials inhibit the allergic response. J Immunol 2007;179:665–72.

Rytting E, Nguyen J, Wang X, Kissel T. Biodegradable polymeric nanocarriers for pulmonary drug delivery. Expert Opin Drug Deliv 2008;5:629–39.

Rzigalinski BA, Meehan K, Davis RM, et al. Radical nanomedicine. Nanomedicine 2006;1:399–412.

Sacconi L, Tolic-Norrelykke IM, Antolini R, Pavone FS. Combined intracellular three-dimensional imaging and selective nanosurgery by a nonlinear microscope. J Biomed Opt 2005;10:14002.

Safinya CR, Raviv U, Needleman DJ, et al. Nanoscale assembly in biological systems: from neuronal cytoskeletal proteins to curvature stabilizing lipids. Adv Mater 2011;23:2260–70.

Santhakumaran LM, Thomas T, Thomas TJ. Enhanced cellular uptake of a triplex-forming oligonucleotide by nanoparticle formation in the presence of polypropylenimine dendrimers. Nucleic Acids Research 2004;32:2102–2112.

Saraste A, Nekolla SG, Schwaiger M. Cardiovascular molecular imaging: an overview. Cardiovasc Res 2009;83:643–52.

Sato K, Hosokawa K, Maeda M. Colorimetric biosensors based on DNA-nanoparticle conjugates. Anal Sci 2007;23:17–20.

Sayes C, et al. The differential cytotoxicity of water-soluble fullerenes. Nano Lett 2004;4:881–7.

Sayes CM, Gobin AM, Ausman KD, et al. Nano-C(60) cytotoxicity is due to lipid peroxidation. Biomaterials 2005;26:7587–95.

Sayes CM, Liang F, Hudson JL, et al. Functionalization density dependence of single-walled carbon nanotubes cytotoxicity in vitro. Toxicol Lett 2006;161:135–42.

Scarberry KE, Dickerson EB, McDonald JF, et al. Magnetic Nanoparticle–Peptide Conjugates for in Vitro and in Vivo Targeting and Extraction of Cancer Cells. J Am Chem Soc 2008;130:10258–62.

Scheerlinck JP, Gloster S, Gamvrellis A, et al. Systemic immune responses in sheep, induced by a novel nano-bead adjuvant. Vaccine 2006;24:1124–31.

Schluep T, Hwang J, Hildebrandt IJ, et al. Pharmacokinetics and tumor dynamics of the nanoparticle IT-101 from PET imaging and tumor histological measurements. Proc Natl Acad Sci U S A 2009;106:11394–9.

Schmieder AH, Winter PM, Caruthers SD, et al. Molecular MR imaging of melanoma angiogenesis with anb3-targeted paramagnetic nanoparticles. Magnetic Resonance in Medicine 2005;53:621–627.

Schroeder A, Heller DA, Winslow MM, et al. Treating metastatic cancer with nanotechnology. Nat Rev Cancer 2012;12:39–50.

Schubert D, Dargusch R, Raitano J, Chan S. Cerium and yttrium oxide nanoparticles are neuroprotective. Biochem Biophys Res Commun 2006;342:86–91.

Seeman NC. An overview of structural DNA nanotechnology. Mol Biotechnol 2007;37:246–57.

Seil JT, Webster TJ. Electrically active nanomaterials as improved neural tissue regeneration scaffolds. Wiley Interdiscip Rev Nanomed Nanobiotechnol 2010;2:635–47.

Sekula-Neuner S, Maier J, Oppong E, et al. Allergen Arrays for Antibody Screening and Immune Cell Activation Profiling Generated by Parallel Lipid Dip-Pen Nanolithography. Small 2012 Jan 26;doi: 10.1002/smll.201101694.

Sengupta S, Eavarone D, Capila I, et al. Temporal targeting of tumour cells and neovasculature with a nanoscale delivery system. Nature 2005;436:568–572.

Serohijos AW, Chen Y, Ding F, et al A structural model reveals energy transduction in dynein. Proc Natl Acad Sci U S A 2006;103:18540–5.

Sha MY, Penn S, Freeman G, Doering WE. Detection of Human Viral RNA via a Combined Fluorescence and SERS Molecular Beacon Assay. NanoBioTechnology 2007;3:23–30.

Sha MY, Walton ID, Norton SM, et al. Multiplexed SNP genotyping using nanobarcode particle technology. Anal Bioanal Chem 2006;384:658–66.

Shah BS, Clark PA, Moioli EK, et al. Labeling of mesenchymal stem cells by bioconjugated quantum dots. Nano Lett 2007;7:3071–9.

Shakeel F, Baboota S, Ahuja A, et al. Skin permeation mechanism and bioavailability enhancement of celecoxib from transdermally applied nanoemulsion. J Nanobiotechnol 2008;6:8.

Shanmukh S, Jones L, Driskell J, et al. Rapid and Sensitive Detection of Respiratory Virus Molecular Signatures Using a Silver Nanorod Array SERS Substrate. Nano Lett 2006;6:2630–6.

Shea TB, Ortiz D, Nicolosi RJ, et al. Nanosphere-mediated delivery of vitamin E increases its efficacy against oxidative stress resulting from exposure to amyloid beta. J Alzheimer's Dis 2005;7:297–301.

Shen X, Ming A, Li X, Zhou Z, Song B. Nanobacteria: A Possible Etiology for Type III Prostatitis. J Urol 2010;184:364–9.

Shen Y, Nakajima M, Yang Z, et al. Design and characterization of nanoknife with buffering beam for in situ single-cell cutting. Nanotechnology 2011;22:305701.

Shi Y, Kim S, Huff TB, et al. Effective repair of traumatically injured spinal cord by nanoscale block copolymer micelles. Nat Nanotech 2010;5:80–7.

Shih WM, Quispe JD, Joyce GF. A 1.7-kilobase single-stranded DNA that folds into a nanoscale octahedron. Nature 2004;427:618–21.

Shilo M, Reuveni T, Motiei M, Popovtzer R. Nanoparticles as computed tomography contrast agents: current status and future perspectives. Nanomedicine 2012;7:257–69.

Shimkunas RA, Robinson E, Lam R, et al. Nanodiamond-insulin complexes as pH-dependent protein delivery vehicles. Biomaterials 2009;30:5720–8.

Shin M. In Vivo Bone Tissue Engineering Using Mesenchymal Stem Cells on a Novel Electrospun Nanofibrous Scaffold. Tissue Engineering 2004;10:33–41.

Shukla GC, Haque F, Tor Y, et al. A boost for the emerging field of RNA nanotechnology. ACS Nano 2011;5:3405–18.

Siegwart DJ, Whitehead KA, Nuhn L, et al. Combinatorial synthesis of chemically diverse core-shell nanoparticles for intracellular delivery. Proc Natl Acad Sci U S A 2011;108:12996–3001.

Silva GA. Nanotechnology applications and approaches for neuroregeneration and drug delivery to the central nervous system. Ann NY Acad Sci 2010;1199:221–30.

Simberg D, Duza T, Park JH, et al. Biomimetic amplification of nanoparticle homing to tumors. Proc Natl Acad Sci U S A 2007;104:932–6.

Simon JA, ESTRASORB Study Group. Estradiol in micellar nanoparticles: the efficacy and safety of a novel transdermal drug-delivery technology in the management of moderate to severe vasomotor symptoms. Menopause 2006;13:222–31.

Singer A, Wanunu M, Morrison W, et al. Nanopore based sequence specific detection of duplex DNA for genomic profiling. Nano Lett 2010;10:738–42.

Singer EM, Smith SS. Nucleoprotein Assemblies for Cellular Biomarker Detection. Nano Lett 2006;6:1184–1189.

Singh P, Destito G, Schneemann A, Manchester M. Canine parvovirus-like particles, a novel nanomaterial for tumor targeting. J Nanobiotechnology 2006a;4:2.

Singh R, Pantarotto D, Lacerda L, et al. Tissue biodistribution and blood clearance rates of intravenously administered carbon nanotube radiotracers. Proc Natl Acad Sci U S A 2006;103:3357–62.

Sirsi SR, Schray RC, Wheatley MA, Lutz GJ. Formulation of polylactide-co-glycolic acid nanospheres for encapsulation and sustained release of poly(ethylene imine)-poly(ethylene glycol) copolymers complexed to oligonucleotides. Journal of Nanobiotechnology 2009;7:1.

Smalley RE. In Bartlett RJ (ed) Comparison of Ab Initio Quantum Chemistry with Experiments for Small Molecules. D. Riedel, Boston, 1985.

So MK, Xu C, Loening AM, Gambhir SS, Rao J. Self-illuminating quantum dot conjugates for in vivo imaging. Nat Biotechnol 2006; 24:339–43.

Soares AF, Carvalho Rde A, Veiga F. Oral administration of peptides and proteins: nanoparticles and cyclodextrins as biocompatible delivery systems. Nanomedicine (Lond) 2007;2:183–202.

Soman NR, Baldwin SL, Hu G, et al. Molecularly targeted nanocarriers deliver the cytolytic peptide melittin specifically to tumor cells in mice, reducing tumor growth. J Clin Invest 2009; 119:2830–42.

Souza GR, Christianson DR, Staquicini FI, et al. Networks of gold nanoparticles and bacteriophage as biological sensors and cell-targeting agents. Proc Natl Acad Sci U S A 2006;103:1215–20.

Spada C, Hassan C, Sturniolo GC, et al. Literature review and recommendations for clinical application of Colon Capsule Endoscopy. Dig Liver Dis 2011;43:251–8.

Speshock JL, Murdock RC, Braydich-Stolle ILK, et al. Interaction of silver nanoparticles with Tacaribe virus. J Nanobiotechnol 2010;8:19

Sprintz M, Benedetti C, Ferrari M. Applied nanotechnology for the management of breakthrough cancer pain. Minerva Anestesiol 2005;71:419–23.

Spyratou E, Mourelatou EA, Makropoulou M. Atomic force microscopy: a tool to study the structure, dynamics and stability of liposomal drug delivery systems. Expert Opin Drug Deliv 2009;6:305–317.

Stabenfeldt SE, Irons HR, Laplaca MC. Stem cells and bioactive scaffolds as a treatment for traumatic brain injury. Curr Stem Cell Res Ther 2011;6:208–20.

Staii C, Johnson AT, Chen M, et al. DNA-Decorated Carbon Nanotubes for Chemical Sensing. Nano Lett 2005;5:1774–1778.

Stampfl A, Maier M, Radykewicz R, et al. Langendorff heart: a model system to study cardiovascular effects of engineered nanoparticles. ACS Nano 2011;5:5345–53.

Star A, Tu E, Niemann J, et al. Label-free detection of DNA hybridization using carbon nanotube network field-effect transistors. PNAS 2006;103:921–6.

Stolz M, Gottardi R, Raiteri R, et al. Early detection of aging cartilage and osteoarthritis in mice and patient samples using atomic force microscopy. Nat Nanotechnol 2009;4:186–92.

Stolz M, Raiteri R, Daniels AU, et al. Dynamic elastic modulus of porcine articular cartilage determined at two different levels of tissue organization by indentation-type atomic force microscopy. Biophys J 2004;86:3269–83.

Stover TC, Sharma A, Robertson GP, Kester M. Systemic delivery of liposomal short-chain ceramide limits solid tumor growth in murine models of breast adenocarcinoma. Clin Cancer Res 2005;11:3465–74.

Straub JA, Chickering DE, Lovely JC, et al. Intravenous hydrophobic drug delivery: a porous particle formulation of paclitaxel (AI-850). Pharm Res 2005;22:347–55.

Strickland CH. Nano-Based Drugs and Medical Devices: FDA's Track Record. Nanotechnology Law & Business 2007;4:511–21.

Subramanian H, Roy HK, Pradhan P, et al. Nanoscale cellular changes in field carcinogenesis detected by partial wave spectroscopy. Cancer Res 2009;69:5357–63.

Suh WH, Suslick KS, Stucky GD, Suh YH. Nanotechnology, nanotoxicology, and neuroscience. Prog Neurobiol 2009;87:133–70.

Suhr J, Victor P, Ci L, et al. Fatigue resistance of aligned carbon nanotube arrays under cyclic compression. Nature Nanotechnology 2006;2:417–421.

Sukhanova A, Devy J, Venteo L, et al. Biocompatible fluorescent nanocrystals for immunolabeling of membrane proteins and cells. Anal Biochem 2004;324:60–7.

Sutton D, Nasongkla N, Blanco E, Gao J. Functionalized Micellar Systems for Cancer Targeted Drug Delivery. Pharmaceutical Research 2007;24:1029–1046.

Swai H, Semete B, Kalombo L, et al. Nanomedicine for respiratory diseases. Wiley Interdiscip Rev Nanomed Nanobiotechnol 2009;1:255–63.

Swami A, Kurupati RK, Pathak A, et al. Biochemical and biophysical research communications 2007;362:835.

Swanson SD, Kukowska-Latallo JF, Patri AK, et al. Targeted gadolinium-loaded dendrimer nanoparticles for tumor-specific magnetic resonance contrast enhancement. Int J Nanomedicine 2008;3:201–10.

Swarnakar NK, Jain AK, Singh RP, et al. Oral bioavailability, therapeutic efficacy and reactive oxygen species scavenging properties of coenzyme Q10-loaded polymeric nanoparticles. Biomaterials 2011;32:6860–74.

Sweeney LG, Wang Z, Loebenberg R, et al. Spray-freeze-dried lipociproflaxacin powder for inhaled aerosol drug delivery. Int J Pharm 2005;305:180–5.

Sykova E, Jendelova P. In vivo tracking of stem cells in brain and spinal cord injury. Prog Brain Res 2007;161C:367–383.

Tabuchi M, Ueda M, Kaji N, Yamasaki Y, et al. Nanospheres for DNA separation chips. Nat Biotechnol 2004;22:337–340.

Tada H, Higuchi H, Wanatabe TM, Ohuchi N: In vivo real-time tracking of single quantum dots conjugated with monoclonal anti-HER2 antibody in tumors of mice. Cancer Res 2007;67:1138–1144.

Tagaram HR, Divittore NA, Barth BM, et al. Nanoliposomal ceramide prevents in vivo growth of hepatocellular carcinoma. Gut 2011;60:695–701.

Tai K, Dao M, Suresh S, et al. Nanoscale heterogeneity promotes energy dissipation in bone. Nat Mater 2007;6:454–62.

Tamerler C, Sarikaya M. Molecular biomimetics: nanotechnology and bionanotechnology using genetically engineered peptides. Philos Transact A Math Phys Eng Sci 2009;367:1705–26.

Tan A, De La Peña H, Seifalian AM. The application of exosomes as a nanoscale cancer vaccine. Int J Nanomedicine 2010;5:889–900.

Tang W, Li Q, Gao S, Shang JK. Arsenic (III,V) removal from aqueous solution by ultrafine α-Fe2O3 nanoparticles synthesized from solvent thermal method. J Hazard Mater 2011a;192:131–8

Tang Y, Heaysman CL, Willis S, Lewis AL. Physical hydrogels with self-assembled nanostructures as drug delivery systems. Expert Opin Drug Deliv 2011;8:1141–59.

Tang BC, Dawson M, Lai SK, et al. Biodegradable polymer nanoparticles that rapidly penetrate the human mucus barrier. PNAS 2010;106;19268–73.

Tang Z, Zhang Z, Wang Y, et al. Self-Assembly of CdTe Nanocrystals into Free-Floating Sheets. Science 2006;314:274–8.

Thakor AS, Luong R, Paulmurugan R, et al. The fate and toxicity of Raman-active silica-gold nanoparticles in mice. Sci Transl Med 2011;3:79ra33.

Thaxton CS, Elghanian R, Thomas AD, et al. Nanoparticle-based bio-barcode assay redefines "undetectable" PSA and biochemical recurrence after radical prostatectomy. Proc Natl Acad Sci USA 2009;106:18437–42.

Thomas CR, Ferris DP, Lee JH, et al. Noninvasive remote-controlled release of drug molecules in vitro using magnetic actuation of mechanized nanoparticles. J Am Chem Soc 2010;132:10623–5.

Thomas K, Aguar P, Kawasaki H, et al. Research Strategies for Safety Evaluation of Nanomaterials, Part VIII: International Efforts to Develop Risk-Based Safety Evaluations for Nanomaterials. Toxicol Sci 2006a;92:23–32.

Thomas T, Thomas K, Sadrieh N, et al. Research strategies for safety evaluation of nanomaterials, part VII: evaluating consumer exposure to nanoscale materials. Toxicol Sci 2006;91:14–9.

Thompson DG, McKenna EO, Pitt A, Graham D. Microscale mesoarrays created by dip-pen nanolithography for screening of protein-protein interactions. Biosens Bioelectron 2011;26:4667–73.

Tok JB, Chuang FY, Kao MC, et al. Metallic Striped Nanowires as Multiplexed Immunoassay Platforms for Pathogen Detection. Angewandte Chemie International 2006;45:6900–4.

Tomalia DA, Baker H, Dewald J, et al. A new class of polymers: starburst-dendritic macromolecules. Polym J 1985;17:117–32.

Tomalia DA, Reyna LA, Svenson S. Dendrimers as multi-purpose nanodevices for oncology drug delivery and diagnostic imaging. Biochem Soc Trans 2007;35(Pt 1):61–7.

Torne S, Darandale S, Vavia P, et al. Cyclodextrin-based nanosponges: effective nanocarrier for Tamoxifen delivery. Pharm Dev Technol 2012 Jan 12. [Epub ahead of print].

Torrecillas R, Moya JS, Diaz LA. Nanotechnology in joint replacement. WIREs Nanomed Nanobiotechnol 2009;1:540–52.

Tosi G, Costantino L, Ruozi B, et al. Polymeric nanoparticles for the drug delivery to the central nervous system. Expert Opinion on Drug Delivery 2008;5:155–74.

Townsend-Nicholson A, Jayasinghe SN. Cell electrospinning: a unique biotechnique for encapsulating living organisms for generating active biological microthreads/scaffolds. Biomacromolecules 2006;7:3364–9.

Truong-Le VL, August JT, Leong KW. Controlled gene delivery by DNA-gelatin nanospheres. Hum Gene Ther 1998;9:1709–17.

Tsurumoto T, Matsumoto T, Yonekura A, Shindo H. Nanobacteria-like particles in human arthritic synovial fluids. J Proteome Res 2006;5:1276–8.

Tung NH, Chikae M, Ukita Y, et al. Sensing technique of silver nanoparticles as labels for immunoassay using liquid electrode plasma atomic emission spectrometry. Anal Chem 2012;84:1210–3.

Uehara H, Kunitomi Y, Ikai A, Osada T. mRNA detection of individual cells with the single cell nanoprobe method compared with in situ hybridization. J Nanobiotechnol 2007;5:7.

Uwatoku T, Shimokawa H, Abe K, et al. Application of nanoparticle technology for the prevention of restenosis after balloon injury in rats. Circ Res 2003;92:e62–9.

Valley JK, Ohta AT, Hsu HY, et al. Optoelectronic Tweezers as a Tool for Parallel Single-Cell Manipulation and Stimulation. IEEE Trans Biomed Circuits Syst 2009;3:424–431.

Van Bockstaele F, Holz JB, Revets H. The development of nanobodies for therapeutic applications. Curr Opin Investig Drugs 2009;10:1212–24.

van Vlerken LE, Duan Z, Seiden MV, Amiji MM. Modulation of intracellular ceramide using polymeric nanoparticles to overcome multidrug resistance in cancer. Cancer Res 2007;67:4843–50.

Vandervoort J, Ludwig A. Ocular drug delivery: nanomedicine applications. Nanomedicine 2007;2:11–21.

Vega E, Egea MA, Valls O, et al. Flurbiprofen loaded biodegradable nanoparticles for ophthalmic administration. J Pharm Sci 2006;95:2393–405.

Veiseh O, Sun C, Fang C, et al. Specific targeting of brain tumors with an optical/magnetic resonance imaging nanoprobe across the blood–brain barrier. Cancer Res 2009;69:6200–7.

Venkatesan N, Yoshimitsu J, Ito Y, et al. Liquid filled nanoparticles as a drug delivery tool for protein therapeutics. Biomaterials 2005;26:7154–63.

Verellen N, Sonnefraud Y, Sobhani H, et al. Fano Resonances in Individual Coherent Plasmonic Nanocavities. Nano Letters 2009;9:1663–7.

Verma A, Uzun O, Hu Y, et al. Surface-structure-regulated cell-membrane penetration by monolayer-protected nanoparticles. Nat Mater 2008;7:588–95.

Vieira DB, Carmona-Ribeiro AM. Cationic nanoparticles for delivery of amphotericin B: preparation, characterization and activity in vitro. Journal of Nanobiotechnology 2008;6:6.

Vij N. Nano-based theranostics for chronic obstructive lung diseases: challenges and therapeutic potential. Expert Opin Drug Deliv 2011;8:1105–9.

Vinogradov SV. Polymeric nanogel formulations of nucleoside analogs. Expert Opin Drug Deliv 2007;4:5–17.

Vo-Dinh T, Zhang Y. Single-cell monitoring using fiberoptic nanosensors. Wiley Interdiscip Rev Nanomed Nanobiotechnol 2011;3:79–85.

Wagner DE, Phillips CL, Ali WM, et al. Toward the development of peptide nanofilaments and nanoropes as smart materials. Proc Natl Acad Sci U S A 2006;102:12656–61.

Wagner E. Programmed drug delivery: nanosystems for tumor targeting. Expert Opinion on Biological Therapy 2007;7:587–593.

Walrant A, Bechara C, Alves ID, Sagan S. Molecular partners for Nanomedicine interaction and cell internalization of cell-penetrating peptides: how identical are they? Nanomedicine 2012;7:133–43.

Wang H, Gu L, Lin Y, et al. Unique Aggregation of Anthrax (Bacillus anthracis) Spores by Sugar-Coated Single-Walled Carbon Nanotubes. J Am Chem Soc 2006;128:13364–5.

Wang P, Jia L, Sanders BG, Kline K. Liposomal or nanoparticle alpha-TEA reduced 66cl-4 murine mammary cancer burden and metastasis. Drug Deliv 2007a;14:497–505.

Wang S, Liu K, Liu J, et al. Highly efficient capture of circulating tumor cells by using nanostructured silicon substrates with integrated chaotic micromixers. Angew Chem Int Ed Engl 2011;50:3084–8.

Wang WU, Chen C, Lin KH, et al. Label-free detection of small-molecule-protein interactions by using nanowire nanosensors. Proc Natl Acad Sci U S A 2005a;102:3208–12.

Wang H, Huff TB, Zweifel DA, et al. In vitro and in vivo two-photon luminescence imaging of single gold nanorods. PNAS 2005c;102:15752–6.

Wang H, Huff TB, Zweifel DA, et al. In vitro and in vivo two-photon luminescence imaging of single gold nanorods. Proc Natl Acad Sci U S A 2005;102:15752–6.

Wang X, Hofmann O, Das R, et al. Integrated thin-film polymer/fullerene photodetectors for on-chip microfluidic chemiluminescence detection. Lab Chip 2007;7:58–63.

Wang X, Inapagolla R, Kannan S, et al. Synthesis, characterization, and in vitro activity of dendrimer-streptokinase conjugates. Bioconjug Chem 2007a;18:791–9.

Wang X, Lou X, Wang Y, et al. QDs-DNA nanosensor for the detection of hepatitis B virus DNA and the single-base mutants. Biosens Bioelectron 2010a;25:1934–40.

Wang X, Ramström O, Yan M. Dye-doped silica nanoparticles as efficient labels for glycans. Chem Commun (Camb) 2011a;47:4261–3.

Warheit DB, Laurence BR, Reed KL, et al. Comparative pulmonary toxicity assessment of single-wall carbon nanotubes in rats. Toxicol Sci 2004;77:117–25.

Warheit DB, Webb TR, Colvin VL, et al. Pulmonary bioassay studies with nanoscale and fine-quartz particles in rats: toxicity is not dependent upon particle size but on surface characteristics. Toxicol Sci 2007;95:270–80.

Watson B, Friend J, Yeo L. Piezoelectric ultrasonic resonant motor with stator diameter less than 250 μm: the Proteus motor. J Micromech Microeng 2009;19:022001.

Webber MJ, Kessler JA, Stupp SI. Emerging peptide nanomedicine to regenerate tissues and organs. J Intern Med 2010;267:71–88.

Wei G, Bhushan B, Torgerson PM, et al. Nanomechanical characterization of human hair using nanoindentation and SEM. Ultramicroscopy 2005;105:248–66.

Wei G, Jin Q, Giannobile WV, Ma PX. Nano-fibrous scaffold for controlled delivery of recombinant human PDGF-BB. J Control Release 2006;112:103–10.

Weigum SE, Floriano PN, Redding SW, et al. Nano-Bio-Chip Sensor Platform for Examination of Oral Exfoliative Cytology. Cancer Prev Res 2010;3;518–28.

Weissig V, Boddapati SV, Jabr L, D'Souza GG. Mitochondria-specific nanotechnology. Nanomedicine 2007;2:275–85.

Weissleder R, Kelly K, Sun EY, et al. Cell-specific targeting of nanoparticles by multivalent attachment of small molecules. Nat Biotechnol 2005;23:1418–23.

Weng KC, Stålgren JJ, Duval DJ, et al. Fluid biomembranes supported on nanoporous aerogel/xerogel substrates. Langmuir 2004;20:7232–9.

Wickline SA, Neubauer AM, Winter P, et al. Applications of nanotechnology to atherosclerosis, thrombosis, and vascular biology. Arterioscler Thromb Vasc Biol 2006;26:435–41.

Williams DN, Ehrman SH, Holoman TR. Evaluation of the microbial growth response to inorganic nanoparticles. J Nanobiotechnol 2006;4:3.

Winter PM, Cai K, Caruthers SD, et al. Emerging nanomedicine opportunities with perfluorocarbon nanoparticles. Expert Rev Med Devices 2007;4:137–45.

Wong-Ekkabut J, Baoukina S, Triampo W, et al. Computer simulation study of fullerene translocation through lipid membranes. Nat Nanotechol 2008;3:363–8.

Wu HB, Huo DF, Jiang XG. Advances in the study of lipid-based cubic liquid crystalline nanoparticles as drug delivery system. Yao Xue Xue Bao 2008;43:450–5.

Wu W, Wieckowski S, Pastorin G, et al. Targeted Delivery of Amphotericin B to Cells by Using Functionalized Carbon Nanotubes. Angew Chem Int Ed Engl 2005;44:6358–62.

Wu Y, SefahK, Liu H, et al. DNA aptamer–micelle as an efficient detection/delivery vehicle toward cancer cells. Proc Natl Acad Sci U S A 2009;107:5–10.

Xenariou S, Griesenbach U, Ferrari S. Using magnetic forces to enhance non-viral gene transfer to airway epithelium in vivo. Gene Ther 2006;13:1445–52.

Xi N, Fung CK, Yang R, et al. Atomic force microscopy as nanorobot. Methods Mol Biol 2011;736:485–503.

Xiao Y, Lubin AA, Baker BR, et al. Single-step electronic detection of femtomolar DNA by target-induced strand displacement in an electrode-bound duplex. PNAS 2006;103: 16677–80.

Xie H, Li YF, Kagawa HK, et al. An intrinsically fluorescent recognition ligand scaffold based on chaperonin protein and semiconductor quantum-dot conjugates. Small 2009;5:1036–42.

Xing Y, Chaudry Q, Shen C, et al. Bioconjugated quantum dots for multiplexed and quantitative immunohistochemistry. Nature Protocols 2007;2:1152–65.

Xing Y, Dai L. Nanodiamonds for nanomedicine. Nanomed 2009;4:207–18.

Xue Q, Wang L, Jiang W.A versatile platform for highly sensitive detection of protein: DNA enriching magnetic nanoparticles based rolling circle amplification immunoassay. Chem Commun (Camb) 2012 Feb 3 [Epub ahead of print].

Yamagishi K, Onuma K, Suzuki T, Okada F, Tagami J, Otsuki M, Senawangse P. Materials chemistry: a synthetic enamel for rapid tooth repair. Nature 2005;433:819.

Yang AH, Moore SD, Schmidt BS, et al. Optical manipulation of nanoparticles and biomolecules in sub-wavelength slot waveguides. Nature 2009;457:71–5.

Yang B, Zhou G, Huang LL. PCR-free MDR1 polymorphism identification by gold nanoparticle probes. Anal Bioanal Chem 2010;397:1937–45.

Yang J, Wang K, Driver J, et al. The use of fullerene substituted phenylalanine amino acid as a passport for peptides through cell membranes. Org Biomol Chem 2007;5:260–6.

Yanik MF, Cinar H, Cinar HN, et al. Neurosurgery: Functional regeneration after laser axotomy. Nature 2004;432:822.

Yao L, Zhao X, Li Q, et al. In vitro and in vivo evaluation of camptothecin nanosuspension: A novel formulation with high antitumor efficacy and low toxicity. Int J Pharm 2012;423:586–8.

Yeh TK, Lu Z, Wientjes MG, Au JL. Formulating paclitaxel in nanoparticles alters its disposition. Pharm Res 2005;22:867–74.

Yen Y, Synold T, Schluep T, et al. First-in-human phase I trial of a cyclodextrin-containing polymer-camptothecin nanoparticle in patients with solid tumors. Journal of Clinical Oncology 2007 June 20 ASCO Annual Meeting Proceedings Part I;25(18S):14078.

Yokoi H, Kinoshita T, Zhang S. Dynamic reassembly of peptide RADA16 nanofiber scaffold. Proc Natl Acad Sci U S A 2005;102:8414–9.

Yoo SI, Yang M, Brender JR, et al. Inhibition of Amyloid Peptide Fibrillation by Inorganic Nanoparticles: Functional Similarities with Proteins. Angew Chem Int Ed Engl 2011;50:5110–5.

Yoshino T, Maeda Y, Matsunag T. Bioengineering of bacterial magnetic particles and their applications in biotechnology. Recent Pat Biotechnol 2010;4:214–25.

Yu KN, Lee SM, Han JY, Park H, et al. Multiplex Targeting, Tracking, and Imaging of Apoptosis by Fluorescent Surface Enhanced Raman Spectroscopic Dots. Bioconjug Chem 2007;18:1155–62.

Yu W, Pirollo KF, Rait A, et al. A sterically stabilized immunolipoplex for systemic administration of a therapeutic gene. Gene Ther 2004;11:1434–40.

Yuan X, Li H, Yuan Y. Preparation of cholesterol-modified chitosan self-aggregated nanoparticles for delivery of drugs to ocular surface. Carbohydr Polym 2006;65;337–45.

Yuan X, Naguib S, Wu Z. Recent advances of siRNA delivery by nanoparticles. Expert Opin Drug Deliv 2011;8:521–36.

Zahavy E, Ber R, Gur D, et al. Application of nanoparticles for the detection and sorting of pathogenic bacteria by flow-cytometry. Adv Exp Med Biol 2012;733:23–36.

Zanello LP, Zhao B, Hu H, Haddon RC. Bone cell proliferation on carbon nanotubes. Nano Lett 2006;6:562–7.

Zavaleta CL, Smith BR, Walton I, et al. Multiplexed imaging of surface enhanced Raman scattering nanotags in living mice using noninvasive Raman spectroscopy. Proc Natl Acad Sci U S A 2009;106:13511–6

Zhang Y, Sun C, Kohler N, et al. Self-Assembled Coatings on Individual Monodisperse Magnetite Nanoparticles for Efficient Intracellular Uptake. Biomedical Microdevices 2004;6:33–40.

Zhang CY, Yeh HC, Kuroki MT, Wang TH. Single-quantum-dot-based DNA nanosensor. Nature Materials 2005;4:826–831.

Zhang L, Granick S. How to Stabilize Phospholipid Liposomes (Using Nanoparticles). Nano Lett 2006;6:694–8.

Zhang J, Lang HP, Huber F, et al. Rapid and label-free nanomechanical detection of biomarker transcripts in human RNA. Nature Nanotechnology 2006a;1:214–20.

Zhang Y, Hong H, Cai W. Imaging with Raman spectroscopy. Curr Pharm Biotechnol 2010; 11:654–61.

Zhang E, Zhang C, Su Y, et al. Newly developed strategies for multifunctional mitochondria-targeted agents in cancer therapy. Drug Discov Today 2011;16:140–6.

Zhang L, Xu J, Mi L, et al. Multifunctional magnetic-plasmonic nanoparticles for fast concentration and sensitive detection of bacteria using SERS. Biosens Bioelectron 2012a;31:130–6.

Zhang LW, Monteiro-Riviere NA. Assessment of quantum dot penetration into intact, tape-stripped, abraded and flexed rat skin. Skin Pharmacol Physiol 2008;21:166–80.

Zhang LW, Yu WW, Colvin VL, Monteiro-Riviere NA. Biological interactions of quantum dot nanoparticles in skin and in human epidermal keratinocytes. Toxicol Appl Pharmacol 2008;228:200–11.

Zhang M, Lenaghan SC, Xia L. Nanofibers and nanoparticles from the insect-capturing adhesive of the Sundew (Drosera) for cell attachment. J Nanobiotech 2010;8:20.

Zhang T, Stilwell JL, Gerion D, et al. Cellular effect of high doses of silica-coated quantum dot profiled with high throughput gene expression analysis and high content cellomics measurements. Nano Lett 2006;6:800–8.

Zhang X, Tung CS, Sowa GZ, et al. Global structure of a three-way junction in a phi29 packaging RNA dimer determined using site-directed spin labeling. J Am Chem Soc 2012;134: 2644–52.

Zhang Y, Hong H, Cai W. Imaging with Raman spectroscopy. Curr Pharm Biotechnol 2010;11:654–61.

Zhang YB, Kanungo M, Ho AJ, et al. Functionalized Carbon Nanotubes for Detecting Viral Proteins. Nano Lett 2007;7:3086–3091.

Zhao B, Hu H, Mandal SK, Haddon RC. A Bone Mimic Based on the Self-Assembly of Hydroxyapatite on Chemically Functionalized Single-Walled Carbon Nanotubes. Chemistry of Materials 2005;17:3235–41.

Zhao Q, Sigalov G, Dimitrov V, et al. Detecting SNPs Using a Synthetic Nanopore. Nano Lett 2007;7:1680–1685.

Zhao Y, Vivero-Escoto JL, Slowing II, et al. Capped mesoporous silica nanoparticles as stimuli-responsive controlled release systems for intracellular drug/gene delivery. Expert Opinion Drug Deliv 2010;7:1013–29.

Zharov VP, Galitovskaya EN, Johnson C, Kelly T. Synergistic enhancement of selective nanophotothermolysis with gold nanoclusters: potential for cancer therapy. Lasers Surg Med 2005;37:219–26.

Zou J, Sood R, Poe D, et al. Manufacturing and in vivo inner ear visualization of MRI traceable liposome nanoparticles encapsulating gadolinium. J Nanobiotechnol 2010;8:32

Zubarev ER, Xu J, Sayyad A, Gibson JD. Amphiphilicity-driven organization of nanoparticles into discrete assemblies. J Am Chem Soc 2006;128:15098–9.

Zwiorek K, Kloeckner J, Wagner E, Coester C. Gelatin nanoparticles as a new and simple gene delivery system. J Pharm Pharm Sci 2005;7:22–8.

Index

K.K. Jain, *The Handbook of Nanomedicine*,
DOI 10.1007/978-1-61779-983-9, © Springer Science+Business Media New York 2012

Printed by Publishers' Graphics LLC